We dedicate this book to our students, and to all those who expand our knowledge of biodiversity and its protection by engaging in conservation biology.

An Introduction to
Conservation Biology

An Introduction to

Conservation Biology

Richard B. Primack
Boston University

Anna A. Sher
University of Denver

Sinauer Associates, Inc. Publishers
Sunderland, MA U.S.A.

Oxford University Press is a department of the University of Oxford.
It furthers the University's objective of excellence in research,
scholarship, and education by publishing worldwide. Oxford is a
registered trade mark of Oxford University Press in the UK and certain
other countries.

Published in the United States of America by Oxford University Press
198 Madison Avenue, New York, NY 10016, United States of America

© 2018 Oxford University Press

Sinauer Associates is an imprint of Oxford University Press.

Address editorial correspondence to:
Sinauer Associates
P.O. Box 407
Sunderland, MA 01375 U.S.A.
publish@sinauer.com

Address orders, sales, license, permissions, and translation inquiries to:
Oxford University Press U.S.A.
2001 Evans Road
Cary, NC 27513 U.S.A.
Orders: 1-800-445-9714

Library of Congress Cataloging-in-Publication Data
Names: Primack, Richard B., 1950- author. | Sher, Anna, author.
Title: Introduction to conservation biology / Richard B. Primack, Anna A.
Sher.
Description: Massachusetts, U.S.A. : Sinauer Associates, Inc. Publishers,
 2016. | Includes bibliographical references and index.
Identifiers: LCCN 2016005729 | ISBN 9781605354736 (pbk. : alk. paper)
Subjects: LCSH: Conservation biology.
Classification: LCC QH75 .P7523 2016 | DDC 333.95/16--dc23
LC record available at http://lccn.loc.gov/2016005729

9 8 7 6 5 4 3 2

Printed in the United States of America

Contents

Chapter 3

The Value of Biodiversity   52

Chapter 4
Threats to Biodiversity   90

Chapter 7

Bringing Species Back from the Brink   234

Chapter 8

Protected Areas   264

Chapter 9

Conservation Outside Protected Areas   304

Chapter 10

Restoration Ecology   336

Chapter 11

The Challenges of Sustainable Development   362

Chapter 12

An Agenda for the Future   392

Preface

The field of Conservation Biology is in a period of rapid growth, collaboration, increased clarity, and optimism. In the last two years, the success of intensive conservation efforts have been seen in the growing populations of wolves and bears in Yellowstone National Park, the establishment of the largest contiguous ocean reserve to date (830,000 km^2) by the U.K., and the discovery of hundreds of new plants and animals across the globe, to only name a few. The public is increasingly engaged with the news of conservation biology, as reflected in the outrage over the 2015 poaching death of Cecil the Lion in Zimbabwe. It can also be seen in the arts, such as the excitement surrounding the 2016 release of *Finding Dory*, an animated movie with strong conservation themes, and by the powerful documentary about orcas, *Blackfish*, which spurred national legislation in 2014 and major changes by SeaWorld in 2016 that support the species. However, the task before us is still vast; as the recent update to the IUCN's list of endangered species has shown, most groups of organisms face greater extinction risk today than ever before as the global human population reaches above 7.25 billion. But if the 2015 Paris Accord on climate change is any indication, the human race is also more united and serious than ever before about confronting these challenges. It is our hope that this book can be an important tool for the development of the next generation.

An Introduction to Conservation Biology is the heir of our two previous, successful volumes: *A Primer of Conservation Biology, Fifth Edition* and *Essentials of Conservation Biology, Sixth Edition*. Since having dedicated myself to this project, I now respect—at an entirely new level—what Richard Primack has done, understanding not only the hours involved but the care it takes to keep content current, coverage across ecosystems and taxa balanced, and to support the text with images and tables that are beautiful, useful, and themselves representative of the diversity of the field at all levels.

My goals have to been to retain the readability of *A Primer of Conservation Biology* with the depth and coverage of *Essentials of Conservation Biology* while upholding the standards of each book to the highest degree. For those who have used either of these texts, *Introduction* should feel both familiar and fresh: three chapters have been added to those offered in the *Primer*, with population biology, conservation tools, restoration ecology, sustainable development, ex situ conservation, and other key elements expanded and updated. There are literally hundreds of new examples, explanations, citations, and figures to enhance learning and excitement for the subject. It is designed to serve a wide variety of courses and students, as a primary textbook or as a supplement.

As Richard has said,

> *It is our goal that readers of this book will be inspired to find out more about the extinction crisis facing species and ecosystems and how they can take action to halt it. I encourage readers to take the field's activist spirit to heart—use the Appendix to find organizations and sources of information on how to help. If readers gain a greater appreciation for the goals, methods, and importance of conservation biology, and if they are moved to make a difference in their everyday lives, this textbook will have served its purpose.*

Acknowledgments

I sincerely appreciate the contributions of everyone who has made this textbook what it is today. Individual chapters were reviewed by J. Michael Reed, Andrew R. Blaustein, Dana Bauer, John J. Cox, Scott Connelly, Michael Reed, Peter Houlihan, Tim Caro, Paige Warren, Janette Wallis, Monique Poulin, Eric Higgs, Federico Cheever, Meg Lowman, and Richard Reading (who also provided several images).

I am grateful for the work of Annie Henry, Eduardo González, and Robert Robinson for their background research for the book and helping to edit material from *Essentials* into the leaner format with new citations and examples. Additional help was provided by Allison Brunner, Brandon Krentz, Jaime Pena, and Matt Herbert, and I appreciate the donation of photos from many fine photographers, including Scott Dressel-Martin of Denver Botanic Gardens, Wright Robinson, and Hector R. Chenge. Andy Sinauer, Rachel Meyers, Martha Lorantos, David McIntyre, Ann Chiara and the rest of the Sinauer staff helped to transform the manuscript into a finished book; I greatly appreciated their encouragement and patience over the past several months.

I would also like to give special thanks to my wife, Fran, and our son Jeremy as well as all the women in my reading and writing groups who have offered support and encouragement along the way. Thanks to my chair, Joseph Angelson, and the Organismal Biologists group at the University of Denver, for their shared enthusiasm for this project. I am indebted to my hundreds of students who have taught me so much about the art and science of teaching.

Finally, I must thank Richard Primack for the honor of being his successor, for his mentorship, and for all of his input at every stage. It has been a deeply rewarding experience.

Anna A. Sher
Denver, Colorado
April 6, 2016

An International Approach

In keeping with the global nature of conservation biology, I feel it is important to make the field accessible to as wide an audience as possible. With the assistance of Marie Scavotto and the staff of Sinauer Associates, I have arranged an active translation program of the *Essentials of Conservation Biology* and the shorter *Primer of Conservation Biology*. The goal has been to create regional or country-specific translations, identifying local scientists to become coauthors and to add case studies, examples, and illustrations from their own countries and regions that would be more relevant to the intended audience. Already, editions of have appeared in Arabic, Brazilian Portuguese, Chinese (four editions), Czech (two editions), Estonian, French (two editions, one with a Madagascar focus), German, Greek, Hungarian, Indonesian (two editions), Italian (two editions), Japanese (two editions), Korean (three editions), Mongolian, Nepal (in English), Romanian (two editions), Russian, Serbia, South Asia (in English), Spanish (two editions; one with a Latin American focus), Turkish, and Vietnamese. New editions for Africa, Bangladesh, Brazil, Germany, Greece, Iran, Laos, Latin America, Madagascar, Pakistan, and Thailand are currently in production. These translations will help conservation biology develop as a discipline with a global scope. At the same time, examples from these translations find their way back into the English language editions, thereby enriching the presentation. I hope that my enthusiastic new coauthor, Anna Sher, will continue this project with our new book, *An Introduction to Conservation Biology*.

Richard Primack
Boston University
April 6, 2016

Media and Supplements

to accompany

An Introduction to
Conservation Biology

eBook

An Introduction to Conservation Biology is available for purchase as an eBook, in several different formats, including VitalSource, Yuzu, BryteWave, and RedShelf. The eBook can be purchased as either a 180-day rental or a permanent (non-expiring) subscription. All major mobile devices are supported. For details on the eBook platforms offered, please visit www.sinauer. com/ebooks.

Instructor's Resource Library

(Available to qualified adopters)

The Instructor's Resource Library for *An Introduction to Conservation Biology* includes all of the textbook's figures (line-art illustrations and photographs) and tables, provided as both high- and low-resolution JPEGs. All have been formatted and optimized for excellent projection quality. Also included are ready-to-use PowerPoint slides of all figures and tables.

An Introduction to
Conservation Biology

1

Defining Conservation Biology

A conservation biologist from the binational Gladys Porter Zoo crew releases Kemp's ridley sea turtles off the coast of Tamaulipas, Mexico.

P opular interest in protecting the world's biological diversity—including its amazing range of species, its complex ecosystems, and the genetic variation within species—has intensified during the last few decades. It has become increasingly evident to both scientists and the general public that we are living in a period of unprecedented losses of **biodiversity*** (Pimm et al. 2014). Around the globe, biological ecosystems—the interacting assemblages of living organisms and their environment that took millions of years to develop—including tropical rain forests, coral reefs, temperate old-growth forests, and prairies, are being devastated. Thousands, if not tens of thousands, of species and millions of unique populations are predicted to go extinct in the coming decades (Barnosky et al. 2011). Unlike the mass extinctions in the geologic past, which followed catastrophes such as asteroid collisions with Earth, today's extinctions have a human face.

During the last 200 years, the human population has exploded. It took more than 160,000 years for the number of *Homo sapiens* to reach 1 billion, an event

***Biological diversity** is often shortened to *biodiversity* (a term credited to biologist E. O. Wilson in 1992); it includes all species, genetic variation, and biological communities and their ecosystem-level interactions.

FIGURE 1.1 The human population in 2016 stood at about 7.25 billion. The World Resources Institute estimates current annual population growth at 1.1%, but even this modest growth rate will add more than 70 million people to the planet in the next year. This number will escalate each year as the increase is compounded. (Data from US Census Bureau, www.census.gov)

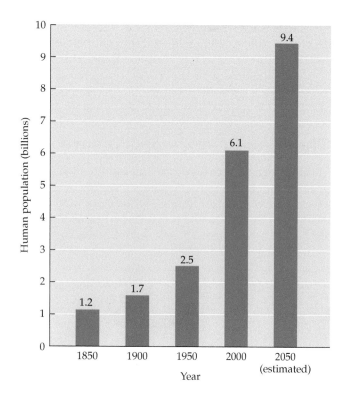

that occurred sometime around the year 1805. Estimates for 2016 put the number of humans at 7.25 billion, with a projected 9.4 billion by 2050 (US Census Bureau); at this size, even a modest rate of population increase adds tens of millions of individuals each year (**FIGURE 1.1**). The threats to biodiversity are accelerating because of the demands of the rapidly increasing human population and its rising material consumption. People deplete natural resources such as firewood, coal, oil, timber, fish, and game, and they convert natural habitats to land dominated by agriculture, cities, housing developments, logging, mining, industrial plants, and other human activities. These changes are not easily reversible, and even aggressive programs to slow population growth do not adequately address the environmental problems we have caused (Bradshaw and Brook 2014).

Worsening the situation is the fact that as countries develop and industrialize, the consumption of resources increases. For example, the average citizen of the United States uses four times more energy than the average global citizen, seven times more than the average Costa Rican citizen, and fifteen times more than the average Indian citizen (US Energy Information Administration 2015). The ever-increasing number of human beings and their intensifying use of natural resources have direct and harmful consequences for the diversity of the living world (Brown et al. 2014).

The threats to biodiversity directly threaten human populations as well because people are dependent on the natural environment for raw materials, food, medicines, and the water they drink. The poorest people are, of course, the ones who experience the greatest hardship from damaged environments because they have fewer reserves of food and less access to medical supplies, transportation, and construction materials.

The New Science of Conservation Biology

The avalanche of species extinctions and the wholesale habitat destruction occurring in the world today is devastating, but there is reason for hope. The last several decades have included many success stories, such as that of the American bald eagle (*Haliaeetus leucocephalus*), which was rescued from near extinction due to a combination of scientific inquiry, public awareness, and political intervention (see Chapter 11). Actions taken—or bypassed—during the next few decades will determine how many of the world's species and natural areas will survive. It is quite likely that people will someday look back on the first decades of the twenty-first century as an extraordinarily exciting time, when collaborations of determined people acting locally and internationally saved many species and even entire ecosystems (Sodhi et al. 2011). Examples of such conservation efforts and positive outcomes are described throughout this book.

Conservation biology is an integrated, multidisciplinary scientific field that has developed in response to the challenge of preserving species and ecosystems. It has three goals:

- To document the full range of biological diversity on Earth
- To investigate human impact on species, genetic variation, and ecosystems
- To develop practical approaches to prevent the extinction of species, maintain genetic diversity within species, and protect and restore biological communities and their associated ecosystem functions

The first two of these goals involve the dispassionate search for factual knowledge that is typical of scientific research. The third goal, however, defines conservation biology as a **normative discipline**—that is, a field that embraces certain values and attempts to apply scientific methods to achieving those values (Lindenmayer and Hunter 2010). Just as medical science values the preservation of life and health, conservation biology values the preservation of species and ecosystems as an ultimate good, and its practitioners intervene to prevent human-caused losses of biodiversity.

Conservation biology arose in the 1980s, when it became clear that the traditional applied disciplines of resource management alone were not comprehensive enough to address the critical threats to biological diversity. The applied disciplines of agriculture, forestry, wildlife management, and

Conservation biology merges applied and theoretical biology by incorporating ideas and expertise from a broad range of scientific fields toward the goal of preserving biodiversity.

fisheries biology have gradually expanded to include a broader range of species and ecosystem processes. Conservation biology complements those applied disciplines and provides a more general theoretical approach to the protection of biological diversity. It differs from these disciplines in its primary goal of long-term preservation of biodiversity.

Like medicine, which applies knowledge gleaned from physiology, anatomy, biochemistry, and genetics to the goal of achieving human health and eliminating illness, conservation biology draws on other academic disciplines, including population biology, taxonomy, ecology, and genetics. Many conservation biologists have come from these ranks. Others come from backgrounds in the applied disciplines, such as forestry and wildlife management. In addition, many leaders in conservation biology have come from zoos and botanical gardens, bringing with them experience in locating rare and endangered species in the wild and then maintaining and propagating them in captivity.

Conservation biology is also closely associated with, but distinct from, **environmentalism**, a widespread movement characterized by political and educational activism with the goal of protecting the natural environment. Conservation biology is a scientific discipline based on biological research whose findings often contribute to the environmental movement (Hall and Fleishman 2010).

Because much of the biodiversity crisis arises from human pressures, conservation biology also incorporates ideas and expertise from a broad range of fields outside of biology (**FIGURE 1.2**) (Reyers et al. 2010). For example, environmental law and policy provide the basis for government protection of rare and endangered species and critical habitats. Environmental ethics provides a rationale for preserving species. Ecological economists provide analyses of the economic value of biological diversity to support arguments for preservation. Climatologists monitor the physical characteristics of the environment and develop models to predict environmental responses to disturbance. Both physical and cultural geography provide information about the relationships among elements of the environment, helping us understand causes and distributions of biodiversity and how humans interact with it. Social sciences, such as anthropology and sociology, provide methods to involve local people in actions to protect their immediate environment. Conservation education links academic study and fieldwork to solve environmental problems, teaching people about science and helping them realize the value of the natural environment. Because conservation biology draws on the ideas and skills of so many separate fields, it can be considered a truly multidisciplinary discipline.

The roots of conservation biology

Religious and philosophical beliefs about the relationship between humans and the natural world are seen by many as the foundation of conserva-

Field experience and research needs

Basic Sciences	Resource Management
Anthropology	Agriculture
Biogeography	Community education
Climatology	and development
Ecology:	Fisheries management
Community ecology	Forestry
Ecosystem ecology	Land-use planning and
Landscape ecology	regulation
Environmental studies:	Management of captive
Ecological economics	populations:
Environmental ethics	Zoos
Environmental law	Aquariums
Ethnobotany	Botanical gardens
Evolutionary biology	Seed banks
Genetics	Management of protected
Population biology	areas
Sociology	Sustainable development
Taxonomy	Wildlife management
Other biological, physical,	Other resource conservation
and social sciences	and management activities

New ideas and approaches

FIGURE 1.2 Conservation biology represents a synthesis of many basic sciences (left) that provide principles and new approaches for the applied fields of resource management (right). The experiences gained in the field, in turn, influence the direction of the basic sciences. (After Temple 1991.)

tion biology (Dudley et al. 2009). Eastern philosophies such as Taoism, Hinduism, and Buddhism revere wilderness for its capacity to provide intense spiritual experiences. These traditions see a direct connection between the natural world and the spiritual world, a connection that breaks down when the natural world is altered or destroyed. Strict adherents to the Jain and Hindu religions in India believe that all killing of animal life is wrong. Islamic, Judaic, and Christian teachings are used by many people to support the idea that people are given the sacred responsibility to be guardians of nature (**FIGURE 1.3**; see Chapter 3, "Enviornmental Ethics"). Many of the leaders of the early Western environmental movement that helped to establish parks and wilderness areas did so because of strong personal convictions that developed from their Christian religious beliefs. Contemporary religious leaders have pointed out that some of the most profound moments in the Bible occur on mountaintops, in the wilderness, or on the banks of a river (Korngold 2008). In Native American tribes of the Pacific Northwest, hunters undergo purification rituals in order to be considered worthy, and the Iroquois consider how their actions would affect the lives of their descendants after seven generations.

FIGURE 1.3 Religious convictions combined with a history of traditional relationships with nature motivate the grassroots conservation organization Kakamega Environmental Education Program (KEEP), established in 1998 to protect one of the last remnants of tropical forest left in Kenya, East Africa. (Photograph by Anna Sher.)

Modern conservation biology in the United States can be traced to the nineteenth-century transcendentalist philosophers Ralph Waldo Emerson and Henry David Thoreau, who saw wild nature as an important element in human moral and spiritual development. Emerson (1836) saw nature as a temple in which people could commune with the spiritual world and achieve spiritual enlightenment. Thoreau was both an advocate for nature and an opponent of materialistic society, writing about his ideas and experiences in *Walden*. His book, published in 1854, has influenced many generations of students and environmentalists. Eminent American wilderness advocate John Muir used the transcendental themes of Emerson and Thoreau in his campaigns to preserve natural areas. According to Muir's **preservationist ethic**, natural areas such as forest groves, mountaintops, and waterfalls have spiritual value that is generally superior to the tangible material gain obtained by their exploitation (Muir 1901).

Subsequent leaders paved the way for conservation biology as an applied academic discipline in the United States. Gifford Pinchot, the first head of the US Forest Service, developed a view of nature known as the **resource conservation ethic** (Ebbin 2009). He defined natural resources as the commodities and qualities found in nature, including timber, fodder, clean water, wildlife, and even beautiful landscapes (Pinchot 1947). The proper use of natural resources, according to the resource conservation ethic, is whatever will further "the greatest good of the greatest number [of people] for the longest time." From the perspective of conservation biology, **sustainable development** is development that best meets present and future human needs without damaging the environment and biodiversity (Czech 2008).

> Preservation of natural resources, ecosystem management, and sustainable development are major themes in conservation biology.

Shortly thereafter, biologist Aldo Leopold published *A Sand County Almanac* (1949), which also illustrated the interrelatedness of living things and their environment, promoting the idea that the most important goal of conservation is to maintain the health of natural ecosystems and ecological processes (Leopold 2004). As a result, Leopold and many others lobbied successfully for certain parts of national forests to be set aside as wilderness areas (Shafer 2001). This idea of considering the ecosystem as a whole, including human populations, is now termed the **land ethic**.

Marine biologist Rachel Carson (**FIGURE 1.4**) is credited with raising public awareness of the complexity of nature with her best-selling books, including *Silent Spring* (1962), which brought attention to the dangers of pesticides and spurred an international environmental movement. An approach known as **ecosystem management**, which combines ideas of Carson, Leopold, and Pinchot, places the highest management priority on cooperation among businesses, conservation organizations, government agencies, private citizens, and other stakeholders to provide for human needs while maintaining the health of wild species and ecosystems.

Depictions of nature in the creative and performing arts have also played an important role in the growing awareness of the value of nature and its preservation in the United States. In the mid-nineteenth century, prolific painters of the Hudson River School were noted for their romantic depictions of "scenes of solitude from which the hand of nature has never been lifted" (Cole 1965). Photographer Ansel Adams

FIGURE 1.4 Rachel Carson (1907–1964) was a marine biologist who, through her popular writing, including *Silent Spring* (1962), helped to found both conservation biology and the environmental movement. (Courtesy of the Beinecke Rare Book and Manuscript Library, Yale University.)

(1902–1984) took breathtaking images of wild America, helping to foster public support for its protection (**FIGURE 1.5**). Popular singer-songwriter John Denver also inspired interest in conservation with such lyrics as:

> *Now he walks in quiet solitude the forest and the streams, seeking grace in every step he takes… His life is filled with wonder but his heart still knows some fear of a simple thing he cannot comprehend: why they try to tear the mountains down … more people, more scars upon the land.*
>
> *Rocky Mountain High*, 1975

Clearly, the arts play a unique role in fostering interest in conservation (Jacobson et al. 2007).

In Europe, expressions of concern for the protection of wildlife began to spread in the nineteenth century when the expansion of agriculture and the use of firearms for hunting led to a marked reduction in wild animals (Galbraith et al. 1998). These dramatic changes stimulated the formation of the British conservation movement, leading to the founding of the Commons, Open Spaces and Footpaths Preservation Society in

FIGURE 1.5 Ansel Adams (1902–1984) showed the public the beauty of wild spaces. (Commissioned by the National Park Service, *The Tetons, Snake River.*)

TABLE 1.1 Some Useful Units of Measurement	
Length	
1 meter (m)	1 m = 39.4 inches = ~3.3 feet
1 kilometer (km)	1 km = 1000 m = 0.62 mile
1 centimeter (cm)	1 cm = 1/100 m = 0.39 inches
1 millimeter (mm)	1 mm = 1/1000 m = 0.039 inches
Area	
1 square meter (m^2)	Area encompassed by a square, each side of which is 1 meter
1 hectare (ha)	1 ha = 10,000 m^2 = 2.47 acres
	100 ha = 1 square kilometer (km^2)
Mass	
1 kilogram (kg)	1 kg = 2.2 pounds
1 gram (g)	1 g = 1/1000 kg = 0.035 ounces
1 milligram (mg)	1 mg = 1/1000 g = 0.000035 ounces
Temperature	
degree Celsius (°C)	°C = 5/9(°F − 32)
	0°C = 32° Fahrenheit (the freezing point of water)
	100°C = 212° Fahrenheit (the boiling point of water)
	20°C = 68° Fahrenheit ("room temperature")

1865, the National Trust for Places of Historic Interest or Natural Beauty in 1895, and the Royal Society for the Protection of Birds in 1899. Altogether, these groups have preserved nearly 1 million hectares (ha) of open land (**TABLE 1.1** provides an explanation of the term *hectare* and other measurements). In contrast to its origins in the United States, biological conservation in Europe has had a more integrated view of human society and ecosystems as a whole, rather than envisioning a dichotomy of man versus nature (Linnell et al. 2015).

Many societies worldwide similarly have strong traditions of nature conservation and land protection. Tropical countries such as Brazil, Costa Rica, and Indonesia have a history of reverence for nature, and their governments have allocated increasing numbers and areas of national parks. The economic value of these protected areas is constantly increasing because of their importance for tourism and the valuable ecosystem services they provide, such as purifying water and absorbing carbon dioxide (see Chapter 3). Many tropical countries have established agencies to regulate the exploration and use of their biodiversity, and these efforts increasingly involve the indigenous peoples

> As demonstrated by the conservation tradition in Europe, habitat degradation and species loss can catalyze long-lasting conservation efforts.

who depend on and have unique knowledge of these ecosystems. Hunting and gathering societies, such as the Penan of Borneo, give thousands of names to individual trees, animals, and places in their surroundings to create a cultural landscape that is vital to the well-being of the tribe. Indeed, traditional societies throughout the world have influenced and enriched modern conservation biology.

A new science is born

By the early 1970s, scientists throughout the world were aware of an accelerating biodiversity crisis, but there was no central forum or organization to address the issue. Scientist Michael Soulé organized the first International Conference on Conservation Biology in 1978 so that wildlife conservationists, zoo managers, and academics could discuss their common interests. At that meeting, Soulé proposed a new interdisciplinary approach that could help save plants and animals from the threat of human-caused extinctions, which he called *conservation biology* (Soulé 1985). Subsequently, Soulé, along with colleagues including Paul Ehrlich of Stanford University and Jared Diamond of the University of California at Los Angeles, began to develop conservation biology as a discipline that would combine the practical experience of wildlife, forestry, fisheries, and national park management with the theories of population biology and biogeography. In 1985, this core of scientists founded the Society for Conservation Biology. This organization, which continues today with an international membership, sponsors conferences and publishes the scientific journal *Conservation Biology*. That journal is now joined by many peer-reviewed publications that feature research in conservation biology, expanding our understanding of the importance of biodiversity and how it can be protected.

Public and policymaker awareness of the value of, and threats to, biodiversity greatly increased following the international Earth Summit held in Rio de Janeiro, Brazil, in 1992 (see Chapter 11). At this meeting, representatives of 178 countries formulated and eventually signed the Convention on Biological Diversity (CBD), which obligates countries to protect their biodiversity but also allows them to obtain a share in the profits of new products developed from that diversity. In 2000, the United Nations General Assembly adopted May 22 as International Biodiversity Day to commemorate the conference, and in 2006, the US Congress designated the third Friday in May as Endangered Species Day. The value of biodiversity was further highlighted by the United Nations designation of 2010 as the International Year of Biodiversity and 2011 as the International Year of Forests. In 2015, the United Nations Climate Change Conference (COP21) was attended by 196 parties and resulted in the Paris Agreement to reduce global greenhouse emissions (**FIGURE 1.6**). Arguably, this increase in understanding and concern would not have been possible without the foundation of a recognized scientific discipline.

> Interdisciplinary approaches, the involvement of local people, and the restoration of important environments and species all attest to progress in the science of conservation biology.

FIGURE 1.6 The heads of delegations at the 2015 United Nations Conference on Climate Change in Paris (also referred to as COP21). President Barak Obama is in the center, speaking to another delegate head. (© Presidencia de la Republica Mexicana/Flickr, licensed under a Creative Commons Attribution license.)

The interdisciplinary approach: A case study with sea turtles

Throughout the world, scientists are using the approaches of conservation biology to address challenging problems, as illustrated by the efforts to save the Kemp's ridley sea turtle (*Lepidochelys kempii*). The Kemp's ridley is the rarest and smallest of the world's sea turtle species, at 70–100 cm (2–3 feet) long and about 45 kg (100 pounds). This critically endangered species is now recovering as a result of international conservation efforts and cooperation between scientists, conservation organizations, government officials, and the interested public.

After its discovery as a distinct species in the late 1800s, it took nearly a century for scientists to determine how and where the Kemp's ridley reproduces (Wibbels and Bevan 2015). They discovered that nearly 95% of Kemp's ridley nesting happens on beaches in the state of Tamaulipas, in the northeastern corner of Mexico, in highly synchronized gatherings of turtles, called *arribada*. This highly concentrated breeding is unusual among turtles and makes the species particularly vulnerable to intensive harvesting. Over many decades, locals collected an estimated 80% of Kemp's ridley eggs from the nesting beaches for eating. Thousands of turtles also drowned in fishing gear, especially in shrimp nets. In 1985, the progressive decline in the population brought it to a low point of only

702 nests worldwide, making the Kemp's ridley the most endangered sea turtle in the world.

Heeding the warning of wildlife biologists that the species was nearing extinction, government officials from Mexico and the United States worked together to help the species recover and establish stable populations. As a first step, nesting beaches were protected as refuges, reserves, and parks. Egg collection was banned. And at sea, shrimp trawlers were required to use turtle excluder devices (TEDs), consisting of a grid of bars with an opening that allows a caught turtle to escape.

In addition to reducing threats, a collaborative group of national and state agencies and conservation organizations in Mexico and the United States has undertaken an ambitious effort to increase nest and hatchling survival and to improve education and appreciation of sea turtle conservation. In the United States, national park authorities began to reestablish a population on Padre Island in Texas, where the species had formerly occurred. From 1978 to 1988, scientists, conservationists, and volunteers collected 22,507 eggs from Mexico, packed them in sand, and transported them to Padre Island National Seashore, which is managed by the US National Park Service (**FIGURE 1.7**). The hatchlings were released on the beach and briefly allowed to swim in the surf before they were captured using aquarium dip nets. The hope was that this brief time on the beach and in the surf would help them imprint on the site so that they would return there to nest as adults. The captured hatchlings were then reared in captivity for 9–11 months as a part of a "head-start" program that allowed the turtles to grow large enough to avoid most predators. (Most sea turtles die as hatchlings.) Scientists carefully monitored growth during this period (Caillouet et al. 1997). Then the one-year-old turtles were released permanently into the Gulf of Mexico (see the chapter opening photo).

Now, each year, the staff at Padre Island, many partner organizations, and over a hundred volunteers patrol the beach during the breeding season, searching for Kemp's ridleys and their nests. When they find nests, teams carefully excavate them and bring the eggs to an incubation facility or a large screen enclosure called a corral. When the young hatchlings are released, it is now a public event that doubles as an education tool—the hope is that the people watching each release will become ad-

FIGURE 1.7 Researchers collect eggs from a Kemp's ridley sea turtle nest. The eggs will either be relocated to a nest within a protected enclosure or brought to an incubation facility. (Courtesy of David Bowman, US Fish and Wildlife Service.)

vocates for the turtles' protection. Outside the national seashore, private conservation organizations also help protect the turtles on their feeding grounds. Together, these conservation activities and associated media coverage expose hundreds of thousands of visitors to information about sea turtle ecology and conservation.

Over a 16-year period, the Kemp's ridley population at Padre Island National Seashore increased dramatically, from 6 nests, 590 eggs, and 369 hatchlings released in 1996 to 209 nests, 20,067 eggs, and 16,577 hatchlings released in 2012. Compared with the low of 702 nests in 1985, researchers and volunteers counted a total of 21,797 nests in 2012. Each female lays two to three clutches of eggs each season, so this number of nests corresponds to at least 7000–9000 mature reproducing females. However, the number declined to only 12,053 nests in 2014, at least partially due to the *Deepwater Horizon* oil spill in 2010 that is believed to have killed hundreds of juvenile turtles (Caillouet et al. 2015). Fortunately, a study of the nests found predation rates to be low and hatchling survival high (Bevan et al. 2014), and the number of nests grew by more than a thousand in the next year (Luis Jaime Peña, pers. comm.).

The Kemp's Ridley Sea Turtle Recovery Plan has set a target of 10,000 nesting females for the population to be considered recovered. Scientific scrutiny, international partnerships, and the participation of volunteers and local communities have brought the Kemp's ridley sea turtle back from the brink of extinction, and all of those involved will continue to seek the answers leading to its complete recovery.

The Ethical Principles of Conservation Biology

Earlier in the chapter, we mentioned that conservation biology is a normative discipline in which certain value judgments are inherent. The field rests on an underlying set of principles that is generally agreed on by practitioners of the discipline (Soulé 1985; "Organizational Values," *sensu* Society for Conservation Biology 2016) and can be summarized as follows:

1. *Biological diversity has intrinsic value*. Species and the biological communities in which they live possess value of their own, regardless of their economic, scientific, or aesthetic value to human society. This value is conferred not just by their evolutionary history and unique ecological role, but also by their very existence. (See Chapter 3 for a more complete discussion of this topic.)

2. *The untimely extinction of populations and species should be prevented*. The ordinary extinction of species and populations as a result of natural processes is an ethically neutral event. In the past, the local loss of a population was usually offset by the establishment of a new population through dispersal. However, as a result of human activity, the loss of

There are ethical reasons why people want to conserve biodiversity, such as belief that species have intrinsic value.

populations and the extinction of species has increased by more than a hundredfold, with no simultaneous increase in the generation of new populations and species (MEA 2005; see Chapter 5).

3. *The diversity of species and the complexity of biological communities should be preserved.* In general, most people agree with this principle simply because they appreciate biodiversity; it has even been suggested that humans may have a genetic predisposition to love biodiversity, called **biophilia** (FIGURE 1.8) (Corral-Verdugo et al. 2009). Many of the most valuable properties of biodiversity are expressed only in natural environments. Although the biodiversity of species may be partially preserved in zoos and botanical gardens, the ecological complexity that exists in natural communities will be lost without the preservation of natural areas (see Chapter 8). Furthermore, biodiversity has been directly linked to ecosystem productivity and stability (Hautier et al. 2015), among other values (see Chapter 3).

4. *Science plays a critical role in our understanding of ecosystems.* It is not enough to simply value diversity and protect natural spaces; objective research is necessary to identify which species and environments are at greatest risk, as well as to understand the nature of these risks and how to mitigate them. Ideally, scientists are involved in all stages of

FIGURE 1.8 People enjoy seeing the diversity of life, as illustrated by the popularity of planting gardens and of public botanical gardens as tourist destinations. (Butchart Gardens, Victoria, BC, Canada © Xuanlu Wang/Shutterstock.)

conservation, including implementation of conservation actions and monitoring results.

5. *Collaboration among scientists, managers, and policy makers is necessary.* In order to achieve the goals of reduced extinction rates and preservation of biological communities, high-quality research findings must be shared with those who create the laws and provide the funding for conservation actions as well as those who must implement them. These actions are most likely to succeed when they are based on scientifically sound information.

Not every conservation biologist accepts every one of these principles, and there is no hard-and-fast requirement to do so. Individuals or organizations that agree with even two or three of these principles are often willing to support conservation efforts. Current progress in protecting species and ecosystems has been achieved in part through partnerships between traditional conservation organizations such as The Nature Conservancy and cattle ranchers, hunting clubs like Ducks Unlimited, and other groups with a vested interest in the health of ecosystems.

Looking to the Future

The field of conservation biology has set itself some imposing—and absolutely critical—tasks: to describe Earth's biological diversity, to protect what remains, and to restore what is degraded. The field is growing in strength, as indicated by increased governmental participation in conservation activities, increased funding of conservation organizations and projects, and an expanding professional society.

In many ways, conservation biology is a crisis discipline. Decisions about selecting national parks, species management, and other aspects of conservation are made every day under severe time pressure (Laurance et al. 2012; Martin et al. 2012). As one of the guiding values mentioned above, biologists and scientists in related fields seek to provide the advice that governments, businesses, and the general public need in order to make crucial decisions, but because of time constraints, scientists are often compelled to make recommendations without thorough investigation. Decisions must be made, with or without scientific input, and conservation biologists must be willing to express opinions and take action based on the best available evidence and informed judgment (Maron et al. 2013). They must also articulate a long-term conservation vision that extends beyond the immediate crisis (Wilhere 2012).

Despite the threats to biodiversity and the limitations of our knowledge, we can detect many positive signs that allow conservation biologists to be cautiously hopeful (Roman et al. 2015). The rate of human population growth has slowed (US Census Bureau), and per capita energy use in the United States, while still high, has been decreasing; in 2014 it was the lowest it has been since the 1960s (World Bank). The number of protected areas around the globe continues to increase, with a dramatic expansion in the number of marine protected areas and a commitment by coastal nations to increase them fivefold by 2020 (Halpern 2014).

These gains are due in part to action spurred by the public. Increasing numbers of social and religious leaders have rallied for the protection of biodiversity. As one powerful example, the first papal encyclical focused solely on the environment was recently released, in which our obligation to "till and keep" the garden of the world (cf. Gen. 2:15) was explained thus:

> *"Tilling" refers to cultivating, … while "keeping" means caring, protecting, overseeing and preserving. This implies a relationship of mutual responsibility between human beings and nature. Among positive experiences in this regard, we might mention, for example, … the binding Convention on International Trade in Endangered Species of Wild Fauna and Flora, which includes on-site visits for verifying effective compliance.*
>
> Pope Francis, *Laudato Si: On Care for Our Common Home,* 2015

Our ability to protect biodiversity has been strengthened by a wide range of local, national, and international efforts. Many endangered species are now recovering as a result of such conservation measures (IUCN 2015). Effective action has resulted from our continuing expansion of knowledge in conservation science, the developing linkages with rural development and social sciences, and our increased ability to restore degraded environments. All of these advances suggest that progress is being made, despite the enormous tasks still ahead.

Summary

■ Human activities are causing the extinction of thousands of species both locally and globally, with threats to species and ecosystems accelerating due to human population growth and the associated demands for resources.

■ Conservation biology is a field that combines basic and applied disciplines with three goals: to describe the full range of biodiversity on Earth, to understand human impact on biodiversity, and to develop practical approaches for preventing species extinctions, maintaining genetic diversity, and protecting and restoring ecosystems.

■ Elements of conservation biology can be found in many cultures, religions, and forms of creative expression. The modern field of conservation biology grew from the ideas of several influential individuals, eventually becoming a recognized scientific discipline with professional societies and academic journals by the 1980s.

■ Conservation biology rests on a number of underlying assumptions that are accepted by most professionals in the discipline: biodiversity has value in and of itself, extinction from human causes should be prevented, diversity at multiple levels should be preserved, science plays a critical role, and scientists must collaborate with nonscientists to achieve our goals.

■ The conservation of biodiversity has become an international undertaking. There

are many successful projects, such as the conservation of Kemp's ridley sea turtles, that indicate that progress can be made.

For Discussion ■

1. How is conservation biology fundamentally different from other branches of biology, such as physiology, genetics, or cell biology? How is it similar to the science of medicine? How is it different from environmentalism?

2. What do you think are the major conservation and environmental problems facing the world today? What are the major problems facing your local community? What ideas for solving these problems can you suggest? (Try answering this question now, and once again when you have completed this book.)

3. Consider the public land management and private conservation organizations

with which you are familiar. Do you think their guiding philosophies are closest to the resource conservation ethic, the preservation ethic, or the land ethic? What factors allow them to be successful or limit their effectiveness? Learn more about these organizations through their publications and websites.

4. How would you characterize your own viewpoint about the conservation of biodiversity and the environment? Which of the religious or philosophical viewpoints of conservation biology stated here do you agree or disagree with? How do you, or could you, put your viewpoint into practice?

Suggested Readings

Barnosky, A. D. and 11 others. 2011. Has Earth's sixth mass extinction already arrived? *Nature* 471: 51–57. Evidence from the fossil records and modern extinction rates suggest that we are on the verge of a major extinction event.

Bevan, E. and 11 others. 2014. *In situ* nest and hatchling survival at Rancho Nuevo, the primary nesting beach of the Kemp's ridley sea turtle, *Lepidochelys kempii. Herpetological Conservation and Biology* 9: 563–577. Research shows that this endangered species has a promising future.

Bradshaw, C. J. A. and B. W. Brook. 2014. Human population reduction is not a quick fix for environmental problems. *Proceedings of the National Academy of Sciences* 111: 16610–16615.

Caillouet, C. W., B. J. Gallaway, and A. M. Landry. 2015. Cause and call for modification of the bi-national recovery plan for the Kemp's ridley sea turtle (*Lepidochelys kempii*)—Second Revision. *Marine Turtle Newsletter* 145: 1–4.

Carson, R. 1962. *Silent Spring*. Houghton Mifflin Company, Boston. Essays written over a period from 1958–1962 on the devastating effects of pesticide use on ecosystems, particularly birds.

Halpern, B. S. 2014. Making marine protected areas work. *Nature* 506: 167–168.

Hautier, Y., D. Tilman, F. Isbell, E. W. Seabloom, E. T. Borer, and P. B. Reich. 2015. Anthropogenic environmental changes affect ecosystem stability via biodiversity. *Science* 348: 336–340. Of several factors explored, it was only those that decreased biodiversity that affected the stability of ecosystem productivity.

Hitzhusen, G. E. and M. E. Tucker. 2013. The potential of religion for Earth Stewardship. *Frontiers in Ecology and the Environment* 11: 368–376. There is a natural alliance between religion and conservation biology that can be a positive force.

Horton, C., T. R. Peterson, P. Banerjee, and M. J. Peterson. 2015. Credibility and advocacy in conservation science. *Conservation Biology* doi: 10.1111/cobi.12558. The nature of conservation biology as a normative science means that there can be a potential conflict between the dispassionate nature of science and the interest in promoting conservation policy.

Kloor, K. 2015. The battle for the soul of Conservation Science. *Issues in Science and Technology* 31: 74. The ongoing debate between those who believe that the field should be primarily guided by "nature for nature's sake" and others who believe that human needs should have greater weight.

Leopold, A. 1949. *A Sand County Almanac*. Oxford University Press, New York. Leopold's evocative essays articulate his "land ethic," defining human duty to conserve the land and the living things that thrive upon it.

Pimm, S. L. and 8 others. 2014. The biodiversity of species and their rates of extinction, distribution, and protection. *Science* 344: 1246752. Most species on the planet are not yet known to science, having restricted geographic ranges and therefore being at risk of going extinct before they are ever known.

Pooley, S. P., J. Andrew-Mendelsohn, and E. J. Milner-Gulland. 2014. Hunting down the chimera of multiple disciplinarity in conservation science. *Conservation Biology* 28: 22–32. Even as conservation biologists attempt to include a wider variety of disciplines, there are methodological and conceptual challenges to such broad approaches.

Roman, J., M. M. Dunphy-Daly, D. W. Johnston, and A. J. Read. 2015. Lifting baselines to address the consequences of conservation success. *Trends in Ecology and Evolution* 30.6: 299–302.

Sodhi, S. N., R. Butler, W. F. Laurance, and L. Gibson. 2011. Conservation successes at micro-, meso- and macroscales. *Trends in Ecology and Evolution* 26: 585–594. There are many examples of successful conservation that can be used to guide future actions.

Soulé, M. E. 1985. "What is conservation biology?" *BioScience* 35: 727–734. Key early paper defining the field, and still relevant today for its emphasis on the intrinsic value of biodiversity.

Wibbels, T. and E. Bevan. 2015. New Riddle in the Kemp's Ridley Saga. In *State of the World's Sea Turtles Report*. Oceanic Society.

KEY JOURNALS IN THE FIELD *Biodiversity and Conservation, Biological Conservation, BioScience, Conservation Biology, Conservation Letters, Ecological Applications, National Geographic, Trends in Ecology and Evolution*

2

What Is Biodiversity?

Coral reefs are built up from the skeletons of billions of tiny individual animals. The intricate coral landscapes create a habitat for a variety of other marine species, including hundreds of different fish and invertebrate creatures.

T he protection of biological diversity is central to conservation biology. Conservation biologists use the term biological diversity, or simply biodiversity, to mean the complete range of species and biological communities on Earth, as well as the genetic variation within those species and all ecosystem processes. By this definition, biodiversity must be considered on at least three levels (**FIGURE 2.1**):

1. *Species diversity*: All the species on Earth, including single-celled bacteria and protists as well as the species of the multicellular kingdoms (plants, fungi, and animals)

2. *Genetic diversity*: The genetic variation within species, both among geographically separate populations and among individuals within single populations

3. *Ecosystem diversity*: The different biological communities and their associations with the chemical and physical environment (the ecosystem)

All three levels of biodiversity are necessary for the continued survival of life as we know it, and all are important to people (Levin 2001; MEA 2005). All of these levels are also currently facing significant threats, to be discussed in Chapters 4 and 5, although threats to species diversity tend to receive the most attention.

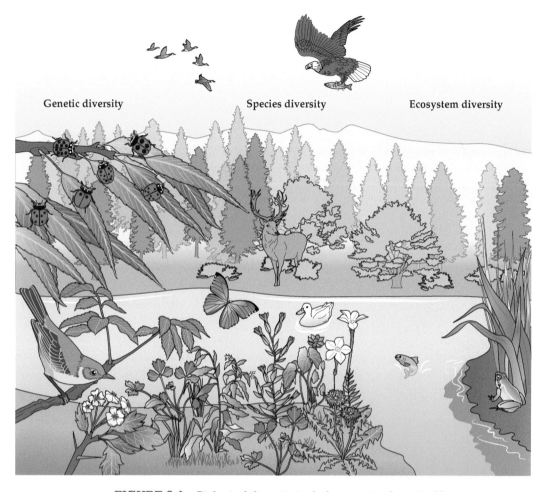

Genetic diversity   Species diversity   Ecosystem diversity

FIGURE 2.1 Biological diversity includes genetic diversity (the genetic variation found within each species), species diversity (the range of species in a given
ecosystem), and ecosystem diversity (the variety of habitat types and ecosystem
processes extending over a given region).

Species diversity reflects the entire range of evolutionary and ecological
adaptations of species to particular environments. It provides people with
resources and resource alternatives; for example, a tropical rain forest or
a temperate swamp with many species produces a wide variety of plant
and animal products that can be used as food, shelter, and medicine. **Genetic diversity** is necessary for any species to maintain reproductive vital-
ity, resistance to disease, and the ability to adapt to changing conditions
(Laikre et al. 2010). Genetic diversity is of particular value in the breeding
programs necessary to sustain and improve modern domesticated plants
and animals and their disease resistance. **Ecosystem diversity** results from
the collective response of species to different environmental conditions.

Biological communities found in deserts, grasslands, wetlands, and forests support the continuity of proper ecosystem functioning, which provides crucial services to people, such as water for drinking and agriculture, flood control, protection from soil erosion, and filtering of air and water. We will now examine each level of biodiversity in turn.

Species Diversity

Recognizing and classifying species is one of the major goals of conservation biology. Identifying the process whereby one species evolves into one or more new species is one of the ongoing accomplishments of modern biology. The origin of new species is normally a slow process, taking place over hundreds, if not thousands, of generations. The evolution of higher taxa, such as new genera and families, is an even slower process, typically lasting hundreds of thousands or even millions of years. In contrast, human activities are destroying the unique species built up by these slow natural processes in only a few decades.

What is a species?

Although seemingly a straightforward concept, how to distinguish a selection of organisms as a species is subject to great scientific discussion, and at least seven different ways of doing this have been proposed (Wiens 2007). The three most commonly used in conservation biology are:

1. *Morphological species*: A group of individuals that appear different from others, that is, that are morphologically distinct. A group that is distinguished exclusively by such visible traits as form or structure may be referred to as a **morphospecies**.
2. *Biological species*: A group of individuals that can potentially breed among themselves in the wild and that do not breed with individuals of other groups.
3. *Evolutionary species*: A group of individuals that share unique similarities in their DNA and hence their evolutionary past.

Because the methods and assumptions used are different, these three approaches to distinguishing species sometimes do not give the same results. Increasingly, DNA sequences and other molecular markers are being used to identify and distinguish species that look almost identical, such as types of bacteria (Papadopoulou 2015).

The **morphological definition of species** is the one most commonly used by **taxonomists**, biologists who specialize in the identification of unknown specimens and the classification of species (**FIGURE 2.2**). The **biological definition of species** is widely accepted, but it is problematic for groups of organisms in which different species readily interbreed, or **hybridize**,

Using morphological and genetic information to identify species is a major activity for taxonomists; accurate identification of a species is a necessary first step in its conservation.

(A)

(B)

FIGURE 2.2 (A) A plant ecologist prepares a museum specimen using a plant press. The flattened and dried plant will later be mounted on heavy paper with a label giving detailed collection information. (B) An ornithologist at the Museum of Comparative Zoology, Harvard University, classifying collections of orioles: black-cowled orioles (*Icterus prosthemelas*), from Mexico, and Baltimore orioles (*Icterus galbula*), which occur throughout eastern North America. (A, photograph by Richard Primack; B, photograph courtesy of Jeremiah Trimble, Museum of Comparative Zoology, Harvard University; © President and Fellows of Harvard College.)

such as plants. Furthermore, the biological definition of species is difficult to use because it requires a knowledge of which individuals actually have the potential to breed with one another. Similarly, the **evolutionary definition of species** requires access to expensive laboratory equipment and so cannot be used in the field. As a result, field biologists must rely on observable attributes, and they may name a group of organisms as a morphospecies until taxonomists can investigate them more carefully to determine if they are a distinct species (Chan et al. 2015).

Ideally, specimens collected in the field are catalogued and stored in one of the world's 6500 natural history museums (for animals and other organisms) or its 300 or more major herbaria (for plants and fungi). These permanent collections form the basis of species descriptions and systems of classification. Each species is given a unique two-part name (a **binomial**), such as *Canis lupus* for the gray wolf. The first part of the name, *Canis*, identifies the genus (the canids, or dogs). The second part of the name, *lupus*, identifies the smaller group within the genus, the species that is the gray wolf. This naming system both separates the gray wolf from and connects it to similar species—such as *Canis latrans*, the coyote, and *Canis rufus*, the red wolf.

FIGURE 2.3   Cope's gray tree frog (*Hyla chrysocelis*, left) is only distinguishable from the gray tree frog (*H. versicolor*, right) by their calls, but they fit both the biological and evolutionary definition of species because they have different numbers of chromosomes (*H. versicolor* is tetraploid; *H. chrysocelis* is diploid) and thus are incapable of interbreeding. (Left photograph, © Jack Glisson/Alamy Stock Photograph; right photograph, David McIntyre.)

Problems in distinguishing and identifying species are more common than many people realize (Frankham et al. 2012). For example, a single species may have several varieties that have observable morphological differences, yet are similar enough to be a single biological or evolutionary species. Different varieties of dogs, such as German shepherds, collies, and beagles, all belong to one species; their genetic differences are actually very small, and they readily interbreed. Alternatively, closely related "sibling" species appear very similar in morphology and physiology, yet are genetically quite distinct (**FIGURE 2.3**).

To further complicate matters, individuals of related but distinct species may occasionally mate and produce **hybrids**, intermediate forms that blur the distinction between species. Hybridization is particularly common among plant species in disturbed habitats. Hybridization in both plants and animals frequently occurs when a few individuals of a rare species are surrounded by large numbers of a closely related species. For example, the endangered California tiger salamander (*Ambystoma californiense*) and the introduced barred tiger salamander (*A. mavortium*) are thought to have evolved from a common ancestor five million years ago, yet they readily mate in California (**FIGURE 2.4**). The hybrid salamanders have a higher fitness and are better able to tolerate environmental pollution than the native species, *A. californiense*, further complicating the conservation of this endangered species (Ryan et al. 2013).

The inability to clearly distinguish one species from another, whether due to similarities of characteristics or to confusion over the correct scientific name, often slows down efforts at species protection. It is difficult to

FIGURE 2.4 The hybrid tiger salamander (left) is larger than its endangered parent species, the California tiger salamander (right), and is increasing in abundance. Note the much larger head of the hybrid salamander. (Photograph courtesy of H. Bradley Shaffer.)

write precise, effective laws to protect a species if scientists and lawmakers are not certain which individuals belong to which species. At the same time, species are going extinct before they are even described. Tens of thousands of new species are being described each year, but even this rate is not fast enough. The key to solving this problem is to train more taxonomists, especially for work in the species-rich tropics (Joppa et al. 2011).

Those conservation biologists primarily concerned with ecosystem function rather than individual species extinction have argued that a better measure than species diversity is **functional diversity**; that is, the diversity of organisms categorized by their ecological roles or traits rather than their taxonomy (Gagic et al. 2015). Functional diversity is an especially important concept in the context of habitat restoration (see Chapter 10). However, if our goal is to prevent untimely extinctions, we cannot avoid the task of identifying and measuring species diversity.

Measuring species diversity

Conservation biologists often want to identify locations of high species diversity. Quantitative definitions of species diversity have been developed by ecologists as a means of comparing the overall diversity of different communities at varying geographic scales (Flohre et al. 2011).

At its simplest level, species diversity can be defined as the number of species, called **species richness**. This number can be determined by several methods and at different geographic scales. Three diversity measurements are based on species richness:

- **Alpha diversity** is the number of species found in a given community, such as a lake or a meadow.
- **Gamma diversity** is the number of species at larger geographic scales that include a number of ecosystems, such as a mountain range or a continent.

- **Beta diversity** links alpha and gamma diversity and represents the *rate of change of species composition as one moves across a large region*. For example, if every lake in a region contained a similar array of fish species, then beta diversity would be low; on the other hand, if the bird species found in one forest were entirely different from the bird species in separate but nearby forests, then beta diversity would be high. There are several ways of calculating beta diversity; a simple measure of beta diversity can be obtained by dividing gamma diversity by alpha diversity.

We can illustrate these three types of diversity with a theoretical example of three mountain ranges (**FIGURE 2.5**). Region 1 has the highest alpha diversity, with more species per mountain on average (six species)

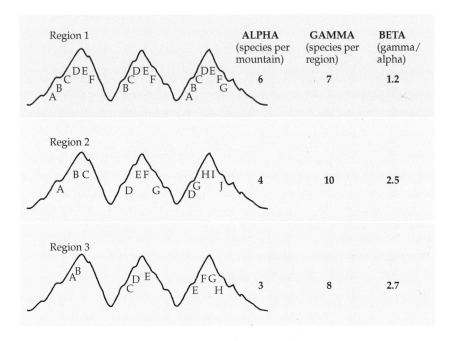

FIGURE 2.5   Biodiversity indexes for three regions, each consisting of three separate mountains. Each letter represents a population of a species; some species are found on only one mountain, while other species are found on two or three mountains. Alpha, gamma, and beta diversity values are shown for each region. If funds were available to protect only one region, Region 2 should be selected because it has the greatest gamma (total) diversity. However, if only one mountain could be protected, a mountain in Region 1 should be selected because these mountains have the highest alpha (local) diversity, that is, the greatest average number of species per mountain. Each mountain in Region 3 has a more distinct assemblage of species than the mountains in the other two regions, as shown by the higher beta diversity. If Region 3 were selected for protection, the relative priority of the individual mountains should then be judged based on how many unique species are found on each mountain.

▶

FIGURE 2.6 If each circle represents a random sample of fish from a pond and colors represent species, both have the same species richness: 5. However, pond sample A is dominated by a single species (6 orange fish out of a total of 10 fish or 60% of the total), while each of the other four species has only 10% of the total. In contrast, pond sample B has perfect evenness, that is, each of the five species has the same number of individuals, or 20% of the total. Therefore we would consider pond B to have greater species diversity. We can further quantify this by calculating H, a measure of diversity, as shown in each table, with pond A having a diversity of 0.53 and pond B having a diversity of 0.70.

> Identifying patterns of species diversity helps conservation biologists establish which locations are most in need of protection.

than the other two regions. Region 2 has the highest gamma diversity, with a total of 10 species. Dividing gamma diversity by alpha diversity shows that Region 3 has a higher beta diversity (2.7) than Region 2 (2.5) or Region 1 (1.2) because all of its species are found on only one mountain each. In practice, indexes of diversity are often highly correlated. The plant communities of the eastern foothills of the Andes, for instance, show high levels of diversity at alpha, beta, and gamma scales.

More complex indexes, such as the **Shannon diversity index** (also called the *Shannon-Wiener index*), Simpson index, and Pielou evenness index, take the relative abundance of different species into account; by these measures, a community dominated by a few species is less diverse than one with a more even distributions of species, even with the same species richness. The Shannon diversity index is calculated as

$$H = -\sum[p_i \times \ln(p_i)]$$

that is, the negative sum of the proportion (*p*) of each species (*i*) multiplied by the natural log (ln) of *p*. In a simple example, let's image two ponds, each of which has five fish species. In pond A, 60% of the individuals are orange carp and each of the remaining four species only represent 10% of the individuals, whereas in pond B there are also five fish species but each of them has equal numbers of individuals, or 20% of the total. Using the Shannon diversity index, pond B will have a greater diversity than pond A (**FIGURE 2.6**). In some cases, one pond may even have a greater number of species but a lower diversity index than another pond if its community is dominated by one or a few particular species. Note that, like the richness values explained above, diversity measures of this type can be calculated at different scales and therefore are useful only as relative, rather than absolute, values. Furthermore, these quantitative definitions of diversity capture only part of the broad definition of biodiversity used by conservation biologists, and new ones continue to be developed (Iknayan et al. 2014; Magurran and McGill 2013). Although each has its limitations, they are useful for comparing regions and highlighting areas that have large numbers of native species requiring conservation protection.

(A)

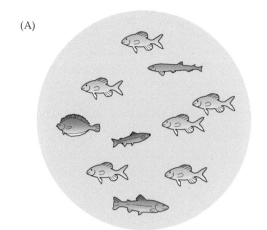

(B)

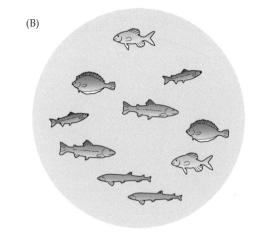

	Number in sample (n)	% in sample (p)	Natu-ral log [ln(p)]	ln(p) × (p)
	6	0.60	−0.51	−0.31
	1	0.10	−2.30	−0.23
	1	0.10	−2.30	−0.23
	1	0.10	−2.30	−0.23
	1	0.10	−2.30	−0.23
Total	10 fish	1 (100%)		−1.23
				H = 1.23

	Number in sample (n)	% in sample (p)	Natu-ral log ln[(p)]	ln(p) × (p)
	2	0.20	−1.61	−0.32
	2	0.20	−1.61	−0.32
	2	0.20	−1.61	−0.32
	2	0.20	−1.61	−0.32
	2	0.20	−1.61	−0.32
Total	10 fish	1 (100%)		−1.61
				H = 1.61

Genetic Diversity

Conservation biology also concerns itself with the preservation of genetic diversity within a species. This level of diversity is important because it provides evolutionary flexibility: when environmental conditions change, a genetically diverse species is more likely to have traits that allow it to adapt. Rare species often have less genetic variation than widespread species and, consequently, are more vulnerable to extinction (Frankham et al. 2009; see Chapter 5).

Genetic diversity is important both within and among populations. A **population** is a group of individuals that mate with one another and produce offspring; species may contain one or many populations.

Genetic diversity arises because individuals have slightly different forms of their **genes** (or **loci**), the units of the chromosomes that code for

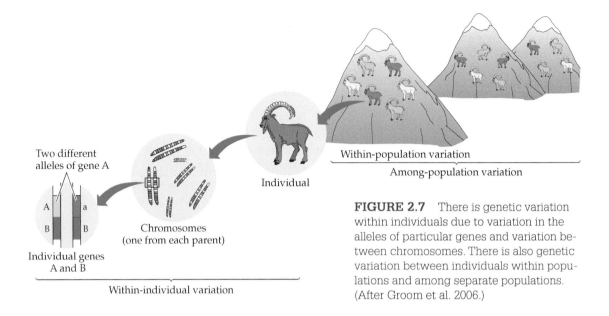

Two different
alleles of gene A

A / a

B / B

Individual genes
A and B

Chromosomes
(one from each parent)

Individual

Within-individual variation

Within-population variation

Among-population variation

FIGURE 2.7 There is genetic variation within individuals due to variation in the alleles of particular genes and variation between chromosomes. There is also genetic variation between individuals within populations and among separate populations. (After Groom et al. 2006.)

specific proteins. These different forms of a gene are known as **alleles**, and the differences originally arise through **mutations**—changes that occur in the deoxyribonucleic acid (DNA) that constitutes an individual's chromosomes. Genetic variation increases when offspring receive unique combinations of genes and chromosomes from their parents via the **recombination** of genes that occurs during sexual reproduction. Genes are exchanged between chromosomes, and new combinations are created when chromosomes from two parents combine to form a genetically unique offspring. Although mutations provide the basic material for genetic variation, the random rearrangement of alleles in different combinations that characterizes sexually reproducing species dramatically increases the potential for genetic variation (**FIGURE 2.7**).

The total array of genes and alleles in a population is the **gene pool** of the population, while the particular combination of alleles that any individual possesses is its **genotype** (Winker 2009). The **phenotype** of an individual represents the morphological, physiological, anatomical, and biochemical characteristics of that individual that result from the expression of its genotype in a particular environment. Examples of phenotypes include eye color and blood type, physical qualities that are determined predominantly by an individual's genotype.

The amount of genetic variation in a population is determined by both the number of genes that have more than one allele (**polymorphic genes**) and the number of alleles for each of these genes. The existence of a polymorphic gene also means that some individuals in the population will be **heterozygous** for the gene; that is, they will receive a different allele of the gene from each parent. On the other hand, some individuals will be **homozygous**: they will receive the same allele from each parent. All these levels of genetic variation contribute to a population's (and therefore a species') ability to adapt to a changing environment.

> Genetic variation within a species can allow the species to adapt to environmental change; genetic variation can also increase the value of domesticated species to people.

Ecosystem Diversity

Ecosystems are diverse, and this diversity is apparent even across a particular landscape. As we climb a mountain, for example, the structure of the vegetation and the kinds of plants and animals present gradually change from those found in a tall forest to those found in a low, moss-filled forest to alpine meadow to cold, barren rock. As we move across the landscape, physical conditions (soil, temperature, precipitation, and so forth) change. One by one, the species present at our starting point drop out, and we encounter new species that were not found there. The landscape as a whole is dynamic and changes in response to the overall environment and the types of human activities that are associated with it.

What are communities and ecosystems?

A **biological community** is defined as the species that occupy a particular locality and the interactions among those species. A biological community, together with its associated physical and chemical environment, is termed an **ecosystem** (**FIGURE 2.8**). Many characteristics of an ecosystem result from ongoing processes, including water cycles, nutrient cycles, and energy capture. These processes occur at geographic scales that range from square meters to hectares and all the way to regional scales involving tens of thousands of square kilometers (see Table 1.1 for definitions of these metric terms). For example, in a temperate forest, rain falls and is absorbed by the soil. Some of that rain evaporates from the surface, some percolates to groundwater reserves, and some is taken up by plants that use it in photosynthesis, converting atmospheric CO_2 into carbohydrates. These plants may then be eaten by animals, which convert the carbohydrates back to energy through respiration, which releases CO_2 back into the environment. Other plants may decompose on the forest floor, releasing nutrients and providing energy for bacteria, fungi, and animals.

The physical environment, especially annual cycles of temperature and precipitation and the characteristics of the land surface, affects the structure and characteristics of a biological community and profoundly influences whether a site will support a forest, grassland, desert, or wetland. In aquatic ecosystems, physical characteristics such as water turbulence and clarity, as well as water chemistry, temperature, and depth, affect the characteristics of the associated **biota** (a region's flora and fauna). In turn, the biological community can alter the physical characteristics of an environment. For example, wind speeds are lower and humidity is higher

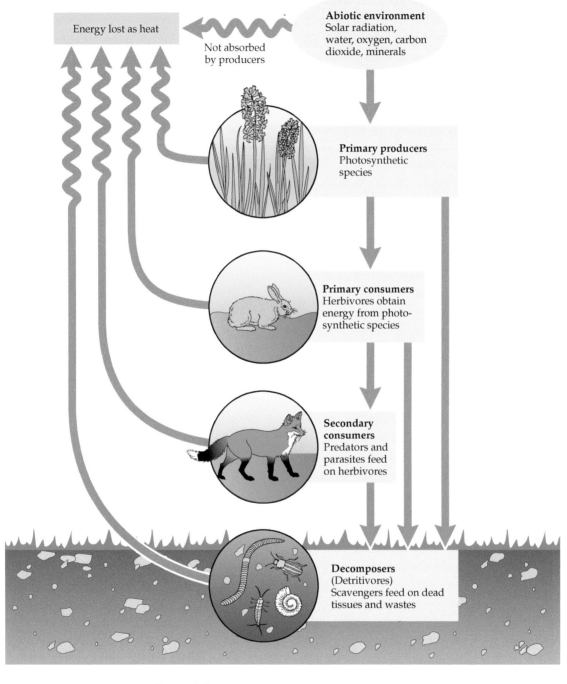

FIGURE 2.8 A model of a field ecosystem, showing its trophic levels and simplified energy pathways.

inside a forest than in a nearby grassland. Marine communities such as kelp forests and coral reefs can affect the physical environment as well by buffering wave action.

Within a biological community, species play different roles and differ in what they require to survive (Roscher et al. 2015). For example, a given plant species might grow best in one type of soil under certain conditions of sunlight and moisture, be pollinated only by certain types of insects, and have its seeds dispersed by certain bird species. Similarly, animal species differ in their requirements, such as the types of food they eat and the types of resting and breeding places they prefer, collectively referred to as **habitat**. Even though a forest may be full of vigorously growing green plants, an insect species that feeds on only one rare and declining plant species may be unable to develop and reproduce because it cannot get the specific food that it requires. Any of these requirements may become a **limiting resource** when it restricts the size of a population.

> Within a community, each species has its own requirements for food, temperature, water, and other resources, any of which may limit its population size and its distribution.

Species interactions within ecosystems

The composition of ecosystems is often affected by **competition** and **predation** (Cain et al. 2014). **Predators** are animals that hunt and eat **prey**, which are the organisms that are eaten. Predation on plants is generally referred to as **herbivory**. Predators of all types may dramatically reduce the densities of certain prey species and even eliminate some species from particular habitats. Indeed, predators may indirectly increase the number of prey species in an ecosystem by keeping the density of each species so low that severe competition for resources does not occur.

In many ecosystems, predators keep the number of individuals of a particular prey species below the number that the resources of the ecosystem can support, a number termed the habitat's **carrying capacity**. If the predators (e.g., wolves; *Canis lupus*) are removed by hunting, poisoning, or some other human activity, the prey population (e.g., deer; *Odocoileus* spp.) may increase to carrying capacity, or it may increase beyond carrying capacity to a point at which crucial resources are overtaxed and the population crashes. In addition, the population size of a species may be controlled by other species that compete with it for the same resources; for example, the population size of terns that nest on a small island may decline or grow if a gull species that uses the same nesting sites becomes abundant or is eliminated from the site.

Community composition is also affected when two species benefit each other in a **mutualism** (**FIGURE 2.9**). Mutualistic species reach higher densities when they occur together than when only one of the species is present. One example of mutualism is the relationship between fruit-eating birds and plants with fleshy fruit containing seeds that are dispersed by birds. Another example is flower-pollinating insects and flowering plants. In some cases these relationships are **symbiotic**, and the species apparently

FIGURE 2.9 This bohemian waxwing (*Bombycilla garrulus*) will disperse the fruit of a mountain ash tree (*Sorbus aucuparia*) by ingesting the fruit and then defecating the seed in some other location far from the parent tree. In this way, both species benefit. (© All Canada Photos/Alamy Stock Photo.)

cannot survive without each other. For example, certain symbiotic algae living inside coral animals are ejected following unusually high water temperatures in tropical areas, leading to the weakening and subsequent death of their associated coral species.

Biological communities can be organized into **trophic levels** that represent the different ways in which species obtain energy from the environment (see Figure 2.8). **Primary producers** make up the *first trophic level*. These organisms obtain their energy directly from the sun via photosynthesis. In terrestrial environments, higher plants, such as flowering plants, gymnosperms, and ferns, are responsible for most photosynthesis, while in aquatic environments, seaweeds, single-celled algae, and cyanobacteria (also called blue-green algae) are the most important. All of these species use solar energy to build the organic molecules they need to live and grow. Because less energy is transferred to each successive trophic level, the greatest biomass (living weight) in a terrestrial ecosystem is usually that of the plants.

The *second trophic level* contains the **herbivores**, which eat primary producers and are thus known as **primary consumers**. The intensity of grazing by herbivores often determines the relative abundance of plant species and even the amount of plant material present.

Carnivores are in the third and higher trophic levels. **Carnivores** are animals that obtain energy by eating other animals. At the third trophic level are **secondary consumers** (e.g., foxes), predators that eat herbivores (e.g., rabbits). At the fourth trophic level are **tertiary consumers** (e.g., bass), predators that eat other predators (e.g., frogs).

Some secondary and higher consumers combine direct predation with scavenging behavior. Others, known as **omnivores**, include both animal and plant foods in their diets. In general, predators occur at lower densities than their prey, and populations at higher trophic levels contain fewer individuals than those at lower trophic levels. A single savanna can support many more zebras than lions.

Parasites and disease-causing organisms, **pathogens**, form an important subclass of predators. Parasites of animals, including mosquitoes, ticks, intestinal worms, and protozoans, as well as microscopic disease-causing organisms such as some bacteria and viruses, do not kill their hosts immediately, if ever. Plants can also be attacked by bacteria, viruses, and a variety of parasites that include fungi, other plants (such as mistletoe), nematode worms, and insects. The effects of parasites range from imperceptibly weakening their hosts to totally debilitating or killing them over time. The spread of parasites and disease from captive or domesticated species, such as dogs, to wild species, such as lions, is a major threat to many rare species (see Chapter 4).

Decomposers and **detritivores** feed on dead plant and animal tissues and wastes (detritus), breaking down complex tissues and organic molecules into the simple chemicals that are the building blocks of primary production. Decomposers release minerals such as nitrates and phosphates back into the soil and water, where they can be taken up again by plants and algae. Decomposers are usually much less conspicuous than herbivores and carnivores, but their role in the ecological community is vital. The most important decomposers are fungi and bacteria, but a wide range of other species play a role in breaking down organic materials. For example, vultures and other scavengers tear apart and feed on dead animals, dung beetles feed on and bury animal dung, and worms break down fallen leaves and other organic matter. Crabs, worms, molluscs, fish, and numerous other organisms eat detritus in aquatic environments. If decomposers were to die off, organic material would accumulate and plant growth would decline greatly (Gessner et al. 2010).

Food chains and food webs

Although species can be organized into the general trophic levels we have just described, their actual requirements or feeding habits within those trophic levels may be quite restricted. For example, a certain aphid species may feed on only one type of plant, and a certain lady beetle species may feed on only one type of aphid. These specific feeding relationships are termed **food chains**. The more common situation in many biological communities, however, is for one species to feed on several other species at the lower trophic level, to compete for food with several species at its own trophic level, and in turn, to be preyed on by several species at the higher trophic level. Consequently, a more accurate description of the organization of biological communities is a **food web**, in which species are linked together through complex feeding relationships (Yodzis 2001) (**FIGURE 2.10**). Species at the same trophic level that use approximately the same environmental resources are considered to be a **guild** of competing species.

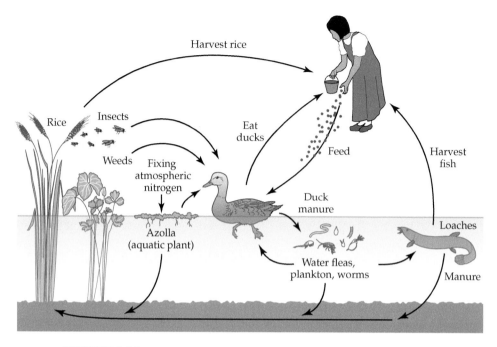

FIGURE 2.10 A simple food web in a traditional agricultural ecosystem. Photosynthetic plants are eaten by people, ducks, and insects. Insects and aquatic invertebrates are eaten by ducks and fish, which are then eaten by people.

Humans can substantially alter the relationships in food webs (Alva-Basurto and Arias-Gonzalez 2014). For example, in urban settings, bird populations may increase due to reduced numbers of predators, reducing insect abundance in the process (Faeth et al. 2005).

Keystone species and resources

Within biological communities, certain species or guilds of species with similar ecological features may determine the ability of many other species to persist in the community (**FIGURE 2.11**). These **keystone species** affect the organization of the community to a far greater degree than we would predict if we considered only their numbers or biomass (Estes et al. 2011). Protecting keystone species is a priority for conservation efforts because loss of a keystone species or guild will lead to losses of numerous other species as well.

Top predators are often considered keystone species because they can markedly influence herbivore populations (Ripple et al. 2014). The elimination of even a small number of individual predators, even though they constitute only a minute fraction of the community biomass, may result in dramatic changes in the vegetation and a great loss in biodiversity, sometimes called

> Keystone species strongly affect the abundance and distribution of other species in an ecosystem. Protecting and restoring keystone species is a conservation priority.

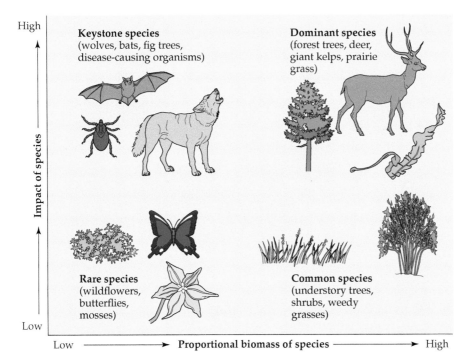

FIGURE 2.11 Keystone species determine the ability of large numbers of other species to persist within a biological community. Although keystone species make up only a small percentage of the total biomass, a community's composition would change radically if one of them were to disappear. Rare species have minimal biomass and seldom have significant effects on the community. Dominant species constitute a large percentage of the biomass and affect many other species in proportion to this large biomass. Some species, however, have a relatively low impact on the community organization despite being both common and heavy in biomass. (After Power et al. 1996.)

a **trophic cascade** (Jorge et al. 2013; Ripple and Beschta 2012). For example, in some places where gray wolves (*Canis lupis*) have been hunted to extinction by humans, deer (*Odocoileus virginiana*) populations have exploded. The deer severely overgraze the habitat, eliminating many herb and shrub species. The loss of these plants, in turn, is detrimental to the deer and to other herbivores, including insects. The reduced plant cover may lead to soil erosion, also contributing to the loss of species that inhabit the soil. When wolves are restored to ecosystems, trophic relationships can sometimes be reestablished (Beyer et al. 2007).

Species that extensively modify the physical environment through their activities, often termed **ecosystem engineers**, are also considered keystone species (Jones et al. 1996; Romero et al. 2015) (**FIGURE 2.12**). Losing keystone species can create a series of linked extinction events, known as an **extinction cascade**, resulting in a degraded ecosystem with much

FIGURE 2.12 The North American beaver (*Castor canadensis*) is considered an ecosystem engineer because the dams it constructs cause overbank flooding, thereby creating new wetland habitat for themselves and other species. (Dam photo, © Richard Hamilton Smith/Corbis; beaver photo, © iStock.com/Musat.)

lower biodiversity at all trophic levels. This may already be happening in tropical forests where overharvesting has drastically reduced the populations of birds and mammals that act as predators, seed dispersers, and herbivores (Naniwadekar et al. 2015). While such a forest appears to be green and healthy at first glance, it is really an "empty forest" in which ecological processes have been irreversibly altered such that many plant and animal species will be eliminated over succeeding decades or centuries (Hollings et al. 2014; Redford 1992). In the marine environment, the loss of key structural species such as sea grasses and seaweeds can lead to the loss of specialized species that inhabit such communities, such as delicate sea dragons and sea horses (Hughes et al. 2009). If the few keystone species in a community being affected by human activities can be identified, they can sometimes be carefully protected or even actively managed to increase their numbers.

Particular habitats may contain **keystone resources**, often physical or structural, that occupy only a small area yet are crucial to many species in the ecosystem (Kelm et al. 2008). For example, deep pools in streams, springs, and ponds may be the only refuge for fish and other aquatic species during the dry season, when water levels drop. For terrestrial animals, these water sources may provide the only available drinking water for a considerable distance. Hollow tree trunks and tree holes are keystone resources as breeding sites for many bird and mammal species and may limit their population sizes (Cockle and Martin 2015). Protecting old hollow trees as a keystone resource is a priority during certain logging activities.

Ecosystem dynamics

An ecosystem in which the processes are functioning normally, whether or not there are human influences, is referred to as a **healthy ecosystem**. In many cases, ecosystems that have lost some of their species will remain healthy because there is often some redundancy in the roles performed by ecologically similar species. Ecosystems that are able to remain in the same state are referred to as **stable ecosystems**. These systems remain stable either because of lack of disturbance or because they have special features that allow them to remain stable in the face of disturbance. Such stability despite disturbance can result from one or both of two features: resistance and resilience. **Resistance** is the ability to maintain the same state even with ongoing disturbance; a river ecosystem that retained its major ecosystem processes after an oil spill would be considered resistant. **Resilience** is the ability to return to an original state quickly after disturbance has occurred; that would be true if, following contamination by an oil spill and the deaths of many animals and plants, a river ecosystem soon returned to its original condition (Bhagwat et al. 2012; Zolli and Healy 2012). As another example, when nonnative fish are introduced into previously fish-free ponds, the number of native animal species declines, indicating low resistance. When the fish die out, however, the number of native species soon recovers, indicating high resilience (Knapp et al. 2005).

Biodiversity Worldwide

Developing a strategy for conserving biodiversity requires a firm grasp of how many species exist on Earth and how those species are distributed across the planet. The answers to both questions can be complex.

How many species exist worldwide?

At present, about 1.5 million species have been described (Costello 2015) (**FIGURE 2.13**). At least two to three times this number of species (primarily insects and other arthropods in the tropics) remain undescribed. Our knowledge of species numbers is imprecise because inconspicuous species have not received their proper share of taxonomic attention. For example, spiders, nematodes (microscopic worms), and fungi living in the soil and insects living in the tropical forest canopy are small and difficult to study.

(A)

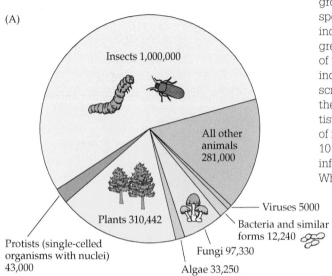

FIGURE 2.13 (A) Approximately 1.73 million species have been identified and described by scientists; the majority of these species are insects and plants. (B) For those groups estimated to contain over 100,000 species, the numbers of described species are indicated by the blue portions of the bars; the green portions are estimates of the number of undescribed species. The vertebrates are included for comparison. The number of undescribed species is particularly speculative for the microorganisms (viruses, bacteria, protists). Most estimates of the possible number of identifiable species range from 5 million to 10 million. (Data from Blackwell 2011, eol.org/info/458, Guiry 2012, Pawlowski et al. 2012, Whitman et al., and Wilson 2010.)

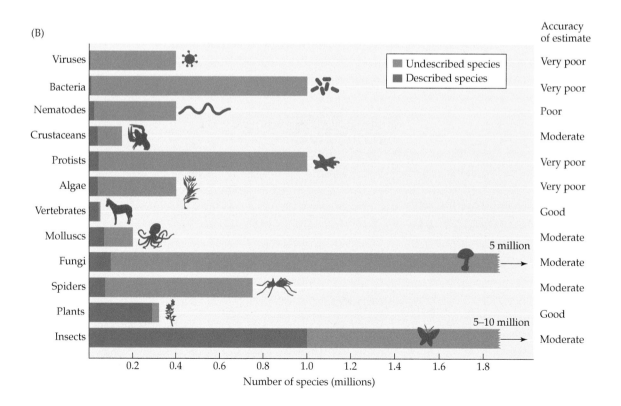

These poorly known groups could number in the hundreds of thousands, or even millions, of species. Our best estimate is that there are between 5 million and 10 million species on Earth (Strain 2011). This number could be many times higher, even up to 100 million species, if it turns out that each species or genus of animal and plant has many unique species of bacteria, protists, and fungi living on or inside it.

Amazingly, about 20,000 new species are described each year. While certain groups of organisms, such as birds, mammals, and temperate flowering plants, are relatively well known, a small but steady number of new species in these groups are being discovered each year (Joppa et al. 2011). Even among a group as well studied as primates, dozens of new monkey species have been found in Brazil, and dozens of new species of lemurs have been discovered in Madagascar—all since 1990. Every decade, 500–600 new species of amphibians are described.

Most of the new animals discovered are insects and other invertebrates; for example, in June 2015, a group of taxonomists and other volunteers discovered seven new species of spiders, including a new genus of tarantula, in Judbarra/Gregory National Park in Australia's Northern Territory during a one-day event called the Bush Blitz. In a **bioblitz**, scientists perform an intensive biological survey of a designated area in a short time with the goal of documenting all living species there (**FIGURE 2.14**). The Bush Blitz is a bioblitz supported by a partnership between the Australian government, museums, and other nonprofit organizations, and since 2010 it has discovered more than 900 new species, including insects, plants, fish, birds, and many other taxonomic groups.

> Many scientists are working to determine the number of species on Earth. The best estimate is that there are about 5 million to 10 million species, about half of them insects.

FIGURE 2.14 Three generations of the Dearborn family look for insect specimens to collect. Acadia National Park, 2009 BioBlitz. (Courtesy of the National Park Service.)

In addition to new species, entire biological communities continue to be discovered, often in extremely remote and inaccessible localities. These communities often consist of inconspicuous species, such as bacteria, protists, and small invertebrates, that have escaped the attention of earlier taxonomists. Specialized exploration techniques have aided in these discoveries, particularly in the deep sea and in the forest canopy. For example, each new deep-sea hydrothermal vent explored reveals dozens, if not hundreds, of species previously unknown to science (Scheckenback et al. 2010). Drilling projects have shown that diverse bacterial communities exist 2.8 km deep in the Earth's crust at densities of up to 100 million bacteria per gram of solid rock (Fisk et al. 1998). In the depths of the northwestern Pacific near Japan, 50 new species of nematodes were found on a single collecting trip (Fadeeva et al. 2015). The unique features of these discoveries not only expand our understanding of the biodiversity of our planet, but also raise questions regarding evolutionary and physiological processes (Nakamura and Takai 2015). These organisms are also being actively investigated as sources of novel chemicals, as medicines, for their potential usefulness in degrading toxic chemicals, and for insight into whether life could exist on other planets.

The bacteria in particular are very poorly known (Azam and Worden 2004) and are thus underrepresented in estimates of the total species on Earth (Dvořák et al. 2015), even as new advances in technology provide an ever more accurate count of species in this group (Zheng et al. 2015). Yet they have an important role to play in ecosystem functioning. Only about 11,000 species of bacteria are currently recognized by microbiologists (Kyrpides et al. 2014) because they are difficult to grow and identify. However, analyses of bacterial DNA indicate that there may be from 6400 to 38,000 species in a single gram of soil and 160 species in a milliliter of seawater (Nee 2003). Such high diversity in small samples suggests that there could be tens of thousands, or even millions, of undescribed bacterial species. The Human Microbiome Project has demonstrated that the human body is occupied by 10 times more bacterial cells than human cells, and that many still undescribed species occupy specific places both inside our bodies and on our skin. Furthermore, the diversity and abundance of species in different parts of the body is related to the health of an individual (Human Microbiome Consortium 2012). The human microbiome is a very active area of current research.

DNA analyses suggest that many thousands of species of bacteria have yet to be described. The marine environment also contains large numbers of species unknown to science.

Considering that about 16,000 new animal species are described each year and that at least 3 million more are waiting to be identified, the task of describing the world's species will not be completed for over 180 years if continued at the present rate. This fact underlines the critical need for taxonomists trained to use the latest molecular technology and for web-based information sharing. International databases such as those of the Global Biodiversity Information Facility (www.gbif.org) and the Encyclopedia of Life project (www.eol.org) will make species names and descriptions more widely and readily available.

Where is the world's biodiversity found?

The most species-rich environments appear to be tropical forests, coral reefs, the deep sea, large tropical lakes and river systems, and regions with Mediterranean climates (Forzza et al. 2012).

TROPICAL FORESTS   Even though the world's tropical forests occupy only 7% of the land area, they contain more than half the world's species, most of which are insects (Corlett and Primack 2010; Gaston and Spicer 2004). Tropical forests also have many species of birds, mammals, amphibians, and plants. Among flowering plants, gymnosperms, and ferns, about 40% of the world's 275,000 species occur in tropical forest areas. Each of the rain forest areas in the Americas, Africa, Madagascar, Southeast Asia, New Guinea, Australia, and various tropical islands has a different biogeographic history, resulting in unique assemblages of species (see Figure 4.5). For example, lemurs are found only in Madagascar, and hummingbirds are found only in the Americas.

OCEANIC DIVERSITY   In the oceans, diversity is spread over a much broader range of phyla and classes than in terrestrial ecosystems. These marine systems contain representatives of 28 of the 35 animal phyla that exist today; one-third of these phyla exist only in the marine environment (Grassle 2001). In contrast, only one phylum is found exclusively in the terrestrial environment. The broad diversity in the ocean may be due to its great age, enormous water volume, degree of isolation of some seas by intervening landmasses, and other factors (**FIGURE 2.15**).

Within oceans, coral reef ecosystems are particularly diverse; although they occupy less than 0.1% of ocean surface area, they are home to one-third of marine fish species (Bowen et al. 2013; Fisher et al. 2015; see the chapter opening photo). The physical structure is created by the corals (invertebrate

> Species diversity is greatest in the tropics, particularly in tropical forests and coral reefs.

FIGURE 2.15 A community of phytoplankton, including various species of both diatoms and dinoflagellates. Microscopic organisms such as these are generally less understood than those more easily seen, and new species are frequently discovered. (© FLPA/Alamy Stock Photo.)

animals), which provide habitat for many other organisms. The photosynthetic algae that live mutualistically inside the corals provide them with abundant carbohydrates. One explanation for the richness of coral reefs is their high primary productivity: 2500 grams of biomass (living matter) per square meter per year, in comparison with 125 $g/m^2/y$ in the open ocean. Extensive niche specialization among coral species and adaptations to varying levels of disturbance may also account for the high species richness found in coral reefs. The world's largest coral reef is Australia's Great Barrier Reef, with an area of 349,000 km^2, which contains over 400 species of corals, 1500 species of fish, 4000 species of molluscs, and 6 species of turtles. Coral reefs also support some 252 species of birds.

MEDITERRANEAN-TYPE COMMUNITIES   Great diversity is found among plant species in Southwest Australia, the Cape region of South Africa, California, central Chile, and the Mediterranean basin, all of which are characterized by a Mediterranean climate of moist winters and hot, dry summers (see Figure 4.5). Of these regions, the Mediterranean basin is the largest in area (2.1 million km^2) and has the most plant species (22,500) (Conservation International and Caley 2008). The Cape Floristic Region of South Africa has an extraordinary concentration of unique plant species (9000) in a relatively small area (78,555 km^2) (**FIGURE 2.16**). The shrub and herb communities in these areas are rich in species, apparently because of their combination of considerable geologic age, complex site characteristics (such as topography and soils), and severe environmental conditions. The high frequency of fire in these areas may also favor rapid speciation and prevent the domination of just a few species.

FIGURE 2.16 The Cape region of South Africa has evolved a unique ecosystem, collectively called the fynbos (meaning "fine-leaved plants" in Dutch), characterized by large numbers of plant species that are found nowhere else. (Photograph by Anna Sher.)

The distribution of species

At various spatial scales, there are concentrations of species in particular places, and there is a rough correspondence in the distribution of species richness between different groups of organisms (Domisch et al. 2015; Simaika et al. 2013). For example, in North America, large-scale patterns of species richness are highly correlated for amphibians, birds, butterflies, mammals, reptiles, land snails, trees, all vascular plants, and tiger beetles; that is, a region with numerous species of one group will tend to have numerous species of the other groups (Ricketts et al. 1999). On a local scale, however, this relationship may break down; for example, amphibians may be most diverse in wet, shady habitats, whereas reptiles may be most diverse in drier, open habitats. At a global scale, each group of living organisms may reach its greatest species richness in a different part of the world because of historical circumstances or the suitability of the site to its needs.

Places with large concentrations of species often have a high percentage of **endemic species**, that is, species that occur there and nowhere else. The countries with the largest numbers of endemic mammals, which represent important targets for conservation efforts, are Indonesia (259), Australia (241), Brazil (183), Madagascar (187), and Mexico (158) (www.iucnredlist. org). Geographically isolated countries and islands also tend to have high percentages of endemic species. For example, most of the species in Madagascar are endemic because it is a large, ancient island on which many unique species have evolved in isolation. Species-rich countries on the mainland, such as Tanzania, have comparatively fewer endemic species because they share many species with neighboring countries.

Almost all groups of organisms show an increase in species diversity toward the tropics (Groombridge and Jenkins 2010) (see Figures 4.5 and 6.20). For example, Thailand has 241 species of mammals, while France has only 104, despite the fact that both countries have roughly the same land area (**TABLE 2.1**). The contrast is particularly striking for trees and other flowering plants: 10 hectares of forest in Amazonian Peru or Brazil might have 300 or more tree species, whereas an equivalent forest area in temperate Europe or the United States would probably contain 30 species or less. Within a given continent, the number of species increases toward the equator.

For many groups of marine species, the greatest diversity of coastal species is found in the tropics, with a particular richness of species in the western Pacific. For open-ocean species, the greatest diversity is found at midlatitudes. Temperature is the most important variable explaining these patterns (Tittensor et al. 2010).

Local variation in climate, sunlight and rainfall, topography, and geologic age also affects patterns of species richness (Zellweger et al. 2015). In terrestrial communities, species richness tends to increase with decreasing elevation, increasing solar radiation, and increasing precipitation; that is, hot, rainy lowland areas have the most species. Species richness can be greater where complex topography and great geologic age provide more environmental variation, which allows genetic isolation, local adaptation, and speciation to occur. Geologically complex areas can produce a variety of soil

TABLE 2.1 Number of Native Mammal Species in Selected Tropical and Temperate Countries Paired for Comparable Size

Tropical country	Area (1000 km^2)	Number of mammal species	Temperate country	Area (1000 km^2)	Number of mammal species
Brazil	8456	648	Canada	9220	202
DRC^a	2268	430	Argentina	2737	374
Mexico	1909	523	Algeria	2382	105
Indonesia	1812	670	Iran	1636	186
Colombia	1039	442	South Africa	1221	297
Venezuela	882	363	Chile	748	143
Thailand	511	311	France	550	123
Philippines	298	207	United Kingdom	242	74
Rwanda	25	184	Belgium	30	70

Source: Data from 2008 IUCN Red List, accessed in 2015.
^{a}DRC = Democratic Republic of the Congo.

conditions with very sharp boundaries between them, leading to multiple communities and plant species adapted to one specific soil type or another.

With better methods of exploration and investigation, we are now able to appreciate the great diversity of the living world. This is truly a golden age of biological exploration. Natural history societies and clubs that combine amateur and professional naturalists contribute to this effort. Yet with this knowledge of biodiversity comes both the awareness of the damaging impact of human activity, which is diminishing it right before our eyes, and the responsibility to protect and restore that biodiversity that still remains.

Summary

- Taxonomists use morphological and genetic information to describe and identify the world's species. Places vary in their species richness, the number of species found in a particular location.

- There is genetic variation among individuals within a species. Genetic variation allows species to adapt to a changing environment, and it is valuable for the continuing improvement of crop plants and domesticated animals.

- Within an ecosystem, species play different roles and have varying requirements for survival. Certain keystone species are important in determining the ability of other species to persist in an ecosystem.

■ It is estimated that there are 5 million to 10 million species, most of which are insects. The majority of the world's species have still not been described and named. Further work is needed to describe microorganisms such as bacteria.

■ The greatest biological diversity is found in tropical regions, with particular concentrations of species in rain forests and coral reefs. The ocean may also have great species diversity but needs further exploration.

For Discussion ■—

1. How many species of birds, trees, and insects can you identify in your neighborhood? How could you learn to identify more? Is it important to be able to identify species in the wild?

2. What are the factors promoting species richness? Why is biological diversity diminished in particular environments? Why aren't species able to overcome these limitations and undergo the process of speciation?

3. Conservation efforts usually target genetic variation, species diversity, biological communities, and ecosystems for protection. What are some other components of natural systems that need to be protected? What do you think is the most important component of biodiversity, and why do you believe it is most important?

Suggested Readings —

Albert, A., K. McKonkey, T. Savini, and M. C. Huynen. 2014. The value of disturbance-tolerant cercopithecine monkeys as seed dispersers in degraded habitats. *Biological Conservation* 170: 300–310. Monkeys are important in dispersing seeds and helping the forest to regenerate.

Chan, Y. F., K.-P. Chiang, J. Chang, Ø. Moestrup, and C.-C. Chung. 2015. Strains of the morphospecies *Ploeotia costata* (Euglenozoa) isolated from the Western North Pacific (Taiwan) reveal substantial genetic differences. *Journal of Eukaryotic Microbiology* 62: 318–326.

Corlett, R. and R. B. Primack. 2010. *Tropical Rainforests: An Ecological and Biogeographical Comparison*, 2nd Ed. Wiley-Blackwell Publishing, Malden, MA. Rain forests on different continents have distinctive assemblages of animal and plant species.

González-Maya, J. F., L.R. Víquez-R, A. Arias-Alzate, J.L. Belant, and G. Ceballos. 2016. Spatial patterns of species richness and functional diversity in Costa Rican terrestrial mammals: implications for conservation. *Diversity and Distributions* 22: 43–56. Understanding the relationship between different measures of diversity can help us understand ecosystem function.

Groombridge, B. and M. D. Jenkins. 2010. *World Atlas of Biodiversity: Earth's Living Resources in the 21st Century*. University of California Press, Berkeley. Great resource, with numerous figures; available on-line.

Hollings, T., M. Jones, N. Mooney, and H. McCallum. 2014. Trophic cascades following the disease-induced decline of an apex predator, the Tasmanian devil. *Conservation Biology* 28: 63–75. Many species are affected when a keystone species is eliminated.

Joppa, L. N., D. L. Roberts, and S. L. Pimm. 2011. The population ecology and social behavior of taxonomists. *Trends in Ecology and Evolution* 26: 551–553. The number of taxonomists and the number of species described per year are steadily increasing.

Laikre, L. and 19 others. 2010. Neglect of genetic diversity in implementation of the Convention on Biological Diversity. *Conservation Biology* 24: 86–88. A greater emphasis on genetic diversity needs to be part of conservation efforts.

Magurran, A. E. 2013. *Measuring Biological Diversity: Frontiers in Measurement and Assessment*. Oxford University Press, Oxford, UK. A widely-cited text that discusses both classic and emerging methods.

Ricklefs, R. E., and F. He. 2016. Region effects influence local tree species diversity. *Proceedings of the National Academy of Sciences USA*, 113: 674–679. Regional species diversity results largely from geologic and geographic properties that affect evolution.

Ripple, W. J. and R. L. Beschta. 2012. Trophic cascades in Yellowstone: The first 15 years after wolf reintroduction. *Biological Conservation* 145: 205–213. Restoring a keystone species has resulted in large changes to this ecosystem.

Strain, D. 2011. 8.7 million: A new estimate for all the complex species on Earth. *Science* 333: 1083. A variety of methods have been developed for estimating the total numbers of species on Earth.

Tittensor, D. P. and 6 others. 2010. Global patterns and predictors of marine biodiversity across taxa. *Nature* 466: 1098–1101. Temperature is the most important factor affecting marine diversity.

3

The Value of Biodiversity

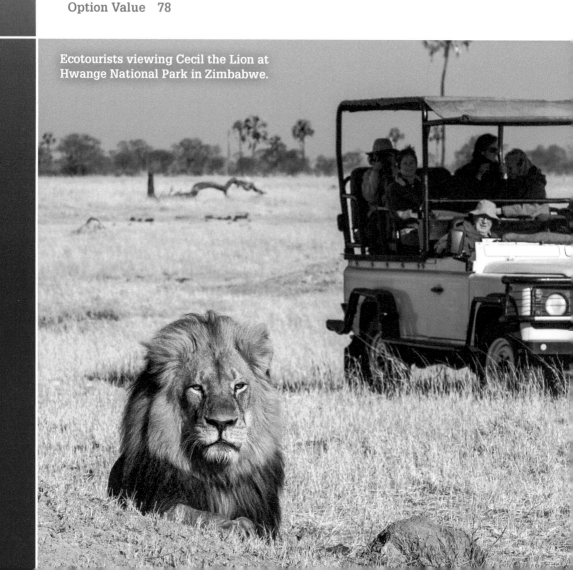

Ecotourists viewing Cecil the Lion at
Hwange National Park in Zimbabwe.

I n July 2015, the economics of biodiversity received international attention when a lion from Hwange National Park in Zimbabwe was illegally shot by an American big-game hunter who had paid poachers $50,000 to do so (CNN 2015). The 13-year-old male, affectionately named "Cecil the Lion," had been an important animal icon and tourism draw for the park; there was an immediate outcry from conservationists and animal lovers around the world when his death was announced. What Cecil and other wild organisms are "worth," both alive and dead, is a critical question in evaluating attitudes and implementing actions that affect biodiversity.

Ultimately, the trend of biodiversity decline will be reversed only if people believe that we are truly losing something of value. But what exactly are we losing? Why should anyone care if a species becomes extinct or an ecosystem is destroyed? What factors induce humans to act in an unsustainable and ultimately destructive manner? What are we willing to spend to protect species, and how will this spending be financed? Governments and communities throughout the world are coming to realize that biodiversity is extremely valuable—indeed, it is essential to human existence.

Most environmental degradation and species losses occur as accidental by-products of human activities. Species are hunted to extinction. Sewage is released into rivers. Low-quality land is cleared for short-term cultivation. **Economics**, the study of the transfer of the production, distribution, and consumption of goods and services, can both help us understand the reasons why people treat the environment in what appears to be a shortsighted manner and provide tools to help protect environmental resources.

One of the most universally accepted tenets of modern economic thought centers on the "voluntary transaction": the idea that a monetary transaction takes place only when it is beneficial to both of the parties involved. For example, a baker who sells his loaves of bread for $40 will find few customers. Likewise, a customer who is willing to pay only 4 cents for a loaf will soon go hungry. A transaction between seller and buyer will occur only when a mutually agreeable price is set that benefits both parties: perhaps $4 for that loaf of bread. Adam Smith, the eighteenth-century philosopher whose ideas are the foundation of much modern economic thought, wrote, "It is not upon the benevolence of the butcher, the baker, or the brewer that we eat our daily bread, but upon his own self-interest" (Smith 1909). The sum of each individual's actions in his or her self-interest results in a more prosperous society.

There are exceptions to Smith's principle, however, that apply directly to environmental issues. For example, Smith assumed that all the costs and benefits of free exchange are accepted and borne by the participants in the transaction. In some cases, however, associated costs (or sometimes benefits) befall individuals not directly involved in the exchange. These hidden costs and benefits are known as negative and positive **externalities**, respectively (**FIGURE 3.1**) (Abson and Termansen 2011). Companies, individuals, or other stakeholders involved in production that results in ecological damage generally do not bear the full cost of their activities, but gain substantial private economic benefits. For example, the company that owns an electric power plant that burns coal and emits toxic fumes benefits from the sale of low-cost electricity, as does the consumer who buys this electricity. Yet the negative externalities of this transaction—decreased air quality and visibility, increased respiratory disease for people and animals, damage to plant life, and a polluted environment—are distributed throughout society.

Even more important and more frequently overlooked is the negative externality of environmental damage done to **open-access resources**, such as water, air, and soil, as a consequence of human economic activity. Open-access resources are those that are collectively owned by society at large (also called **common-property resources**) or owned by no one. They are available for everyone to use and are essentially free. When there are no regulations on their use, then people, industries, and governments use and damage these resources without paying more than a minimal cost, or sometimes nothing at all. This situation, which has been referred to as the **tragedy of the commons** (a phrase popularized by Garett Hardin's 1968 essay of that name on human population control), means that the

FIGURE 3.1   Smog in the Andes, looking eastward over Santiago, Chile. Air pollution is responsible for 3.3 million premature human deaths (and unknown numbers of wildlife) worldwide per year, a devastating externality of the combined emissions from industry, transportation, and agriculture. (© Matt Mawson/Corbis.)

value of the open-access resource is gradually lost to all of society. The unregulated dumping of industrial sewage into a river as a by-product of manufacturing is a common example. The externalities of this activity are degraded drinking water and an increase in disease, loss of opportunity to bathe and swim in the water, fewer fish that are safe to eat, and the loss of many species unable to survive in the polluted river. The factory owner gains free disposal of sewage, but the society pays the price in terms of lost products and services.

When externalities are not identified and managed, or there is inadequate regulation of common-property resources, certain economic activities make the society as a whole *less* prosperous, not more prosperous. When an economic system fails, and thus the balance of supply and demand for products or services is lost, economists call this **market failure** (**FIGURE 3.2**). Avoiding market failure is arguably the primary goal of government policies that regulate economic activities that can affect the environment.

Ecological and Environmental Economics

A major problem for conservation biology is that natural resources have often been undervalued by modern society. Thus, the costs of environmental damage have been ignored, the depletion of natural resource stocks disre-

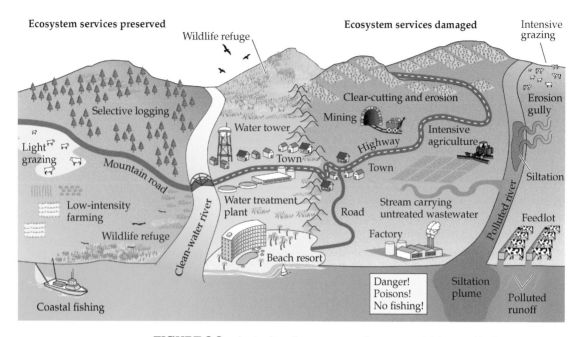

FIGURE 3.2 Agricultural ecosystems, forestry activities, and industries are usually valued by the products that they produce. In many cases, these activities have negative externalities in that they erode soil, degrade water quality, and harm aquatic life (right side of figure). Farming, forestry, and other human activities could also be valued on the basis of their public benefits, such as soil retention and maintaining water quality and fish populations, and their owners might receive subsidies for these benefits (left side of figure).

garded, and the future value of resources discounted (MEA 2005). Because the underlying causes of environmental damage are so often economic in nature, the solution must incorporate economic principles (Kubiszewski et al. 2013). In an effort to account for all costs of economic transactions, including environmental costs, two closely related research areas have evolved—environmental economics and ecological economics—that integrate economics, environmental science, ecology, and public policy and that include valuations of biodiversity in economic analyses (Common and Stagl 2005). **Environmental economics** is a subdiscipline of economics that places a value on components of the environment. Its modern form can be traced to a popular 1972 book, *The Limits to Growth*, by environmental scientist Donella Meadows and coauthors; they used system dynamics modeling to predict potential future balances of the human population, pollution, and agricultural production (Hanley et al. 2015). **Ecological economics**, which is more closely allied to conservation biology, seeks to integrate the thinking of ecologists and economists into a transdiscipline aimed at developing a sustainable world (Sachs 2008). One of the core agenda

Arguments for the protection of biodiversity are often strengthened by evidence provided by ecological economics.

items of ecological economics is to develop methods to value biodiversity by integrating economic valuation with ecology, environmental science, sociology, and ethics and to use those new valuation methods to design better public policies related to conservation and environmental issues (Reyers et al. 2013). The fundamental challenge facing conservation biologists is to ensure that all the costs of economic activity, as well as all the benefits, are understood and taken into account when decisions that will affect biodiversity are made (Hoeinghaus et al. 2009; Junk et al. 2014).

Cost–benefit analysis

Economic methods are now being used to review development projects and evaluate their potential environmental effects *before* the projects proceed. **Environmental impact assessments**, in particular, consider the present and future effects of projects on the environment. "The environment" is often broadly defined to include not only harvestable natural resources but also air and water quality, the quality of life for local people, and biodiversity. In its most comprehensive form, **cost–benefit analysis** compares the values gained against the costs of a project or resource use (Maron et al. 2013; Newbold and Siikamäki 2009). In practice, though, cost–benefit analyses are notoriously difficult to calculate accurately because benefits and costs change over time and are difficult to measure. Today, there is an increasing tendency by governments, conservation groups, and economists to apply the **precautionary principle**. That is, it may be better not to approve a project that has risk associated with it and to err on the side of doing no harm to the environment, rather than doing harm unintentionally or unexpectedly, as by building wind turbines where they could harm endangered birds (Braunisch et al. 2015). The precautionary principle is a key feature of many national and international policies and agreements regarding environmental management, even though its interpretation can be vague and variable (Foster et al. 2000).

It would be highly beneficial to apply cost–benefit analysis to many of the basic industries and practices of modern society. Many environment-damaging economic activities appear to be profitable even when they are actually losing money because governments subsidize the industries involved in them with tax breaks, direct payments or price supports, cheap fossil fuels, free water, and road networks—sometimes referred to as **perverse subsidies** (Myers et al. 2007). The elimination of such subsidies that are harmful to biodiversity by 2020 is one of the explicit targets of the Convention on Biological Diversity (CBD Decision X/2; Dobson 2005). Subsidies in agriculture and fisheries can be as high as 20%–30% of the production value of those industries (MEA 2005). Without these subsidies, many environmentally damaging or expensive activities—such as farming in areas with high labor, energy, and water costs; overfishing in the ocean; and inefficient and highly polluting energy use—would be reduced (Merckx and Pereira 2015).

Attempts have been made to include the loss of natural resources in calculations of gross domestic product (GDP) and other indexes of national

Unsustainable activities such as clear-cut logging, strip mining, and over-fishing may cause a country's apparent productivity to increase temporarily, but are generally destructive to long-term economic well-being.

production. The problem with GDP is that it measures economic activity in a country without accounting for all the costs of unsustainable activities (such as overfishing of coastal waters and poorly managed strip mining), which cause the GDP to increase, even though they may be destructive to a country's long-term economic well-being. In actuality, the economic costs associated with environmental damage can be considerable, and they often offset the gains attained through agricultural and industrial development. In the United Kingdom, hidden environmental costs in agriculture, including soil erosion and water pollution, are estimated to be worth about $2.6 billion per year, or 9% of the value of the country's agriculture (MEA 2005).

A system that accounts for natural resource depletion, pollution, and unequal income distribution in measures of national production is the Index of Sustainable Economic Welfare (ISEW), the updated version of which is called the Genuine Progress Indicator (GPI; www.progress.org). This index includes factors such as the loss of farmlands, the loss of wetlands, the impact of acid rain, the number of people living in poverty, and the effects of pollution on human health. According to the GPI, the world economy reached a peak around 1978 and has been slowly declining since then, even though the standard GDP index showed a dramatic gain (Kubiszewski et al. 2013). The GPI suggests what conservation biologists have long feared: Many modern economies are achieving their growth only through the unsustainable consumption of natural resources and environmental degradation. As these resources run out and as humans suffer the effects of pollution, the true economic situation will continue to deteriorate.

A third measure of national productivity is the Environmental Performance Index (EPI), which uses 20 environmental indicators to rank countries according to the health of, and threats to, their ecosystems; the vulnerability of their human population to adverse environmental conditions; the ability of their society to protect the environment; and their participation in global environmental protection efforts (epi.yale.edu). In general, developing countries with a low GDP per person also have low EPI scores, including such large countries as China and India, as **FIGURE 3.3** shows. Higher-income countries, such as the United States, Japan, and Germany, tend to have much higher EPI scores. There is a concern among many economists and businesspeople that a country that rigorously protects its environment, as shown by a high EPI, might not be competitive in the world economy, as measured by a competitiveness index that includes worker productivity and a country's ability to grow and prosper. But environmental sustainability is not linked to a country's economic competitiveness. Countries such as Finland have an economy that is both sustainable and competitive, whereas Belgium is competitive but ranks poorly in

New measures of national productivity take environmental sustainability into account. These measures include both the benefits and the costs of human activities.

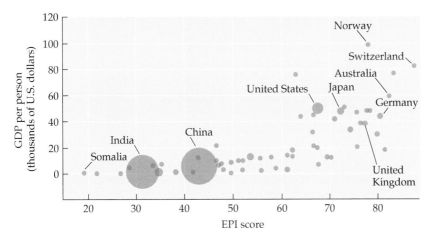

FIGURE 3.3   Wealthy countries with higher gross domestic product (GDP) per person tend to have higher scores on the Environmental Performance Index (EPI), as measured by various indicators: health and stress level of ecosystems, human vulnerability to environmental change, ability of the society and institutions to cope with environmental changes, and cooperation in international environmental initiatives. The size of the human population is indicated by the size of the circle. (After epi.yale.edu/epi/data-explorer.)

sustainability. The rapidly growing economies of China and India are intermediate in competitiveness, but rank low in environmental sustainability.

Financing conservation

Another important aspect of environmental economics is the cost of conservation. Especially when a species is already rare or endangered, it is not enough to simply do no harm. People must intervene to protect, manage, and otherwise support its health and survival. But such interventions can be expensive. In an extreme example, whooping crane conservation is estimated to cost $2–2.5 million per year, with a total projected cost of $48 million (US Fish and Wildlife Service). The effort increased this species from 16 individuals in 1941 to 603 as of 2015. Although the rescue of the whooping crane is a success story, there are many other less charismatic species that have less financial support than they need. Thus, in addition to cost–benefit analysis, it is important to do **cost-effectiveness analysis** as well, or to ask, "Where do we get the most with our conservation dollar?" (Cannon 1996).

We must also concern ourselves with where and how we get these funds. Conservation is financed in many ways, including by the sometimes-controversial source of hunting (Crosmary et al. 2015). Some countries, such as Namibia, depend on sales of expensive trophy licenses, even for the hunting of endangered species, to support conservation (Rust 2015). In the United States, most state wildlife conservation efforts are primarily funded by the sale of hunting licenses, tags, and stamps, and

$200 million a year in hunters' federal excise taxes is distributed to states, primarily for wildlife management (US Fish and Wildlife Service). This is considered a utilitarian approach to conservation: it assumes that species and their habitats will be conserved if people find them valuable, even if only to kill for sport. In Great Britain, landowners were much more likely to plant and protect woodland for nonagricultural purposes if they hunted game there (Oldfield et al. 2003). However, it has also been argued that using hunting as a foundation for conservation is morally wrong and will inevitably lead to poaching and decreased biodiversity (Selier et al. 2014).

Ultimately, only that which is perceived to have value will be saved, so the basis on which we assign this value is of utmost importance. It will also inevitably raise questions regarding moral and ethical values.

What are species worth?

There are many classification systems used to evaluate the benefits we receive from our natural environment (Wallace 2007). As yet there is no universally accepted framework for assigning value to biodiversity, but a variety of approaches have been proposed. Among the most useful is the framework used by McNeely et al. (1990) and Barbier et al. (1994), in which economic values are first divided into **use values** and **non-use values**. Use values of biodiversity are divided between **direct use values** (also known in other frameworks as **commodity values** and **private goods**) and **indirect use values**. Direct use values are assigned to products harvested by people, such as timber, seafood, and medicinal plants from the wild, while indirect use values are assigned to benefits provided by biodiversity that do not involve harvesting or destroying the resource. Indirect use values provide current benefits to people, such as recreation, education, scientific research, and scenic amenities, and include the benefits of ecosystem services, such as water quality, pollution control, natural pollination and pest control, ecosystem productivity, soil protection, and regulation of climate. **Option value** is determined by the prospect for possible future benefits for human society, such as new medicines, possible future food sources, and future genetic resources. **Existence value** is the non-use value that can be assigned to biodiversity—for example, economists can attempt to measure how much people are willing to pay to protect a species from going extinct or an ecosystem from being destroyed.

Ecosystem services

The many and varied environmental benefits provided by biodiversity and ecosystems in general to humans are collectively referred to as **ecosystem services** (Ehrlich and Ehrlich 1982), which are typically divided into four categories:

Provisioning services are the material or energy outputs of an ecosystem, including food, fresh water, and raw materials. The worth of biological outputs to humans will be discussed in detail below under "Direct use values," whereas non-living products such as clean water give the organisms that supply it indirect economic value.

Regulating services are services provided by the ecosystem acting as regulators of the quality of the air and soil. Forests provide many of these by regulating local climate, removing pollutants from the atmosphere, and holding soil with their roots that would otherwise blow or wash away. Other examples can be found in the section "Indirect use values."

Habitat/supporting services refers to the role ecosystems play in supporting biodiversity, including genetic diversity that humans depend on for cultivating crops and livestock. Other examples can be found in "Species relationships and environmental monitors." These supportive services mean that they also provide option use value.

Cultural services include inspiration for art, design, music and other cultural expression, aesthetics, intellectual stimulation, and spiritual value. Examples of these can be found under "Amenity value," "Education and scientific value," and "Existence Value."

These categories were first defined in the Millennium Ecosystem Assessment (2005), a project initiated by the United Nations in 2000 to evaluate the impact of ecosystem change on human well-being and determine which actions were needed to protect these services. Below we will discuss these services in terms of the value that the organisms (both living and dead) themselves have for humans and how this value is quantified in financial terms.

Economic Use Values

Direct use values

Direct use values can often be readily calculated by observing the activities of representative groups of people, by monitoring collection points for natural products, and by examining import and export statistics. Direct use values are further divided into **consumptive use value**, for goods that are consumed locally, and **productive use value**, for products that are sold in markets.

Consumptive use value

People living close to the land often derive a considerable proportion of the goods they require for their livelihood from the surrounding environment (**FIGURE 3.4**). These goods, such as fuelwood and wild meat, are consumed locally and are therefore assigned consumptive use value (Davidar et al. 2008). These goods do not appear in the GDP of countries because they are neither bought nor sold beyond the village or local region and do not appear in the national or international marketplace. However, if rural people are unable to obtain these products (as might occur following environmental degradation, overexploitation of natural resources, or even creation of a protected reserve), their standard of living will decline, possibly to the

(A)

(B)

FIGURE 3.4 Examples of natural products with consumptive use value: (A) Women in India return to their village with loads of wood. Fuelwood is one of the most important natural products consumed by local people, particularly in Africa and southern Asia. The value of these products can be estimated based on what these people would have to spend to purchase rather than harvest them. (B) Along a river in India, fishermen catch small fish to eat. (A, photograph © Robert Harding World Imagery/Corbis; B, photograph courtesy of Sandesh Kadur.)

point where they are forced to relocate. Consumptive use value can be assigned to a product by considering how much people would have to pay if they had to buy an equivalent product when their local source was no longer available. This valuation is sometimes referred to as a **replacement cost approach**.

Studies of traditional societies in the developing world show how extensively these people use their natural environment to supply themselves with fuelwood (Figure 3.4A), meat, vegetables, fruit, medicine, rope and string, and building materials (Angelsen et al. 2014). About 80% of the world's population still relies principally on traditional medicines derived from plants and animals as their primary source of treatment (Shanley and Luz 2003).

One of the crucial requirements of rural people is protein, which they obtain by hunting and collecting wild animals for meat. In some places, this meat is called **bushmeat**. In many areas of Africa, bushmeat constitutes a significant portion of the protein in the average person's diet—about 40% in Botswana and about 80% in the Democratic Republic of the Congo (formerly Zaire; Powell et al. 2013). Bushmeat extraction rates for Africa are undeniably unsustainable, perhaps by a factor of six. This wild meat includes not only birds, mammals, and fish, but spiders, snails, caterpillars, and insects. In certain areas of Africa, because of overharvesting of larger animals, insects may constitute the majority of the dietary protein and supply critical vitamins.

In areas along coasts, rivers, and lakes, wild fish represent an important source of protein (Figure 3.4B). Throughout the world, 130 million tons of fish, crustaceans, and molluscs, mainly wild species, are harvested each year, 100 million tons from the oceans and 30 million tons from freshwater (Chivian and Bernstein 2008). Much of this catch is consumed locally. In coastal areas, fishing is often the most important source of employment, and seafood is the most widely consumed protein. Even though fish farming is increasing rapidly, much of the feed used is fish meal derived from wild-caught fish (Gross 2008).

Although dependency on local natural products is primarily associated with the developing world, there are rural areas of the United States, Canada, Europe, and other developed countries where hundreds of thousands of people are dependent on fuelwood for heating, on wild game and seafood for their protein needs, and on intact ecosystems for clean drinking water and sewage treatment. Many of these people would be unable to survive in these locations if they had to pay for these necessities.

> Consumptive use value can be calculated by considering how much people would have to pay to buy an equivalent product if their local source were no longer available.

Productive use value

Resources that are harvested from the wild and sold in national or international commercial markets are assigned productive use value (**FIGURE 3.5**). In standard economics, these products are valued at the price paid at the first point of sale minus the costs incurred up to that point, whereas

FIGURE 3.5 Examples of productive use value. (A) A wide variety of marine animals are collected in the wild or produced by aquaculture and then sold as seafood, as shown by this market in South Korea. (B) The productive use value of the trees in this forest in Canada is simply calculated as their worth on the market. (A, photograph by Richard B. Primack; B, photograph © Lloyd Sutton/ Alamy Stock Photo.)

(A)

(B)

other methods value the resource at the final retail price of the products. For example, the bark and leaves from wild shrubs and trees of the common witch hazel (*Hamamelis virginiana* and related species) are used to make a variety of astringent herbal products, including aftershave lotions, insect-bite creams, and hemorrhoid preparations. The final retail price of the medicine, which includes the values of all inputs (labor, energy, other materials, transportation, and marketing, as well as witch hazel bark and leaves), is vastly greater than the purchase price of the witch hazel raw materials.

The productive use value of natural resources is significant, even in industrial nations. It has been estimated that approximately 4.5% of the US GDP depends in some way on wild species (Prescott-Allen and Prescott-Allen 1986). This translates to about $780 billion (out of a GDP of $17.4 trillion) for the year 2014. The percentage is far higher for developing countries that have less industry and a higher percentage of their population living in rural areas. The international trade in wildlife, fisheries, and timber products harvested from the wild has been estimated to be $332 billion (Engler 2008, cited in Barber-Meyer 2010). However, it is difficult to accurately calculate the total value of wild-harvested products because of the unknown contribution of "invisible trades" due to low detection rates, underreporting, and non-reporting, especially of illegal products (Phelps and Webb 2015).

The range of products obtained from the natural environment and sold in the marketplace is enormous: these products include fuelwood, construction timber, fish and shellfish, medicinal plants, wild fruits and vegetables, wild meat and skins, fibers, rattan (a vine used to make furniture and other household articles), honey, beeswax, natural dyes, seaweed, animal fodder, natural perfumes, and plant gums and resins (Baskin 1997; Chivian and Bernstein 2008). Additionally, there are large international industries associated with collecting tropical cacti, orchids, and other plants for the horticultural industry and birds, mammals, amphibians, and reptiles for zoos and private collections. The value of ornamental fishes in the aquarium trade is estimated at $1 billion per year, with wild-caught fish representing about 20% of the total.

> A wide variety of natural resources are sold commercially and have enormous total market value. Their value can be considered the productive value of biodiversity.

FOREST PRODUCTS   Wood is one of the most significant products obtained from natural environments, with an export value of about $231 billion per year (www.fao.org/forestry/statistics/80938). The total value of timber and other wood products is far greater—perhaps about $400 billion per year—because most wood is used locally and is not exported. In tropical countries such as Indonesia, Brazil, and Malaysia, timber products earn billions of dollars per year (Corlett and Primack 2010). Non-wood products from forests, including bushmeat, fruits, gums and resins, rattan, and medicinal plants, also have a large productive use value. These non-wood products are sometimes erroneously called "minor forest products"; in reality, they are often very important economically and may even rival the value of wood.

THE NATURAL PHARMACY   Effective drugs are needed to keep people healthy, and they represent an enormous industry, with worldwide sales of about $300 billion per year (Chivian and Bernstein 2008). The natural world is an important source of medicines currently in use as well as possible future medicines. All 20 of the pharmaceutical products most frequently used in the United States are based on chemicals that were first identified in natural organisms. More than 25% of the prescriptions filled in the United States contain active ingredients derived directly from plants, and many of the most important antibiotics, including penicillin and tetracycline, are derived from fungi or microorganisms (Waterman et al. 2016).

Many modern medicines were first discovered in a wild species used in traditional medicine, and then produced synthetically by chemists (Chivian and Bernstein 2008; Cox 2001). For example, the use of coca (*Erythroxylum coca*) by natives of the Andean highlands eventually led to the development of synthetic derivatives such as Novocain, procaine, and lidocaine, commonly used as local anesthetics in dentistry and surgery. The rose periwinkle (*Catharanthus roseus*) from Madagascar (**TABLE 3.1**) is the

TABLE 3.1   Twenty Drugs from the Plant World First Discovered in Traditional Medical Practice

Drug	Medical use	Plant source	Common name
Ajmaline	Treats heart arrhythmia	*Rauwolfia* spp.	Rauwolfia
Aspirin	Analgesic, anti-inflammatory	*Spiraea ulmaria*	Meadowsweet
Atropine	Dilates eyes during examination	*Atropa belladonna*	Belladonna
Caffeine	Stimulant	*Camellia sinensis*	Tea plant
Cocaine	Ophthalmic analgesic	*Erythroxylum coca*	Coca plant
Codeine	Analgesic, antitussive	*Papaver somniferum*	Opium poppy
Digitoxin	Cardiac stimulant	*Digitalis purpurea*	Foxglove
Ephedrine	Bronchodilator	*Ephedra sinica*	Ephedra plant
Ipecac	Emetic	*Cephaelis ipecachuanha*	Ipecac plant
Morphine	Analgesic	*Papaver somniferum*	Opium poppy
Pseudoephedrine	Decongestant	*Ephedra sinica*	Ephedra plant
Quinine	Antimalarial prophylactic	*Cinchona pubescens*	Chinchona
Reserpine	Treats hypertension	*Rauwolfia serpentina*	Rauwolfia
Sennoside A, B	Laxative	*Cassia angustifolia*	Senna
Scopolamine	Treats motion sickness	*Datura stramonium*	Thorn apple
THC	Antiemetic	*Cannabis sativa*	Marijuana
Toxiferine	Relaxes muscles during surgery	*Strychnos guianensis*	Strychnos plant
Tubocurarine	Muscle relaxant	*Chondrodendron tomentosum*	Curare
Vincristine	Treats pediatric leukemia	*Catharanthus roseus*	Rose periwinkle
Warfarin	Anticoagulant	*Melilotus* spp.	Sweet clover

Sources: Balick and Cox 1996; Chivian and Bernstein 2008.

source of two potent drugs that have increased the rate of survival of childhood leukemia from 10% to 90%. Venomous animals such as rattlesnakes, bees, and cone snails have been especially rich sources of chemicals with valuable medical and biological applications. An enzyme derived from a heat-tolerant bacterium (*Thermus aquaticus*) collected from hot springs at Yellowstone National Park forms a key component in the polymerase chain reaction used to amplify DNA in the biotechnology industry and in biological research (**FIGURE 3.6**). This enzyme is also used in the medical field to detect human diseases. The industries using this enzyme have generated hundreds of billions of dollars of value and employ hundreds of thousands of people. How many more such valuable species will be discovered in the years ahead—and how many will go extinct before they are discovered?

FIGURE 3.6   Specialized bacteria, such as these bright orange bacteria at Prism Lake, grow abundantly in Yellowstone National Park's hot mineral springs and have contributed essential enzymes for the high temperature reactions used in the biotechnology industry. (Photograph by Richard Primack.)

Indirect use values

Many organisms provide benefits to humans only when alive and in their natural ecosystem, such as plants growing on a hillside that prevent soil erosion. These resources have **nonconsumptive use value** because they are not consumed. Indirect use values are nonconsumptive use values that can be assigned to all aspects of biodiversity that provide economic benefits without being harvested or destroyed. Because these benefits are not goods or services in the usual economic sense, they do not typically appear in the statistics of national economies, such as the GDP. They are often called **public goods** because they belong to society in general, without private ownership. However, these benefits may be crucial to the continued availability of the natural products on which economies depend. If natural ecosystems are not available to provide these benefits, substitute sources must be found—often at great expense—or local and even regional economies may face a decline in prosperity or even collapse.

Human societies are totally dependent on the free services that we obtain from natural ecosystems because we could not pay to replace these ecosystems if they were permanently degraded or destroyed. Especially important in this regard are wetland ecosystems because of their role in water purification and nutrient recycling as well as their enormous importance in flood control (Horwitz and Finlayson 2011; see the section "Water and soil protection," p. 70).

Economists are actively improving calculations of the indirect use value of ecosystem services at regional and global levels (Bateman et al. 2013; Reyers et al. 2013). However, many ecological economists sharply disagree about how calculations of indirect use value should be done, or even whether they should be done at all (Peterson et al. 2010). One such calculation suggests that the annual value of ecosystem services worldwide is as much as $145 trillion (Costanza et al. 2014), exceeding the current $78 trillion annual value of the world's economy (data.worldbank.org). Using different approaches, other ecological economists have come up with much lower estimates, but those estimates have still amounted to trillions of dollars a year. The disparity in these various estimates indicates that much more work needs to be done on this topic.

The great variety of environmental services that ecosystems provide can be assigned particular types of indirect use value. The following sections discuss some of the specific indirect use values derived from biodiversity. Later in the chapter, we will consider two other nonconsumptive ways of valuing biodiversity: option value, the value that biodiversity may have in the future, and existence value, the amount that people are willing to pay to protect biodiversity (or other environmental goods or services) even if they never expect to experience it.

Ecosystem productivity

All life on Earth is made possible by the energy of the sun, which is converted into usable energy through photosynthesis in plants and algae. Humans depend on the energy stored in plants for many direct uses, such as food, fuelwood, and hay and other fodder for animals. This plant material is also the starting point for innumerable food chains, from which people harvest many animal products. Humans appropriate approximately half of the productivity of the terrestrial environment to meet their needs for natural resources (MEA 2005), and most of the remaining half performs services that have indirect use value to humans, including the production of oxygen (O_2) by plants through the process of photosynthesis.

The destruction of the vegetation in an area through overgrazing by domestic animals or overharvesting of timber will destroy the system's ability to perform these functions (**FIGURE 3.7**). Eventually, it will lead to losses of plant biodiversity and of the associated production of plant biomass, loss of the animals that live in that area, and losses of natural resources and ecosystem services for people.

Likewise, coastal estuaries are areas of rapid plant and algal growth that provide the starting point for food chains leading to commercial stocks of fish and shellfish. When these coastal areas are filled in for development, their value to society is lost. Even when degraded or damaged wetland ecosystems are re-built or restored—usually at great expense—they often do not function as well as they initially did and almost certainly do not contain their original species composition or species richness.

Scientists are actively investigating how the loss of species from biological communities affects ecosystem processes such as the total growth of plants, the ability of plants to absorb atmospheric CO_2, and the ability of communities to adapt to global climate change (King et al. 2012). Many studies of natural and experimental grassland communi-

> Ecosystems with reduced species diversity are less able to adapt to the altered conditions associated with rising carbon dioxide levels and global climate change.

FIGURE 3.7 Although many grassland systems are adapted to grazing, overgrazing reduces ecosystem services and can lead to decreased ecosystem productivity and increased soil erosion. (Photograph by Anna Sher.)

(A)

(B)

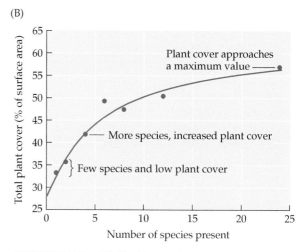

FIGURE 3.8 (A) Healthy prairie ecosystems are naturally diverse, with many species of grasses and forbs. (B) Varying numbers of grassland species were grown in experimental plots. The plots containing the most species had the greatest overall amount of growth as measured by the total plant cover (the percentage of the total surface area occupied by plants). (A, photograph © Clint Farlinger/Alamy Stock Photo; B, after Tillman 1999.)

ties confirm that as species are lost, overall productivity declines, and the community is less flexible in responding to environmental disturbances such as drought (Hautier et al. 2015) (**FIGURE 3.8**).

Water and soil protection

Biological communities are of vital importance in protecting watersheds, buffering ecosystems against extremes of flood and drought, and maintaining water quality (Thorp et al. 2010). Plant foliage and dead leaves intercept the rain and reduce its impact on the soil. Plant roots and soil organisms aerate the soil, increasing its capacity to absorb water. This increased water-holding capacity reduces the flooding that would otherwise occur after heavy rains and allows a slow release of water for days and weeks after the rains have ceased.

When logging, farming, and other human activities disturb vegetation, the rates of soil erosion, and even occurrences of landslides, increase rapidly, decreasing the value of the land for some human activities. Damage to the soil limits the ability of plant life to recover from disturbance and can render the land useless for agriculture. Erosion and flooding also contaminate drinking water supplies for humans in the communities along rivers, leading to an increase in human health problems. Soil erosion increases sediment loads entering the reservoirs behind dams, causing a loss of electrical output, and it creates sandbars and islands, which reduces the navigability of rivers and ports.

Floods are currently the most common natural disaster in the world, killing thousands of people each year, and losses of wetland and floodplain ecosystems have contributed to these disasters. In the industrial nations of the world, wetlands protection has become a priority in order to prevent flooding of developed areas (**FIGURE 3.9**). In certain locations, wetlands are estimated to have a value of $6000 per hectare per year in flood damage reduction and other ecosystem services, which is three times the value of farmland developed on the same site (MEA 2005). The conversion of floodplain habitat to farmland along the Mississippi, Missouri, and Red Rivers in North America and the Rhine River in Europe is considered a major factor in the massive, damaging floods along those rivers in past years. The most dramatic example of such flooding is the devastating flooding of New Orleans in 2005 after Hurricane Katrina struck the Mississippi delta, which

FIGURE 3.9 Wetlands perform many vital functions for humans. The trees, their root mats, and other vegetation act like a living sponge that traps and slowly releases rainwater and snowmelt, while creating resistance that slows water flows, thus preventing damage from floodwater. Aquatic plants also clean water by taking up excess nitrogen, phosphorus, heavy metals, and other substances that can be harmful to humans. (© Patricia Hofmeester/Shutterstock.)

> Wetland ecosystem services whose value is typically not accounted for in the current market system include waste treatment, water purification, and flood control—all of which are essential to healthy human societies.

has undergone heavy conversion of wetlands for urban, industrial, and agricultural development. The risk of flooding would be substantially reduced if even a small proportion of the wetlands along these rivers were restored to their original condition.

Wetlands also perform important functions for filtering excess nutrients and toxins in water. The government of New York City paid $1.5 billion in the late 1980s to county and town governments in rural New York State to maintain forests on the watersheds surrounding its reservoirs and to improve agricultural practices in order to protect the city's water supply. Water filtration plants doing the same job would have cost $8–9 billion (www.nyc.gov/watershed). The nonconsumptive use value of the US national forests alone for protecting the nation's water supply has been estimated at $4 billion per year.

Aquatic ecosystems such as swamps, lakes, rivers, floodplains, tidal marshes, mangroves, estuaries, the continental shelf, and the open ocean are capable of breaking down and immobilizing toxic pollutants, such as the heavy metals and pesticides that have been released into the environment by human activities (Balmford et al. 2002). When such aquatic communities, especially the bacteria and other microorganisms they contain, are damaged by a combination of sewage overload and habitat destruction, alternative systems have to be developed. These contrived systems, such as waste treatment facilities and giant landfills, cost tens of billions of dollars. In regions that cannot afford to build such facilities, people's quality of life can be severely harmed.

Climate regulation

Plant communities are important in moderating local, regional, and even global climate conditions (West et al. 2011). At the local level, trees provide shade and evaporate water from their leaf surfaces during photosynthesis, reducing the local temperature in hot weather. This cooling effect reduces the need for fans and air conditioners and increases people's comfort and work efficiency. Trees are also locally important because they act as windbreaks for agricultural fields, reducing soil erosion by wind and reducing heat loss from buildings in cold weather.

At the regional level, plants capture water that falls as rain and then transpire it back into the atmosphere, from which it can fall as rain again. The loss of vegetation from large forested regions such as the Amazon basin and western Africa may result in a reduction of average annual rainfall or greatly altered weather patterns over large areas.

In both terrestrial and aquatic environments, plant growth is tied to the carbon cycle. A reduction in plant life results in reduced uptake of CO_2, contributing to the rising CO_2 levels that lead to global warming (McKinley et al. 2011; Pan et al. 2011). Environmental economists also recognize the value of intact and restored forests in retaining carbon and absorbing atmospheric CO_2 (Butler et al. 2009). As countries and corporations reduce their

CO_2 emissions as part of the worldwide effort to address global climate change, they are paying to protect and restore forests and other ecosystems (Venter et al. 2010).

Species relationships and environmental monitors

Many of the species harvested by people for their direct consumptive use value depend on other wild species for their continued existence. For example, the wild game and fish harvested by people are dependent on wild insects and plants for their food. Thus, a decline in a wild species of little immediate value to humans may result in a corresponding decline in a harvested species that is economically important.

Crop plants benefit from wild insects, birds, and bats (Kross et al. 2012; Wanger et al. 2014). Predatory insects such as praying mantises, as well as many bird and bat species, feed on pest insect species that attack crops, increasing crop yields and reducing the need to spray pesticides (Boyles et al. 2011; Kross et al. 2012). Insects, birds, and bats also act as pollinators for numerous crop species (**FIGURE 3.10**). About 150 species of crop plants in the United States require insect pollination of their flowers, which is often performed by a combination of wild insects and domesticated honeybees (Garibaldi et al. 2013; Vanbergen and the Insect Pollinator Initiative 2013). The global value of these pollinators in increasing crop yield has been estimated at $200 billion per year. The value of wild insect pollinators will increase in the near future if they take over the pollination role of honeybees, whose populations are declining in many places because of disease and pests. Many useful wild plant species depend on fruit-

> Relationships between species are often essential for preserving biodiversity and providing value to people. For example, many insects pollinate the crops on which people depend for food.

(A)

(B)

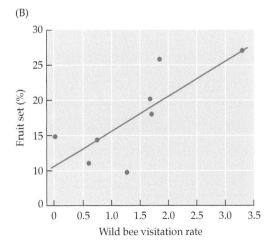

FIGURE 3.10 (A) Bees provide ecosystem services for humans by pollinating food crops such as this peach tree; without pollination, the tree will not produce fruit. The indirect economic value of these insects can be estimated by what it would cost to pay humans to hand-pollinate the trees—something that is sometimes necessary when bee populations decline. (B) Higher visitation rates by wild bees (the number of bees observed in 15-minute intervals foraging on 1000 flowers) increase the fruit set of sweet cherries (the percentage of flowers that develop into fruits). (A, photograph by David McIntyre; B, after Holzschuh et al. 2012.)

eating birds, primates, and other animals to act as seed dispersers. Where these animals have been overharvested, fruits remain uneaten, seeds are not dispersed, and plant species head toward local extinction (Sethi and Howe 2009). It should be noted, however, that there is redundancy in guilds of similar species, and the service of one natural predator, pollinator, or seed disperser may be carried out equally well by another species.

One of the most economically significant relationships in ecosystems is the one between many forest trees and crop plants and the soil organisms that provide them with essential nutrients (Beattie and Ehrlich 2010). Fungi and bacteria break down dead plant and animal matter in the soil, which they use as their energy source. In the process, the fungi and bacteria release mineral nutrients such as nitrogen into the soil. These nutrients are used by plants for further growth. The poor growth and dieback of many trees observed in certain areas of North America and Europe is attributable in part to the deleterious effects of acid rain and air pollution on soil fungi that help supply the trees with mineral nutrients and water.

Amenity value

Ecosystems provide many recreational services for humans; for instance, they furnish a place to enjoy nonconsumptive activities such as hiking, photography, and birdwatching (Buckley 2009) (**FIGURE 3.11**). This expe-

FIGURE 3.11 A picnic area inside a national park in Turkey. It is argued that humans have a need to be near natural features such as this lake and forest; their use for recreation gives these natural features amenity value, a type of indirect use value. (Photograph by Richard Primack.)

rience of nature is not only enjoyable, but also leads to improved health for the participants (Donovan et al. 2013). The monetary value of these activities, sometimes called their **amenity value**, can be considerable and can have a major impact on local economies. In the United States, more than 250 million people visit national parks each year. People in the United States spend around 7 billion hours per year enjoying nature at national parks, state parks, wildlife refuges, and other protected public lands (Siika-mäki 2011). If we estimate that these nature experiences have a value of $12 per hour (the amount people might spend at a movie or dinner), then US protected areas have an estimated value of $84 billion per year for just this one type of use! Recreation represents over 75% of the value of US national forests, far greater than the value of the wood being extracted (Groom et al. 2006). Even sportfishing and hunting, which in theory are consumptive uses, are in practice both consumptive and nonconsumptive because the food value of the animals caught by fishermen and hunters is insignificant compared with the time and money spent on these activities. In national and international sites known for their conservation value or exceptional scenic beauty, such as Yellowstone National Park, nonconsumptive recreational value often dwarfs the value generated or captured by all other economic enterprises there, including ranching, mining, and logging (Power and Barrett 2001).

Ecotourism is a special category of recreation that involves people visiting places and spending money wholly or in part to experience unusual biological communities (such as rain forests, African savannas, coral reefs, deserts, the Galápagos Islands, or the Everglades) and to view particular "flagship" species (such as elephants on safari trips; Balmford et al. 2009; www.ecotourism.org) (**FIGURE 3.12**). Tourism, valued at $600 billion dollars per year, is among the world's largest industries (on the scale of the petroleum and motor vehicle industries), and ecotourism currently represents about 20% of the tourism industry.

Ecotourism has traditionally been a key industry in East African countries such as Kenya and Tanzania, and it has also become important in Latin America, including Costa Rica and Belize, and many other parts of the world. Tourism associated with the Great Barrier Reef in Australia is estimated to be worth $5.5 billion per year and employs more than 50,000 people, which is 36 times more than the commercial fishing industry in Australia (Catlin et al. 2013). In addition to international tourism, the rapidly growing middle classes in developing countries, such as China and India, are increasingly traveling within their own countries to visit national parks and nature reserves (Karanth and DeFries 2011).

The revenue provided by ecotourism has the potential to provide one of the most immediate justifications for protecting biodiversity, particularly when ecotourism activities are integrated into overall management plans (Vianna et al. 2012). In integrated conservation and development projects (ICDPs), local communities develop accommodations, expertise in nature guiding, local handicraft outlets, and other sources of income; the income

(A)

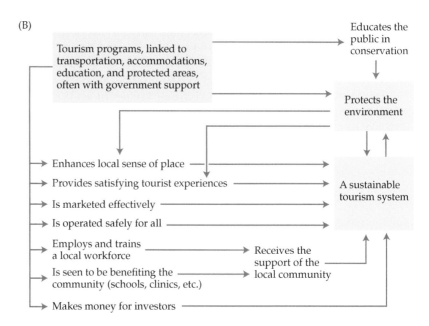

(B)

FIGURE 3.12   (A) Ecotourism can provide an economic justification for protecting biodiversity and can also provide benefits to people living nearby. (B) The diagram illustrates some of the main elements in a successful ecotourism program. (A, photograph © Andrew Parkinson/Corbis; B, after Braithwaite 2001.)

from ecotourism allows the local people to give up unsustainable or destructive hunting, fishing, or grazing practices (see Chapter 9). The local community benefits from the learning of new skills, employment opportunities, greater protection for the environment, and the development of additional community infrastructure such as schools, roads, medical clinics, and stores.

> The rapidly developing ecotourism industry can provide income to protect biodiversity, but possible costs must be weighed along with benefits.

A danger of ecotourism is that the tourists themselves will unwittingly damage the sites they visit—by trampling wildflowers, breaking coral, or disrupting nesting bird colonies, for instance—thereby contributing to the degradation and disturbance of sensitive areas (Nash 2009; Schlacher and Thompson 2012; Shutt et al. 2014). Tourists might also indirectly damage sites by creating a demand for fuelwood for heating and cooking, thus contributing to deforestation. In addition, the presence, affluence, and demands of tourists can transform traditional human societies in tourist areas by changing employment opportunities (Dahles 2005). As local people increasingly enter a cash-based economy, their values, customs, and relationship to nature may be lost along the way. A final potential danger of this industry is that ecotourist facilities may provide a sanitized fantasy experience rather than helping visitors understand the serious social and environmental problems that endanger biodiversity.

Educational and scientific value

Many books, television programs, movies, and websites produced for educational and entertainment purposes are based on nature themes (Osterlind 2005). These natural history materials are continually incorporated into school curricula and are worth billions of dollars per year. To take one example, recent movies with penguins as main characters or themes have had revenues estimated at around $1.6 billion. They represent a nonconsumptive use value of biodiversity because they use nature only as intellectual content. A considerable number of professional scientists, as well as highly motivated amateurs, are engaged in making ecological observations and preparing educational materials. In rural areas, their activities often take place in scientific field stations, which can become sources of training and employment for local people. While these scientific activities provide economic benefits to the communities surrounding field stations, their real value lies in their ability to increase human knowledge, enhance education, and enrich the human experience.

Multiple uses of a single resource: A case study

Horseshoe crabs (*Limulus polyphemus*) provide an example of the diverse values that can be provided by just one species. They are usually noticed only as clumsy creatures that seem to move with difficulty in shallow seawater (**FIGURE 3.13**). In the United States, commercial fishermen harvest these animals in large quantities for use as cheap fishing bait. In recent years, however, ecologists have realized that horseshoe crab eggs

FIGURE 3.13 Horseshoe crabs gather in great numbers to spawn in shallow coastal waters. These aggregations have significant value to people. (© Prisma Bildagentur AG/ Alamy.)

Sometimes a specific species or ecosystem can provide a variety of goods and services to human society. Compromises are often needed to balance competing uses.

and juveniles are extremely important as a food source for shorebirds and coastal fish, which have a major role in local tourism related to birdwatching and sportfishing. Without horseshoe crabs, shorebird and sport fish populations decline in abundance (Niles 2009). Additionally, the blood of horseshoe crabs is collected to make limulus amoebocyte lysate (LAL), a highly valuable chemical used to detect bacterial contamination in injection-administered medications and vaccines (Odell et al. 2005). This chemical cannot be manufactured synthetically, and horseshoe crabs are its only source. Without this natural source of LAL, our ability to determine the purity of injected medicines would be compromised.

Currently, commercial fishing and sportfishing interests, environmental groups, birdwatching groups, and the biomedical industry are competing for control of horseshoe crabs along US coastlines. Each group can make a good argument for its own right to use or protect horseshoe crabs. Hopefully, the final result will be a working compromise that allows the crabs a place in a functioning ecosystem and still provides for the needs of people living in the area and elsewhere in the world.

The Long-Term View: Option Value

The potential of biodiversity to provide an economic benefit to human society at some point in the future is its option value. As the needs of society change, so must the methods of satisfying those needs, and such methods often lie in previously untapped animal or plant species. For example, the continued genetic improvement of cultivated plants is necessary not only for increased yield, but also to guard against pesticide-resistant insects and more virulent strains of fungi, viruses, and bacteria (Sairam et al. 2005).

We are continually searching the biological communities of the world for new plants, animals, fungi, and microorganisms that can be used to fight human diseases or to provide some other economic value, an activity referred to as **bioprospecting** (Lawrence et al. 2010). These searches are generally carried out by government research institutes and pharmaceutical companies. The US National Cancer Institute has been carrying out a program to test extracts of thousands of wild species for their effectiveness in controlling cancer cells and HIV (the virus that causes AIDS). To facilitate the search for new medicines and to profit financially from new products, GlaxoSmithKline, a British pharmaceutical corporation, and the Brazilian government signed a contract in 1999 worth $3 million allowing the company to sample, screen, and investigate approximately 40,000 plants, fungi, and bacteria from Brazil, with part of the royalties going to support scientific research and local community-based conservation and development projects. Similar agreements involving international corporations have been signed in many other countries as well. Another approach has been to target plants and other natural products used in traditional medicine for screening, often in collaboration with local healers and rural villagers. Programs such as these provide financial incentives for countries to protect their natural resources and the biodiversity knowledge of their indigenous inhabitants.

The search for valuable natural products is wide-ranging: entomologists search for insects that can be used as biological control agents, microbiologists search for bacteria that can assist in biochemical manufacturing processes, and wildlife biologists search for species that can potentially produce animal protein more efficiently and with less environmental damage than existing domesticated species (Chivian and Bernstein 2008). The growing biotechnology industry is finding new ways to reduce pollution, to develop better industrial processes, and to fight diseases threatening human health. The gene-splicing techniques of molecular biology are allowing unique, valuable genes found in one species to be transferred to another species. Both newly discovered and well-known species are often found to have exactly those properties needed to address some significant human problem. If biodiversity is reduced, the ability of scientists to locate and utilize a broad range of species will also be reduced.

> A question currently being debated among conservation biologists, governments, ecological economists, corporations, and local individuals is, Who owns the commercial rights to the world's biodiversity?

A question currently being debated among conservation biologists, governments, environmental economists, and corporations is, Who owns the commercial development rights to the world's biodiversity? The answer to this is complex, especially in the modern era. An excellent example is provided by the immunosuppressant drug cyclosporine, which occurs naturally in the fungus *Tolypocladium inflatum*. The Swiss company Sandoz (which later merged with another company to become Novartis) developed cyclosporine into a family of drugs with sales of $1.2 billion per year (Bull 2004). Cyclosporine is a key medicine used following transplant surgery to prevent organ rejection. The fungus was found in a sample of soil that was collected in Norway, without permission, by a Sandoz biologist on vacation and was later screened for biological activity. Norway has not yet received any payment for the use of this fungus in drug production. Similarly, the US National Park Service has not received any payments for the bacteria collected at Yellowstone National Park that led to breakthroughs in DNA technology. Such past and present unauthorized bioprospecting for commercial purposes is now often termed **biopiracy**. Many developing countries have reacted to this situation by passing laws that require permits for collecting biological material for research and commercial purposes, and they impose criminal penalties and fines for the violation of such laws

(Bhatti et al. 2009). People collecting samples without the needed permits have been arrested for violating the law.

Both developing and developed countries now frequently demand a share in the commercial activities derived from the biodiversity contained within their borders, and rightly so. Local people in developing countries who possess knowledge of species, protect them, and show them to scientists should also share in the profits from any use of them. Writing treaties and developing procedures to guarantee participation in this process will be a major diplomatic challenge in the coming years.

While most species may have little or no direct economic value and little option value, a small proportion may have enormous potential to supply medical treatments, to support a new industry, or to prevent the collapse of a major agricultural crop. Other species or sets of species may provide other kinds of future values even if they don't provide them now. If just one of these species goes extinct before it is discovered, it could be a tremendous loss to the global economy, even if the majority of the world's species are preserved. As Aldo Leopold commented in *Round River*,

> *If the biota, in the course of aeons, has built something we like but do not understand, then who but a fool would discard seemingly useless parts? To keep every cog and wheel is the first precaution of intelligent tinkering.*

The diversity of the world's species can be compared to a manual on how to keep the Earth running effectively. The loss of a species is like tearing a page out of the manual. If we ever need the information from that page to save ourselves and the Earth's other species, the information will have been irretrievably lost.

Existence Value

Many people throughout the world care about wildlife, plants, and entire ecosystems and want to see them protected. Their concern may be associated with a desire to someday visit the habitat of a unique species and see it in the wild; alternatively, concerned individuals may not expect, need, or even desire to see a species personally or experience the habitat in which it lives. For this reason, existence value is considered a non-use value: people value the resource without any intention to use it now or in the future. In economic terms, existence value is the amount that people are willing to pay to prevent species from going extinct, habitats from being destroyed, and genetic variation from being lost (Zander and Garnett 2011). A related idea is **beneficiary value**, or **bequest value**: how much people are willing to pay to protect something of value for their own children and descendants, or for future generations.

Particular species—the so-called charismatic megafauna, such as pandas, whales, lions, and many birds—elicit strong

> People, governments, and organizations annually contribute large sums of money to ensure the continuing existence of certain species and ecosystems.

FIGURE 3.14   Whale watching can make a significant contribution to a local economy, and participants often later contribute money to organizations promoting whale conservation. Here, people greet a California gray whale (*Eschrichtius robustus*) in Magdalena Bay, on the Pacific side of the Baja Peninsula. Such meetings can enrich human lives. (© Robert Harding Specialist Stock/Corbis.)

responses in people (**FIGURE 3.14**). Special groups have been formed to appreciate and protect butterflies and other insects, wildflowers, and fungi. People place value on wildlife in a direct way by joining and contributing billions of dollars to organizations that protect species (Wallmo and Lew 2012). In the United States, billions of dollars are contributed each year to conservation and environmental organizations, with The Nature Conservancy ($871 million in 2013), the World Wildlife Fund ($245 million), Ducks Unlimited ($184 million), and the Sierra Club ($98 million) high on the list. Citizens also show their concern by directing their governments to spend money on conservation programs and to purchase land for habitat. For example, the government of the United States has spent millions of dollars to protect a single rare species, the brown pelican (*Pelecanus occidentalis*), which was protected initially under the US Endangered Species Act and is now considered recovered.

Existence value can also be attached to ecosystems, such as temperate old-growth forests, tropical rain forests, coral reefs, and prairie remnants, and to areas of scenic beauty. Growing numbers of people and organizations contribute large sums of money annually to ensure the continuing existence of these habitats, and the environment is increasingly an important issue during national elections. Further, people want environmental education included in public school curricula (www.neefusa.org; www.epa.gov/education).

In summary, ecological economics has helped to draw attention to the wide range of goods and services provided by biodiversity. That attention has enabled scientists to account for environmental impacts that were previously left out of the equation. When analyses of large-scale development projects have finally been completed, some projects that initially appeared to be successful have been seen to actually be running at an economic loss. For example, to evaluate the success of an irrigation project using water diverted from a tropical wetland ecosystem, the short-term benefits (improved crop yields) must be weighed against the environmental costs. **FIGURE 3.15** shows the total economic value of a tropical wetland ecosystem, including its use value, option value, and existence value. When the wetland ecosystem is damaged by the removal of water, the ecosystem's ability to provide the goods and services shown in the figure are curtailed,

Total Economic Value of a Tropical Wetland Ecosystem

Use Values

Direct Use Values
Fish and meat
Fuelwood
Timber and other
  building materials
Medicinal plants
Edible wild fruits
  and plants
Animal fodder

Indirect Use Values
Flood control
Soil fertility
Pollution control
Drinking water
Transportation
Recreation and tourism
  (e.g., birdwatching)
Education
Biological services
  (e.g., pest control, pollination)

Option Value
Future products:
  Medicines
  Genetic resources
  Biological insights
  Food sources
  Building supplies
  Water supplies

Existence Value
Protection of biological
  diversity
Maintaining culture of
  local people
Continuing ecological and
  evolutionary processes

FIGURE 3.15 Evaluation of the success of a development project must incorporate the full range of its environmental impacts. This figure shows the total economic value of a tropical wetland ecosystem, including direct and indirect use value, option value, and existence value. A development project such as an irrigation project lowers the value of the wetland ecosystem when water is removed for crop irrigation. When that lowered value is taken into account, the irrigation project may represent an economic loss. (After Groom et al. 2006; based on data in Emerton 1999.)

their value greatly diminishes, and the economic success of the project is called into question. It is only by incorporating the value of the wetland into this equation that an accurate view of the total project can be gained.

Environmental Ethics

In most modern societies, people attempt to protect biodiversity, environmental quality, and human well-being through regulations, incentives, fines, environmental monitoring, and assessments. A complementary approach is to change the fundamental values of our materialistic society. **Environmental ethics**, a vigorous and growing discipline within philosophy, articulates the ethical value of the natural world (Alexander 2009; Minteer and Collins 2008). As a corollary, it challenges the materialistic values that tend to dominate modern societies. If contemporary societies de-emphasized the pursuit of wealth and instead focused on furthering genuine human well-being, the preservation of the natural environment and the maintenance of biodiversity would probably become honored practices, rather than occasional afterthoughts (Mills 2003).

Ethical values of biodiversity

Environmental ethics provides virtues and values that make sense to people today. At a time when there are unprecedented threats to the environment, ethical arguments can and do convince people to conserve biodiversity. Ethical arguments are also important because, although economic arguments by themselves provide a basis for valuing some species and ecosystems, economic valuation can also provide grounds for extinguishing species, or for saving one species and not another (Redford and Adams 2009; Rolston 2012). According to conventional economic thinking, a species with low population numbers, an unattractive appearance, no immediate use to people, and no relationship to any species of economic importance will be given a low value. Halting profitable developments or making costly attempts to preserve these species may not have any obvious economic justification. In fact, in many circumstances, economic cost–benefit analyses will support the destruction of endangered species that stand in the way of "progress."

> Ethical arguments can complement economic and biological arguments for protecting biodiversity. Such ethical arguments are readily understood by many people.

Despite any economic justification, however, many people would make a case against species extinctions on ethical grounds, arguing that the conscious destruction of a natural species is morally wrong, even if it is economically profitable. Similar arguments can be advanced for protecting unique ecosystems and genetic variation. Ethical arguments for preserving biodiversity resonate with people because they appeal to our nobler instincts or to belief in a divine creation, which do play a role in societal decision making (Bhagwat et al. 2011).

The following arguments, based on the intrinsic value of species and on our duties to other people, are important to conservation biology because they provide a rationale for protecting all species, including rare species and species of no obvious economic value.

EACH SPECIES HAS A RIGHT TO EXIST   All species represent unique biological solutions to the problem of survival. For this reason, the survival of each species must be respected, regardless of its importance to humans. This statement is true whether the species is large or small, simple or complex, ancient or recently evolved; whether it is economically important or of little immediate economic value to humans; and whether it is loved or hated by humans. Each species has value for its own sake—an **intrinsic value** unrelated to human needs or desires (Sagoff 2008). This argument suggests not only that we have no right to destroy any species, but also that we have a moral responsibility to actively protect species from going extinct as the result of our activities, as articulated by the deep ecology movement described later in this section. This argument also recognizes that humans are part of the larger biotic community and reminds us that we are not the center of the universe (**FIGURE 3.16**).

> An argument can be made that people have a responsibility to protect species and other aspects of biodiversity because of their intrinsic value, not because of human needs.

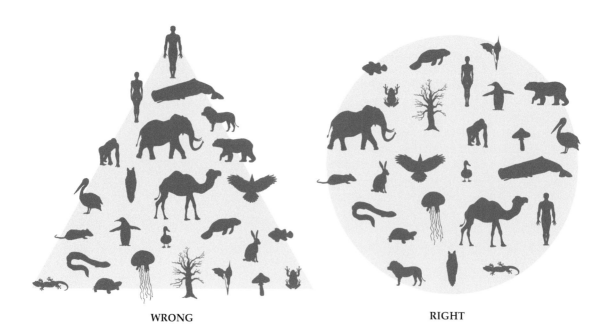

WRONG   RIGHT

FIGURE 3.16 According to certain principles of environmental ethics, it is wrong for humans to act as if they are at the top of the living world and have the right to exploit and damage other species; rather, there is a moral imperative for humans to behave as if all species have an equal right to exist. (Silhouettes © Shutterstock.)

ALL SPECIES ARE INTERDEPENDENT   Species interact in complex ways in natural communities. The loss of one species may have far-reaching consequences for other members of the community (as described in Chapter 2). Other species may become extinct in response, or the entire ecosystem may become destabilized as the result of cascades of species extinctions. For these reasons, if we value some parts of nature, we should protect all of nature (Leopold 1949). We are obligated to conserve the system as a whole because that is the appropriate survival unit (Diamond 2005).

> All species and ecosystems are interdependent, and so all parts of nature should be protected. It is in the long-term survival interest of people to protect all of biodiversity.

PEOPLE HAVE A RESPONSIBILITY TO ACT AS STEWARDS OF THE EARTH   Many religious adherents find it wrong to destroy species because they are God's creations (Moseley 2009; see Chapter 1). If God created the world, then presumably the species God created have value. Within the Jewish, Islamic, and Christian traditions, human responsibility for protecting animal species is explicitly described in the Bible as part of the covenant with God. For example, Muslim clerics in Malaysia recently joined Indonesia in a *fatwa* (a religious edict) against hunting endangered species; Asrorun Niam Sholeh of the Indonesian Ulema Council stated, "As Muslims, we have a duty to maintain the ecological balance" (Yi 2016). Other major religions, including Hinduism, Buddhism, and Islam strongly support the preservation of nonhuman life. As Mahatma Gandhi (1869–1948) taught, "The good man is the friend of all living things."

PEOPLE HAVE A RESPONSIBILITY TO FUTURE GENERATIONS   If in our daily living we degrade the natural resources of the Earth and cause species to become extinct, future generations will pay the price in terms of a lower standard of living and quality of life (Gardiner et al. 2010). As species are lost and wild lands developed, children are deprived of one of the most exciting experiences in growing up: the wonder of seeing "new" animals and plants in the wild. To remind us to act more responsibly, we might imagine that we are borrowing the Earth from future generations who expect to get it back in good condition.

RESPECT FOR HUMAN LIFE AND HUMAN DIVERSITY IS COMPATIBLE WITH A RESPECT FOR BIODIVERSITY   Some people worry that recognizing intrinsic value in nature requires taking resources and opportunities away from human beings. But respect for and protection of biodiversity can be linked to greater opportunities and better health for people (Jacob et al. 2009). Some of the most exciting developments in conservation biology involve supporting the economic development of disadvantaged rural people in ways that are linked to the protection of biodiversity. In developed countries, the **environmental justice** movement seeks to empower poor and politically weak people, who are often members of minority groups, to protect their own environments; in the process, their well-being and the protection of biodiversity are enhanced (Robinson 2011).

PEOPLE BENEFIT FROM AESTHETIC AND RECREATIONAL ENJOY-
MENT OF BIODIVERSITY   Throughout history, poets, writers, painters,
and musicians of all cultures have drawn inspiration from wild nature
(Swanson et al. 2008; Thoreau 1854). Nearly everyone enjoys wildlife and
landscapes aesthetically, and this joy increases the quality of our lives.
Nature-related activities are important in childhood development (Carson
1965; Luck et al. 2011) and may even improve human health (Donovan et al.
2013). The presence and abundance of trees in urban settings has frequently
been associated with lower incidence of crime (e.g., Gilstad-Hayden et al.
2015). Recreational activities such as hiking, canoeing, and mountain climb-
ing are physically, intellectually, and emotionally satisfying. People spend
tens of billions of dollars annually in these pursuits, proof enough that
they value them highly. A loss of biodiversity diminishes this experience.
What if there were no more migratory birds or no more meadows filled
with wildflowers and butterflies? Would we still enjoy nature as much?

PEOPLE BENEFIT FROM THE KNOWLEDGE THE NATURAL WORLD
PROVIDES   Three of the central mysteries in the world of science are:
(1) how life originated, (2) how the diversity of life interacts to form complex ecosystems, and (3) how humans evolved. Thousands of biologists are
working on these questions and are coming ever closer to the answers. New
techniques of molecular biology allow greater insight into the relationships
of living species as well as some extinct species, which are known to us
only from fossils. When species become extinct and ecosystems are damaged, however, important clues are lost, and the mysteries become harder to
solve. For example, if *Homo sapiens'* closest living relatives—chimpanzees,
bonobos, gorillas, and orangutans—disappear from the wild, we will lose
important clues to human physical and social evolution.

Deep ecology

One well-developed environmental philosophy that supports environmental activism is known as **deep ecology** (Naess 2008). Deep ecology
builds on the basic premise of biocentric equality, which expresses "the
intuition…that all things in the biosphere have an equal right
to live and blossom and to reach their own individual forms of
unfolding" (Devall and Sessions 1985). Humans have a right to
live and thrive, as do the other organisms with whom we share
the planet (see Figure 3.16). Deep ecologists oppose what they
see as the dominant worldview, which places human concerns
above all and views human happiness in materialistic terms.
Deep ecologists see acceptance of the intrinsic value of nature
less as a limitation than as an opportunity to live better lives.

Deep ecology is an
environmental philosophy
that advocates placing
greater value on
protecting biodiversity
through changes in
personal attitude,
lifestyle, and even
societies.

Paul Shepard (1925–1996) introduced to the deep ecology
movement the idea that we should achieve this ideal by returning to a more primitive state; that civilization has made us immature and out of sync with our environment. He argued that

because we evolved in close contact with nature, this contact is necessary for our emotional and psychological well-being. A related idea, called "nature deficit disorder," was popularized by author Richard Louv to describe the wide variety of problems people suffer when deprived of interactions with the natural world (Louv 2005).

Because present human activities are destroying the Earth's biodiversity, our existing political, economic, technological, and ideological structures must change. The changes that deep ecology calls for entail enhancing the quality of life for all people, emphasizing improvements in environmental quality, aesthetics, culture, and spirituality rather than higher levels of material consumption. The philosophy of deep ecology includes an obligation to work to implement needed programs through political activism and a commitment to personal lifestyle changes, in the process transforming the institutions in which we work, study, pray, and shop (Bearzi 2009). Deep ecology and other ethical and religious perspectives urge professional biologists, ecologists, and all concerned people (such as you?) to escape from their narrow, everyday concerns and act and live "as if nature mattered" (Naess 1989).

Summary

- Ecological economics is developing methods for valuing biodiversity and, in the process, providing arguments for its protection. Direct use values are assigned to products harvested from the wild, such as timber, fuelwood, fish, wild animals, edible plants, and medicinal plants. Direct use values can be further divided into consumptive use values, for products that are used locally, and productive use values, for products harvested in the wild and later sold in markets.

- Indirect use values can be assigned to aspects of biodiversity that provide economic benefits to people but are not harvested during their use. Nonconsumptive use values include ecosystem productivity, protection of soil and water resources, positive interactions of wild species with commercial crops, and regulation of climate. Biodiversity also provides value to recreation, education, and ecotourism activities.

- The option value of biodiversity is its potential to provide future benefits to human society, such as new medicines, industrial products, and crops. Biodiversity also has existence value, which is the amount of money people and their governments are willing to pay to protect species and ecosystems without any plans for their direct or indirect use.

- Environmental ethics appeals to religious and secular value systems to justify preserving biodiversity. The most central ethical argument asserts that people must protect species and other aspects of biodiversity because they have intrinsic value, unrelated to human needs. Further, biodiversity must be protected because human well-being is linked to a healthy and intact environment.

For Discussion ■—

1. Find a recent large development project in your area, such as a dam, office park, shopping mall, highway, or housing development, and learn all you can about it. Estimate the costs and benefits of this project in terms of biological diversity, economic prosperity, and human health. Who pays the costs and who receives the benefits? Consider other projects carried out in the past and determine their impact on the surrounding ecosystem and human community.

2. Consider the natural resources that people use near where you live. Can you place an economic value on those resources? If you can't think of any products harvested directly, consider basic ecosystem services such as flood control, freshwater provisioning, and soil retention.

3. Imagine that the only known population of a dragonfly species will be destroyed unless money can be raised to purchase the pond where it lives and the surrounding land. How could you assign a monetary value to this species?

4. Do living creatures, species, biological communities, and physical entities, such as rivers, lakes, and mountains, have rights? Can we treat them any way we please? Where should we draw the line of moral responsibility?

Suggested Readings —

Bateman, I. J., A. R. Harwood, G. M. Mace, R. T. Watson, D. J. Abson, B. Andrews, and 19 others. 2013. Bringing ecosystem services into economic decision-making: land use in the United Kingdom. *Science* 341: 45–50.

Braunisch, V., J. Coppes, S. Bachle, and R. Suchant. 2015. Underpinning the precautionary principle with evidence: A spatial concept for guiding wind power development in endangered species' habitats. *Journal for Nature Conservation* 24: 31–40.

Cannon, J. R. 1996. Whooping crane recovery: a case study in public and private cooperation in the conservation of endangered species. *Conservation Biology* 10: 813–821.

Chan, K. M., P. Balvanera, K. Benessaiah, M. Chapman, et al. 2016. Opinion: Why protect nature? Rethinking values and the environment. Proceedings of the National Academy of Sciences 113: 1462–1465. Considering the value of nature in terms of relational values (instrumental) as distinct from those that are of the object itself (intrinsic).

Costanza, R. R. de Groot, P. Sutton, S. van der Ploeg, and 4 others. 2014. Changes in the global value of ecosystem services. *Global Environmental Change* 26: 152–158. Global ecosystem services are valued between $125 trillion and $145 trillion per year, with significant modern losses due to land conversion and other factors.

Ehrlich, P. R. and A. H. Ehrlich. 1982. *Extinction: The Causes and Consequences of the Disappearance of Species*. Gollancz, London.

Foster, K. R., P. Vecchia, and M. H. Repacholi. 2000. Science and the precautionary principle. *Science* 288: 979–981. Despite its popular use in government policies and international agreements, the precautionary principle is problematic due to its variable interpretation.

Hardin, G. 1968. The tragedy of the commons. *Science* 162: 1243–1248. This often-cited work suggests that population control is the only solution to the overuse of common-property resources.

Helm, D. 2015. *Natural Capital: Valuing the Planet*. Yale University Press. 296 pp. This author attempts to quantify the value of various environmental resources and argues from an economist's perspective that the environment should be at the center of any economy, rather than an afterthought.

Merckx, T. and H. M. Pereira. 2015. Reshaping agri-environmental subsidies: From marginal farming to large-scale rewilding. *Basic and Applied Ecology* 16: 95–103. Re-purposing European Union subsidies that make it profitable to farm marginal land will benefit biodiversity.

4

Threats to Biodiversity

Suburban development, like that seen here outside of
Las Vegas, decreases and degrades native habitat.

Maintaining a healthy environment means preserving all of its components in good condition—ecosystems, biological communities, species, populations, and genetic variation. If species, ecosystems, and populations are adapted to local environmental conditions, why are they being lost? Why don't they tend to persist in the same place over time? Why can't they adapt to a changing environment? These questions have a single, simple answer: massive anthropogenic disturbances (that is, disturbances caused by human activities) have altered, degraded, and destroyed the landscape on a vast scale, destroying populations, species, and even whole ecosystems.

There are seven major threats to biodiversity: habitat destruction, habitat fragmentation, habitat degradation (including pollution), global climate change, the overexploitation of species for human use, the invasion of nonnative species, and the spread of disease (**FIGURE 4.1**). Most threatened species face at least two of these threats, which may interact to speed their way toward extinction and hinder efforts to protect them (Forister et al. 2010). Typically, these threats develop so rapidly, and on such a large scale, that species are not able to adapt genetically to the changes or disperse to a more hospitable location. Moreover, multiple threats

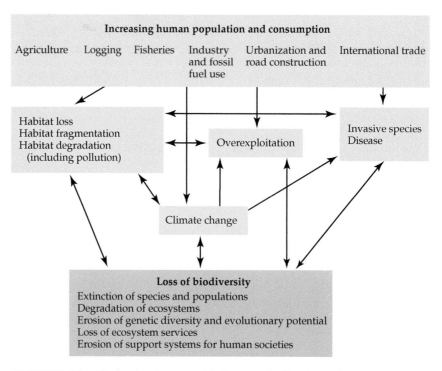

FIGURE 4.1 The major threats to biodiversity (yellow boxes) are the result of human activities. These seven factors can interact synergistically to speed up the loss of biodiversity. (After Groom et al. 2006.)

may interact additively, or even synergistically, such that their combined impact on a species or an ecosystem is greater than their individual effects.

Human Population Growth and Its Impact

The seven major threats to biodiversity are all caused by the ever-increasing use of the world's natural resources by an expanding human population. Up until the last 300 years, the rate of human population growth had been relatively slow, with the birthrate only slightly exceeding the mortality rate. The greatest destruction of ecosystems has occurred since the Industrial Revolution, during which time the human population exploded from 1 billion in 1850 to 7.25 billion in 2016 (see Figure 1.1). The human population could reach a maximum of 9–10 billion by the end of the twenty-first century. Humans have increased in such numbers because birthrates have remained high while mortality rates have declined—a result of both modern medical achieve-

> The major threats to biodiversity—habitat destruction, habitat fragmentation, pollution, global climate change, overexploitation of resources, invasive species, and the spread of disease—are all rooted in the expanding human population.

TABLE 4.1	Three Ways Humans Dominate the Global Ecosystem

1. Land surface
   Human land use, mainly for agriculture, and our need for resources, especially forest products, have transformed as much as half of the Earth's ice-free land surface.

2. Nitrogen cycle
   Each year, human activities, such as cultivating nitrogen-fixing crops, using nitrogen fertilizers, and burning fossil fuels, release more nitrogen into terrestrial systems than do natural biological and physical processes.

3. Atmospheric carbon cycle
   By the middle of the twenty-first century, human use of fossil fuels and the cutting down of forests will result in a doubling of the concentration of carbon dioxide in the Earth's atmosphere.

Sources: Data from MEA 2005 and Kulkarni et al. 2008.

ments (specifically, the control of disease) and more reliable food supplies. Population growth has slowed in the industrialized countries of the world, as well as in some developing countries in Asia and Latin America, but it is still high in other areas, particularly in tropical Africa. If these countries implemented immediate and effective programs of population control, the human population could possibly peak at "only" 9.4 billion in 2050 and then gradually decline.

Humans and their activities dominate ecosystems worldwide (**TABLE 4.1**). People use large amounts of natural resources, such as fuelwood, timber, wild meat, and wild plants, and convert vast areas of natural habitat into agricultural and residential lands. Agricultural systems and other human activities now occupy one-fourth of the Earth's land surface (Krausmann et al. 2013). All else being equal, more people equals greater human impact, more land clearing for agriculture, and less biodiversity (Allendorf and Allendorf 2012; Godfray et al. 2010). For example, nitrogen pollution is greatest in rivers flowing through landscapes with high human population densities, and rates of deforestation are greatest in countries with the highest rates of human population growth. Therefore, some scientists have argued strongly that controlling the size of the human population is the key to protecting biodiversity (Ehrlich et al. 2012; O'Neill et al. 2012; but see Bradshaw and Brook 2014).

Healthy ecosystems can persist close to areas with high population densities, even large cities, as long as human activities are regulated by local customs or government officials. The sacred groves of trees that are preserved next to villages in Africa, India, and China are examples of locally managed biological communities. When this regulation breaks down during war, political unrest, or other periods of social instability, the result is usually a scramble to collect and sell resources that had been used sustainably for generations. The higher the human population density, and the larger the city, the more closely human activities must be regulated, because the potential for both destruction and conservation is greater (Gaston 2010). For example, larger cities have been found to produce proportionally greater carbon dioxide (CO_2) emissions than smaller ones (Fragkias et al. 2013).

People in industrialized countries (and the wealthy minority in developing countries) consume a disproportionately large share of the world's energy, minerals, wood products, and food (Mills Busa 2013), and therefore have disproportionate effects on the environment. Each year, the United States, which has 5% of the world's human population, uses roughly 25% of the world's natural resources. And each year, the average US citizen uses 28 times more energy and 79 times more paper products than does the average citizen of India (*Encyclopedia of the Nations* 2009; Randolph and Masters 2008).

The impact (*I*) of any human population on the environment is roughly captured by the formula $I = PAT$ where *P* is the number of people, *A* is the average income, and *T* is the level of technology (Davidson and Andrews 2013; Elrich and Holdren 1971). It is important to recognize that the impact of a population is often felt over a great distance; for example, citizens of Germany, Canada, and Japan affect the environment in other countries through their use of foods, luxury goods, and other materials produced elsewhere (Berger et al. 2013). The increasing interconnectedness of resource and labor markets is termed **globalization**. The fish eaten quietly at home in Washington, DC, may have come from Alaskan waters, where its capture may have contributed to the population decline of sea lions, seals, and sea otters; the chocolate cake and coffee consumed at the end of a meal in Italy or France were made with cacao and coffee beans that might have grown in plantations carved out of rain forests in western Africa, Indonesia, or Brazil. Residents of industrialized countries also affect other countries through the production of waste, including greenhouse gases (with China, first among the top 10 CO_2 emitters, producing more than the remaining 9 combined; PBL Netherlands Environmental Assessment Agency 2012).

> The enormous consumption of resources in an increasingly globalized world is not sustainable in the long term.

This linkage has been captured in the idea of the **ecological footprint**, defined as the per capita influence a group of people has on both the surrounding environment and locations across the globe (Holden and Hoyer 2005; Wackernagel and Rees 1996; **FIGURE 4.2**). The ecological footprint per person is high in developed countries such as the United States and Canada and relatively low in developing countries such as China and India.

A modern city in a developed country typically has an ecological footprint that is hundreds of times its area. For example, the city of Toronto, Canada, occupies an area of 630 km², but each of its citizens requires the environmental services of 7.7 ha (0.077 km²) to provide food, water, and waste disposal sites. With a population of 2.4 million people, Toronto has

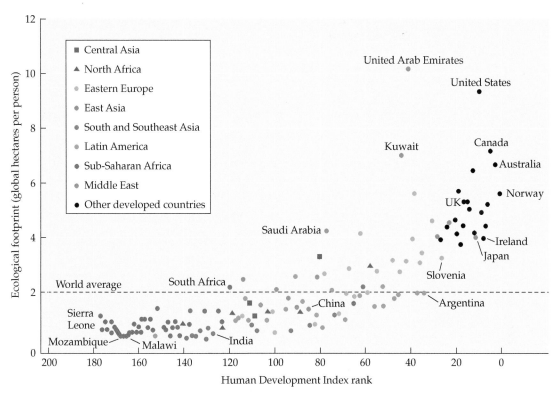

FIGURE 4.2 An ecological footprint for a nation is arrived at by calculations that estimate the number of global hectares needed to support an average
citizen of that nation. Although the methods used to arrive at these calculations
can be debated, the overall message is clear. When plotted against the Human Development Index, which reflects living standards, ecological footprints
graphically illustrate the disproportionate use of natural resources by people in
developed nations. However, the *total* impacts (not shown in this graph) of developing countries such as China, with 1.3 billion people, are also huge because
of their large populations. (Data from Global Footprint Network and the United
Nations Development Programme 2009.)

an ecological footprint of 185,000 km^2, an area equal to the state of New
Jersey or the country of Syria and more than 300 times its own area. This
excessive consumption of resources is not sustainable in the long term.
Unfortunately, this pattern of consumption is now being adopted by the
expanding middle class in the developing world, including the large, rapidly developing countries of China and India, increasing the probability
of massive environmental disruption (Feng et al. 2009; Grumbine 2007).
In fact, the developing countries now generate more greenhouse gases
and consume more of certain natural resources than the developed countries; China, in particular, has emerged as a rapidly growing industrial
powerhouse that not only exports manufactured goods but also imports

resources from around the world. The affluent citizens of developed countries must confront their excessive consumption of resources and reevaluate their lifestyles while at the same time offering aid to curb population growth, protect biodiversity, and assist industries in the developing world to grow in a responsible way.

An alternative view is that development can have a positive effect on biodiversity because wealthier countries can better afford to establish and maintain their national parks and other natural areas. Across the world, countries with a gross domestic product (GDP) of $10,000 per capita per year or greater have stable or increasing forest areas, while countries with a GDP of less than $10,000 per capita per year have declining forest areas. Furthermore, developed countries are often characterized by increasing urbanization and consequent reduced impacts on rural areas.

Habitat Destruction

The primary cause of the reduction in biodiversity, including variation at the genetic, species, and ecosystem levels, is the habitat loss that inevitably results from the expansion of human populations and activities (**FIGURE 4.3**). For the next few decades, land-use change will continue to be the main factor affecting biodiversity in terrestrial ecosystems, probably followed by overexploitation, climate change, and the introduction of invasive species (IUCN 2004). Consequently, the most important means of protecting

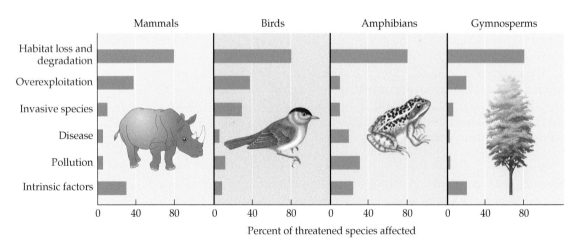

FIGURE 4.3 Habitat loss and degradation are the greatest threats to the world's species, followed by overexploitation and intrinsic factors, which include poor dispersal ability, low reproductive success, and high juvenile mortality. Groups of species face different threats: birds are more threatened by invasive species, whereas amphibians are more affected by disease and pollution. Percentages add up to more than 100% because many species face multiple threats. (After IUCN 2004.)

biodiversity is habitat preservation. *Habitat loss* does not necessarily mean wholesale habitat destruction—habitat fragmentation and habitat damage associated with pollution can also mean that the habitat is effectively "lost" to species that cannot tolerate these changes, even though, to the casual onlooker, the habitat appears intact.

> The main threat to biodiversity is habitat destruction.

In many areas of the world, particularly on islands and in locations where the human population density is high, most of the original habitat has been destroyed (Hambler et al. 2011). Fully 98% of the land suitable for agriculture has already been transformed by human activity (Sanderson et al. 2002), and 53.5% of Earth's surface has been modified by humans (Hooke et al. 2012). Because the world's population, as well as its standard of living, will continue to increase, the world's farmers will need to increase agricultural output by 30%–50% over the next 30 years. Thus, the need to protect biodiversity will be forced to compete directly against the need for new agricultural lands and the intensification of agriculture on existing lands.

Habitat disturbance has been particularly severe throughout Europe; in southern and eastern Asia, including the Philippines, China, and Japan; in southeastern and southwestern Australia; in New Zealand; in Madagascar; in western Africa; on the southeastern and northern coasts of South America; in Central America; in the Caribbean; and in central and eastern North America (**FIGURE 4.4**). In many of these regions, more than 50% of the natural habitats have been disturbed or removed. Only 15% of the land area in Europe remains unmodified by human activities, and in some regions of Europe, the percentage is even lower. In Germany or the United Kingdom, for example, one can hardly find any habitat that has not been modified by humans at one time or another.

The principal human activities that threaten the habitats of endangered species, in order of decreasing importance, are agriculture (affecting 38% of endangered species), commercial developments (35%), water projects (30%), outdoor recreation (27%), livestock grazing (22%), pollution (20%), infrastructure and roads (17%), disruption of fire ecology (13%), and logging (12%) (Stein et al. 2000; Wilcove and Master 2005).

As a result of farming, logging, and other human activities, very little **frontier forest**—intact blocks of undisturbed forest large enough to support all aspects of biodiversity—remains in many countries; the global decline in frontier forest is estimated as approximately 0.5% per year during the past decade (Hansen et al. 2013). In the Mediterranean region, which has been densely populated by people for thousands of years, only 10% of the original forest cover remains. An important point to remember here is that wildlife individuals and populations are lost in approximate proportion to the amount of habitat that has been lost; even though the Mediterranean forest still exists in places, approximately 90% of the individuals and populations of birds, butterflies, wildflowers, frogs, and mosses that once existed are no longer there.

(A) Biomes current and future

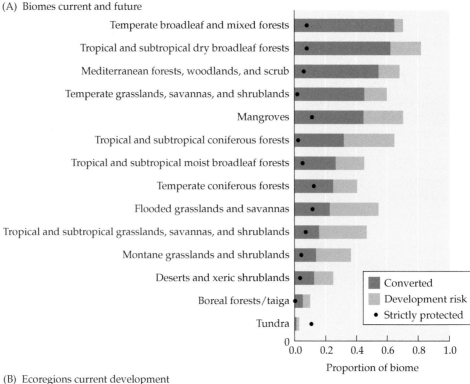

(B) Ecoregions current development

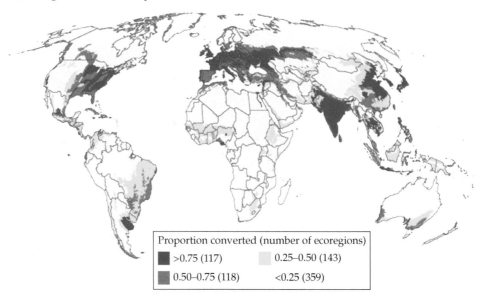

Proportion converted (number of ecoregions)

■ >0.75 (117) □ 0.25–0.50 (143)

■ 0.50–0.75 (118) <0.25 (359)

FIGURE 4.4 Many of the world's major biomes have already had a large proportion of their area converted to human uses. (A) Current and future conversion. (B) Current conversion. (After Oakleaf et al. 2015.)

Tropical rain forests

The destruction of tropical rain forests has come to be synonymous with the rapid loss of species. The tropics occur between approximately 23.5° north and south of the equator, and *rain forests* (sometimes written rainforests) are those characterized by a closed, evergreen canopy often more than 25 m tall with abundant, woody vines and generally more than 2000 mm rain per year (Turner 2001). Tropical rain forests occupy 7% of the Earth's land surface (**FIGURE 4.5**), but they are estimated to contain over 50% of its species (Bradshaw et al. 2009; Corlett and Primack 2010; see Chapter 2). Many tropical rain forest species are important to local economies and have the potential for greater use by the entire world population (see Chapter 3). Tropical rain forests also have local significance as home to numerous indigenous cultures, regional importance in protecting watersheds and moderating climate, and global importance as sinks to absorb some of the excess CO_2 that is produced by the burning of fossil fuels. The original extent of tropical rain forests and related moist forests has been estimated at 17 million km^2 based on current patterns of rainfall and temperature. A combination of ground surveys, aerial photos, and remote-sensing data from satellites has shown that only 11 million km^2 remain (Corlett and Primack

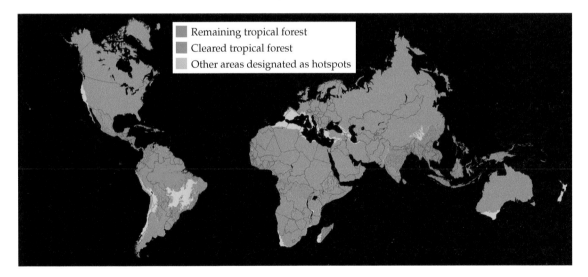

Remaining tropical forest
Cleared tropical forest
Other areas designated as hotspots

FIGURE 4.5 The current extent of tropical forests, and the areas that have been cleared of tropical forests. Note the extensive amount of land that has been deforested in northern and southeastern South America, India, Southeast Asia, Madagascar, and western Africa. The map also shows hotspots of biodiversity, a subject that will be treated in further detail in Chapter 7. Many of the biodiversity hotspots in the temperate zone have a Mediterranean climate, such as southwestern Australia, South Africa, California, Chile, and the Mediterranean basin. This map is a Fuller Projection, which distorts the sizes and shapes of continents less than typical world maps. (Map created by Clinton Jenkins; originally appeared in Pimm and Jenkins 2005.)

2010). More than 60% of the recent loss has occurred in the Neotropics, with Brazil alone accounting for almost half. Another third has occurred in Asia, with Indonesia second to Brazil in the absolute rate of tropical forest loss. Africa has contributed only 5.4% to the total area lost, reflecting the current absence of industrial-scale agricultural clearance there. Strikingly, 55% of all recent tropical forest losses occurred within only 6% of the total tropical forest area, forming an "arc of deforestation" in the south and southeast of the Brazilian Amazon, in much of Malaysia, and in Sumatra and parts of Kalimantan in Indonesia. These areas are experiencing rapid deforestation; it is estimated that in Malaysia, 2% of the forest area is lost per year (Hansen et al. 2013). Meanwhile, the rate of deforestation in the Brazilian Amazon appears to be slowing down due to changes in government policy and stricter enforcement of logging regulations.

In relative terms, the deforestation rate is greatest in Asia, averaging about 1.2% per year, while in absolute terms, tropical America has the greatest amount of deforestation because of its larger total area. If the current rates of loss continue, there will be little tropical forest left after the year 2050, except in the relatively small national parks and remote, rugged, or infertile areas of the Amazon basin, Congo River basin, and New Guinea. The move to establish large new parks in many tropical countries is cause for some hope; however, these parks will need to be well funded and managed to be effective in preserving biodiversity, as described in Chapter 8. In many cases, these parks are only "paper parks" with few employees or facilities.

On a global scale, much of rain forest destruction may still result from small-scale cultivation of crops by poor farmers, who are often forced onto remote forest lands by poverty or sometimes moved there by government-sponsored resettlement programs (Peres and Schneider 2012). Much of this farming is **shifting cultivation**, a kind of subsistence farming sometimes referred to as slash-and-burn, or swidden, agriculture, in which trees are cut down and then burned. The cleared patches are farmed for two or three seasons, after which soil fertility has usually diminished and soils have eroded to the point where crop production is so low that the patches are then abandoned and a new area is cleared (Phua et al. 2008). Although these patches may recover with time, studies show that shifting cultivation has a negative effect on both plants and animals (Mukul and Herbohn 2016). Shifting cultivation is often practiced because farmers are unwilling or unable to spend the time and money necessary to develop more permanent forms of agriculture on land that they do not own and may not occupy for very long. Rain forests are also destroyed by fuelwood production, mostly to supply local villagers with wood for cooking fires. More than two billion people cook their food with firewood, so their impact is significant. Increasing human populations in poor tropical countries will cause further loss of tropical forests in coming decades.

In an increasing proportion of the tropics, however, clearance by peasant farmers to meet subsistence needs is now dwarfed by clearance by large landowners and commercial interests to create pasture for cattle ranching or to plant cash crops, such as oil palms, soybeans, and rubber trees (Rosa

FIGURE 4.6 Complex and diverse tropical forests give way to an African tea plantation, a sea of green that supports almost no biodiversity—not only because it is a monoculture, but also because of the liberal use of pesticides. Conversion of tropical forests to agriculture is considered the most common cause of loss of these ecosystems. This plantation and others like it were initially established to be buffer zones around protected remnant forests; however, over time, the plantations facilitated further encroachment and conversion by the growing local human population. In this region of western Kenya, almost 60% of the forest cover was lost between 1984 and 2009, with 90% of this loss attributable to agriculture (Cordeiro et al. 2015). (Photograph by Anna Sher.)

et al. 2012; **FIGURE 4.6**). Shifting cultivation is often less detrimental to biodiversity than these large-scale, often pesticide-laden plantations in which large areas are maintained under a uniform crop cover (Mukul and Herbohn 2016). Commercial agriculture displaces poor farmers and justifies the expansion of roads. In addition, large areas of rain forest are damaged during commercial logging operations, most of which are poorly managed selective logging. These logged forests are prone to widespread fires due to the large numbers of branches and dead trees on the ground. In many cases, logging operations precede conversion of tropical forest land to agriculture and ranching. The relative importance of these enterprises varies by geographical region: logging is a significant activity in tropical Asia and America, cattle ranching is most prominent in tropical America, and farming is more important for the rapidly expanding population in tropical Africa (Corlett and Primack 2010). As tropical forests are affected by people in any of these ways, their diversity of species often rapidly declines (Gibson et al. 2011).

The destruction of tropical rain forests is frequently caused by demand in industrialized countries for cheap agricultural products, such as rubber, palm oil, cocoa, soybeans, orange juice, and beef, and for low-cost wood products (Nepstad et al. 2006). At present, most consumers in industrialized countries are not aware of how their food choices affect land use. Many people would be surprised to learn how widely palm oil is used in processed food and consumer products, not to mention where it comes from. It is also true that increasing proportions of these agricultural and wood products are consumed within the countries that produce them or exported to rapidly industrializing countries, such as China and India.

The story of Indonesian Borneo and Sumatra in Southeast Asia illustrates how rapid and serious rain forest destruction can be. Between 1990 and 2005, an incredible 42% of the lowland forest of these two large islands was cleared. Most of the clearing was due to logging, both legal and illegal, and the development of cash crops, especially oil palms (Hansen et al. 2009; Lee et al. 2014).

Other threatened habitats

The plight of the tropical rain forests is perhaps the most widely publicized case of habitat destruction, but many other habitats are also in grave danger.

TROPICAL DECIDUOUS FORESTS A month of less than 100 mm is considered dry in tropical forests, and trees that experience drought regularly for long periods adapt by being *deciduous* (i.e., dropping their leaves). The land occupied by tropical deciduous forests is more suitable for agriculture and cattle ranching than is the land occupied by tropical rain forests. These forests are also easier than rain forests to clear and burn. Moderate rainfall in the range of 250–2000 mm per year allows mineral nutrients to be retained in the soil, from which they can be taken up by plants. Consequently, human population density is five times greater in deciduous forest areas of Central America than in adjacent rain forest areas. Today, the Pacific coast of Central America has less than 2% of its original deciduous forest remaining (WWF and McGinley 2009), and less than 3% remains in Madagascar, which is home to the endangered lemurs, an endemic group of primates (see Figure 5.10).

GRASSLANDS Temperate grassland is another habitat type that has been almost completely destroyed by human activities. It is relatively easy to convert large areas of grassland to farmland or cattle ranches. Between 1800 and 1950, as much as 98% of North America's tallgrass prairie was converted to farmland. The remaining area of prairie is fragmented and widely scattered across the landscape. Increasing prices for agricultural products are currently driving the conversion of even marginal grasslands to farmland (Wright and Wimberly 2013). Worldwide, only 4% of temperate grasslands are protected (see Chapter 8).

> Between 1800 and 1950, as much as 98% of North America's tallgrass prairie was converted to farmland.

FRESHWATER HABITATS Wetlands and other freshwater aquatic habitat are critical habitats for fish, aquatic invertebrates, amphibians, and birds.

They are also a resource for flood control, water filtration, and power production (as described in Chapter 3). Freshwater systems are often filled in or drained for development, or are otherwise altered by dams, channelization of watercourses, and chemical pollution (Mitsch and Gosselink 2015). Over half of the wetland ecosystems that existed in the early twentieth century have been lost in North America, Europe, Australia, and China (Moreno-Mateos et al. 2012). In the United States, 98% of the country's 5.2 million km of streams have been degraded in some way to the point that they are no longer considered wild or scenic. More importantly, their ecosystem functions and services are lost, including their ability to serve as dispersal routes for aquatic animals and plants (see Chapter 3). Destruction of wetlands and streams has been equally severe in other parts of the industrialized world, such as Europe and Japan. About 60%–70% of wetlands in Europe have been lost. Only 2 of Japan's 30,000 rivers can be considered wild, without dams or some other major modification.

In the last few decades, major threats to wetlands and other aquatic environments in developing countries have included massive development projects involving drainage, irrigation, and dams, organized by governments and often financed by international aid agencies. The Three Gorges Dam on the Yangtze River of China is a recent example (Sun et al. 2012; Yang et al. 2014). The dam is the largest hydroelectric power plant in the world, generating much-needed clean and renewable energy. However, the dam and reservoir have displaced more than 1 million people, destroyed untold numbers of ecosystems and archaeological sites, and altered the river and delta systems, with unknown ecological consequences. The economic benefits of such projects are important, but the rights of local people and the value of ecosystems are often not adequately considered.

MARINE COASTAL AREAS   Human populations are increasingly concentrated in marine coastal areas. Twenty percent of marine coastal areas have already been degraded, filled in, or highly modified by human activity, despite their importance in the harvesting of fish, shellfish, seaweeds, and other marine products (see Figure 3.5A). Coastal wetlands are also threatened by pollution, dredging, sedimentation, destructive fishing practices, invasive species, and now rising temperatures. Human impacts on these habitats are less well studied than in the terrestrial environment, but they are probably equally severe, especially in shallow coastal areas. Two coastal habitats of special note are mangroves and coral reefs.

MANGROVES   Mangrove forests are among the most important wetland communities in tropical areas (Polidoro et al. 2010), and it is estimated that 2%–8% of global mangrove cover is lost each year (Miththapala 2008). Composed of species that are among the few woody plants able to tolerate salt water, mangrove forests occupy coastal areas with saline or brackish water, typically where there are muddy bottoms. Such habitats, like salt marshes in the temperate zone, are extremely important breeding grounds and feeding areas for shrimp and fish. In Australia, for example, two-thirds of the species caught by commercial fishermen depend to some degree on the mangrove ecosystem. Despite their great economic value and their utility for protecting coastal areas from storms and tsunamis, mangroves are often cleared for rice cultivation and commercial shrimp and prawn hatcheries, particularly in Southeast Asia. Mangroves have also been severely degraded by overcollection of wood for fuel, construction poles, and timber throughout the region; it has been argued that there are too few incentives for local users and managers to stop this overexploitation (Máñez et al. 2014). Over half of the world's mangrove ecosystems have already been destroyed (Twilley and Day 2012). Today, almost 40% of mangrove-dependent animal species are considered to be at high risk of extinction (Daru et al. 2013).

CORAL REEFS Tropical coral reefs (see Chapter 2 opener) contain an estimated one-third of the ocean's fish species in only 0.2% of its surface area (**FIGURE 4.7**). At least 38% of all coral reefs have already been destroyed (Butchart et al. 2010). A further 20% have been degraded by overfishing, overharvesting, pollution, and the introduction of invasive species (MEA 2005). The most severe destruction is taking place in the Philippines, where a staggering 90% of the reefs are dead or dying. In China, coral reefs have declined by 80% over the past 30 years (Hughes et al. 2013). The main culprits are pollution, which either kills the corals directly or allows excessive

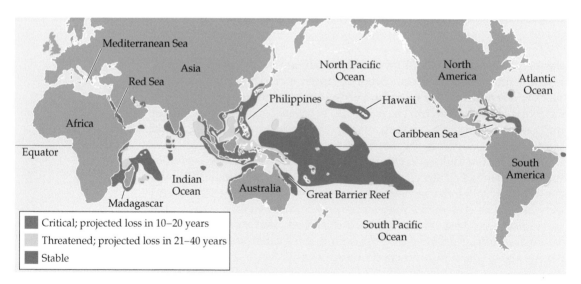

FIGURE 4.7 Extensive areas of coral are likely to be damaged or destroyed by human activities over the next 40 years unless conservation measures can be implemented. (After Bryant et al. 1998, with updates from Burke et al. 2011.)

growth of algae; sedimentation resulting from deforestation, agriculture, and coastal development; overharvesting of fish, clams, and other animals; and finally, stunning with dynamite or cyanide to collect the few remaining living creatures for food and the aquarium trade. The effects of global climate change, including ocean warming, ocean acidification, and the increasing intensity of tropical storms fueled by warmer surface waters, are also playing a major role in the rapid degradation of coral reefs (Comeau et al. 2015; Schmidt et al. 2014). Worst of all is the combination of local and global impacts, which makes it difficult or impossible for coral reefs to bounce back from natural disturbances, such as hurricanes, to which they may otherwise be relatively resilient.

Further losses of coral reefs are expected within the next 40 years, especially in tropical East Asia, the areas around Madagascar and East Africa, and throughout the Caribbean. Elkhorn and staghorn corals, which were formerly common in the Caribbean and gave three-dimensional structure to the reef community, have already become rare in many locations (Carpenter et al. 2009; Mora 2009).

Over the last 15 years, scientists have discovered extensive reefs of corals living in cold water at depths of 300 m or more, many of which are in the temperate zone of the North Atlantic. These coral reefs are rich in species, many of which are new to science. Yet just as these communities are being explored for the first time, they are being destroyed by trawlers, which drag nets across the seafloor to catch fish. The trawlers destroy the very coral reefs that protect and provide food for young fish. The damage to these cold-water reefs by careless harvesting is costing the industry its resource base in the long run.

Desertification

Many ecosystems in seasonally dry climates are degraded by human activities into human-made deserts, a process known as *desertification* (Lavauden 1927; Okin et al. 2009). These dryland ecosystems include grasslands, scrub, and tropical deciduous forests as well as temperate shrublands, such as those found in the Mediterranean region, southwestern Australia, South Africa, central Chile, and California. Naturally occurring dry areas cover about 41% of the world's land area and are home to about one billion people. Approximately 10%–20% of these drylands are at least moderately degraded, and more than 25% of the productive capacity of their plant growth has been lost (Sayre et al. 2013). These areas may initially support agriculture, but their repeated cultivation, especially during dry and windy years, often leads to soil erosion and loss of water-holding capacity in the soil. The land may also be chronically overgrazed by domesticated livestock, and woody plants may be cut down for fuel. Frequent fires during long dry periods often damage the remaining vegetation. The result is the progressive and largely irreversible degradation of the ecosystem and the loss of soil cover. Ultimately, formerly productive farmland and pastures take on the appearance of a desert. Desertification has been ongoing for thousands of years in the Mediterranean region and was known even to ancient Greek observers.

Worldwide, 9 million km^2 of arid lands have been converted to human-made deserts. These areas are not functional desert ecosystems, but wastelands, lacking the flora and fauna characteristic of natural deserts. The process of desertification is most severe in the Sahel region of Africa, just south of the Sahara, where most of the native large mammal species are threatened with extinction. The human dimension of the problem is illustrated by the fact that the Sahel region is estimated to have 2.5 times more people (100 million currently) than the land can sustainably support. Further desertification appears to be almost inevitable, especially given the higher temperatures and lower rainfall associated with predictions of future climate change (Verstraete et al. 2009). In such areas, the solution will be programs involving the implementation of sustainable agricultural practices, the elimination of poverty, increased political stability, and control of the human population.

Habitat Fragmentation

In addition to being destroyed outright, habitats that formerly occupied wide, unbroken areas are often divided into pieces by roads, fields, towns, and a broad range of other human constructs. **Habitat fragmentation** is the process whereby a large, continuous area of habitat is both reduced in area and divided into two or more fragments (**FIGURE 4.8**). When habitat is destroyed, a patchwork of habitat fragments may be left behind. These fragments are often isolated from one another by a highly modified or degraded landscape, and their edges experience altered conditions, referred to as **edge effects,** such as increased wind, fire, species invasion, or predation (Murcia 1995; Porensky and Young 2013). The fragments are often on the least desirable land for human uses, such as steep slopes, poor soils, and inaccessible areas.

Fragmentation almost always occurs during a severe reduction in habitat area, but it can also occur when habitat area is reduced to only a minor degree by roads, railroads, canals, power lines, fences, oil pipelines, fire lanes, or other barriers to the free movement of species. In many ways, habitat fragments resemble islands of original habitat in an inhospitable, human-dominated landscape. Habitat fragmentation is a serious threat to biodiversity, as species are often unable to survive under the altered conditions found in fragments.

Habitat fragments differ from the original habitat in three important ways:

1. Fragments have a greater amount of edge per area of habitat (and thus a greater exposure to edge effects).
2. The center of a habitat fragment is closer to an edge.
3. When a formerly continuous habitat hosting a large population is divided into fragments, each fragment hosts a smaller population.

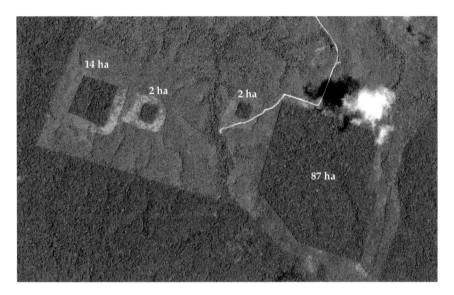

FIGURE 4.8   The Biological Dynamics of Forest Fragmentation (BDFF) study in Brazil is the largest and longest-running experiment of tropical forest fragmentation; forest fragments of different sizes were created 1980 when the land was cleared for cattle ranches and farms. Shown in this aerial photograph are forest fragments of 2 ha, 14 ha, and 87 ha. These fragments were surveyed for species composition, microclimate, and other ecosystem characteristics before the forest was cleared, and have been regularly surveyed over the past 36 years. In addition, there have been forest blocks of equivalent size inside intact forest that have been surveyed as controls. In the photograph, note that the fragments vary in their distance from the intact forest. An important factor in the experiment is that land use around the fragments has changed over time, with many of the cattle pastures being abandoned, and undergoing succession to secondary forests. For example, the left side of the 14 ha fragment is secondary vegetation while the right side has recently been cleared. See Figure 4.10 for some of the key results. (© DigitalGlobe.)

Threats posed by habitat fragmentation

LIMITS TO DISPERSAL AND COLONIZATION   Fragmentation may limit a species' potential for dispersal and colonization by creating barriers to normal movements (Stouffer et al. 2011). In an undisturbed environment, seeds, spores, and animals move passively or actively across the landscape. Over time, populations of a species may build up and go extinct on a local scale as the species disperses from one suitable site to another and the biological community undergoes succession. At a landscape level, a series of populations exhibiting this pattern of extinction and recolonization is sometimes referred to as a *metapopulation* (see Figure 6.12).

When a habitat is fragmented, the potential for dispersal and colonization is often reduced. Many bird, mammal, and insect species of the forest interior will not cross even short stretches of open ground (Laurance et al. 2009). If they do venture into the open, they may find predators waiting on the forest edge to catch and eat them. Agricultural fields 100 m wide may represent an impassable barrier to the dispersal of many invertebrate species. Roads, too, may be significant barriers to animal movement. Many species avoid crossing roads, which represent an environment totally different from the habitat they are leaving. For animals that do attempt to cross roads, motor vehicles are a major source of mortality (Beebee 2013), as has been observed in the endangered Florida panther (Kroll 2015). To deal with such problems, highway officials are building animal underpasses, overpasses, and other improvements to minimize animal mortality.

As species go extinct within individual fragments through natural successional and metapopulation processes, new species will be unable to arrive because of barriers to dispersal colonization, and the number of species present in the habitat fragment will decline over time. Extinction will be most rapid and severe in small habitat fragments.

> The barriers that fragment a habitat reduce the ability of animals to forage, find mates, disperse, and colonize new locations. Habitat fragmentation often creates small subpopulations that are vulnerable to local extinction.

RESTRICTED ACCESS TO FOOD AND MATES   Many animal species need to move freely across the landscape, either as individuals or in social groups, to feed on widely scattered resources (Becker et al. 2010). A given resource may be needed for only a few weeks each year, or even only once in a few years, but when a habitat is fragmented, species confined to a single habitat fragment may be unable to migrate over their normal home range in search of that scarce resource. Gibbons and other primates, for example, typically remain in forests and forage widely for fruits. Finding scattered trees with abundant fruit crops may be crucial during episodes of fruit scarcity. Clearings and roads that break up the forest canopy may prevent these primates from reaching nearby fruiting trees because the primates are unable or unwilling to descend to the ground and cross the intervening open landscape. Fences may prevent the natural migration of large grazing animals such as wildebeest and bison, forcing them to overgraze unsuitable habitat, which eventually leads to starvation and further degradation of the habitat (Gates et al. 2012).

Barriers to dispersal can also restrict the ability of widely scattered species to find mates, leading to a loss of reproductive potential for many animal species. Plants may have reduced seed production if butterflies and bees are less able to migrate among habitat fragments to pollinate flowers.

CREATION OF SMALLER POPULATIONS   Habitat fragmentation may precipitate population decline and extinction by dividing an existing widespread population into two or more subpopulations in restricted areas. These smaller populations are then more vulnerable to inbreeding depression, genetic drift, and other problems associated with small population size (see Chapter 5). While a large area may support a single large population, it is possible that none of the smaller subpopulations in the fragments will be sufficiently large to persist, even if the total area is the same. Connecting the fragments with properly designed movement corridors may be the key to maintaining populations.

INTERSPECIES INTERACTIONS   Habitat fragmentation increases the vulnerability of the fragments to invasion by nonnative and native pest species (Aguirre-Acosta et al. 2014; Flory and Clay 2009). Road edges themselves may represent dispersal routes for invasive species. The forest edge represents a high-energy, high-nutrient, disturbed environment in which many pest species of plants and animals can increase in number and then disperse into the interior of the fragment.

Omnivorous animals may increase in population size along forest edges, where they can eat foods, including eggs and nestlings of birds, from both undisturbed and disturbed habitats. In the coniferous forests of Finland, predation on nests was found to be higher in edges of clear-cut areas than in either the forest interior or forest corridors, and sometimes higher than in the clear-cut area (Huhta and Jokimäki 2015). The invading predators include not only native omnivores—such as raccoons, skunks, and blue jays in North America—but also introduced domesticated cats. In a study done in the southeastern United States, successful hunting cats were found to capture an average of 2.4 prey items during 7 days of roaming, with Carolina anoles (*Anolis carolinensis*) being the most common prey species (Loyd et al. 2013). In rural Poland, an examination of both scat and stomach contents and prey brought home by cats found mammals to be the most common prey, followed by birds (Krauze et al. 2012).

Nest-parasitizing cowbirds, which live in fields and edge habitats, use habitat edges as invasion points, flying up to 15 km into forest interiors, where they lay their eggs in the nests of forest songbirds (Cox et al. 2012; Lloyd et al. 2005). The combination of habitat fragmentation, increased nest predation, and destruction of tropical wintering habitats is probably responsible for the dramatic decline of certain migratory songbird species of North America, such as the cerulean warbler (*Dendroica cerulea*), particularly in the eastern half of the United States (Valiela and Martinetto 2007; With 2015). Populations of deer and other herbivores can also build up in edge areas, where plant growth is lush, eventually overgrazing the vegetation and selectively eliminating certain rare and endangered plant species for several kilometers into the forest interior.

Habitat fragmentation also puts wild populations of animals in closer proximity to domesticated animals. Diseases of domesticated animals can then spread more readily to wild species, which often have no immunity to them. There is also the potential for diseases to spread from wild species to domesticated plants, animals, and even people, once the level of contact increases. Fragmented forest habitats characteristic of suburban development often have high densities of white-footed mice and black-legged ticks with high rates of infection with Lyme disease, along with a corresponding increase in Lyme disease in people living in those areas (Tran and Waller 2013).

Edge effects

Habitat fragmentation greatly increases the amount of edge relative to the amount of interior habitat. A simple example will illustrate these characteristics and the problems they can cause. Consider a square conservation reserve 1000 m (1 km) on each side (**FIGURE 4.9**). The total area of the reserve is 1 km² (100 ha). The perimeter (or edge) of the reserve totals 4000 m. A point in the middle of the reserve is 500 m from the nearest perimeter. If the principal edge effect for birds in the reserve is predation by domesticated cats and introduced rats, which forage 100 m into the forest from the perimeter of the reserve and prevent forest birds from successfully raising their young, then only the reserve's interior—64 ha—is available to the birds for breeding. If the reserve is divided into four fragments by a road and a railroad, each of which is 495 m in area, the nesting habitat is further reduced to 8.7 ha in each fragment, for a total of 34.8 ha. Even though the road and railroad remove only 2% of the reserve area, they reduce the habitat available to the birds by about half. The implications of edge effects can be seen in the decreased

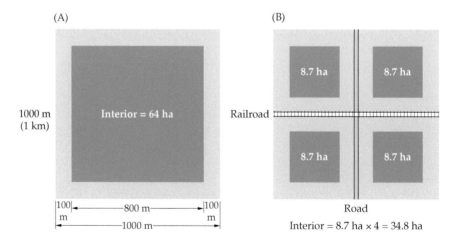

FIGURE 4.9 A hypothetical example shows how habitat area is reduced by fragmentation and edge effects. (A) A 1 km² protected area. Assuming that edge effects (gray) penetrate 100 m into the reserve, approximately 64 ha are available as usable habitat for nesting birds. (B) The bisection of the reserve by a road and a railroad, although taking up little actual area, extends the edge effects so that almost half the breeding habitat is destroyed. Edge effects are proportionately greater when forest fragments are irregular in shape, as is usually the case.

ability of birds to live and breed in small forest fragments compared with larger blocks of forest habitat.

The microenvironment at a forest fragment edge is different from that in the forest interior. Some of the more important differences are greater fluctuations in levels of light, temperature, humidity, and wind (Laurance et al. 2011). These edge effects are most evident up to 100 m inside the forest, although certain effects are detectable up to 400 m from the forest edge (**FIGURE 4.10**). Because so many plant and animal species are precisely adapted to certain levels of temperature, humidity, and light, changes in those conditions eliminate many species from forest fragments. Shade-tolerant wildflower species of temperate forests, late-successional tree species of tropical forests, and

> Habitat fragmentation increases edge effects— changes in light, humidity, temperature, and wind that may be less favorable for many species living there.

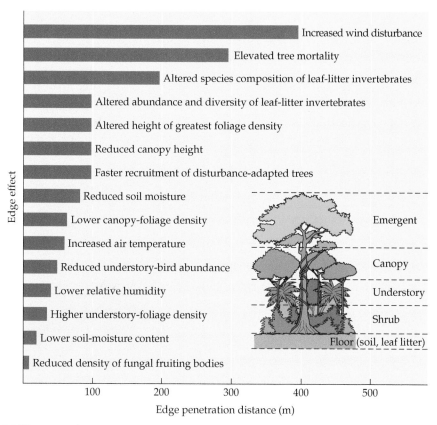

FIGURE 4.10 Edge effects in the Amazon rain forest as measured in the experiment shown in Figure 4.8. The bars indicate how far into the forest fragment the specified effect occurs. For example, trees growing within 300 m of an edge have a higher mortality rate, and the average height of trees in the forest canopy (see drawing) is reduced within 100 m of the edge. (After Laurance et al. 2002.)

humidity-sensitive animals such as amphibians are often rapidly eliminated by habitat fragmentation.

When a forest is fragmented, increased wind, lower humidity, and higher temperatures make fires more likely (Uriarte et al. 2012). Fires may spread into habitat fragments from nearby agricultural fields that are being burned regularly, as in sugarcane harvesting, or from the irregular activities of farmers practicing slash-and-burn agriculture. Forest fragments may be particularly susceptible to fire damage when wood has accumulated at the edge of the forest where trees have died or have been blown down by the wind. In Borneo and the Brazilian Amazon, millions of hectares of tropical moist forest burned during unusually dry periods in 1997 and 1998. A combination of factors contributed to these environmental disasters: forest fragmentation due to road construction, farming, the accumulation of brush following selective logging, and human-caused fires.

Environmental Degradation and Pollution

Even when a habitat is unaffected by overt destruction or fragmentation, the ecosystems and species in that habitat can be profoundly affected by human activities. Ecosystems can be damaged and species driven to extinction by external factors that do not obviously change the structure of the dominant plants or other features in the community. Sometimes this damage and its cause(s) are quite obvious (**FIGURE 4.11**). For example, keeping too many cattle in a grassland community gradually damages it, often eliminating many native species and favoring invasive species that can tolerate grazing and trampling. On the other hand, out of sight from the public, fishing trawlers drag across an estimated 15 million km^2 of ocean floor each year, an area 150 times greater than the area of forest cleared in the same period. The trawling destroys delicate creatures such as anemones and sponges, reduces species diversity and biomass, and alters community structure (Hinz et al. 2009; Maynou and Cartes 2012).

Other types of environmental damage are not visually apparent even though they occur all around us, every day, in nearly every part of the world. The most subtle and universal form of environmental degradation is pollution, commonly caused by pesticides, herbicides, sewage, fertilizers from agricultural fields, industrial chemicals and wastes, emissions from factories and automobiles, and sediment deposits from eroded hillsides. The general effects of pollution on water quality, air quality, and even the global climate are cause for great concern, not only because of their threats to biodiversity, but also because of their effects on human health and agriculture (Dearborn and Kark 2010). Although environmental pollution is sometimes highly visible and dramatic, as in the case of massive oil spills, the subtle, unseen forms of pollution are probably the most threatening—primarily because they are so insidious.

Pollution of the air, water, and soil by chemicals, wastes, and the by-products of energy production destroys habitats in insidious ways.

FIGURE 4.11   This 80-mile-long spill of toxic materials from the Gold King Mine, which extended through three states and two Native American reservations in 2015, prompted a state of emergency in the southwestern United States. The extent of the harm done to wildlife and people by the arsenic, lead, mercury, and cadmium released is not fully known. An EPA report said that human error, in the form of a botched cleanup effort, was to blame for the 3-million-gallon spill (*Denver Post* 2016). (Animas River at Bakers bridge near Durango, Colorado, USA, taken August 6, 2015. © Whit Richardson/Alamy Stock Photo.)

Pesticide pollution

The dangers of pesticides were brought to the world's attention in 1962 by Rachel Carson's influential book *Silent Spring* (see Figure 1.4). Carson described a process known as **biomagnification**, through which dichloro-diphenyl-trichloro-ethane (DDT) and other organochlorine pesticides become concentrated as they ascend the food chain (Kohler and Triebskorn 2013; Weis and Cleveland 2008; **FIGURE 4.12**). These pesticides, used on crop plants to kill insects and sprayed on water bodies to kill mosquito larvae, were harming wildlife populations, especially birds, such as hawks and eagles, that eat large amounts of insects, fish, or other animals exposed to DDT and its by-products.

Recognition of this situation in the 1970s led many industrialized countries to ban the use of DDT and other chemically related pesticides. The ban eventually allowed the partial recovery of many bird populations (**FIGURE 4.13**). Nevertheless, massive use of pesticides—even DDT—persists because of their benefits to people. For example, DDT is still highly effective in controlling mosquitoes, which, through the malaria they spread, are still a significant cause of death in tropical regions. These benefits must be

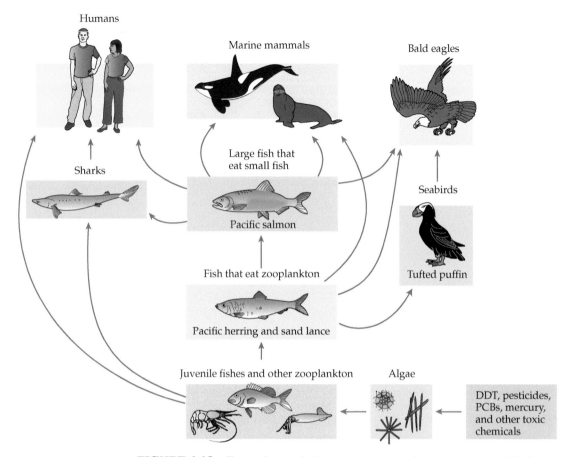

FIGURE 4.12 Toxic chemicals become successively concentrated at higher levels in the food chain, leading to health problems for humans, marine mammals, seabirds, and raptors. (After Groom et al. 2006.)

weighed against harm not only to endangered animal species, but also to people, particularly the workers who handle these chemicals in the field and the consumers of the agricultural products, such as crops and even chicken eggs, exposed to these chemicals (Bouwman et al. 2015). Even plants, animals, and people living far from where the chemicals are applied can be harmed; high concentrations of pesticides are found even in the tissues of polar bears in northern Norway and Russia, where they have a harmful effect on bear health (Elliott and Elliott 2013).

Water pollution

Water pollution has negative consequences for all species: it destroys important food sources and contaminates drinking water with chemicals that can cause immediate and long-term harm to the health of humans and other species that come into contact with the polluted water (Feist et al. 2011) (see Figure 4.11). In the broader picture, water pollution often severely

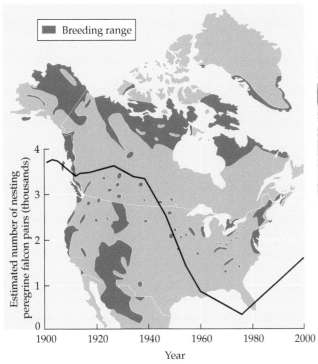

FIGURE 4.13   Peregrine falcons are now breeding in many areas across North America. Their population declined dramatically when DDT use began in the 1940s and then recovered after DDT was banned in 1972 (Hoffman and Smith 2003). (After Canadian Wildlife Service and Connecticut Department of Environmental Protection; photograph courtesy of the US Fish and Wildlife Service.)

damages aquatic ecosystems. Rivers, lakes, and oceans are sometimes used as open sewers for industrial wastes and residential sewage. And higher densities of people almost always mean greater levels of water pollution. Pesticides, herbicides, petroleum products, heavy metals (such as mercury, lead, and zinc), detergents, toxic chemicals such as polychlorinated biphenyls (PCBs), and industrial wastes directly kill organisms such as insect larvae, fish, amphibians, and even marine mammals living in aquatic environments. Pollution is a threat to 90% of the endangered fishes and freshwater mussels in the United States. An increasing source of pollution in coastal areas is the discharge of nutrients and chemicals from shrimp and salmon farms. Medicines used by people or given to domesticated animals can enter the aquatic environment through sewage, either because waste treatment plants cannot not remove them or because they leak into wells (Schaider et al. 2014). These biologically active chemicals, especially hormones, can have an adverse effect on the physiology, behavior, and reproduction of fish and other animals that ingest them (Brodin et al. 2013).

In contrast to wastes in the terrestrial environment, which have primarily local effects, toxic wastes in aquatic environments diffuse over a wide area. Toxic chemicals, even at very low concentrations in the water, can be lethal to aquatic organisms through biomagnification. Many aquatic environments are naturally low in essential minerals, such as nitrates and phosphates, and aquatic plant and animal species have adapted to their natural absence by developing the ability to process large volumes of water and concentrate these minerals. When these species process polluted water, they concentrate toxic chemicals along with the essential minerals, and the toxins may eventually poison them. Species that feed on these aquatic species ingest the toxic chemicals they have concentrated. One of the most serious consequences for humans is the accumulation of mercury and other toxins by long-lived predatory fishes, such as swordfish and sharks, and its effect on the nervous systems of people who eat these types of fish frequently (Jaeger et al. 2009).

Even essential minerals that are beneficial to plant and animal life can become harmful pollutants at high concentrations (McWilliams 2013). Anthropogenic releases of human sewage, agricultural fertilizers, detergents, and industrial wastes often add large amounts of nitrates and phosphates to aquatic systems, initiating the process of **eutrophication**. Humans release as much nitrate into the environment as is produced by all natural processes, and the anthropogenic release of nitrogen is expected to keep increasing as the human population continues to increase. Even small amounts of these nutrients can stimulate growth, and high concentrations often result in thick "blooms" of algae at the surfaces of ponds and lakes (**FIGURE 4.14**). These algal blooms may be so dense that they outcompete other plankton species and shade bottom-dwelling plant species. As the algal mat becomes thicker, its lower layers sink to the bottom and die. The bacteria and fungi that decompose the dying algae multiply in response to this added sustenance and consequently absorb all the oxygen in the water. Without oxygen, much of the remaining animal life dies off, sometimes visibly in the form of masses of dead fish floating on the water's surface. The result is a greatly impoverished and simplified community, a *dead zone* consisting of only those species that are tolerant of polluted water and low oxygen levels.

This process of eutrophication can also affect marine systems, particularly coastal areas and bodies of water in confined areas, such as the Gulf of Mexico, the Mediterranean, the North and Baltic Seas in Europe, and the enclosed seas of Japan which have large anthropogenic inputs of nutrients (Greene et al. 2009). In warm tropical waters, eutrophication favors algae, which grow over coral reefs and completely change

FIGURE 4.14 Large algal blooms such as this one on a pond in Oulu, Finland, are often the result of human activities, and when extensive can have significant negative effects on an entire ecosystem. (© RJH_CATALOG/Alamy Stock Photo.)

the reef community (Díaz-Ortega and Hernández-Delgado 2014). The key to stopping eutrophication and its negative effects is to reduce the release of excess nutrients through improved sewage treatment and better farming practices, including reduced applications of fertilizer and establishment of buffer zones between fields and waterways.

Sediments eroded from logged or farmed hillsides can also harm aquatic ecosystems. The sediment covers submerged plant leaves and other green surfaces with a muddy film that reduces light availability and can block O_2 and CO_2 exchanges, thus decreasing the rate of photosynthesis. Increasing water turbidity (a decrease in water clarity) reduces the depth at which photosynthesis can occur and may prevent animal species from seeing, feeding, and living in the water. Sediment loads are particularly harmful to the many coral species that require crystal-clear waters to survive.

Air pollution

In the past, people assumed that the atmosphere was so vast that materials they released into the air would be widely dispersed and their effects would be minimal. But today, several types of air pollution are so widespread that they have damaged whole ecosystems. These same pollutants also have severe effects on human health, demonstrating again the common interests shared by people and nature.

ACID RAIN   **Acid rain** is produced when industries such as smelting operations and coal- and oil-fired power plants release huge quantities of nitrogen oxides and sulfur oxides into the air, where those chemicals combine with moisture in the atmosphere to produce nitric and sulfuric acids. These acids are incorporated into cloud systems and dramatically lower the pH (the standard measure of acidity) of rainwater, leading to the weakening and death of trees over wide areas. Acid rain also lowers the pH of soil moisture and water bodies and increases the concentration of toxic metals such as aluminum in the soil and water.

> Acid rain and other effects of air pollution are increasing rapidly in Asian countries as they industrialize. Acid rain is particularly harmful to freshwater species.

Increased acidity alone damages many plant and animal species; as the acidity of water bodies increases, many fish either fail to spawn or die outright. Increased acidity, along with both aquatic and terrestrial pollution, is a contributing factor to the dramatic decline of many amphibian populations throughout the world (Brühl 2013; Hayes et al. 2010). Most amphibian species depend on bodies of water for at least part of their life cycle, and a decline in water pH causes a corresponding increase in the mortality of eggs and young animals. Acidity also inhibits the microbial process of decomposition, lowering the rate of mineral recycling and ecosystem productivity. Many ponds and lakes in industrialized countries have lost large portions of their animal communities as a result of acid rain. These damaged water bodies are often in supposedly pristine areas hundreds of kilometers from major sources of urban and industrial pollution, such as the North American Rocky Mountains and Scandinavia. While the acidity of rain is decreasing in many areas because of better pollution control, acid rain remains a serious problem. In developing countries such as China, acid rain is becoming a greater problem as the country powers its rapid industrial development with fuels that are high in sulfur.

OZONE PRODUCTION AND NITROGEN DEPOSITION   Automobiles, power plants, and industrial activities release hydrocarbons and nitrogen oxides as waste products. In the presence of sunlight, these chemicals react with the atmosphere to produce ozone and other secondary chemicals, collectively called *photochemical smog*. Although ozone in the upper atmosphere is important in filtering out ultraviolet radiation, high concentrations of ozone at ground level damage plant tissues and make them brittle, harming biological communities and reducing agricultural productivity. Ozone and smog are detrimental to people and other animals when inhaled, so both people and biological communities benefit from air-pollution controls. Smog can be so severe that people may avoid outdoor activities. Severe smog is becoming an increasing problem in China and other Asian countries.

When airborne nitrogen compounds are deposited in terrestrial environments by rain and dust, ecosystems throughout the world are damaged and altered by potentially toxic concentrations of nitrogen, and many species are unable to survive in the altered conditions (Bobbink et al. 2010). In particular, the combination of nitrogen deposition and acid rain is responsible for a decline in the density of soil fungi that form beneficial relationships with trees (see Chapter 3).

TOXIC METALS   Leaded gasoline (still used in many developing countries, despite its clear danger to human health), coal burned for heat and power, mining and smelting operations, and other industrial activities release large quantities of lead, zinc, mercury, and other toxic metals into the atmosphere (Wade 2013). These elements are directly poisonous to plant and animal life and can cause permanent injury to children (see Figure 4.11). The effects of these toxic metals are particularly evident in areas surrounding large smelting operations, where there may be measurable negative effects on life for miles around.

Enforcement of local and national policies and regulations may sometimes reduce levels of air pollution; eliminating lead from gasoline is one such example. Concentrations of air pollutants are declining in certain areas of North America and Europe, but they continue to rise in many other areas of the world that are undergoing industrialization. Because air and water pollutants do not observe political boundaries, international agreements addressing these issues are important, as we will see in Chapter 11. Lowering levels of air and water pollution is necessary for the health of both human populations and biodiversity.

Global Climate Change

Carbon dioxide, methane, and other trace gases in the atmosphere allow light energy to pass through and warm the surface of the Earth. However these gases, as well as water vapor (in the form of clouds), are able to trap the energy radiating from the Earth as heat, slowing the rate at which heat leaves the Earth's surface and radiates back into space. These gases are called **greenhouse gases** because they function much like the glass in a greenhouse, which is transparent to sunlight but traps energy inside the greenhouse once it is transformed into heat. The similar warming effect of Earth's atmospheric gases is called the **greenhouse effect**. We can imagine these gases as "blankets" over the Earth's surface: the denser the concentration of gases, the more heat is trapped near the Earth, and the higher the planet's surface temperature.

The greenhouse effect allows life to flourish on Earth—without it, the temperature at the Earth's surface would fall dramatically. Today, however, as a result of human activities, concentrations of greenhouse gases are increasing so much that they are already affecting the Earth's climate (Gore 2006; IPCC 2014). The term *global warming* is used to describe the rise in temperatures resulting from the increase in greenhouse gases, and *global climate change* refers to the complete set of climate characteristics that are changing now and will continue to change in the future because of this increase, including patterns of precipitation and wind.

During the past 130 years, global atmospheric concentrations of carbon dioxide, methane, and other trace gases have been steadily increasing, primarily as a result of the burning of fossil fuels—coal, oil, and natural gas (Climate Central 2012; IPPC 2013). Clearing of forests by logging and for agriculture also contribute to rising concentrations of CO_2. Through all these activities, humans currently release about 70 million tons of CO_2 into the atmosphere every day. The concentration of CO_2 in the atmosphere has increased from 290 parts per million (ppm) to around 400 ppm over the last 100 years, and it is projected to reach 580 ppm at some point in the latter half of the twenty-first century.

There is broad scientific agreement among the Intergovernmental Panel on Climate Change (IPCC), a study group of leading scientists organized by the United Nations, that the increased levels of greenhouse gases have affected the world's climate and ecosystems already and that these effects will increase in the future (IPCC 2013). An extensive review of the evidence supports the conclusion that global surface temperatures have increased by 0.8°C (1.4°F) during the last 110 years (**FIGURE 4.15**). In fact, August 2015 was the hottest month in Earth's recorded history; the average global temperature was 15.6°C (60.1°F), 0.88°C (1.58°F) above the twentieth-century average, while temperatures in the Pacific Ocean were 2°C (3.6°F) above the 1981–2010 average in the eastern half of the equatorial Pacific Ocean (NOAA 2015). The El Niño (a periodic warming of Pacific temperatures) during winter of 2015–2016 is one of the most severe in at least 50 years, creating record highs in several cities during December (NOAA 2016). Temperatures at high latitudes, such as in Siberia, Alaska, and Canada, have increased more than in other regions.

There is a broad consensus among scientists that increased atmospheric concentrations of carbon dioxide and other greenhouse gases produced by human activities have already resulted in warmer temperatures and will continue to affect Earth's climate in the coming decades.

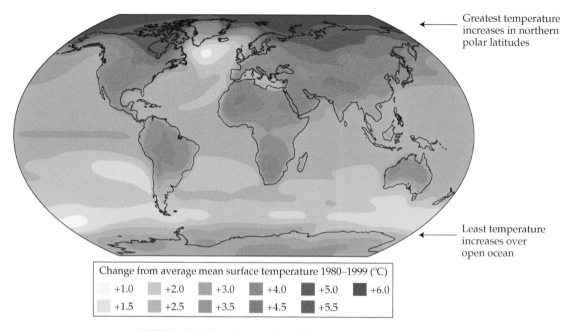

Greatest temperature increases in northern polar latitudes

Least temperature increases over open ocean

Change from average mean surface temperature 1980–1999 (°C)

+1.0 +2.0 +3.0 +4.0 +5.0 +6.0
+1.5 +2.5 +3.5 +4.5 +5.5

FIGURE 4.15 Over the last 130 years, atmospheric CO_2 concentrations have increased dramatically as a result of human activities, and global temperatures have risen as a result. Here, the average annual temperature used for comparison is that for the period from 1980 to 1999; temperature changes are reported in terms of difference (anomaly) from this average annual temperature. Global annual temperatures were colder than average prior to 1980, when annual temperatures began to be warmer than average. (After Karl 2006, updated from NOAA 2015.)

Climatologists predict that the world climate will warm by an additional 1°C–3°C (1.8°F–5.4°F) by 2100 as a result of increased atmospheric concentrations of CO_2 and other gases. The increase could be even greater, in the range of 3°C–6°C (5.4°F–10.8°F), if CO_2 levels rise faster than predicted; conversely, it could be slightly less if all countries reduce their emissions of greenhouse gases in the very near future. The increase in temperature will be greatest at high latitudes and over large continents (IPCC 2013; **FIGURE 4.16**). Rainfall has already started to increase on a

FIGURE 4.16 (A) All climate models predict that the greatest warming will ▶ take place in the northern polar regions. The graphs show satellite data-derived measurements of the extent of polar sea ice from 1979 to 2013 for the Arctic (top), Antarctica (middle), and the global combined total (bottom). Straight lines represent the overall trend of observed data. Slight gains in polar ice in the south do not compensate for dramatic losses in the north. (B) Polar ice caps are already melting at alarming rates, as these walrus crowded onto an ice floe in the Bering Sea off Alaska seem to attest. (A, Graphs by Josh Stevens, NASA Earth Observatory; B, photograph by Budd Christman, courtesy of NOAA.)

(A)

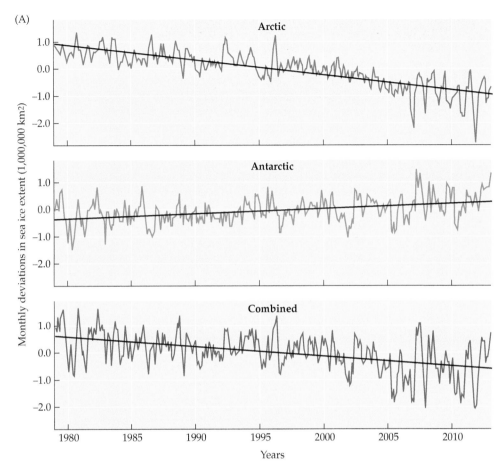

(B)

global scale and will continue to do so, but changes in rainfall will vary by region, with some regions showing decreases. There will also probably be an increase in extreme weather events, such as hurricanes, floods, snowstorms, and regional drought, associated with global warming. In tropical deciduous forests and savannas, warmer conditions will result in an increased incidence of fire. In coastal areas, storms will cause increased destruction of cities and other human settlements and will severely damage coastal vegetation, including beaches and coral reefs. The series of hurricanes that devastated the southern United States in 2005, including Hurricane Katrina, could be an indication of what the future may bring.

As a result of global climate change, climate regions in the northern and southern temperate zones will be shifted toward the poles. This warming is expected to create a snowball effect in which thawing tundra and melting permafrost, the layer of soil that typically stays frozen throughout the year, will increase the rate of CO_2 release through respiration by soil microbes, but more research is needed to understand its extent (Schuur et al. 2015). However, the effects of warming are already being observed both at high altitudes and at high latitudes. Alpine plants are found growing higher on mountains and blooming earlier in the spring; Rocky Mountain species are blooming as much as a month earlier now than a century ago (Munson and Sher 2015), which puts them at potential risk of frost damage and disconnects them from their pollinators (Inouye et al. 2002). Migrating birds have been observed spending longer times at their summer breeding grounds (Bussière et al. 2015). In the coming century, global climate change is predicted to have profound effects on Arctic, boreal, and alpine ecosystems as a result of warmer temperatures and a longer growing season.

> As rainfall patterns change and most regions become warmer, many terrestrial plant and animal species may not be able to adapt quickly enough to survive. Climate change may also have huge impacts on marine ecosystems and coastal areas occupied by people.

The effects of global climate change on temperature and rainfall are also expected to have dramatic effects on tropical ecosystems (IPCC 2014). Many tropical species and biological communities appear to have narrow tolerances for temperature and rainfall, so even small changes in the climate could have major effects on species composition, cycles of plant reproduction, patterns of migration, and susceptibility to fire (Corlett 2011). Major contractions in the area of rain forest are quite likely. In particular, cool-adapted species that live atop tropical mountains could be highly vulnerable to increasing temperatures; as bands of vegetation move higher on mountains, the species at the top will have nowhere to go (Şekercioğlu et al. 2012).

Ocean acidification, warming, and rising sea level

Evidence indicates that ocean water temperatures have increased over the last 40 years by about 0.44°C. As a consequence, certain marine species are expanding their ranges farther from the equator, and coral reefs and other marine habitats are threatened by rising seawater temperatures (Pandolfi et al. 2011). In the coastal waters off California, warm-water southern species are increasing in abundance while cold-water northern species are declining. Zooplankton are declining in some areas because of warmer seawater temperatures, with dire consequences for the marine animals that feed on them (Robinson et al. 2008). Abnormally high water temperatures in oceans across the globe in 1998 caused corals to sicken and expel the symbiotic algae that would normally live inside their bodies and provide them with essential carbohydrates. These "bleached" corals then suffered a massive dieback, with an estimated 70% coral mortality in Indian Ocean reefs in 1998. Although reefs in some places have been able to recover from that event (Gilmour et al. 2013), subsequent global bleaching events in 2010 and 2015 have taken a significant toll (**FIGURE 4.17**).

Increased anthropogenic CO_2 in the atmosphere has also caused another problem: as much as a third of it is absorbed by our oceans, causing a reduction in the pH (Doney et al. 2009). Ocean acidification is associated with many problems, most notably the decrease in calcium carbonate saturation states that affect shell-forming marine animals.

Warming temperatures are already causing mountain glaciers to melt and the polar ice caps to shrink, and this process will continue and ac-

FIGURE 4.17 The 2015 coral bleaching event had dramatic effects on this reef in American Samoa. The photo on the left was taken in December 2014, and the one on the right was taken in February 2015. The 2015 event was the third global bleaching event in recorded history, the others taking place in 2010 and 1998; all three events coincided with periods of abnormally high ocean temperatures. The photos were taken by the XL Catlin Seaview Survey, a group documenting bleaching across the globe. Although fluctuations in ocean temperature are normal, the geographic pattern and intensity of these recent warming events have been very unusual.

celerate. As a result of this melting and the thermal expansion of ocean water, sea level is currently rising at a rate of about 3 cm (just over an inch) per decade. By the year 2100, sea level is predicted by the IPCC to rise by 40–60 cm (16–24 in.), and possibly by as much as 100 cm (3 ft) (IPCC 2013; Jones 2013). Many scientists, however, consider these predictions to be too conservative and believe that a rise of up to 130 cm (4 ft) is possible. In regions with low elevational gradients, a 4-ft rise in sea level could affect areas thousands of meters inland; massive flooding of coastal cities and communities will certainly occur. Much of the current land area of low-lying countries such as Bangladesh could be underwater within 100 years. Somewhere between 25% and 80% of coastal wetlands could be altered by rising sea level. Many coastal cities, such as Miami and New York, will have to build expensive seawalls or be flooded by the rising waters. There is evidence that this process has already begun; sea level has already risen by about 20 cm (8 in.) over the last 100 years (IPCC 2013).

The overall effect of global warming

Global climate change has the potential to radically restructure ecosystems by changing the ranges of many species. The pace of this change could overwhelm the natural dispersal abilities of species. There is mounting evidence that this process has already begun (**TABLE 4.2**). The distributions of bird and plant species are moving poleward, and reproduction is oc-

---

TABLE 4.2　Some Effects of Global Warming

1. Increased temperatures and incidence of heat waves
   Examples: 2015 was the hottest year worldwide in the historical record so far; previously, the warmest year was 2014. An August 2003 heat wave in France killed over 10,000 people as temperatures reached 40°C (104°F).

2. Melting of glaciers and polar ice
   Examples: Arctic Ocean summer ice has declined in area by 15% over the past 25 years. Since 1850, glaciers in the European Alps have disappeared from more than 30%–40% of their former range.

3. Rising sea level
   Example: Since 1938, one-third of the coastal marshes in a wildlife refuge in the Chesapeake Bay have been submerged by rising seawater.

4. Earlier spring activity
   Example: One-third of English birds are now laying eggs earlier in the year, and two-thirds of temperate plant species are now flowering earlier than they did several decades ago.

5. Shifts in species ranges
   Example: Two-thirds of European butterfly species studied are now found 35 to 250 km farther north than recorded several decades ago.

6. Population declines
   Example: Adélie penguin populations have declined over the past 25 years as their Antarctic sea ice habitat melts away.

Sources: Data from Union of Concerned Scientists (www.ucsusa.org) and NASA.

curring earlier in the spring (Chen et al. 2011; Willis et al. 2008). Increasing temperatures are also associated with summer drought and rising mortality rates of trees in Europe and with devastating outbreaks of tree-killing beetles in the Rocky Mountains (Carnicer et al. 2011). Because the implications of global climate change are so far-reaching, biological communities, ecosystem functions, and climate need to be carefully monitored over the coming decades (Nelson et al. 2013). Global climate change will also have an enormous effect on human populations in coastal areas affected by rising sea level and increased hurricane impacts, as well as in areas that are already experiencing drought stress and desertification. In many areas of the world, crop yields will decline because of less favorable growing conditions (Hannah et al. 2013). Crop yields are predicted to decline by an average of 8% in tropical areas of Africa and South Asia where chronic, severe hunger is already an enormous problem, and by 30% or more in certain populous countries such as Brazil and Indonesia. The disadvantaged people of the world will be least able to adjust to these changes and will suffer the consequences disproportionately. However, all countries of the world will be affected, and it is time for people and their governments to recognize the urgent need to address global climate change.

It is likely that, as the climate changes, many existing protected areas will no longer preserve the rare and endangered species that currently live in them (Hole et al. 2011). We need to establish new conservation areas now to protect sites that will be suitable for these species in the future, such as sites with large elevational and latitudinal gradients. Potential future migration routes, such as north–south river valleys, also need to be identified and established now (Nuñez et al. 2013). If species are in danger of going extinct in the wild because of global climate change, the last remaining individuals may have to be maintained in captivity. Another strategy that we need to consider is to move isolated populations of rare and endangered species to new localities at higher elevations and closer to the poles, where they can survive and thrive. This approach has been termed *assisted colonization*. There is considerable debate within the conservation community about whether assisted colonization represents a valid strategy or whether it is too problematic because of the potential for transplanted species to become invasive species in their new ranges. Even if global climate change is not as severe as predicted, establishing new protected areas can only help to protect biodiversity.

Although the prospect of global climate change is cause for great concern, it should not divert our attention from the massive habitat destruction that is the principal current cause of species extinctions. Many conservation biologists believe that preserving intact ecosystems, restoring degraded ones, and increasing the connectivity of existing protected areas are the most important and immediate priorities for conservation, especially in the marine environment. These protected areas will also facilitate the migration of species as they adjust their ranges in response to a changing climate. Regardless, it is imperative that we reduce our use of fossil fuels and protect and replant forests in order to decrease levels of greenhouse gases (Kintisch 2013).

Overexploitation

People have always hunted and harvested the food and other resources they need to survive. As long as human populations were small and their methods of collection unsophisticated, people could sustainably harvest and hunt the plants and animals in their environment. As human populations have increased, however, our use of the environment has escalated, and our methods of harvesting have become dramatically more efficient. In many areas, this has led to an almost complete depletion of large animals from many biological communities and the creation of strangely "empty" habitats (Redford 1992).

Technological advances mean that, even in the developing world, guns are used instead of blowpipes, spears, or arrows for hunting in tropical forests and savannas. Small-scale local fishermen now have outboard motors on their boats, which allow them to harvest wider areas more rapidly. Powerful motorized fishing boats and enormous factory ships harvest fish from the world's oceans and sell them on the global market (**FIGURE 4.18**). Even in preindustrial societies, intense exploitation, particularly for meat,

FIGURE 4.18 Intensive harvesting has reached crisis levels in many of the world's fisheries. These bluefin tuna (*Thunnus thynnus*) are being transferred from a fishing trawler to a factory ship, aboard which huge quantities of fish are efficiently processed for human consumption. Such efficiency can result in massive overfishing. (Photograph © Images & Stories/Alamy.)

has led to the decline and extinction of local species of birds, mammals, and reptiles (Doughty 2013).

Some traditional societies imposed restrictions on themselves to prevent overexploitation of jointly owned common property or natural resources (Cinner and Aswani 2007). For example, the rights to specific harvesting territories were rigidly controlled, and hunting and harvesting in certain areas were banned. There were often prohibitions against harvesting female, juvenile, and undersized animals. Certain seasons of the year and times of the day were closed for harvesting. Certain efficient methods of harvesting were not allowed. (Interestingly enough, these restrictions, which allowed some traditional societies to exploit communal resources on a long-term, sustainable basis, are almost identical to some of the fishing restrictions regulators have imposed on or proposed for many fisheries in industrialized nations.)

> Today's vast human population and improved technology have resulted in unsustainable harvest levels of many biological resources.

Such self-imposed restrictions on using common-property resources are often less effective today. In much of the world, resources are exploited opportunistically without restraint. In economic terms, a regulated common-property resource sometimes becomes an open-access resource and available to everyone without regulation. The lack of restraint applies to both ends of the economic scale—the poor and hungry as well as the rich and greedy. If a market exists for a product, local people will search their environment to find and sell it. Sometimes traditional groups will sell the rights to a resource, such as a forest or mining area, for cash. In rural areas, the traditional controls that regulate the extraction of natural products have generally weakened. Whole villages are mobilized to systematically remove every usable animal and plant from an area of forest. Where there has been substantial human migration, civil unrest, or war, controls of any type may no longer exist. In countries beset with civil conflict, such as Somalia, Cambodia, the former Yugoslavia, the Democratic Republic of the Congo, and Afghanistan, firearms have come into the hands of rural people. The breakdown of food distribution networks in countries such as these leaves the resources of the natural environment vulnerable to whoever can exploit them (Loucks et al. 2009).

In some areas, populations of large primates, such as gorillas and chimpanzees, as well as ungulates and other mammals may be reduced by 80% or more by hunting. Populations of certain species may be eliminated altogether, especially those that occur within a few kilometers of a road (Lindsey et al. 2013). In many places, hunters are extracting animals at a rate six or more times greater than the resource base can sustain. The result is a forest with a mostly intact plant community that is lacking its animal community (Galetti and Dirzo 2013). Without large animals, many plant species lack effective seed dispersal and decline in abundance.

The decline in animal populations caused by intensive hunting for food, which has been termed the *bushmeat crisis*, is a major concern for wildlife officials and conservation biologists, especially throughout Africa (www.bushmeat.org; Linder and Oates 2011). Furthermore, eating primate bushmeat increases the possibility that new diseases will be transmitted to human populations; both HIV and the Ebola virus have been linked to such practices (Calmy et al. 2015). Solutions involve restricting the sale and transport of bushmeat, restricting the sale of firearms and ammunition, closing roads following logging, extending legal protection to key endangered species, establishing protected reserves where hunting is not allowed, and most importantly, providing alternative protein sources to reduce the demand for bushmeat (Moro et al. 2013).

International wildlife trade

International trade in wildlife is valued at over $240 billion per year, and trade within developing countries is rapidly increasing as well (Goss and Cumming 2013; Hinsley et al. 2015). Major exporters are primarily located in the developing world, often in the tropics. Most major importers are in the developed countries and East Asia; they include Canada, China, the European Union, Japan, Singapore, Taiwan, and the United States.

The legal and illegal trade in wildlife is responsible for the decline of many species (Nijman et al. 2011). For example, in 2015, 1338 African rhinos were killed—the highest number in one year since the current poaching epidemic began in 2008 (IUCN 2016). Most of these killings were motivated by the international market for rhino horn, which is believed by some to have medicinal properties. Overharvesting of butterflies by insect collectors; of parrots by aviculturists; of orchids, cacti, and other plants by horticulturists; of marine molluscs by shell collectors (see Figure 5.5); and of tropical fishes by aquarium hobbyists are further examples of the targeting of whole biological communities to supply an enormous international demand (**TABLE 4.3**; Raghavan et al. 2013).

It has been estimated that 1 billion tropical fish are sold annually worldwide for the aquarium market, and many times that number are killed during collection and shipping (Whittington and Chong 2007). The international trade in other live animals is similarly large: 1 million reptiles and 250,000 birds per year. More than 400 endangered bird species are hunted or traded (Regueira and Bernard 2012). Or consider primates: as many as 70,000 live primates are exported each year (Nijman et al. 2011); China is the largest exporter of live primates and the United States is the largest importer (primarily for research). In an attempt to regulate and restrict this trade, many declining species are listed as protected under the Convention on International Trade in Endangered Species (CITES; see Chapter 6). Listing species with CITES has often protected species or groups of species from further exploitation.

A striking example of overexploitation is the worldwide trade in frog legs; each year Indonesia exports the legs of roughly 100 million frogs to western European countries for luxury meals. There is no information on how this intensive harvesting affects frog populations, forest ecology, or agriculture, and perhaps not surprisingly, the names of the frog species on the shipping labels are often wrong, which adds to the difficulty in quantifying the extent of the problem (Warkentin et al. 2009; www.traffic.org). Adding

TABLE 4.3 Major Groups Targeted by the Worldwide Trade in Wildlife

Group	Total individuals traded in 2013 (alive or dead)	Comments
Primates	637,000	Mostly used for biomedical research; also for pets, zoos, circuses, and private collections.
Birds	3.3 million	Zoos and pets. Mostly perching birds, but also legal and illegal trade of about 80,000 parrots.
Reptiles	15.6 million	Zoos, pets, and raw skins. Also 10–15 million raw skins. Reptiles are used in some 50 million manufactured products. Mainly come from the wild, but increasingly from farms.
Ornamental fish	3.7 million	Most saltwater tropical fish come from wild reefs and may be caught by illegal methods that damage other wildlife and the surrounding coral reef.
Reef corals	2 million	Reefs are being destructively mined to provide aquarium decor and coral jewelry.
Orchids	9–10 million	70% of the species listd by CITES are orchids. Approximately 10% of the international trade comes from the wild.
Cacti	7–8 million	Approximately 15% of traded cacti come from the wild; smuggling is a major problem.

Sources: Data from CITES Trade Database, Appendix 2 for 2013 (as compiled in *National Geographic* 2015); WRI 2005; Karesh 2005; and Nijman et al. 2011.

to the stress placed on frog populations is the use of some species, such as the Lake Titicaca frog (*Telmatobius coleus*), as medicine (**FIGURE 4.19**). This critically endangered frog is endemic to Lake Titicaca, which straddles Peru and Bolivia (De la Riva and Reichle 2014). Harvesting of these frogs is believed to have decreased the population by 80%, but vigorous conservation and education efforts seem to be helping (Reading et al. 2011).

Yet another example is the enormous demand for sea horses (*Hippocampus* spp.) in China. The Chinese use dried sea horses in their traditional medicine because they resemble dragons and are believed to have a variety of healing powers. About 54 tons of sea horses are consumed in China per year—roughly 19 million

FIGURE 4.19 The Lake Titicaca frog (*Telmatobius coleus*) has been harvested to the point of extinction from this South American lake for use as an exotic dish and as a treatment for asthma and other ailments. The frogs are blended whole and then sold in Peru and Bolivia as drinks, frog extract (as shown here), canned food, and flour. Also added into the product is maca (*Lepidium meyenii*), a radish-like plant from the Andes that used to be endangered but is now grown commercially. (Photograph courtesy of Walter Silva [ATFSS-Lima].)

animals. Sea horse populations throughout the world are being decimated to supply this ever-increasing demand, with the result that international trade in sea horses is increasingly monitored and regulated by international treaty (Vincent et al. 2014).

Although most international trade in wildlife is legal and therefore can be regulated, an estimated $10 billion per year is not. A black market links poor local people, corrupt customs officials, rogue dealers and criminal gangs, and wealthy buyers who do not question the sources from which they buy. This trade has many of the same characteristics, the same practices, and sometimes the same players as the illegal trade in drugs and weapons, and it is extremely widespread and highly profitable. Confronting those who perpetuate such illegal activities has become a major and dangerous job for international law enforcement agencies.

Commercial harvesting

> Species can often recover when they are protected from overexploitation.

Governments and industries often claim that they can avoid the overharvesting of wild species by applying modern scientific management. As part of this approach, an extensive body of literature has developed in wildlife and fisheries management and in forestry to describe the *maximum sustainable yield*: the greatest amount of a resource, that can be harvested each year and replaced through population growth without detriment to the population. In many real-world situations, however, industry representatives and government officials managing commercial harvesting operations may lack the key biological information that is needed to make accurate calculations. Not surprisingly, attempts to harvest at high levels can lead to abrupt species declines.

For many marine species, direct exploitation is less important than the indirect effects of commercial fishing (Burgess et al. 2013). Many marine vertebrates and invertebrates are caught incidentally during fishing operations; most of these organisms, referred to as *bycatch*, are killed or injured in the process. Between 25% and 75% of the catch in fishing operations is dumped back into the sea. The declines of skates, rays, and seabirds of 148 species have all been linked to their wholesale death as bycatch (Zydelis et al. 2009). The huge number of sea turtles and dolphins killed by commercial fishing boats as bycatch resulted in a massive public outcry and led to the development of improved nets to reduce these accidental catches. The development of improved nets and hooks, as well as other methods to reduce bycatch, is an active area of current fisheries research (Riskas et al. 2016).

One of the most heated debates over the harvesting of wild marine species has involved the hunting of whales. The debate is due in part to the strong emotional attachment to whales that many people in Western countries have (see Figure 3.14). After recognizing that many whale species had been hunted to dangerously low levels, the International Whaling Commission finally banned all commercial whaling in 1986. Despite that ban, certain species remain at densities far below their original numbers.

Those species include the blue whale (*Balaenoptera musculus*) and the northern right whale (*Eubalaena glacialis*), which have been protected since 1967 and 1935, respectively. The densities of other species, such as the gray whale (*Eschrichtius robustus*), appear to have recovered, however (**TABLE 4.4**). The slow recovery of some species may be due to continued hunting, both legal and illegal. Whale hunting by the Japanese fleet continues under the dubious claim that additional scientific data are needed to assess the status of whale populations.

Finding the best methods to protect and manage the remaining individuals in overharvested populations is a priority for conservation biologists. As we will see in Chapter 8, projects linking the conservation of biodiversity to local economic development represent one possible approach. In some cases, this linkage may be made possible by acknowledging the sustainable harvesting of a natural resource with a special certification that allows producers to receive a higher price for their product. Certified timber products and seafoods are already entering the market, but it

---

TABLE 4.4   Worldwide Populations of Whale Species Harvested by Humans

Species	Numbers prior to whaling[a]	Present numbers	Primary diet items	IUCN status[b]
Baleen whales				
Blue	200,000	10,000–25,000	Plankton	Endangered
Bowhead	56,000	25,500	Plankton	Least concern
Fin	475,000	60,000	Plankton, fish	Endangered
Gray (Pacific stock)	23,000	15,000–22,000	Crustaceans	Least concern
Humpback	150,000	60,000	Plankton, fish	Least concern
Minke	140,000	1,000,000	Plankton, fish	Least concern
Northern right	Unknown	1,300	Plankton	Endangered
Sei	250,000	54,000	Plankton, fish	Endangered
Southern right	100,000	10,000	Plankton	Least concern
Toothed whales				
Beluga	Unknown	200,000	Fish, squid, crustaceans	Near threatened
Narwhal	Unknown	50,000	Fish, squid, crustaceans	Near threatened
Sperm	1,100,000	360,000	Fish, squid	Vulnerable

Sources: American Cetacean Society (www.acsonline.org); IUCN Red List.

[a]Preexploitation population numbers are highly speculative; genetic evidence suggests the populations might have been even greater (Roman and Palumbi 2003; Alter et al. 2007).

[b]Status is determined by a combination of numbers, threats, and trends. For example, numbers of Southern right whales are low but are increasing.

> ◼———————
> Certifying timber, sea-
> food, and other products
> as sustainable may
> be a way to prevent
> overharvesting.
> ———————◼

remains to be seen whether they will have a significant positive effect on biodiversity, particularly among increasingly affluent consumers in China and other Asian countries. It is also unknown whether the regulations associated with certification will be enforced in practice and will restrain the ongoing threats (Christian et al. 2013).

National parks, nature reserves, marine sanctuaries, and other protected areas can also be established to conserve over-harvested species. In some cases, cooperative actions involving international organizations, individual countries, and nongovernmental conservation organizations are needed to prevent overharvesting (Osterblom and Bodin 2012). When harvesting can be reduced or stopped by the enforcement of international regulations such as CITES and comparable national regulations, species may be able to recover. Elephants, sea otters, sea turtles, seals, and certain whale species provide hopeful examples of species that have recovered—once overexploitation was stopped—in certain places in the world (Lotze et al. 2011; Magera et al. 2013).

Invasive Species

Species that become established and proliferate in new (i.e., nonhistorical) ranges where they cause environmental harm are considered **invasive** (Mack et al. 2000). Not unlike chemical pollution and other risk factors mentioned above, invasive species have negative effects on ecosystems because the native organisms have not evolved adaptations to deal with the new conditions they impose. In most cases, the conditions imposed by an invasive species are novel because it has only recently been introduced to that ecosystem. However, native species can also become invasive, usually in response to human alterations to their ecosystem, such as the availability of human food waste supporting great increases in corvids (crows and ravens) and their predation on other bird's nests in coniferous forests (Carey et al. 2012). Invasive species represent threats to 42% of the endangered species in the United States and have had particularly severe impacts on bird and plant species (Pimentel et al. 2005). The effects of invasive species have been estimated to cost countries from 1.4% up to 12% of their GDP, amounting to billions of dollars per year (Marbuah et al. 2014). Globally, over half of all recent animal extinctions are attributable in whole or in part to the effects of invasive species (Clavero and García-Berthou 2005).

Species invasions have occurred by a variety of means:

- *European colonization.* European settlers arriving at new colonies released hundreds of European bird and mammal species into places like New Zealand, Australia, North America, and South Africa to make the countryside seem familiar and to provide game for hunting. Numerous species of fish (trout, bass, carp, etc.) have been widely released to provide food and recreation. In some cases, food sources introduced to support game species have disrupted food webs and led to declines in populations of game species.

- *Agriculture, horticulture, aquaculture.* Large numbers of plant species have been introduced and grown as crops, ornamentals, pasture grasses, or soil stabilizers. Many of these species have escaped from cultivation and have become established in local ecosystems. As aquaculture develops, there is a constant danger of more plant species escaping and becoming invasive in marine and freshwater environments (Xu et al. 2014).

- *Accidental transport.* Species are often transported unintentionally (Hulme 2015). For example, weed seeds are accidentally harvested with commercial seeds and sown in new localities; rats, snakes, and insects stow away aboard ships and airplanes; and disease-causing microbes, parasitic organisms, and insects travel along with their host species, particularly in the leaves and roots of plants and the soil of potted plants (Liebhold et al. 2012). Around 70% of the nonnative forest pest insects in the United States arrived on imported living plants. Seeds, insects, and microorganisms on shoes, clothing, and luggage can be transported across the world in a few days by people traveling by plane. Ships frequently carry organisms in their ballast tanks, releasing vast numbers of bacteria, viruses, algae, invertebrates, and small fish into new locations. Large ships may hold up to 150,000 tons of ballast water. Governments are now developing regulations to reduce the transport of species in ballast water, such as requiring ships to exchange their ballast water 320 km (200 miles) offshore in deep water before approaching a port (Costello et al. 2007).

- *Biological control.* When a nonnative species becomes invasive, a common solution is to release an animal species from its original range that will consume the pest and hopefully control its numbers. While biological control can be dramatically successful, there are cases in which a biological control agent itself has become invasive, attacking native species along with (or instead of) the intended target species (Elkinton et al. 2006). For example, an herbivorous weevil (*Larinus planus*) introduced into North America to control invasive Eurasian thistles (*Carduus* spp.) has been found to attack populations of rare native North American thistles (*Cirsium* spp.; Havens et al. 2012; Louda et al. 1997). In order to minimize the probability of such effects, species being considered as biological control agents are tested before release to determine whether they will restrict their feeding to the intended target species.

> Invasive species may displace native species through competition for limiting resources, prey on native species to the point of extinction, or alter the habitat so that natives are no longer able to persist.

Threats posed by invasive species

The negative effects of invasive species can be both direct and indirect. Abundant research has documented these for many types of organisms and may cascade through an ecosystem. When invasive plant species dominate a community, the diversity and abundance of native plant species and the insects that feed on them may show a corresponding decline (Heleno et al. 2009). Evidence also indicates that invasive plants can even reduce the diversity of soil microbe species (Callaway et al. 2004). Mechanisms whereby invasive species cause harm include the following:

- *Competition for resources.* Through their dramatic expansion, invasive organisms may use up or block access to resources needed by native organisms, including food, space, light, and even mates. Nonnative plants can compete with native plants for pollinators or overwhelm them with nonnative pollen (Beans and Roach 2015). Introduced European earthworm species are currently outcompeting native species in soil communities across North America, with negative effects on certain ground-nesting songbirds as well as potentially enormous, but as yet unknown, consequences for underground biological communities (Loss and Blair 2011).

- *Predation and parasitism.* Extinction risk from invasion is often associated with predation. Introduced rats are a primary management concern on many islands; when they were successfully controlled in areas of New Zealand, many populations of native insects (Ruscoe et al. 2013) and birds (O'Donnell and Hoare 2012) showed dramatic increases. American bullfrogs (*Lithobates catesbeianus*) are voracious predators and invasive in many parts of the world; a study in Argentina found them to consume 40 different prey taxa, including several species of native frogs and crustaceans (Quiroga et al. 2015).

- *Changes in ecosystem processes.* Both cheatgrass (*Bromus tectorum*) and tamarisk (*Tamarix* spp.), two of the most pervasive weeds in the western United States, indirectly kill native plants by providing fuel that promotes wildfires (Drus 2013). Soil microbial communities are also different under tamarisk trees, lacking important symbionts for native plants (Meinhardt and Gehring 2013). The presence of domesticated cats has been shown to affect a wide range of ecological processes, including bird migration patterns, seed dispersal, and breeding and parental behaviors (Medina et al. 2014).

- *Alteration of abiotic conditions.* Some invasive plants have been found to alter soil properties, including moisture, acidity, and enzymes, which can facilitate more invasions or otherwise harm native species (Kuebbing et al. 2014). Both invasive plants and animals can alter the structural environment (Gutiérrez et al. 2014): animals through behaviors such as burrowing and plants because of their particular growth forms, which may differ in shape or density from those of natives. Such changes to the environment may then affect ecosystem processes.

Invasive species may have more than one of these effects on an ecosystem at once, particularly those that are in the middle of the food chain, as insects and other invertebrates usually are. Insects introduced both deliberately, such as European honeybees (*Apis mellifera*), and accidentally, such as fire ants (*Solenopsis invicta*) and gypsy moths (*Lymantria dispar*), can build up huge populations, both competing with native animals and preying on native plants and animals. At some localities in the southern United States, the diversity of insect species declined by 40% following the invasion of nonnative fire ants, and there was a similarly large decline in native birds (**FIGURE 4.20**).

Invasive species on oceanic islands

The isolation of oceanic island habitats encourages the evolution of a unique assemblage of endemic species (see Chapter 7), but it also leaves those species particularly vulnerable to depredations by invading species. Many plants that grow on islands with few herbivores do not produce the bad-tasting, tough vegetative tissue that discourages herbivores, nor do they have the ability to resprout rapidly following damage. Some island birds that have evolved in the absence of predators have lost the power of flight and simply build their nests on the ground. Such species often

FIGURE 4.20 The abundance of northern bobwhites (*Colinas virginianus*) in Texas has been declining over a 20-year period following the arrival of the nonnative red fire ant (*Solenopsis invicta*). The fire ants may directly attack and disturb bobwhites, particularly at the nestling stage, and may compete with them for food items, such as insects. (After Allen et al. 1995; bobwhite photograph courtesy of Steve Maslowski/US Fish and Wildlife Service; fire ant photograph courtesy of Richard Nowitz/USDA ARS.)

succumb rapidly when invasive herbivores or predators are introduced. Mammals and other vertebrates introduced to islands often prey efficiently on endemic animal species and have grazed some native plant species to extinction (Garzón-Machado et al. 2010). Moreover, island species often have no natural immunity to mainland diseases. When nonnative domesticated species (e.g., chickens, ducks) arrive, they frequently carry pathogens or parasites that, though relatively harmless to the carriers, devastate the native populations (e.g., wild birds).

The introduction of just one nonnative species to an island may cause the local extinction of numerous native species. The brown tree snake (*Boiga irregularis*) has been introduced onto a number of Pacific islands, where it is devastating endemic bird populations. The snake eats eggs, nestlings, and adult birds. On Guam alone, the brown tree snake has driven 10 of 13 forest bird species extinct (Perry and Vice 2009). The government spends $3 million per year on attempts to control the brown tree snake population, including aerial drops of dead mice injected with acetaminophen (a poison for snakes), so far without success.

A quarter of the native bird species of New Zealand have gone extinct since the arrival of humans on the islands and the resulting introduction of many invasive species to which the birds (and other native organisms) were not adapted (Russell et al. 2015a). New Zealand's national symbol, the kiwi (*Apteryx* spp.), is an endemic flightless bird that did not evolve with mammalian predators such as dogs (*Canis lupus familiaris*) and stoats (*Mustela ermine*, a type of weasel). It is estimated that New Zealand is losing 2% of the kiwi population each year to these predators. In 2014, the Predator Free New Zealand program was launched with the goal of extirpating all invasive mammals from the country, beginning with its smallest islands.

Invasive species in aquatic habitats

The diversity and abundance of both freshwater and saltwater species is frequently found to decline in association with aquatic invaders, which include plants, fishes, and invertebrates (Gallardo et al. 2016). Whereas invasions in freshwater environments are often more readily noticed, a widespread survey found that there are 329 invasive marine species and that 84% of marine areas worldwide are affected by at least one of them (Molnar et al. 2008). Freshwater ecosystems are somewhat similar to oceanic islands in that they are isolated habitats and are thus at increased risk of invasion. There has been a long history of introductions of nonnative commercial and sport fishes into lakes, such as the introduction of the Nile perch (*Lates niloticus*) into Lake Victoria in East Africa, which was followed by the extinction of numerous endemic cichlid fishes. Often the introduced nonnative fishes are larger and more aggressive than the native fish fauna, and they may eventually drive the local fish to extinction. But once these invasive species are removed from aquatic habitats, the native species are sometimes able to recover (Vredenburg 2004).

Aggressive aquatic invaders also include plants and invertebrates. One of the most alarming invasions in North America was the arrival in 1988 of the zebra mussel (*Dreissena polymorpha*) in the Great Lakes (Strayer 2009). This small, striped native of the Caspian Sea apparently was a stowaway in the ballast tanks of a European cargo ship (**FIGURE 4.21**). Within 2 years, zebra mussels had reached almost unbelievable densities of 700,000 individuals per square meter in parts of Lake Erie, encrusting every hard surface and choking out native mussel species in the process. As they continue to spread throughout the waters of the United States, these non-native molluscs are causing enormous economic damage, estimated at $1 billion per year, to fisheries, dams, power plants, water treatment facilities, and boats, as well as devastating the aquatic communities they encounter (Pimentel et al. 2005). Both the zebra mussel and, more recently, the quagga mussel (*Dreissena rostriformis bugensis*) have become significant problems not only in the United States, but in Europe as well (**FIGURE 4.22**).

One-third of the worst invasive species in aquatic environments are aquarium and ornamental species, which are traded worldwide to the tune of $25 billion per year (Keller and Lodge 2007). Notable examples include water hyacinth (*Eichornia crassipes*), an ornamental plant that creates huge floating mats (**FIGURE 4.23**). These mats deprive submerged plants of light; use up oxygen in the water, thereby killing fish and turtles; are barriers to animals that fish; and even act as vectors of disease by creating habitat for mosquitoes that carry malaria and for a snail that hosts a flatworm that causes disease in humans. Another such invader, *Caulerpa taxifolia*, is a species of marine algae that has blanketed the seafloor, covering native algae, sea grass, and sessile animals, in the Mediterranean Sea. Both water hyacinth and *Caulerpa* are listed by the **International Union for the Conservation of Nature (IUCN)** as among the world's top 100 worst invasive species.

The ability of species to become invasive

The great majority of introduced species do not survive outside of their native ranges, and of those that do, only a small fraction (perhaps less than 1%) are capable of increasing and spreading in their new locations. Why are certain species able to invade and dominate new habitats and displace native species so easily? These species may be better suited to taking advantage of disturbed conditions than native species (Dukes et al. 2011), particularly if the disturbance is new to that system and alters the way physical or chemical resources are made

FIGURE 4.21 This current meter was retrieved from Lake Michigan after a crust of thousands of zebra mussels made it inoperable. Such encroachment is typical of the tiny molluscs, which also encrust and destroy native mussel species and other organisms. (Photograph by M. McCormick, courtesy of NOAA/Great Lakes Environmental Research Laboratory.)

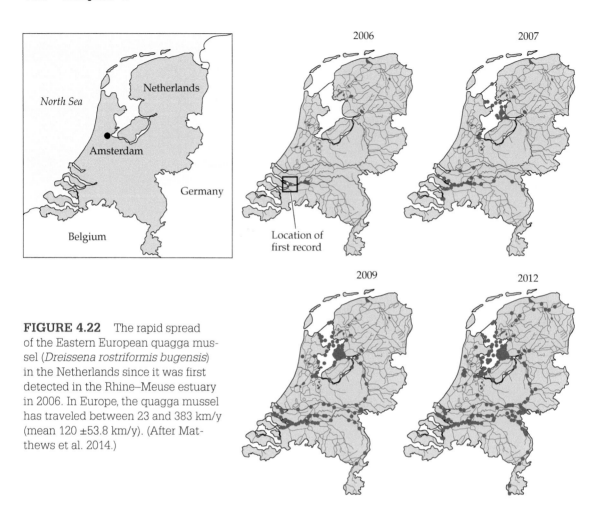

FIGURE 4.22 The rapid spread of the Eastern European quagga mussel (*Dreissena rostriformis bugensis*) in the Netherlands since it was first detected in the Rhine–Meuse estuary in 2006. In Europe, the quagga mussel has traveled between 23 and 383 km/y (mean 120 ±53.8 km/y). (After Matthews et al. 2014.)

available (Sher and Hyatt 1999). Human activity that causes disturbances may create unusual environmental conditions, such as higher or lower mineral nutrient levels, increased or decreased incidence of fire, or enhanced or lowered light availability, to which nonnative species are sometimes better adapted than native species. In fact, the highest concentrations of invasive species are often found in those habitats that have been most altered by human activity. Many of the threats to biodiversity mentioned above can cause or exacerbate invasion; for example, ocean acidification was found to increase predation by invasive snails on native oysters by 48% (Sanford et al. 2014). When habitats are altered by global climate change, they become even more vulnerable to invasion (Bradley et al. 2012).

As mentioned above, native species can become invasive within their home ranges, usually when they are suited to the ways in which humans have altered the environment, and may be almost as much of a concern as nonnative invasive species (Carey et al. 2012). Within North America,

FIGURE 4.23   Water hyacinth (*Eichhornia crassipes*), which is native to South America, has become a significant ecological and economic problem in many other parts of the world, including North America, Africa, and Asia. (© Stephanie Jackson/Alamy Stock Photo.)

fragmentation of forests, suburban development, and easy access to garbage have allowed the numbers of coyotes, red foxes, and certain gull species to increase. Native jellyfishes have increased in abundance in the Gulf of Mexico because they use oil rigs and human structures for spawning and feed on plankton blooms stimulated by nitrogen pollution (Duarte et al. 2013). As these aggressive species increase, they do so at the expense of other local native species, such as the juvenile stages of commercially harvested fish, representing a further challenge to the management of vulnerable native species and protected areas.

Invasive species often thrive where human activities have changed the environment.

Another explanation for why some introduced species become invasive is the **predator release hypothesis**, which attributes their rapid proliferation in their new habitat to the absence of the specialized natural predators and parasites that would otherwise control the invaders' population growth (Davis 2009). In Australia, for example, introduced rabbits spread uncontrollably, grazing native plants to the point of extinction, because there were no effective checks on their numbers. Australian control efforts have focused in part on introducing specific diseases as a biological control for rabbit populations.

In one of the key generalizations of this field, the species that are most likely to become invasive and have significant effects in a new location are those species that have already been shown to do so somewhere else (Ricciardi 2003). A special class of invasive species is made up of those introduced species that have close relatives in the native biota (Laikre et al. 2010). When invasive species hybridize with the native species and varieties, unique genotypes may be eliminated from local populations and taxonomic boundaries may become obscured (see Chapter 2)—a process called *genetic swamping*. This appears to be the fate of native trout species when confronted by introduced species. In the southwestern United States, the Apache trout (*Oncorhynchus apache*) has had its range reduced by habitat destruction and competition with introduced species. The species has also hybridized extensively with rainbow trout (*O. mykiss*), an introduced sport fish, blurring its identity as a distinct species.

Control of invasive species

Invasive species are considered to be the most serious threat facing the biota of the US national park system. While the effects of habitat degradation, fragmentation, and pollution can potentially be corrected and reversed in a matter of years or decades as long as the original species are present, well-established invasive species may be impossible to remove from ecosystems. They may have built up such large numbers and become so widely dispersed and so thoroughly integrated into an ecosystem that eliminating them may be extraordinarily difficult and expensive (Rinella et al. 2009).

The threats posed by invasive species are so severe that reducing the rate of their introduction needs to become a greater priority for conservation efforts (Liebhold et al. 2012). Governments must pass and enforce laws and customs restrictions prohibiting the transport and introduction of nonnative species. In some cases, this may require restrictions and inspections related to the movement of soil, wood, plants, animals, and other items across international borders and even through checkpoints within countries. Currently, vast sums are spent controlling widespread outbreaks of invasives, but inexpensive, prompt control and eradication efforts at the time of first sighting can stop a species from getting established in the first place (McConnachie et al. 2012; Van Wilgen et al. 2012). Training citizens and protected areas staff to monitor vulnerable habitats for the appearance of known invasive species and promptly implementing intensive control efforts can be an effective way to stop the establishment and early spread of a potential new invader. This strategy may require a cooperative effort on the part of multiple levels of government and private landowners.

A variety of strategies for controlling and eliminating invasive species exists, including three general approaches to their removal: pesticides (i.e., chemical poisons), mechanical culling (physical removal, such as pulling weeds and trapping or shooting animals), and biological control (use of

Governments must act to prevent the introduction of new invasive species, to monitor the arrival and spread of invasives, and to eradicate new populations of invasives.

one organism to control another). Each of these approaches, used alone or in combination with the others, has both benefits and limitations. In many cases, use of pesticides is the most effective means of controlling undesirable species, but the chemicals can be expensive and may hurt nontarget, desirable species. Mechanical approaches usually have the fewest nontarget effects but are the most labor-intensive. Biological control, such as use of a specialist insect or pathogen, can be the cheapest approach to managing invaders and is usually the most gentle on the ecosystem as a whole, but biological controls will not usually eradicate the species in question and may take time to show any effect. An extensive public education program is often necessary so that people are aware of why invasive species need to be removed or killed, especially when they are charismatic mammals such as cats, dogs, and rabbits (Oppel et al. 2011). An emphasis on "compassion for the ecosystem" is one approach that has been suggested to help explain why lethal controls are sometimes warranted (Russell et al. 2015b).

Habitat manipulation, changing agricultural practices, and the use of crop strains that are more resistant to pests are other methods frequently employed to prevent invasions or lessen their effects. Using a combination of these techniques for the long-term management of pests is called **integrated pest management** (**IPM**).

Even though the effects of invasive species are generally considered to be negative, they may provide some benefits as well, especially when the habitat is so degraded that the original species are unlikely to reestablish themselves, even once the invasive species are removed. Invasive plant species can sometimes stabilize eroding lands, provide nectar for native insects, and supply nesting sites for birds and mammals (Bateman et al. 2013; Shackelford et al. 2013). In such situations, the trade-offs need to be evaluated to determine whether the potential benefits of removal will outweigh the overall costs. Increasingly, biologists are recognizing the value of accepting "novel ecosystems" (see Chapter 10) in which a mixture of native and nonnative species is best suited to the new conditions created by human activities (Hobbs et al. 2013).

GMOs and conservation biology

A special topic of concern for conservation biologists is the increasing use of **genetically modified organisms** (**GMOs**) in agriculture, forestry, aquaculture, and toxic waste cleanup. GMOs are organisms to whose genetic code scientists have added genes from a different ("source") species using the techniques of recombinant DNA technology. Such gene transfers can be done not only across species, but across taxonomic domains, as when a bacterial gene that produces a chemical toxic to insects is transferred into a crop species such as corn. Enormous amounts of cropland—especially in the United States, Argentina, China, and Canada—have already been planted with GMOs, mainly soybeans, corn (maize), cotton, and oilseed rape (canola). Genetically modified animals are under development, with salmon and pigs showing commercial potential.

Humans have been genetically modifying domesticated crop and animal species since the dawn of civilization by means of selective breeding, hybridization, and other forms of artificial selection. However, many species being investigated as potential sources of transferable genes, including viruses, bacteria, insects, fungi, and shellfish, have not previously been used in breeding programs. Fear of "crossovers" between unrelated species has resulted in some governments implementing special controls on this type of research and its commercial applications. There is concern among some people, especially in Europe, that genetically modified crop species will hybridize with related species, leading to invasion by new, aggressive weeds and virulent diseases (Bagla 2010). Additionally, the use of GMOs could potentially harm noncrop species, such as insects, birds, and soil organisms that live in or near agricultural fields. Further, some people want assurances that eating food from GMO crops will not harm their health or cause unusual allergic reactions.

It is clear that GMOs have the ability to increase crop production to feed a growing human population, to produce new and cheaper medicines, and to reduce the use of pesticides on agricultural fields and the pesticide pollution associated with such use (Lu et al. 2012). (On the other hand, one hugely popular GMO—the Roundup Ready soybean—has been genetically engineered by the manufacturer of the herbicide glyphosate, commonly sold as the weed killer Roundup, so that the crop can be treated with more—not less—Roundup.)

In summary, the benefits of GMOs need to be examined and weighed against the unknown potential risks. The best approach involves proceeding cautiously, investigating GMOs thoroughly before commercial releases are authorized, and monitoring environmental and health effects after release.

Disease

The increased transmission of disease as a result of human activities is a major threat to many endangered species and ecosystems. Pathogens (disease-causing organisms) such as bacteria, viruses, fungi, and protists can have major effects on vulnerable species and even the structure of entire ecosystems. Human activities can lead to increased populations of many pathogens, leading to outbreaks of animal and human disease. In addition, interaction with humans and their domesticated animals exposes wild animals to diseases never previously encountered that can reduce the size and density of wild populations (Jones et al. 2008; **FIGURE 4.24**).

Disease may be the single greatest threat to some rare species. The decline of numerous frog populations in visually pristine montane habitats across the world is apparently due to the introduction of a nonnative fungal disease. The last population of black-footed ferrets (*Mustela nigripes*) known to exist on its own in the wild was destroyed by the canine distemper virus in

> Increased incidence of infectious disease threatens wild and domesticated species as well as humans. Transfer of disease between species is a subject of special concern.

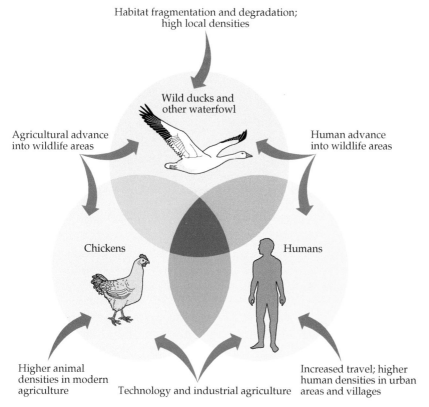

Habitat fragmentation and degradation; high local densities

Agricultural advance into wildlife areas

Human advance into wildlife areas

Wild ducks and other waterfowl

Chickens

Humans

Higher animal densities in modern agriculture

Technology and industrial agriculture

Increased travel; higher human densities in urban areas and villages

FIGURE 4.24   Infectious diseases—such as rabies, Lyme disease, influenza, bird flu, hantavirus, and canine distemper—spread among wildlife populations, domesticated animals, and humans as a result of increasing population densities and the advance of agriculture and human settlements into wild areas. The diagram illustrates the infection and transmission routes of bird flu, which is caused by a virus that infects wild waterfowl, chickens, and humans. The shaded areas of overlap indicate that the disease can be shared among the three groups. Green arrows indicate factors contributing to higher rates of infection; blue arrows indicate factors contributing to the spread of disease among the three groups. (After Daszak et al. 2000.)

1987, though a few healthy individuals were caught for a captive breeding program. One of the main challenges of managing the captive breeding program has been protecting the captive ferrets from canine distemper, human viruses, and other diseases; this is being done through rigorous quarantine measures and subdivision of the captive colony into geographically separate groups.

White-nose syndrome is a fungal disease that is currently killing millions of bats across the eastern United States (Thogmartin et al. 2013). In some caves, 90% of the bats have died. The disease is characterized by a powdery white fuzz on a bat's snout and other membranous areas

FIGURE 4.25 A little brown bat (*Myotis lucifugus*) infected with white-nose syndrome is being examined by a researcher. This fungal disease, which first appeared in the United States in 2006, affects a variety of bat species and has killed up to 90% of the bats in many caves. (Photograph courtesy of Ryan Von Linden/New York Department of Environmental Conservation.)

(**FIGURE 4.25**). Bats die when the fungus causes skin irritation and the bat wakes from hibernation in midwinter, when its diet of flying insects is not available, instead of in spring, depleting its energy reserves and subsequently starving to death. Discovered in one cave in New York State in 2006, the disease has spread rapidly across the region, probably during bat migration. It is possible that cave explorers or bat researchers accidentally introduced the fungus to the United States as a contaminant on their clothes, boots, or equipment following a visit to a European bat cave. At this point, the only effective way to prevent its spread to new colonies is to close bat caves to all human visitors except for scientists who sterilize their clothes and equipment before entering.

Three basic principles of epidemiology have obvious practical implications for limiting disease in the management of endangered species. First, a high rate of contact between host and pathogen or parasite is one factor that encourages the spread of disease. In general, as host population density increases, so does risk of disease. In addition, a high density of the infective stages of a parasite in the environment of the host population can lead to increased incidence of disease. In natural situations, the rate of infection is typically reduced when animals migrate away from their droppings, saliva, old skin, dead conspecifics, and other sources of infection. However, in unnaturally confined situations, such as habitat fragments, zoos, or even nature reserves, the animals remain in contact with these potential sources of infection, and disease transmission increases. Furthermore, at higher densities, animals have abnormally frequent contact with one another, and once one animal becomes infected, the parasite can rapidly spread throughout the entire population.

Second, indirect effects of habitat destruction can increase an organism's susceptibility to disease. When a host population is crowded into a smaller area because of habitat destruction, its members often face lowered habitat quality and food availability, lowered nutritional status, and less resistance to infection in addition to higher contact rates (Zylberberg et al. 2013). Furthermore, crowding can lead to social stress within a population, which also lowers the animals' resistance to disease. Pollution may make individuals more susceptible to infection by pathogens, particularly in aquatic environments (Harvell et al. 2004). It has even been observed that biodiversity regulates disease by diluting the number of suitable host species or by constraining the sizes of host populations through predation and competition (Ostfeld 2009). For example, the increased incidence of Lyme disease and other tick-borne pathogens in some areas has been linked not only to the local abundance of certain host rodent species, as mentioned above, but also to the overall loss of local species diversity (**FIGURE 4.26**).

Third, in many conservation areas, zoos, national parks, and newly cleared agricultural areas, species come into contact with other species that they would rarely or never encounter in the wild—including humans and domesticated animals—so infections such as rabies, influenza, distemper, hantavirus, and bird flu can spread from one species to another. Infectious diseases can spread between wildlife populations, domesticated animals, and humans as a result of increasing human population densities and the advance of agriculture and human settlements into wildlife areas. The

(A)

(B)

(C)

FIGURE 4.26   (A) The white-footed mouse (*Peromyscus leucopus*), one of the main hosts for the bacterium that causes Lyme disease, increases in abundance in habitat fragments created by suburban development. (B) Field biologists sampling mice for the presence of infectious diseases, such as plague. (C) A black-legged tick, which can be up to 3 mm (0.12 in.) long, can transfer the Lyme disease bacterium to a human after acquiring it from an infected animal host. (A, © Rob and Ann Simpson/Visuals Unlimited, Inc.; B, from Crowl et al. 2008; C, courtesy of Michael L. Levin/CDC.)

> Steps must be taken
> to prevent the spread
> of disease in captive
> animals and to ensure
> that new diseases are not
> accidentally introduced
> into wild populations.

human immunodeficiency virus (HIV) and the deadly Ebola virus both appear to have spread from wildlife populations to humans and to domesticated animals. Such examples are likely to become more common as a result of anthropogenic changes to the environment, the increase in international travel, and globalization of the economy.

In zoos, colonies of animals are often caged together in small areas, and similar species are often housed close to one another. Consequently, if one animal becomes infected, the pathogen can spread rapidly to other animals and to related species. Once they are infected with an exotic disease, captive animals cannot be returned to the wild without threatening the entire wild population. Furthermore, a species that is both common and fairly resistant to a disease can act as a reservoir for the disease, which can then infect populations of susceptible species. For example, apparently healthy African elephants can transmit a fatal herpesvirus to related Asian elephants when they are kept together in zoos. Diseases can spread very rapidly between captive species kept in crowded conditions.

A Concluding Remark

This chapter has described seven major categories of threats faced by species and ecosystems. A study of 181 threatened and endangered species (Lawler et al. 2002) found that more than 85% of them faced at least four types of threats, the most common of which were related to habitat destruction and degradation, interactions with invasive species, and overexploitation. When the threats to biodiversity are well understood, protection and recovery efforts have the best chance for success.

Summary

- The major threats to biodiversity are habitat destruction, fragmentation, degradation (which includes pollution), climate change, overexploitation, invasive species, and disease. All of these threats result from the use of the world's natural resources by an increasing human population.

- Habitat destruction threatens rain forests, wetlands, coral reefs, and other species-rich communities.

- Habitat fragmentation is the process whereby a large, continuous area of habitat is both reduced and divided into two or more fragments. Habitat fragmentation can lead to the rapid loss of some of the remaining species because it creates barriers to the normal processes of dispersal, colonization, and foraging. Particular fragments may contain altered environmental conditions that make them less suitable for the original species.

- Environmental pollution eliminates many species from ecosystems even where the structure of the community is not obviously disturbed. Environmental pollution results in pesticide biomagnification; contamination of water with industrial wastes, sewage, and fertilizers; and air pollution resulting in acid rain, excess nitrogen deposition, photochemical smog, and high ozone levels.

- Global climate change, including warmer temperatures and changing precipitation patterns, is already occurring because of the large amounts of carbon dioxide and other greenhouse gases produced by the burning of fossil fuels and deforestation. Predicted temperature increases may be so rapid in coming decades that many species will be unable to adjust their ranges and will become extinct.

- Overexploitation is driving many species to extinction and can consequently undermine entire ecosystems. Overexploitation is caused by increasingly efficient methods of harvesting and marketing, increasing demand for products, and increased access to remote areas.

- Humans have deliberately and accidentally moved thousands of species to new regions of the world. Some of these nonnative species have become invasive, greatly increasing their numbers at the expense of native species.

- Levels of disease often increase when animals are confined to nature reserves, zoos, or habitat fragments and cannot disperse over wide areas. In zoos and botanical gardens, diseases sometimes spread between related species of animals and plants. Diseases may also spread between domesticated species and wild species, and even between humans and both wild and domesticated animals.

- Species may be threatened by a combination of factors, all of which must be addressed in a comprehensive conservation plan.

For Discussion ■

1. Human population growth is often blamed for the loss of biological diversity. Is this valid? What other factors are responsible, and how do we weigh their relative importance? Is it possible to find a balance between providing for increasing numbers of people and protecting biodiversity?

2. Consider the most damaged and the most pristine habitats near where you live or go to school. Why have some habitats been preserved and others fragmented and degraded?

3. Learn about one endangered species in detail. What is the full range of immediate threats to this species? How do these immediate threats connect to larger social, economic, political, and legal issues?

Suggested Readings

Bussière, E., L. G. Underhill, and R. Altwegg. 2015. Patterns of bird migration phenology in South Africa suggest northern hemisphere climate as the most consistent driver of change. *Global Change Biology* 21: 2179–2190. The warming climate in breeding grounds has been found to have impacts on the timing of bird movements.

Clavero, M. and E. García-Berthou, E. 2005. Invasive species are a leading cause of animal extinctions. *Trends in Ecology and Evolution*, 20(3), 110. A frequently cited documentation of the impact of invasive animals.

Gallardo, B., M. Clavero, M. I. Sánchez, and M. Vilá. 2016. Global ecological impacts of invasive species in aquatic ecosystems. *Global Change Biology*. 22: 151–163. A review and analysis of 733 evaluations of the effects of aquatic invaders.

Gibson, L., A. J. Lynam, C. J. Bradshaw, F. He, D. P. Bickford, D. S Woodruff, and 2 others. 2013. Near-complete extinction of native small mammal fauna 25 years after forest fragmentation. *Science* 341: 1508–1510. Forest fragments gradually lose many of their original species.

Gutiérrez, J. L. and C. Bernstein. 2014. Ecosystem impacts of invasive species. BIOLIEF 2011. 2nd World Conference on Biological Invasion and Ecosystem Functioning, Mar del Plata, Argentina, 21–24 November 2011. *Acta Oecologica* 54 :1–138.

Intergovernmental Panel on Climate Change (IPCC). 2013. *Climate Change 2013: The Physical Science Basis. Contribution of Working Group I to the Fifth Assessment Report of the Intergovernmental Panel on Climate Change*. T. F. Stocker and 9 others (eds). Cambridge University Press, Cambridge, UK. Comprehensive presentation of the evidence for global climate change, along with predictions for the coming decades.

Larson, C. 2016. Shell trade pushes giant clams to the brink. *Science* 351: 323–324. *Tridacna gigas* faces extinction from over harvesting.

Lovich, J. E., C. B. Yackulic, J. Freilich, M. Agha, M. Austin, K. P. Meyer, and 4 others. 2014. Climatic variation and tortoise survival: Has a desert species met its match? *Biological Conservation* 169: 214–224. Prolonged drought associated with climate change is impacting endangered desert species.

Primack, R. B. and A. J. Miller-Rushing. 2012. Uncovering, collecting, and analyzing records to investigate the ecological impacts of climate change: A template from Thoreau's Concord. *BioScience* 62: 170–181. We can see the effects of climate change happening all around us if we find the right tools and baseline studies.

Robertson, G., C. Moreno, J. A. Arata, S. G. Candy, K. Lawton, J. Valencia, and 4 others. 2014. Black-browed albatross numbers in Chile increase in response to reduced mortality in fisheries. *Biological Conservation* 169: 319–333. Seabird numbers increased after the Chilean fishing industry switched to a system of fast-sinking baited hooks that were less accessible to birds.

Siraj, A. S., M. Santos-Vega, M. J. Bouma, D. Yadeta, D. R. Carrascal, and M. Pascual. 2014. Altitudinal changes in malaria incidence in highlands of Ethiopia and Colombia. *Science* 343: 1154–1158. Malaria spreads to higher altitudes in warmer years and is predicted to have a wider impact with a warming climate.

Wilson, M. C., X. Y. Chen, R. T. Corlett, R. K. Didham, et al. 2016. Habitat fragmentation and biodiversity conservation: key findings and future challenges. *Landscape Ecology* 31(2): 219–227. Many of the impacts of habitat fragmentation are indirect.

5 Extinction Is Forever

The Tasmanian tiger-wolf (*Thylacinus cynocephalus*) was a marsupial that lived in Australia and on the island of Tasmania. After competition with introduced dogs led to its extinction on the mainland, the remaining populations in Tasmania were destroyed by a combination of habitat loss, hunting, and disease. By 1933, the species was extinct in the wild, and the last known individual died in captivity in 1936.

T
he diversity of species found on the Earth has been increasing since life first originated. The increase has not been steady; rather, it has been characterized by periods of high rates of speciation followed by periods of minimal change and episodes of mass extinction (Raup and Sepkoski 1982; Ward 2004). In addition to this overall increase, there have been five episodes of mass extinction in the fossil record, occurring at intervals ranging from 60 million to 155 million years (**FIGURE 5.1**). These episodes, which ended the Ordovician, Devonian, Permian, Triassic, and Cretaceous periods, could be called *natural mass extinctions*. The most famous is the extinction of the dinosaurs during the Late Cretaceous, 65 million years ago, after which mammals achieved dominance in terrestrial communities. The most massive extinction took place at the end of the Permian, 250 million years ago, when about 95% of all marine animal species and half of all animal families are estimated to have gone extinct (Wake and Vredenburg 2008). As David Raup (1988) observed, "If these estimates are even reasonably accurate, global biology (for higher organisms at least) had an extremely close brush with total destruction." It is quite likely that some massive disturbance—such as widespread volcanic eruptions, a collision with an asteroid, or both—caused the dramatic change in the Earth's climate that resulted in

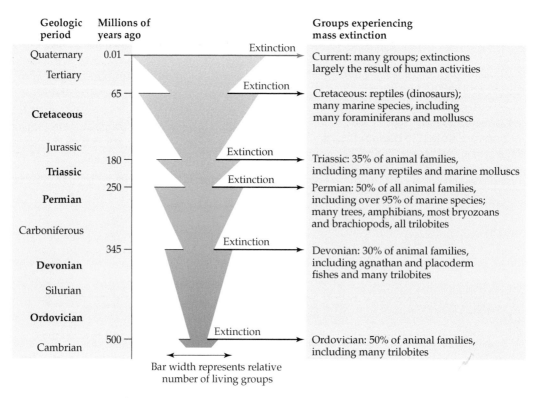

Geologic period	Millions of years ago		Groups experiencing mass extinction

Bar width represents relative
number of living groups

FIGURE 5.1 Relative numbers of animal families over geologic time. Although the total number of species groups on Earth has increased over the eons, a large percentage of these groups disappeared during each of five episodes of natural mass extinction (named in boldface at left). The most dramatic period of loss occurred about 250 million years ago, at the end of the Permian period. A sixth mass extinction episode (red arrow at top of figure) began during the present geologic period and will continue for decades to come.

the end of so many species. Another speculation is that there might have been a massive release of methane gas from beneath the ocean floor—a "big burp," if you will. Such an event not only would have released toxic plumes but almost certainly would have affected the climate, since methane is an even more potent greenhouse gas than carbon dioxide. It took about 80–100 million years of evolution for Earth's biota to regain the number of families lost during the Permian extinction.

Species go extinct even in the absence of violent disturbance. One species may outcompete another for a vital resource, or predators may drive prey species to extinction. Extinction is as much a natural process as speciation is. If extinction is a natural process, however, why is the current loss of species of such concern? The answer lies in the relative rates of extinction and speciation as well as in the causes of extinction. Speciation is typically a slow process, occurring through the gradual accumulation of mutations and shifts in allele frequencies over thousands, if not millions, of years. As long as the rate of speciation equals or exceeds the rate of extinction, biodiversity will remain constant or increase. In past geologic periods, the loss of existing species was eventually balanced and then exceeded by the evolution of new species. However, current extinction rates are much greater than speciation rates, and more than 99% of modern species extinctions are linked to human activity (Pimm and Jenkins 2005; Wake and Vrendenberg 2008). We are presently in the midst of a **sixth extinction episode**, caused by human activities rather than a natural disaster (Ceballos et al. 2015).

The current rate of species loss is unprecedented, unique, and irreversible. Ninety-nine percent of current extinctions are caused in some way by human activities.

Extinction is a top concern of conservation biology because the number of threatened species is so large. About 21% of the world's remaining bird species are threatened with extinction. **TABLE 5.1** shows certain animal groups for which the danger is particularly severe, including turtles, manatees, rhinoceroses, and penguins. The danger is even

TABLE 5.1 Numbers of Species Threatened with Extinction in Major Groups of Animals and Plants

Group	Approximate number of species	Number of species threatened with extinction	Percentage of species threatened with extinction
Vertebrates			
Fishes	28,000	2,084	7[a]
Amphibians	6,338	2,303	36
Reptiles	9,400	1,018	11[a]
Crocodiles	23	10	43
Turtles	228	170	75
Birds	10,052	2,097	21
Penguins	18	11	61
Mammals	5,499	1,462	27
Primates	418	227	54
Manatees, dugongs	5	4	80
Horses, tapirs, rhinos	16	14	88
Plants			
Gymnosperms	18,000	521	3[a]
Angiosperms (flowering plants)	260,000	9,614	4[a]
Palms	361	275	82
Fungi	100,000	3	0

Source: Data from IUCN 2011 (www.iucnredlist.org). Data include the categories critically endangered, endangered, vulnerable, and near threatened.

[a]Low percentages reflect inadequate data due to the small number of species evaluated. For example, 11% of reptiles are listed as endangered, but only about one-third of species have been evaluated. Among reptile species that have been evaluated, 31% are considered endangered.

greater for other large groups: 27% of mammal species and 36% of amphibian species are threatened (IUCN 2013). Some other groups of species, such as fishes (9%) and reptiles (12%), face lower threat levels. Some plant species are also at risk; palms are especially vulnerable. Most groups of plants, fungi, fishes, and insects are not well known, and for these groups the extinction risk cannot be accurately determined. Extinction rates are likely to be much higher once threats faced by poorly known groups of species are assessed (Hoffmann et al. 2010; McClenachan et al. 2012).

The Meaning of "Extinct"

The word *extinct* has many nuances, and its meaning can vary somewhat depending on the context.

- A species is considered **extinct** when, after a thorough search, no member of the species is found alive anywhere in the world: the Tasmanian tiger-wolf (*Thylacinus cynocephalus*), for example, is extinct (see the chapter opening photo).
- If individuals of a species remain alive only in captivity or in other human-controlled situations, the species is said to be **extinct in the wild**: The St. Helena ebony tree (*Trochetiopsis ebenus*) (**FIGURE 5.2**) is nearly extinct in the wild, although it grows well under cultivation.
- A species is **locally extinct**, or **extirpated**, when it is no longer found in a specific area it once inhabited, but is still found elsewhere in the wild: "The purple bladderwort can no longer be found in Fairhaven Bay, but it can still be found in other nearby locations." This is in contrast to a species that no longer lives anywhere in the world, sometimes referred to as being **globally extinct**.

FIGURE 5.2 The St. Helena ebony tree (*Trochetiopsis ebenus*) is endemic to the island of St. Helena in the South Atlantic Ocean. The wild population has been reduced to just two individuals on the side of a cliff. Because of these low numbers, this species will almost certainly go extinct in the wild. The species will remain alive in cultivation. (© fotoFlora/Alamy.)

- Some conservation biologists speak of a species as being **functionally extinct** (or **ecologically extinct**) if it persists at such reduced numbers that its effects on the other species in its community are negligible: "Tigers are functionally extinct because so few remain in the wild that their effect on prey populations is insignificant." Species that are functionally extinct are generally at great risk of becoming globally extinct.

In addition to the various nuances of the term *extinct*, conservation biologists work with a variety of categories, including *endangered*, *vulnerable*, and *threatened*, that more specifically describe the status of species. These categories are discussed in detail in Chapter 6.

The current, human-caused mass extinction

The global diversity of species reached an all-time high in the present geologic period. Many groups of organisms—such as insects, vertebrates, and flowering plants—reached their greatest diversity about 30,000 years ago. Since that time, however, species richness has slowly decreased as one species—*Homo sapiens*—has asserted its dominance. In our need to consume natural resources, humans have increasingly altered terrestrial and aquatic environments at the expense of other species.

The first noticeable effects of human activity on extinction rates can be seen in the elimination of large mammals from Australia and North and South America at the time humans first colonized these continents tens of thousands of years ago. Shortly after humans arrived, approximately 80% of the *megafauna*—mammals weighing more than 44 kg (100 pounds)—became extinct. These extinctions were probably caused directly by hunting and indirectly by the burning and clearing of forests and grasslands and the introduction of invasive species and new diseases. On all continents, paleontologists and archaeologists have found an extensive record of prehistoric human alteration and destruction of habitat coinciding with high rates of species extinctions (Sandom et al. 2014). For example, deliberate burning of savannas, presumably to encourage plant growth for browsing wildlife and thereby to improve hunting, has been occurring for 50,000 years in Africa.

Extinction rates during the last 2000 years are best known for terrestrial vertebrates, especially birds and mammals, because these species are conspicuous and are therefore well studied. Extinction rates for the other 99% of the world's species are just rough guesses at present because scientists have not systematically searched the remote sites where many of these species occur to determine whether they still exist. For example, a species of giant tortoise (*Chelonoidis elephantopus*) known from only one island in the Galápagos was thought to have gone extinct shortly after Charles Darwin visited in 1835, but it was recently rediscovered on another island and its identity confirmed using DNA sequencing methods.

On the other hand, some species presumed to be **extant** (still living) may actually be extinct; the Yangtze river dolphin (*Lipotes vexillifer*), for example, has not officially been designated as extinct, but no individuals could be found during an intensive survey in 2006 (Fisher and Blomberg 2012), and many of the previous sightings are probably not reliable (Turvey 2008).

Sometimes it is difficult to determine whether a species is truly extinct. In 2004, for example, ornithologists in North America announced sightings of an ivory-billed woodpecker (*Campephilus principalis*) in an Arkansas swamp forest—decades after this bird was believed to have gone extinct. Since then, however, intensive efforts to find and conclusively identify existing individuals of the species have been unsuccessful (Gotelli et al. 2012; Scott et al. 2008; Solow et al. 2012).

One set of estimates based on the best available evidence indicates that 79 species of mammals have gone extinct and another 126 are critically endangered—a total 3.7% of all known species (Pimm et al. 2014). Amphibians have fared far worse (8.1%). The majority of human-caused extinctions have occurred in the last 150 years. The extinction rate for birds was about 0–5 species every 25 years during the period from 1500 to 1725, but it rose to 8–12 species every 25 years during the period from 1750 to 1850. After 1850, the extinction rate rose again to more than 16 species every 25 years. This increase in the rate of extinction reflects the growing intensity of the threat to biodiversity.

> Island species have had higher rates of extinction than mainland species. Freshwater species are more vulnerable to extinction than marine species.

One trend to note is that all of the earliest extinctions documented by humans were on islands. In fact, 89% of known bird extinctions since 1500 have been on islands, even though greater than 80% of bird species occur on continents (Butchart et al. 2006). It has been determined from archeological findings that 20% of bird species worldwide have gone extinct since the arrival of humans on Pacific islands 30,000 years ago, due to both hunting and the introduction of new predators (Steadman 1995). In contrast, extinctions of birds in mainland areas were first observed about 1800, and they have been increasing since then (**FIGURE 5.3**). In the future, mainland species will account for an increasing proportion of extinctions. Studying island extinctions has taught us much about population dynamics that is applicable to mainland conservation efforts.

Fortunately, extinction rates appear to have declined since 1950, in part due to deliberate efforts to protect rare species in danger of going extinct. These numbers can be misleading, however, because of the current practice of not declaring a species extinct until decades after any individuals of the species have been found. In coming years, numerous species will be declared to have gone extinct during the past half century. In the last decade, a number of species that were not found despite intensive searches were declared extinct, including the Monteverde golden toad of Costa Rica (*Bufo periglenes*) and the Alaotra grebe (*Tachybaptus rufolavatus*) of Madagascar. Many species not yet listed as extinct—and some species that have never been documented at all—have been reduced to such low numbers by human activities that their ability to persist is

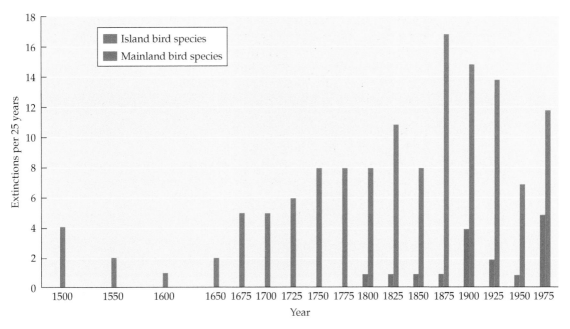

FIGURE 5.3 Rates of extinctions of bird species during 25-year intervals since 1500. Initially, extinctions were almost exclusively of island species, but extinctions of mainland species have increased since 1800. (After Butchart et al. 2006.)

uncertain. In Britain, the country with the best-kept biological records, extinction rates for major groups of species have been in the range of 1%–5% per century, but are predicted to increase in the present century (Hambler et al. 2011).

A count of extant species may also mask the true scale of the current extinction crisis. For many species, a few individuals in scattered small populations might persist for years, decades, or centuries (isolated individuals of woody plants, in particular, can persist for hundreds of years), but their ultimate fate is global extinction, either because they are functionally extinct or because negative environmental effects have yet to be fully felt. Remaining individuals of species that are doomed to extinction have been called "the living dead" or "committed to extinction" (Hanski et al. 1996; Janzen 2001). There are certainly many species in this category in the remaining fragments of species-rich tropical forests. The presumed eventual loss of species following habitat destruction and fragmentation is called the *extinction debt* (Tilman et al. 1994; Dullinger et al. 2013). For example, it is estimated that 80% of forest-dependent vertebrate species will disappear because of destruction that has already occurred (Wearn et al. 2012).

Many species today are represented only by scattered populations with reduced numbers of individuals. Although these isolated populations could persist for years or decades, the ultimate fate of such species is extinction.

FIGURE 5.4 This tropical checkered-skipper butterfly (*Pyrgus oileus*) is one of the remaining extant species on Barro Colorado Island, Panama. Fully 23 known species of butterflies have gone locally extinct there since the 1930s, although the actual number is probably much higher, given how many cryptic butterfly species exist. (© Rick and Nora Bowers/Alamy Stock Photo.)

Local extinctions

In addition to the global extinctions that are a primary focus of conservation biology, many species are experiencing a series of local extinctions, or extirpations, across their range (Leidner and Neel 2011). Where habitats are degraded and destroyed, populations of plants and animals go extinct. For example, 23 species of butterflies have gone locally extinct on Barro Colorado Island, Panama, since the 1930s, representing 20% of the common butterfly species there (Basset et al. 2015) (**FIGURE 5.4**). Many carnivores have experienced local extinctions in recent history across the globe (Ripple et al. 2014). Biological communities are impoverished by such local extinctions.

Concord, Massachusetts, was first assessed for wildflower species in the 1850s by the famous naturalist and philosopher Henry David Thoreau. Twenty-seven percent of the native species seen by Thoreau and other nineteenth-century Concord botanists could not be found when the area was surveyed 150 years later (Miller-Rushing and Primack 2012; Primack et al. 2009). A further 36% of the species now persist in only one or two populations and are therefore vulnerable to extinction. In some cases, only a few individual plants remain of species that were formerly common. Certain groups, such as orchids and lilies, have shown particularly severe losses. A combination of forest succession, invasive species, air and water pollution, grazing by deer, habitat destruction and fragmentation, and now climate change has contributed to these species losses in Concord. In other examples, according to surveys by state Natural Heritage programs, 4%–8% of the plant species formerly found in Hawaii, New York, and Pennsylvania can no longer be found in these states. Due to deforestation over the last 200 years, 34% of bird species and 26% of vascular plant species in Singapore have become locally extinct (Brook et al. 2003). And in a survey of one part of the Indonesian island of Sumatra, only 3 of 12 populations of Asian elephants known from the 1980s were still present 20 years later (Hedges et al. 2005).

It is estimated that there are an average of 200 populations per species (Hughes and Roughgarden 2000). While some species have just a few populations, other species may have thousands of populations. The loss of populations is roughly equal to the proportion of a habitat that is lost, so the world's populations are being lost at a far higher rate than its species. When 90% of an extensive grassland ecosystem is destroyed, 90% of the populations of plants, animals, and fungi there will also be lost. Tropical rain forests contain at least half of the world's species, and they are being lost at the rate of about 1% per year. This represents a loss of 5 million populations per year (1% of the 500 million tropical forest populations), or about 13,500 populations per day.

> Species-rich tropical rain forests are being lost at a rate of 1% a year, a rate believed to result in the destruction of more than 13,500 biological populations each day. Population losses eventually result in species extinctions.

These large numbers of local extinctions serve as important biological warning signs that something is wrong in the environment. Action is needed to prevent further local extinctions as well as global extinctions. The loss of local populations not only represents a loss of biodiversity but also diminishes the value of an area for nature enjoyment, scientific research, and the provision of crucial materials to local people in subsistence economies.

Extinction rates in aquatic environments

In contrast with the large amount of information we have on extinct terrestrial species, only about 14 species—4 marine mammals, 5 marine birds, 1 fish, and 4 molluscs—are known to have gone extinct in the world's vast oceans during historic times (Régnier et al. 2009). This number of extinctions is almost certainly an underestimate, because marine species are not nearly as well known as terrestrial species, but it also may reflect a greater resiliency of marine species to disturbance. Regardless, the significance of these losses may be greater than the numbers suggest. Many marine mammals are top predators, and their loss could have major effects on marine communities. Some marine species are the sole species of their genus, family, or even order, so the extinction of even a few of them could represent a serious loss to global biodiversity.

The oceans were once considered so enormous that it seemed unlikely that marine species could go extinct; many people still share this viewpoint. However, as marine coastal waters become more polluted and species are harvested more intensely, even the vast oceans will not provide safety from extinction (McCauley et al. 2015) (**FIGURE 5.5**). Many species of whales and large fishes have declined by 90% or more because of overharvesting and other human activities and are in danger of extinction.

Freshwater species appear to have a higher extinction rate than marine species. The modern extinction rate for North American freshwater fishes is conservatively estimated to be 877 times greater than the background extinction rate (Burkhead 2012). The fishes of California are particularly

FIGURE 5.5 (A) The endangered white abalone (*Haliotis sorenseni*), a type of mollusc, in its natural habitat on the coast of California. Several abalone species have reached dangerously low numbers due to overharvesting by humans, both for their muscular foot, which is a delicacy, and for their beautiful shells, from which jewelry is made (B). Shown here is an antique pin in the shape of the now extinct Huia bird from New Zealand. Abalones are also threatened by withering shell syndrome, caused by the bacterium *Candidatus xenohaliotis californiensis*, for which a cure has yet to be found. Due to their low numbers, some scientists consider them functionally extinct (Haw 2013). (A, Courtesy of NOAA; B, © Marc Tielemans/Alamy Stock Photo.)

(A)

(B)

vulnerable because of the state's scarcity of water and its intense development: 6% of California's 129 native fish species are already extinct, and 51% are in danger of extinction. Large numbers of fishes and aquatic invertebrates, such as molluscs, are in danger of extinction because of dams, pollution, irrigation projects, overharvesting, invasive species, disease, and general habitat damage.

Measuring Extinction

Quantifying rates of extinction and likelihood of extinction is critical for understanding the magnitude of the problem and how best to address it.

Calculating extinction rates requires information about many facets of organisms and their environments, including the extent to which humans are playing a role.

Background extinction rates

To better understand how calamitous the present extinction rates are, we can compare them with the natural extinction rates that would prevail regardless of human activity. What is the natural rate of extinction in the absence of human influence? Natural **background extinction rates** can be estimated by looking at the fossil record. On average, an individual species lasts about 1 million–10 million years before it goes extinct or evolves into a new species (Mace et al. 2005; Pimm and Jenkins 2005). Since there are perhaps 10 million species on the Earth today, we can predict that 1–10 of the world's species would be lost per year as a result of a natural extinction rate of 0.0001%–0.00001% per year. These estimates are derived from studies of wide-ranging marine animals, so they may be lower than natural extinction rates for species with narrower distributions, which are more vulnerable to habitat disturbance; however, they do appear to be applicable to terrestrial mammals. The current observed rate of extinction for birds and mammals is 1% per century (or 0.01% per year), which is 100–1000 times greater than would be predicted based on background rates of extinction. Putting it another way, about 100 species of birds and mammals were observed to go extinct between 1850 and 1950, but the natural rate of extinction would have predicted that, at most, only 1 species would have gone extinct. Therefore, the other 99 extinctions can be attributed to the effects of human activities.

Extinction rate predictions and the island biogeography model

Studies of island communities have led to general rules on the distribution of biodiversity, synthesized as the island biogeography model[1] by Mac-Arthur and Wilson (1967). The **island biogeography model** was built to explain the species–area relationship: islands with large areas have more species than do islands with smaller areas (**FIGURE 5.6**). This rule makes intuitive sense because a large island will tend to have a greater variety of local environments and community types than a small island. In addition, large islands allow a larger number of separate populations per species and larger sizes per individual population, increasing the likelihood of speciation and decreasing the probability of local extinction of newly evolved as well as recently arrived species.

---

[1]A scientific model is a physical, conceptual, or mathematical representation of something observed in nature that can help us understand it and in some cases make predictions. Models are used to illustrate and explore specific aspects of a real object or phenomenon and so are necessarily simplifications of reality.

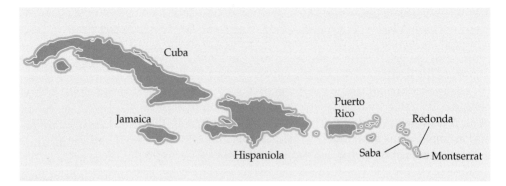

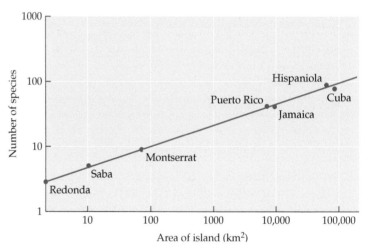

FIGURE 5.6 The number of species on an island can be predicted from the island's area. In this figure, the number of species of reptiles and amphibians is shown for seven islands in the West Indies. The number of species on large islands such as Cuba and Hispaniola far exceeds that on the tiny islands of Saba and Redonda. (After Wilson 1989.)

The **species–area relationship** can be summarized by the empirical formula

$$S = CA^Z$$

where S is the number of species on an island, A is the area of the island, and C and Z are constants. The exponent Z determines the slope of the relationship between species and area. The values for C and Z depend on the types of islands being compared (tropical vs. temperate, dry vs. wet, etc.) and the types of species involved (birds vs. reptiles, etc.). Z values are typically about 0.25, with a range from 0.15 to 0.35. Island species with restricted ranges, such as reptiles and amphibians, tend to have Z values near 0.35, while widespread mainland species tend to have Z values closer to 0.15. Values of C tend to be high in groups, such as insects, that contain many species, and they tend to be lower in groups, such as birds, that contain relatively few species (Connor and McCoy 2001).

Imagine the simplest situation, in which $C = 1$ and $Z = 0.25$, for raptorial birds on a hypothetical archipelago:

$$S = (1)A^{0.25}$$

The formula predicts that islands of 10, 100, 1000, and 10,000 km^2 in area would have 2, 3, 6, and 10 species, respectively. It is important to note that a tenfold increase in island area does not result in a tenfold increase in the number of species; with this equation, each tenfold increase in island area increases the number of species by a factor of approximately 2.

The island biogeography model has been empirically validated and is accepted by most biologists (Quammen 1996; Triantis et al. 2008; Chen and He 2009). For numerous groups of plants and animals, it has been found to describe the observed species richness reasonably well, explaining about half of the variation in numbers of species. For example, actual data from three Caribbean islands conform to this relationship: with increasing area, Nevis (93 km^2), Puerto Rico (8959 km^2), and Cuba (114,524 km^2) have 2, 10, and 57 species of anolis lizards, respectively; with a C of 0.5 and a Z of 0.35, the islands would be predicted to have 2, 12, and 30 species, respectively.

In their classic text, MacArthur and Wilson (1967) also hypothesized that the number of species occurring on an island represents a dynamic equilibrium between colonization by (and evolution of) new species and extinctions of existing species. Starting with an unoccupied island, the number of species will increase over time, since more species will be arriving (or evolving) than will be going extinct, until the rates of extinction and immigration are balanced (**FIGURE 5.7**). Species establishment rates will be higher for large islands than for small islands because large islands represent larger targets for dispersing or-

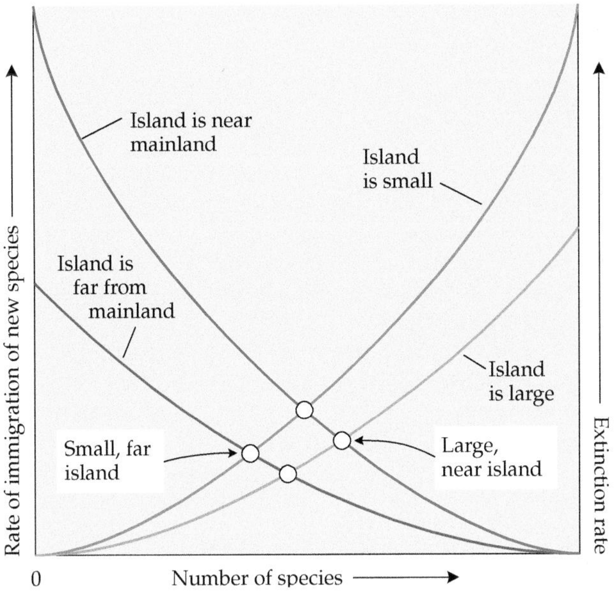

> The island biogeography model can be used to predict how many species will go extinct due to habitat loss. The model can also be used to predict how many species will remain in protected areas of different sizes.

FIGURE 5.7   The island biogeography model describes the relationship between the rates of colonization and extinction in islands. The immigration rates (blue and red curves) on unoccupied islands are initially high, as species with good dispersal abilities rapidly take advantage of the available open habitats. The immigration rates slow as the number of species increases and sites become occupied. The extinction rates (green and gold curves) increase with the number of species on the island; the more species on an island, the greater the likelihood that a species will go extinct at any time interval. Colonization rates will be highest for islands near a mainland population source, since species can disperse over shorter distances more easily than longer ones. Extinction rates are highest on small islands, where both population sizes and habitat diversity are low. The number of species present on an island reaches equilibrium when the colonization rate equals the extinction rate (circles). The equilibrium number of species is greatest on large islands near the mainland and lowest on small islands far from the mainland. (After McArthur and Wilson 1967.)

ganisms and are more likely to have open habitat suitable for colonization. Extinction rates will be lower on large islands than on small islands because large islands will have greater habitat diversity and greater numbers of populations. Furthermore, rates of immigration of new species will be higher for islands near the mainland than for islands farther away, since mainland species will be able to disperse to near islands more easily than to distant islands. The model predicts that for any group of organisms, such as birds or orchids, the number of species found on a large island near a continent will be greater than that on a small island far from a continent.

Extinction rates and habitat loss

In addition to describing species diversity in natural systems, species–area relationships have been used to predict the number and percentage of species that would become extinct if habitats were destroyed (Rompré et al. 2009). The calculation assumes that reducing the area of natural habitat on an island would effectively result in a smaller island that would support fewer species. This model has great utility because it can be extended to nature reserves or other natural areas on the mainland that are surrounded by damaged habitat (Chittaro et al. 2010). The reserves can be viewed as "habitat islands" in a "sea" of unsuitable habitat. The model predicts that when 50% of an island (or habitat island) is destroyed, approximately 10% of the species occurring on the island will be eliminated (**FIGURE 5.8**). If these species are **endemic** to that island—that is, if they occur there and nowhere else—they will become extinct. When 90% of the habitat is destroyed, 50% of the species will be

FIGURE 5.8 According to the island biogeography model, the number of species present in an area increases asymptotically—that is, it rises sharply and then levels off, as shown by the red curve in this example. The shape of the curve differs from region to region and among different species groups, but this model gives a general indication of the relationship between habitat loss and species loss. Here, if the area of habitat is reduced by 50%, then 10% of the species in the group will be expected to disappear; if the habitat is reduced by 90%, half the species will be lost. Stating this in another way, a system of protected areas covering 10% of a country could be expected to include 50% of the country's species.

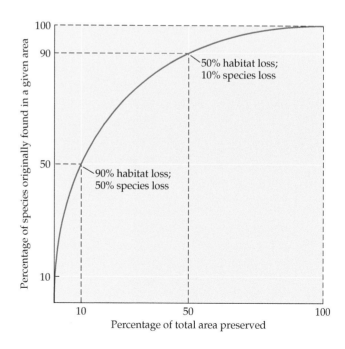

lost; and when 99% of the habitat is gone, about 75% of the original species will be lost. The island of Singapore can be used as an example. Over the last 180 years, over 99% of its original forest cover has been removed; the model estimates that about 70% of its forest species should have been lost. In fact, between 1923 and 1998, more than 90% of Singapore's native birds were lost, with higher rates of loss for large ground birds and for insectivorous birds of the forest canopy (Castelletta et al. 2000) (**FIGURE 5.9**).

Predictions of extinction rates based on habitat loss vary considerably because each species–area relationship is unique. Using the conservative estimate that 1% of the world's rain forest is being destroyed each year, Wilson (1989) estimated that 0.2%–0.3% of all rain forest species will be lost per year. Assuming a total of 5 million rain forest species worldwide, that would result in a loss of 10,000 to 15,000 species per year, or 34 species per day.

FIGURE 5.9   The banded kingfisher (*Lacedo pulchella*) is locally extinct in Singapore. (© PanuRuangjan/Getty Images.)

This estimate predicts that species extinctions by 2050 will be up to 35% in tropical Africa, 20% in tropical Asia, 15% in the Neotropics, and 8%–10% elsewhere (MEA 2005). Extinction rates might in fact be higher because the highest rates of deforestation are occurring in countries with large concentrations of rare species, and because large forest areas that remain are increasingly being fragmented by roads and development projects (Arima et al. 2015; Walker et al. 2013).

And yet, these predictions of high extinction rates do not seem to be coming true in the best-studied groups of birds and mammals. For example, there are 17 large areas of the world, including the Amazon basin, the trans-Himalayan region, northern Canada, and the island of New Guinea, where there have been no recorded extinctions of vertebrates in the last 200 years (Sanjayan et al. 2012). Extinction rates may be lower than predicted in these areas because threatened species are being given extra protection, and perhaps because hotspots particularly rich in endemic species are being targeted for conservation. Lower extinction rates could also be observed because species that are committed to extinction are able to persist at low numbers for several generations. Regardless of which estimate gives the most reasonable prediction, all estimates indicate that tens of thousands—if not hundreds of thousands—of species are headed for extinction within the next 50 years (McCallum 2015; Pimm et al 2015).

The time required for a given species to go extinct following a reduction in area or fragmentation of its range is a vital question in conservation biology, and the island biogeography model makes no prediction about how long it will take. Small populations of some species may persist for decades or even centuries in habitat fragments, even though they are committed to extinction (Sharma et al. 2014). Of the species that will eventually be lost, the best estimates predict that half will be lost in 50 years from a 1000-ha fragment, while half will be lost in 100 years from a 10,000-ha fragment (Brooks et al. 1999). Certain forest mammals in Australia have an expected persistence time of less than 10 years in small habitat fragments under 10 ha, 50 years in 40–80-ha fragments, and 100 years in 300-ha fragments (Laurance et al. 2008b). In situations in which there is widespread habitat destruction followed by recovery, such as in New England and Puerto Rico over the last several centuries, species may be able to survive in small numbers in isolated fragments and then reoccupy adjacent recovering habitat. Even though 98% of the forests of eastern North America were cut down, the clearing took place in a patchwork fashion over hundreds of years, so forest always covered half of the area, providing refuges for mobile animal species such as birds.

This ability of species to persist for several generations and several decades in habitat fragments may also be why there have not been more observed species losses. This conclusion has two important implications for conservation. First, many species will go extinct in coming decades as their populations continue to decline in these fragmented habitats. And second, the persistence of species in fragmented habitats provides a narrow window in which conservation actions have the potential to rescue declining species from extinction, as described in later chapters.

Vulnerability to Extinction

Populations (and ultimately species) that are declining in number are likely to go extinct unless the cause of decline is identified and corrected (Martin et al. 2012; Peery et al. 2004). As Charles Darwin pointed out more than 150 years ago in *On the Origin of Species* (1859):

> *To admit that species generally become rare before they become extinct, to feel no surprise at the rarity of the species, and yet to marvel greatly when the species ceases to exist, is much the same as to admit that sickness in the individual is the forerunner of death—to feel no surprise at sickness, but when the sick man dies, to wonder and to suspect that he died of some deed of violence.*

Past decline has been found to be the number one predictor of future decline (Di Marco et al. 2015). However, once a population is actively declining, it may be too late to prevent its extinction (see "The extinction vortex" on p. 188). Thus, conservation biologists have sought to determine what features of species or populations might be predictive of future decline or extinction. Some groups of species clearly are more vulnerable to extinction than others. In Europe, a higher proportion of reptile species are threatened with extinction than plants, birds, mammals, or fishes (Dullinger et al. 2013). Ecologists have observed that across taxonomic lines, there are features that increase a species' risk of extinction, and through statistical modeling, they have identified those features that are most predictive of extinction (Di Marco et al. 2015; Purvis et al. 2000):

- *Narrow geographic range.* This feature, which defines endemic species, has been found to be the most predictive of extinction or population decline (Botts et al. 2013; Di Marco et al. 2015). This conclusion is intuitive: if the whole range is affected by human activity or a natural disaster, the species may become extinct (Hanna and Cardillo 2013). Bird species on oceanic islands and fish species confined to a single lake or watershed are good examples of species with limited ranges that are especially vulnerable to global climate change. Many tropical bird species with narrow ranges will face increasing threats of extinction due to climate change in the coming decades (Şekercioğlu et al. 2012). Furthermore, a narrow range often (but not always) encompasses only one or a few populations or a small population size.

- *Only one or a few populations.* Any one population of a species may become extinct as a result of chance factors, such as earthquake, fire, an outbreak of disease, or human activity. Species with many populations are less vulnerable to extinction than are species with only one or a few populations. This feature is linked to the previous feature because species with few populations also tend to have narrow geographic ranges (**FIGURE 5.10**).

> Species most vulnerable to extinction have the following characteristics: narrow geographic range, only one or a few populations, small populations, declining population size, and being hunted or harvested by people.

- *Small population size.* As we will see in the next section, small populations are more likely to go locally extinct than large populations because of their greater vulnerability to loss of genetic diversity and to demographic and environmental variation. Species that characteristically have small population sizes, such as large predators and extreme specialists, are more likely to become extinct than species that typically have large populations (Bulman et al. 2007). At the extreme are species whose numbers have declined to just a few individuals. A special category is species with a widely fluctuating population size in which the population is sometimes small.

- *Island habitat.* As mentioned above, the highest species extinction rates during historic times have occurred on islands. This is not surprising, given that species on islands often have limited range areas, small population sizes, and small numbers of populations and are more likely to be endemic (Régnier et al. 2009). Of the terrestrial animal and plant species known to have gone extinct from 1600 to the present, almost half were island species, even though islands represent only a tiny fraction of the Earth's land surface. Island species usually have evolved and undergone speciation with a limited number of competitors, predators, and pathogens, which makes them particularly vulner-

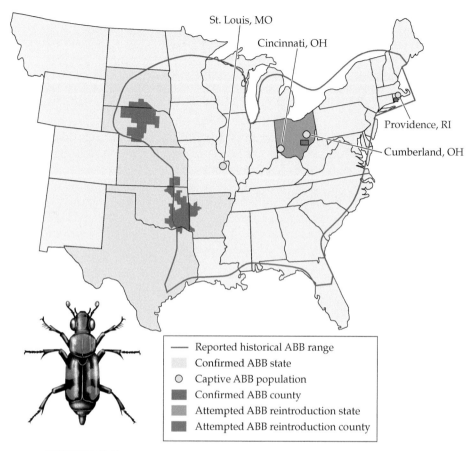

St. Louis, MO
Cincinnati, OH
Providence, RI
Cumberland, OH

— Reported historical ABB range
  Confirmed ABB state
○ Captive ABB population
■ Confirmed ABB county
  Attempted ABB reintroduction state
■ Attempted ABB reintroduction county

FIGURE 5.10 Although the American burying beetle (*Nicrophorus america-nus*) was once widespread in the eastern and central United States (outlined in red), its range is now greatly reduced, which makes it vulnerable to extinction. It is found in the wild in only two separate areas of the central United States and on Block Island in Rhode Island. Intensive efforts have been initiated to determine the cause of this decline and develop a recovery plan. The species is also being bred in captivity. (After O'Meilla 2004, with updates from the US Fish and Wildlife Service.)

able to these threats if they are introduced. Species extinction rates peak soon after humans occupy an island and then decline after the most vulnerable species are eliminated. In general, the longer an island has been occupied by people, the greater its percentage of extinct biota. In Madagascar, 72% of the 9000 plant species are endemic, and 189 species are threatened with extinction. The lemurs are also endemic to Madagascar, and most species of these unique primates are threatened (Schwitzer et al. 2014) (**FIGURE 5.11**).

- *Hunting or harvesting by people.* Overharvesting can rapidly reduce the population size of a species. If hunting and harvesting are not regulated, either by law or by local customs, the species can be driven to extinction. Utility has often been the prelude to extinction, as was the case with the dodo, now a symbol of modern extinctions.

Some additional features of species or populations have been found to put them at risk:

- *Large home range.* A species in which individuals or social groups need to forage over wide areas is prone to die off when part of its range is damaged or fragmented by human activity.

- *Large body size.* Large animals tend to have large individual ranges, have low reproductive rates, require more food, and be hunted by humans. Top carnivores, especially, are often killed by humans because they compete with humans for wild game, sometimes damage livestock, and are hunted for sport. Larger animals are generally more vulnerable to the effects of habitat fragmentation (Riverá Ortíz et al. 2014). Within groups of species, the largest species is often the most prone to extinction; that is, the largest carnivore, the largest lemur, the largest whale will go extinct first.

- *Slow reproduction.* In some analyses, a low number of offspring in each reproductive event was one of the most predictive features of a species becoming endangered (Purvis et al. 2000). This characteristic often overlaps with large body size: elephants, whales, giraffes, rhinos, and gorillas all usually have only one or two young at a time. Other traits associated with slow reproduction are slow growth rates, late sexual maturity, long gestation periods, and long intervals between reproductive events. These traits mean that a species cannot compensate easily for increased mortality rates (McArthur and Wilson 1967).

- *Limited dispersal ability.* Species unable to adapt physiologically, genetically, or behaviorally to changing environments must either migrate to more suitable habitat or face extinction. The rapid pace of anthropogenic changes often precludes evolutionary adaptation, leaving migration as a species' only alternative. Terrestrial species that are unable to cross roads, farmland, or disturbed habitats are doomed to extinction as their original habitat deteriorates. Dispersal is important in the aquatic environment as well,

FIGURE 5.11 More than 30 species of lemurs are all endemic to Madagascar, a large island off the eastern coast of Africa. Their endemism, that they live on an island, their small population sizes and numbers of populations, and even the fact that they are primates (a group with many endangered species), puts them at higher anticipated risk of extinction. Shown here is a Coquerel's sifaka and its young (*Propithecus coquereli*). (© Mint Images/ Frans Lanting/Getty Images.)

where dams, point sources of pollution, channelization, and sedimentation can limit movement. Limited dispersal ability may explain in part why freshwater fauna such as mussels and snails are more likely to be extinct or threatened with extinction than are dragonfly species, which are strong fliers.

- *Seasonal migration.* Species that migrate seasonally depend on two or more distinct habitat types. If either one of those habitat types is damaged, the species may be unable to persist. The billion songbirds of 120 species that migrate each year between the northern United States and the Neotropics depend on suitable habitat in both locations to survive and breed. In addition, if barriers to dispersal are created by roads, fences, or dams between the needed habitats, a species may be unable to complete its life cycle (Wilcove and Wikelski 2008). Salmon species that are blocked by dams from swimming up rivers and spawning are striking examples of this problem (**FIGURE 5.12**).

- *Little genetic variation.* Genetic variation within a population can sometimes allow a species to adapt to a changing environment (Forsman 2014); (see the section "Loss of genetic diversity" on p. 172). Species with little or no genetic variation may have a greater tendency to become extinct when a new disease, a new predator, or some other factor alters their environment. However, one study determined that for mammals, genetic variation, while a contributing factor, was not as important as other, ecological factors, such as body size, for extinction risk (Polishchuk et al. 2015).

- *Specialized niche requirements.* Once a habitat has been altered, it may no longer be suitable for specialized species (Tingley et al. 2013). For

FIGURE 5.12   Aggregations of salmon (*Oncorhynchus* spp.) migrate upstream to spawning pools. Impediments to their movement (such as dams), or anything else (such as harvesting) that threatens these aggregations, will severely threaten populations. (© Thomas Kline/Design Pics/Getty Images.)

example, wetland plants that require very specific and regular changes in water level may be rapidly eliminated when human activity affects the hydrology of an area. Species with highly specific dietary requirements are also at risk. For instance, there are species of mites that feed only on the feathers of a single bird species. If the bird species goes extinct, so do its associated feather mite species.

- *Low tolerance for disturbance.* Many species are adapted to stable, pristine environments where disturbance is minimal, such as old stands of tropical rain forests or the interiors of rich temperate deciduous forests. When these forests are logged, grazed, burned, or otherwise altered, these species are unable to tolerate the changed microclimatic conditions (more light, less moisture, greater temperature variation) or an influx of invasive species.

- *Permanent or temporary aggregations.* Species that group together in specific places are highly vulnerable to local extinction (Reed 1999) (see Figure 5.12). Herds of bison, flocks of passenger pigeons, and schools of spawning ocean fish all represent aggregations that have been exploited and overharvested by people. Many species of social animals may be unable to persist when population size or density falls below a certain number because they may be unable to forage, find mates, or defend themselves.

- *No prior contact with people.* Species that have experienced prior human disturbance and persisted have a lower current extinction risk than species encountering people—along with the nonnative species associated with them—for the first time (Balmford 1996).

- *Close relatives that are recently extinct or threatened with extinction.* Some groups of species, such as primates, cranes, sea turtles, and orchids, are particularly vulnerable to extinction. The characteristics that make certain species in these groups vulnerable are often shared by related species.

These characteristics of extinction-prone species are not independent; rather, they tend to group together. For example, many orchid species have specialized habitat requirements, have specialized relationships with pollinators, and are overharvested by collectors; all of those characteristics lead to small, declining populations and eventually to extinction. A high percentage of seabirds are also in danger of extinction because they have low reproductive rates; they form dense breeding aggregations, often in small areas, where their eggs and nestlings are prone to attack by introduced predators; they are killed by oil pollution and as bycatch during commercial fishing operations; and their eggs are overharvested by people (Munilla et al. 2007).

By using these features to identify extinction-prone species, conservation biologists can anticipate the need for managing their populations. Those species that are most vulnerable to extinction may have the full range of characteristics, like the animal David Ehrenfeld (1970) imagined:

... a large predator with a narrow habitat tolerance, long gestation period, and few young per litter [that is] hunted for a natural product and/or for sport, but is not subject to efficient game management. It has a restricted distribution but travels across international boundaries. It is intolerant of man, reproduces in aggregates, and has nonadaptive behavioral idiosyncrasies.

There is another great gap in our knowledge of threatened species: most threatened species that have been identified so far are members of the best-studied groups, such as birds and mammals, highlighting the point that only when we are knowledgeable about a species can we recognize the dangers it faces. Other groups of species, such as beetles, ocean fish, and fungi, are much less well known. A lack of knowledge about these groups should not be taken to mean that the species are not threatened with extinction; rather, it should be seen as an argument for studying those species. The conservation status of amphibians, for example, was relatively unknown until 15 years ago, when intensive study revealed that a high proportion of species were in danger of extinction.

Problems of Small Populations

As mentioned above, species with small populations are in increased danger of going extinct. Small populations are subject to rapid decline in numbers and local extinction for three main reasons:

1. Loss of genetic diversity and related problems of inbreeding depression and genetic drift
2. Demographic fluctuations due to random variations in birth and death rates
3. Environmental fluctuations due to variation in predation, competition, disease, and food supply as well as natural catastrophes that occur at irregular intervals, such as fires, floods, storms, or droughts

We'll now examine in detail each of these causes for decline in small populations.

Loss of genetic diversity

As described in Chapter 2, a population's ability to adapt to a changing environment depends on genetic variation, which occurs as a result of different individuals having different alleles of the same gene. Individuals with certain alleles or combinations of alleles may have just the characteristics needed to survive and reproduce under new conditions (Allendorf and Luikart 2007; Frankham 2005). Within a population, the frequency of a given allele can range from common to very rare. New alleles arise in a population either by random mutations or through the migration of individuals from other populations. Small populations often have very low genetic diversity, which compromises their ability to respond to environmental changes. Species of yew trees (*Taxus* spp.) in the central Himalayan region between Nepal and China have been reduced to small populations that have such low genetic variation as to be considered unsustainable (Poudel et al. 2014).

In small populations, allele frequencies may change significantly from one generation to the next simply because of chance—based on which individuals happen to survive to sexual maturity, mate, and leave offspring. This random process of allele frequency change, known as **genetic drift**, is a separate process from changes in allele frequency caused by natural selection (Hedrick 2005). When an allele occurs at a low frequency in the gene pool of a small population, it has a significant probability of being lost in each generation (**FIGURE 5.13**). For example, if a rare allele occurs in 5% of all the gene copies present in a population of 1000 individuals, then 100 copies of the allele are present (1000 individuals × 2 gene copies per individual × 0.05 allele frequency), and the allele will probably remain in the population for many generations. However, in a population of 10 individuals, only 1 copy of the allele is present (10 individuals × 2 gene copies per individual × 0.05 allele frequency), and it is possible that the allele will be lost from the population by chance in the next generation.

A problem related to the loss of alleles is the reduction of heterozygosity—that is, the proportion of individuals with two different alleles of a

> Because of genetic drift, small populations lose genetic variation more rapidly than large populations. Some small populations may lack any genetic variation.

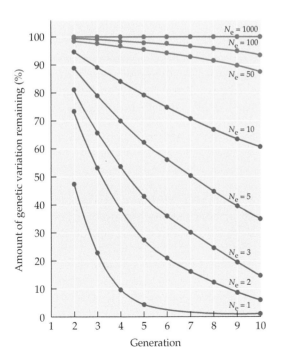

FIGURE 5.13 The rate at which genetic variation is lost through genetic drift varies with population size. This graph shows the average percentage of genetic variation remaining after 10 generations in theoretical populations of various effective population sizes (N_e). After 10 generations, a population with a size of 10 loses approximately 40% of its genetic variation, a population of 5 loses 65%, and a population of 2 loses 95%. Blue lines indicate large populations; red lines indicate small populations. (After Groom et al 2006.)

particular gene. Considering the general case of an isolated population in which there are two alleles of each gene in the gene pool, Wright (1931) proposed a formula to express the proportion of the original heterozygosity (H_0; which in this case is 1) remaining after each generation (H_1). The formula includes the **effective population size** (N_e)—the size of the population as estimated by the number of its breeding individuals[2]:

$$H_1 = [1 - 1/(2N_e)]\, H_0$$

According to this equation, a population of 50 breeding individuals would retain 99% of its original heterozygosity after 1 generation:

$$H_1 = (1 - 1/100)\, 1 = 1.00 - 0.01 = 0.99$$

The proportion of the original heterozygosity remaining after t generations (H_t/H_0) decreases over time, and the decrease is greater for smaller populations[3]:

$$H_t/H_0 = (H_1)^t$$

For our population of 50 individuals, then, the remaining heterozygosity would be 98% after 2 generations (0.99×0.99), 97% after 3 generations ($0.99 \times 0.99 \times 0.99$), and 90% after 10 generations (0.99^{10}). However, a population of 10 individuals would retain only 95% of its original heterozygosity after 1 generation, 90% after 2 generations, 86% after 3 generations, and 60% after 10 generations. Loss of heterozygosity has been directly linked to extinction risk (**FIGURE 5.14**).

  This formula demonstrates that significant losses of genetic variation can occur in isolated small populations. Such small populations are often found on islands and in fragmented landscapes (Vranckx et al. 2012). However, the amount of genetic variation within a small population can increase over time through two means: regular mutation of genes and migration of even a few individuals from other populations. Mutation rates found in nature vary between 1 in 10,000 and 1 in 1 million per gene per generation; mutations may therefore make up for genetic drift in large populations and, to a lesser extent, in small populations. However, mutations alone are not sufficient to counter genetic drift in populations of 100 individuals or fewer. Fortunately, even a low frequency of movement of individuals between populations minimizes the loss of genetic variation associated with small population size (Weiser et al. 2013). If even 1 or 2 immigrants arrive each generation in an isolated population of about 100 individuals, the effect of genetic drift will be greatly reduced. With 4–10 immigrants arriving per generation, the effects of genetic drift are negligible. Gene flow from neighboring populations appears to be the

---

[2]Factors that affect N_e, the effective population size, are discussed in detail beginning on p. 180.

[3]This is a simplification of the following equation: $H_t/H_0 = [1 - 1/(2N_e)]^t$

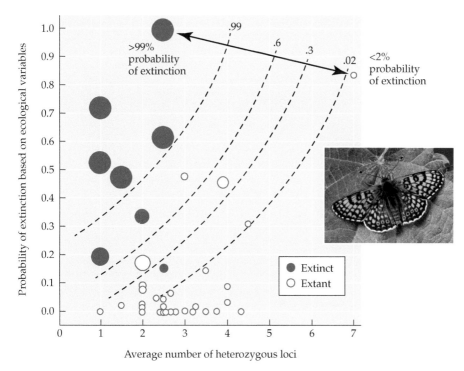

FIGURE 5.14   Whether or not a population of the Glanville fritillary butterfly (*Melitaea cinxia*) went extinct was a function of both ecological variables and the average number of heterozygous loci; the more heterozygosity in a population, the lower its likelihood of extinction (as shown by size of the circle). Populations that had low heterozygosity and were under high environmental pressure (higher numbers on the *y*-axis) had a 100% probability of extinction, but populations subjected to high environmental pressure but which also had high heterozygosity had a low probability of extinction (as shown by the arrow). (After Saccheri et al. 1998; photograph © Andrew Darrington/Alamy Stock Photo.)

major factor preventing the loss of genetic diversity in small populations of Galápagos finches (Grant and Grant 2008). In addition, genetic variation that increases fitness will tend to be retained longer in a population, even when there is genetic drift.

Field data confirm that a lower effective population size leads to a more rapid loss of alleles from a population (Evans and Sheldon 2008). For example, in a broad survey of 89 bird species, there was a strong tendency for abundant birds to have more heterozygosity than species with small populations. Almost all species with populations over 10 million have over 60% heterozygosity, in contrast to most species with fewer than 100,000 individuals, which have less than 60% heterozygosity.

Unfortunately, rare and endangered species often have small, isolated populations and thus suffer rapid losses of genetic diversity. In 170 paired comparisons, threatened taxa with narrow ranges had an average of 35% lower genetic diversity than taxonomically related nonthreatened species with wide distributions (Spielman et al. 2004). In some cases, entire species lacked genetic variation. In the evolutionarily isolated Wollemi pine (*Wollemia nobilis*) of Australia, only 40 plants occur in two nearby populations. As might be predicted, an extensive investigation failed to find any genetic variation in this species (Peakall et al. 2003).

Consequences of reduced genetic diversity

Small populations subjected to genetic drift are susceptible to a number of deleterious genetic effects, such as inbreeding depression, outbreeding depression, and loss of evolutionary flexibility. These effects may contribute to a decline in population size, leading to an even greater loss of genetic diversity, a loss of fitness, and a greater probability of extinction (Frankham et al. 2009).

INBREEDING DEPRESSION   Matings between parents and their offspring, siblings, or cousins and self-fertilization in hermaphroditic species may result in **inbreeding depression**, a condition that occurs when an individual receives two identical copies of a defective allele from each of its parents. Inbreeding depression is characterized by higher mortality of offspring, fewer offspring, or offspring that are weak or sterile or have low mating success (Frankham et al. 2009) (**FIGURE 5.15**). These factors result in even fewer individuals in the next generation, leading to more pronounced inbreeding depression.

Evidence for the existence of inbreeding depression comes from studies of human populations (in which there are records of marriages between close relatives for many generations), captive and wild animal populations, and cultivated plants (Frankham et al. 2014). In a wide range of captive mammal populations, matings among close relatives resulted, on average, in offspring with a 33% higher mortality rate than in non-inbred animals. This lower fitness resulting from inbreeding is sometimes referred to as a "cost of inbreeding." Inbreeding depression can be a severe problem in small captive populations in zoos and in domestic livestock breeding programs. Deleterious effects of inbreeding in the wild have also been demonstrated (Crnokrak and Roff 1999): of over 150 studies, 90% showed inbreeding to be detrimental. The scarlet gilia (*Ipomopsis aggregata*) provides an example: plants that come from populations with fewer than 100 individuals produce smaller seeds with a lower rate of germination and exhibit greater susceptibility to environmental stress than do plants from larger populations (Heschel and Paige 1995). In another example, genetic analysis of lions from two protected areas in Nigeria revealed evidence of inbreeding and a lack of gene flow between the populations; in this case, trapping and moving lions as a conservation approach may be necessary to avoid the expected deleterious effects (Tende et al. 2014).

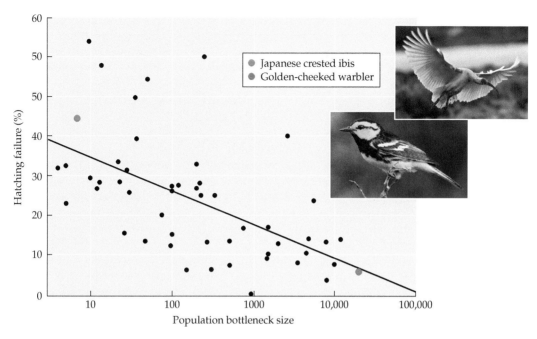

FIGURE 5.15 Among 51 species of birds, the rate of hatching failure is highest for those with the smallest population sizes (as expressed by the bottleneck population size—the lowest size recorded for the population). Hatching failure is a function of inbreeding depression. The *x*-axis is on a log scale. The Japanese crested ibis (*Nipponia nippon*) (top photograph) represents one extreme, with fewer than 10 individuals in one year and hatching failure of about 45%; at the other extreme is the golden-cheeked warbler (*Dendroica chrysoparia*) (bottom photograph), with a population size always over 10,000 individuals and hatching failure of about 5%. (After Heber and Briskie 2010; Japanese crested ibis photograph © Jed Weingarten/Getty Images; golden-cheeked warbler photograph © Rolf Nussbaumer Photography/Alamy.)

OUTBREEDING DEPRESSION   Individuals of different species rarely mate in the wild; there are strong ecological, behavioral, physiological, and morphological isolating mechanisms that ensure that mating occurs only between individuals of the same species. However, when a species is rare or its habitat is damaged, outbreeding—mating between individuals of different populations or species—may occur. Individuals unable to find mates of their own species may mate with individuals of related species. The resulting offspring sometimes exhibit outbreeding depression, a condition that results in weakness, sterility, or lack of adaptability to the environment (Waller 2015). Outbreeding depression may be caused by incompatibility of the chromosomes and enzyme systems that are inherited from the different parents. To use an example from artificial selection, domestic horses and donkeys are commonly bred to produce mules. Although mules are not physically weak (on the contrary, they are quite strong, which is why humans find them useful), they are almost always sterile.

Outbreeding depression can also result from matings between different subspecies, or even matings between divergent populations of the same species. Such matings might occur in a captive breeding program or when individuals from different populations are kept together in captivity. In such cases, the offspring of parents with such different genotypes are unlikely to have the precise mixture of genes that allows individuals to survive and reproduce successfully in a particular set of local conditions (Frankham et al. 2009). For example, when the ibex (*Capra ibex*) population of Slovakia went extinct, ibex from Austria, Turkey, and the Sinai were brought in to start a new population. These different subspecies mated and produced hybrids that bore their young in the harsh conditions of winter rather than in the spring, and consequently their offspring had a low survival rate (Templeton 1986). Outbreeding depression caused by the pairing of individuals from the extremes of the species' geographic range meant failure for the experiment. However, many other studies of animals have failed to demonstrate outbreeding depression, or have even shown that some hybrids are more vigorous than their parent species (McClelland and Naish 2007), a condition known as *hybrid vigor*. Thus, outbreeding depression may be less of a concern for animals than inbreeding depression, the negative effects of which are well documented.

Outbreeding depression may be considerably more significant in plants, in which the arrival of pollen on the receptive stigma of a flower is to some degree a matter of the chance movement of pollen by wind or animal vectors. A rare plant species growing near a closely related common species may be overwhelmed by the pollen of the common species and fail to produce seeds (Willi et al. 2007). Even when hybrids are produced by matings between a common and a rare species, the genetic identity of the rare species is lost as its small gene pool is mixed into the much larger gene pool of the common species. The seriousness of this threat is illustrated by the fact that more than 90% of California's threatened and endangered plants occur in close proximity to other species in the same genus with which the rare plants could possibly hybridize. Such losses of identity can also take place in gardens when individuals from different parts of a species' range are grown next to one another.

LOSS OF EVOLUTIONARY FLEXIBILITY   Genetic diversity is extremely important to a species' long-term survival. Rare alleles and unusual combinations of alleles that are harmless (or even slightly harmful) but confer no immediate advantage on the few individuals who carry them may turn out to be uniquely suited for a future set of environmental conditions. If such alleles and combinations do become advantageous, their frequency in the population will increase rapidly through natural selection because the individuals who carry them will be those most likely to survive and reproduce successfully, passing on the formerly rare alleles to their offspring.

Loss of genetic diversity in a small population may limit its ability to respond to new conditions and long-term changes in the environment, such as pollution, new diseases, or global climate change (Willi et al. 2007). According to the *fundamental theorem of natural selection*, the rate of evolutionary change in a population is directly related to the amount of genetic variation in that population. A small population is less likely than a large population to possess the genetic variation necessary for adaptation to environmental changes and is therefore more likely to go extinct. In many plant populations, for example, a few individuals have alleles that promote tolerance for high concentrations of toxic metals such as zinc and lead, even when these metals are not present. If toxic metals become abundant in the environment because of pollution, individuals with these alleles will be better able to adapt to them and will grow, survive, and reproduce better than typical individuals; consequently, the frequency of these alleles in the population will increase dramatically. However, if the population has become small and the genotypes for metal tolerance have been lost, the population could go extinct.

Factors that determine effective population size

Earlier in this section, we mentioned the effective population size, which is the size of a population as estimated by the number of its breeding individuals. The effective population size is lower than the total population size because many individuals do not reproduce, due to factors such as inability to find a mate, being too old or too young to mate, poor health, sterility, malnutrition, and small body size (Hare et al. 2011). Many of these factors are initiated or aggravated by habitat degradation and fragmentation. Furthermore, many plant, fungus, bacteria, and protist species have seeds, spores, or other structures that remain dormant in the soil unless stable conditions for germination appear. These individuals could be counted as members of the population though they are obviously not part of the breeding population.

> The effective population size, N_e, will be much smaller than the total population size, N, when there is a small proportion of individuals reproducing, great variation in reproductive output, an unequal sex ratio, or wide fluctuations in population size.

Because of these factors, the effective size of a population (N_e) is often substantially smaller than its actual size (N). Because the rate of loss of genetic diversity depends on effective population size, loss of genetic diversity can be quite severe even in a large population. For example, consider a population of 1000 alligators consisting of 990 immature animals and only 10 mature breeding animals: 5 males and 5 females. The effective size of this population is 10, not 1000. In a population of a rare oak species, there might be 20 mature trees, 500 saplings, and 2000 seedlings, resulting in a population size of 2520, but an effective population size of only 20.

In addition, the effective population size is often even lower than the actual number of breeding individuals because of an unequal sex ratio, variation in reproductive output, or large annual changes in population size (Jamieson 2011).

UNEQUAL SEX RATIO   A population may consist of unequal numbers of males and females due to chance, selective mortality, or the harvesting of one sex by people. If, for example, a population of a goose species that is monogamous (in which one male and one female form a long-lasting pair bond) consists of 20 males and 6 females, then only 12 individuals—6 males and 6 females—will be mating. In this case, the effective population size is 12, not 26. In other animal species, social systems may prevent many individuals from mating even though they are physiologically capable of doing so. Among elephant seals, many ungulates, and many primates, for example, a single dominant male usually mates with a large number of females and prevents other males from mating with them (**FIGURE 5.16**), whereas among African wild dogs and hyenas, the dominant female in the pack often bears all of the pups.

The effect of unequal numbers of breeding males and females on N_e can be described by this formula:

$$N_e = [4(N_f N_m)]/(N_f + N_m)$$

where N_f and N_m are the numbers of adult breeding females and adult breeding males, respectively, in the population. In general, as the sex ratio of breeding individuals becomes increasingly unequal, the ratio of the effective population size to the number of breeding individuals (N_e/N) goes down. This occurs because only a few individuals of one sex are making a disproportionately large contribution to the genetic makeup of the next generation. In the case of Asian elephants (*Elephas maximus*), for example, males are hunted by poachers for their tusks. At the Periyar Tiger Reserve in India in 1997, there were 1166 elephants, of which 709 were adults. Of these adults, 704 were female and 5 were male (Ramakrishnan et al. 1998).

FIGURE 5.16 Two large male elk compete to mate with the many smaller females surrounding them. The effective population size is reduced because only one male is providing genetic input to many females. (© Lee Foster/Alamy Stock Photo.)

If all of the adults were breeding, this sex ratio would result in an effective population size of only 20.

In many fish and reptile species, sex is affected by temperature. As global climate change increases water and air temperatures in many places, the sex ratios of these species may become skewed, lowering their effective population sizes. In Switzerland, for example, grayling (*Thymallus thymallus*) populations that used to have highly variable sex ratios centered around 65% males before 1990 now consistently have 80%–90% males. The effective population size for this fish species will be far lower than the number of individuals in the population.

VARIATION IN REPRODUCTIVE OUTPUT   In many species, the number of offspring varies substantially among individuals. This variation is particularly pronounced in highly fecund species, such as plants and fishes (see Hedrick 2005), in which many or even most individuals produce a few offspring while others produce huge numbers. Unequal production of offspring leads to a substantial reduction in N_e because a few individuals in the present generation will be disproportionately represented in the gene pool of the next generation. In general, the greater the variation in reproductive output, the more the effective population size is lowered. For a variety of species in the wild, Frankham (1995) estimated that variation in offspring number reduces effective population size by a factor of 54%. In the many annual plant populations that consist of large numbers of tiny plants producing one or a few seeds and a few gigantic individuals producing thousands of seeds, N_e may be reduced even more.

POPULATION FLUCTUATIONS AND BOTTLENECKS   In some species, population size varies dramatically from generation to generation. Particularly good examples of this phenomenon are butterflies, annual plants, and amphibians. In a population with extreme size fluctuations, the effective population size is much nearer the lowest than the highest number of individuals, and it tends to be determined by the years in which the population has the smallest numbers.

The effective size of a fluctuating population can be calculated over a period of t years using the number of individuals (N) breeding in any one year:

$$N_e = t/(1/N_1 + 1/N_2 + \ldots + 1/N_t)$$

Consider a butterfly population, monitored for 5 years, that has 10, 20, 100, 20, and 10 breeding individuals in those successive 5 years. In this case,

$$N_e = 5/(1/10 + 1/20 + 1/100 + 1/20 + 1/10) = 5/(31/100) = 5(100/31) = 16.1$$

The effective population size over the course of 5 years is above the lowest population size (10), but well below the maximum (100) and the average (32) population size.

As these calculations suggest, a single year of drastically reduced population numbers will substantially lower the value of N_e. This principle applies to a phenomenon known as a **population bottleneck**, which occurs when a population is greatly reduced in size. Such a population will lose rare alleles if no individuals possessing those alleles survive and reproduce (Jamieson 2011). With fewer alleles present and a decline in heterozygosity, the overall fitness of the individuals in the population may decline.

A special category of population bottleneck, known as the **founder effect**, occurs when a few individuals leave one population and establish a new population. The new population often has less genetic diversity than the larger, original population. For example, the wolf population in Sweden was established by 5 individuals (Laikre et al. 2013). Similarly, if a population is fragmented by human activities, each of the resulting small subpopulations may lose genetic variation and go extinct. Such is the fate of many fish populations fragmented by dams (Wofford et al. 2005). Bottlenecks can also occur when captive populations are established using relatively few individuals.

The lions (*Panthera leo*) of Ngorongoro Crater in Tanzania provide a well-studied example of a population bottleneck (Munson et al. 2008). The lion population in the crater consisted of 60–75 individuals until an outbreak of biting flies in 1962 reduced the adult population to 9 females and 1 male (**FIGURE 5.17**). Two years later, 7 additional males immigrated to the crater, and no new immigration was observed for the next 49 years. The dramatic reduction in population size, the isolation of the population, and variation in reproductive success among individuals have apparently created a population bottleneck, leading to inbreeding depression. In comparison with the large Serengeti lion population nearby, the crater lions showed reduced genetic diversity, high levels of sperm abnormalities, reduced reproductive rates, increased cub mortality, and higher rates of disease (Munson et al. 2008). After reaching a peak of 125 animals in the 1980s, the population declined again. By 2003, following an outbreak of canine distemper virus that had spread from domestic dogs kept by people living just outside the crater area, the population had dropped to 34 lions. For the next decade the population size stayed between 50 and 60 individuals, but in 2013 a coalition of four males entered the Crater. Genetic testing will confirm whether these lions are actually "fresh blood," but the fact that 16 of their 20 cubs have survived—a very high rate for this population—is a promising sign that these individuals introduced some new alleles; the current population size is 70 lions (updates from Packer 2016, pers. comm.).

Population bottlenecks do not always lead to greatly reduced heterozygosity, however. The effects of population bottlenecks are most evident when the breeding population remains below 10 individuals for several generations. If a population expands rapidly in size after a temporary bottleneck, heterozygosity may be restored even though the number of alleles present is severely reduced. An example of this phenomenon is

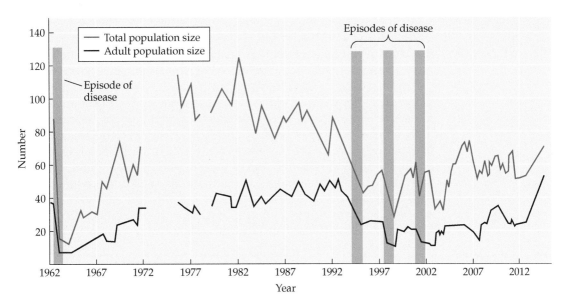

FIGURE 5.17   The Ngorongoro Crater lion population consisted of about 90 individuals in 1961 before crashing in 1962. Since that time, the population reached a peak of 125 individuals in 1983 before collapsing to 34 individuals (fewer than 20 of which were adults). Small population size, an isolated location, lack of immigration since 1964, and the impact of disease have contributed to the loss of genetic variation caused by a population bottleneck. A lack of census data for certain years is the cause of gaps in the lines. The four green bars represent episodes of disease outbreak. (After Munson et al. 2008, with updates from C. Packer.)

the high level of heterozygosity found in the greater one-horned rhinoceros (*Rhinoceros unicornis*) in Chitwan National Park, Nepal, even after the population passed through a bottleneck. Population size declined from 800 individuals to fewer than 100 individuals; fewer than 30 were breeding. With an effective population size of 30 individuals for one generation, the population would have lost only 1.7% of its heterozygosity. As a result of strict protection of the species by park guards, the population recovered to 400 individuals. The Mauritius kestrel (*Falco punctatus*), which represents an even more extreme case, experienced a long population decline that resulted in only one breeding pair remaining in 1974. An intensive conservation program has allowed the population to recover to about 300 adult birds today. A study comparing the present birds with preserved museum specimens and kestrel species living elsewhere has found that the Mauritius kestrel lost only about half of its genetic variation after passing through this bottleneck (Ewing et al. 2008).

MANAGING FOR GENETIC VARIATION   These examples demonstrate that effective population size is often substantially less than the total number of individuals in a population. Particularly where a combination of factors, such as fluctuating population size, numerous nonreproductive individuals, and an unequal sex ratio, are involved, the effective population size may be far lower than the number of individuals alive in a good year.

A review of a wide range of wildlife studies revealed that effective population size averaged only 11% of total population size; that is, a population of 300 animals, seemingly enough to maintain the population, might have an effective size of only 33, which would indicate that it was in serious danger of extinction (Frankham 2005). For highly fecund species, such as fishes, seaweeds, and many invertebrates, the effective population size may be less than 1% of the total population size (Frankham et al. 2009). Consequently, management aimed toward simply maintaining large populations may not prevent the loss of genetic variation unless the effective population size is also large. In the case of captive populations of rare and endangered species, genetic variation may be effectively maintained by controlling breeding, perhaps by subdividing the population, periodically removing dominant males to allow subdominant males the opportunity to mate, and periodically transporting a few selected individuals among subpopulations.

Other factors that affect the persistence of small populations

Random fluctuations in birth and death rates, disruption of social behavior by lowered population density, and environmental stochasticity all contribute to further decrease in the size of populations, often leading to local extinction.

Random variation, or **stochasticity**, in the biological or physical environment can cause variation in the population size of a species. For example, the population of an endangered butterfly species might be affected by fluctuations in the abundance of its food plants or its predators. Weather might also strongly influence the butterfly population: in an average year, the weather may be warm enough for caterpillars to feed and grow, whereas a cold year might cause many caterpillars to become inactive and consequently starve. Such environmental stochasticity affects all individuals in the population and is linked to demographic stochasticity (or demographic variation), which is variation in birth and death rates among individuals and across years within a given population.

Demographic stochasticity

In an ideal, stable environment, a population would increase until it reached the carrying capacity (K) of the environment, at which point the average birthrate (b) per individual would equal the average death rate (d) and there would be no net change in population size. In any real population, however, individuals do not usually produce the average number of offspring: they may leave no offspring, somewhat fewer than the average, or more than the average. For example, in an ideal, stable giant panda (*Ailuropoda melanoleuca*) population, each female would produce an average of two surviving offspring in her lifetime, but field studies show that rates of reproduction among individual females vary widely around that number. However, as long as the population size is large, the average birthrate provides an accurate description of the population. Similarly, the average death rate in a population can be determined only by examining large numbers of individuals because some individuals die young and other individuals live a relatively long time. But as long as the population size is large, the average death rate provides an accurate and relatively stable description of the population.

Population size may fluctuate over time because of changes in the environment or other factors without ever approaching a stable value. Random fluctuations upward in population size are eventually bounded by the carrying capacity of the environment, after which the population may fluctuate downward again. In general, once population size drops below about 50 individuals, individual variation in birth and death rates begins to cause the population size to fluctuate randomly up or down (Schleuning and Matthies 2009). Variation in population size due to random variation in reproduction and mortality rates is known as **demographic variation** or **demographic stochasticity**. If population size fluctuates downward in any one year because of a higher than average number of deaths or a lower than average number of births, the resulting smaller population will be even more susceptible to demographic fluctuations in subsequent years.

Consequently, once a population decreases because of habitat destruction and fragmentation, demographic variation becomes important, and the population has a higher probability of declining more and even going extinct due to chance alone (in a year with low reproduction and high mortality) (Melbourne and Hastings 2008). Species with highly variable birth and death rates, such as annual plants and short-lived insects, may be particularly susceptible to population extinction due to demographic stochasticity. The chance of extinction is also greater in species that have low birthrates, such as elephants, because these species take longer to recover from chance reductions in population size.

As a simple example, imagine a population of three hermaphroditic individuals; each lives for 1 year, attempts to find a mate and reproduce, and then dies. Assume that each individual has a 33% probability of producing zero offspring, one offspring, or two offspring, resulting in an average birthrate of 1 per individual; in this case, there is theoretically a stable population. When these individuals attempt to reproduce, however, there is a 1-in-27 chance ($0.33 \times 0.33 \times 0.33$) that no offspring will be produced in the next generation and the population will go extinct. Consider also that there is a 1-in-9 chance that only one offspring will be produced in the next generation ($0.33 \times 0.33 \times 0.33 \times 3$); because this individual will not be able to find a mate, the population will be doomed to extinction in the following generation. There is also a 22% chance that the population will decline to two individuals in the next generation. Thus, random variation in birthrates can lead to demographic stochasticity and extinction in small populations.

Similarly, random fluctuations in the death rate can lead to fluctuations in population size that could eliminate the population altogether.

When populations drop below a critical number, deviations from an equal sex ratio may occur, leading to a declining birthrate and a further decrease in population size. For example, imagine a population of four birds that includes two mating pairs of males and females, in which each female produces an average of two surviving offspring in her lifetime. There is a 1-in-8 chance that only male or only female birds will be produced in the next generation, in which case no eggs will be laid to produce the following generation. There is a 50% (8-in-16) chance that there will be either three males and one female or three females and one male in the next generation, in which case only one pair of birds will mate, and the population will decline. This scenario is illustrated by the now extinct dusky seaside sparrow (*Ammodramus maritimus nigrescens*); the last five individuals were males, so there was no opportunity to establish a captive breeding program. Such demographic effects are also seen in the Spanish imperial eagle (*Aquila adalberti*), in which only mature birds breed when the population is large, but immature birds are more likely to breed when the population is small. Such immature birds are more likely to produce predominantly male offspring, contributing to further population decline and increasing the probability of local extinction (Ferrer et al. 2009). For these eagles, a management strategy involving supplemental feedings was able to increase the population size, keep mature males from leaving the site, and restore the sex ratio.

Many small populations are demographically unstable because social interactions (especially mating) can be disrupted once population density falls below a certain level. This interaction among population size, population density, population growth rate, and behavior is sometimes referred to as the **Allee effect** (Allee 1931; Bonsall et al. 2014). Herds of grazing mammals and flocks of birds may be unable to find food and defend themselves against predators when their numbers fall below a certain level. Animals that hunt in packs, such as wild dogs and lions, may need a certain number of individuals to hunt effectively.

Perhaps the most significant aspect of the Allee effect for small populations involves reproductive behavior: many species that live in widely dispersed populations, such as bears, spiders, and tigers, have difficulty finding mates once the population density drops below a certain point (**FIGURE 5.18**). Even among plant species, as population size and density decrease, the distance between individual plants increases. Pollinating animals may fail to visit isolated, scattered plants, resulting in insufficient transfer of compatible pollen and a subsequent decline in seed production. In such cases, the birthrate will decline, population density will become lower yet, problems such as unequal sex ratio will worsen, and birthrates will drop even more. Once the birthrate falls to zero, extinction is guaranteed. Detecting and anticipating Allee effects are necessary for the management and recovery of endangered species.

FIGURE 5.18 Bears, such as this Kamchatka brown bear (*Ursus arctos beringianus*) in eastern Siberia, cannot reproduce when densities decrease below a certain number because they are typically solitary and may have difficulty finding each other to mate. (© Steve Winter/Getty Images.)

Environmental stochasticity and catastrophes

Random variation in the biological and physical environment, known as **environmental stochasticity**, can also cause variation in the size of a population. For example, a population of an endangered rabbit species might be affected by fluctuations in the population of a deer species that eats the same types of plants, in the population of a fox species that feeds on the rabbits, and in the populations of parasites and disease-causing organisms that affect the rabbits. Variation in the physical environment might also strongly influence the rabbit population: rainfall during an average year might encourage plant growth and allow the rabbit population to increase, while dry years might limit plant growth and cause rabbits to starve. Environmental stochasticity affects all individuals in the population, unlike demographic stochasticity, which causes variation among individuals within the population.

Natural catastrophes that occur at unpredictable intervals, such as droughts, storms, earthquakes, and fires, along with cyclical die-offs in the surrounding biological community, can cause dramatic fluctuations in population levels. Natural catastrophes can kill part of a population or even eliminate an entire population from an area. Numerous examples exist of die-offs in populations of large mammals; in many cases, 70%–90% of the population dies (Young 1994). For a wide range of vertebrates, the probability of catastrophes is about 15% per generation (Reed et al. 2003). Even though the probability of a natural catastrophe in any one year is low, over the course of decades and centuries, natural catastrophes—including extended periods of unseasonable weather, excessive or insufficient rainfall, and events such as hurricanes and earthquakes—have a high likelihood of occurring.

Modeling efforts by Menges (1992) and others have shown that random environmental variation is generally more important than random demographic variation in increasing the probability of extinction in populations of small to moderate size. Environmental variation can substantially increase the risk of extinction even in populations that showed positive population growth under the assumption of a stable environment (Mangel and Tier 1994). In general, introducing environmental variation into population models, in effect making them more realistic, results in populations with lower growth rates, lower population sizes, and higher probabilities of extinction.

Imagine a rabbit population of 100 individuals in which the average birthrate is 0.2 and an average of 20 rabbits are eaten each year by foxes. On average, the population will maintain its numbers at exactly 100 individuals, with 20 rabbits born each year and 20 rabbits eaten each year. However, if there are 3 successive years in which the foxes eat 40 rabbits per year, the population size will decline to 80 rabbits, 56 rabbits, and 27 rabbits in years 1, 2, and 3, respectively. If there are then 3 years of no fox predation, the rabbit population will increase to 32, 38, and 46 individuals in years 4, 5, and 6, respectively. Even though the same average rate of predation (20 rabbits per year) occurred over this 6-year period, variation in year-to-year predation rates caused the rabbit population size to decline by more than 50%. At a population size of 46 individuals, the rabbit population will probably go extinct within the next 5–10 years if it is subjected to the average rate of 20 rabbits eaten by foxes per year.

The interaction between initial population size and environmental variation was demonstrated using the biennial herb garlic mustard (*Alliaria petiolata*), an invasive plant in the United States, as an experimental subject (Drayton and Primack 1999). Populations of various sizes were assigned at random either to be left alone as controls or to be experimentally eradicated by removal of every flowering plant in each of the 4 years of the study; removal of all plants could be considered a natural catastrophe. Overall, the probability of an experimental population's going extinct over the 4-year period was 43% for small populations (fewer than 10 individuals initially), 9% for medium-sized populations (10–50 individuals), and 7% for large populations (more than 50 individuals). For control populations, the probability of going extinct for small, medium, and large populations was 11%, 0%, and 0%, respectively. Large numbers of dormant seeds in the soil apparently allowed most experimental populations to persist even when every flowering plant was removed in 4 successive years. However, small populations were far more susceptible to extinction than large populations.

The extinction vortex

The smaller a population becomes, the more vulnerable it is to the combined effects of low genetic diversity, demographic variation, and environmental stochasticity that tend to lower reproduction, increase mortality rates, and so reduce its size even more, driving the population to extinction (**FIGURE 5.19**). This tendency of small populations to decline toward extinction has been likened to a vortex, a whirling mass of gas or liquid spiraling inward: the closer an object gets to the center of the vortex, the faster

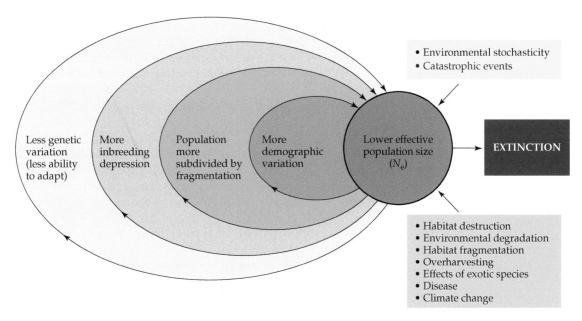

FIGURE 5.19 Once a population drops below a certain size, it enters an extinction vortex in which factors that affect small populations tend to drive its size progressively lower. This downward spiral often leads to the local extinction of species. (After Gilpin and Soulé 1986 and Guerrant 1992.)

it moves. At the center of an **extinction vortex** is oblivion: the local extinction of the species. Once caught in such a vortex, it is difficult for a species to resist the pull toward extinction (Palomares et al. 2012). Small populations often require a careful program of population and habitat management, as described in later chapters, to increase their growth rates and allow them to escape from the extinction vortex.

> Intensive management is often required to prevent small populations from declining further in size and going extinct.

Summary

- Many extant species are on the brink of extinction, with 21% of bird species, 27% of mammal species, and 36% of amphibian species believed to be threatened.

- Island species have had a higher rate of extinction than mainland species. Among aquatic species, freshwater species apparently have a higher extinction rate than marine species.

- Current rates of extinction are between 100 and 1000 times greater than background levels. More than 99% of modern species extinctions are attributable to human activity.

- The island biogeography model is used to predict the numbers of species that will persist in new protected areas and the numbers that will go extinct elsewhere due to habitat destruction and other human activities.

- Those species that are most vulnerable to extinction have particular features, including a narrow range, one or only a few populations, small population size, declining population size, and economic value to humans, which leads to overexploitation.

- Rare species are more prone to extinction than common ones. A species can be considered rare if it has one of the following characteristics: it occupies a narrow geographical range, it occupies only one or a few specialized habitats, or it is always found in small populations. Isolated habitats such as islands, lakes, and mountaintops may have locally endemic species with narrow distributions.

- Small populations are vulnerable to further declines in size and eventual extinction due to genetic, demographic, and environmental factors, including those that are stochastic (randomly occurring). Intensive management of small populations may be required to prevent their extinction. Effective population size measures the number of breeding individuals, which is likely smaller than the total population and is affected by sex ratios, age structure, and other issues.

For Discussion ■

1. Why should conservation biologists, or anyone else, care if a species goes locally extinct if it is still found somewhere else?

2. Consider one of the species that has gone extinct in the last two centuries. Did it fall into one of the categories of extinction vulnerability listed in this chapter? What other biological or ecological traits might it have had that contributed to its extinction? Did it have one dominant feature that was predictive of extinction, or a combination of such features?

3. Find out about a species that is currently endangered in the wild. How might this species be affected by the problems of small populations? Address genetic, physiological, behavioral, and ecological aspects as appropriate.

Suggested Readings

Bell, C. D., J. M. Blumenthal, A. C. Broderick, and B. J. Godley. 2010. Investigating potential for depensation in marine turtles: How low can you go? *Conservation Biology* 24: 226–235. High gene flow may explain why certain marine turtles can persist at low population densities.

Bonsall, M. B., C. A. Dooley, A. Kasparson, T. Brereton, D. B. Roy, and J. A. Thomas. 2014. Allee effects and the spatial dynamics of a locally endangered butterfly, the high brown fritillary. *Ecological Applications* 24: 108–120. Using a modeling approach, the authors demonstrate that Allee effects have a major effect on most populations of this butterfly.

Fisher, D. O. and S. P. Blomberg. 2012. Inferring extinction of mammals from sighting records, threats, and biological traits. *Conservation Biology* 26: 57–67. Using past observations, a model can estimate the probability that an extinct species is still alive and that an endangered species is actually extinct.

Frankham, R., J. D. Ballou, and D. A. Briscoe. 2009. *Introduction to Conservation Genetics*, 2nd Ed. Cambridge University Press, Cambridge, UK. Excellent introduction to the importance of genetics to conservation.

Hedrick, P. 2005. Large variance in reproductive success and the N_e/N ratio. *Evolution* 59: 1596–1599. A variety of factors can result in reduced effective population size.

Morton, T. A. L., A. Thorn, J. M. Reed, R. Van Driesche, R. A. Casagrande, and F. S. Chew. 2015. Modeling the decline and potential recovery of a native butterfly following serial invasions by exotic species. *Biological Invasion* 17:1683–1695. A stochastic model is used to predict rates of loss of heterozygosity in a desirable trait that allows it to adapt to an exotic host.

Palomares, F., J. A. Godoy, J. V. López-Bao, A. Rodriguez, S. Roques, M. Casas-Marce and 2 others. 2012. Possible extinction vortex for a population of Iberian lynx on the verge of extirpation. *Conservation Biology* 26: 689–697. The Iberian lynx exhibits many of the characteristics that drive small populations toward extinction.

Pe'er, G., M. A. Tsianou, K. W. Franz, Y. G. Matsinos, A. D. Mazaris, D. Storch, and 5 others. 2014. Toward better application of minimum area requirements in conservation planning. *Biological Conservation* 170: 92–102. Minimum area requirements have practical application in efforts to protect populations.

Polishchuk, L. V., K. Y. Popadin, M. A. Baranova, and A. S. Kondrashov. 2015. A genetic component of extinction risk in mammals. *Oikos* 124(8): 983–993.

Wake, D. B. and V. T. Vredenburg. 2008. Are we in the midst of the sixth mass extinction? A view from the world of amphibians. *Proceedings of the National Academy of Sciences USA* 105: 11466–11473. As a result of intensive study, amphibians are now recognized as among the most threatened animal groups.

Waller, D. M. 2015. Genetic rescue: a safe or risky bet? *Molecular Ecology* 24: 2595–2597. doi: 10.1111/mec.13220. Human-facilitated outcrossing between populations to increase genetic diversity can sometimes lead to outbreeding depression.

Wedekind, C. G. Evanno, T. Székely, M. Pompini, O. Darbellay, and J. Guthruf. 2013. Persistent unequal sex ratio in a population of grayling (Salmonidae) and possible role of temperature increase. *Conservation Biology* 27: 229–234. Case study of the linkage between temperature and sex ratio in fish.

6

Conserving Populations and Species

This Colorado hookless cactus (*Sclerocactus glaucus*) is listed as threatened under the Endangered Species Act; researchers tag individuals within transects to track population demographics over time.

A t any point in time, a population of any species can naturally be stable, increasing, decreasing, or fluctuating in number. In general, widespread human disturbance destabilizes populations of many native species, often sending them into sharp decline. But how can this disturbance be measured, and what actions should be taken to prevent or reverse it? How can conservation biologists determine whether a specific plan to manage a rare or endangered species has a good chance of succeeding? This chapter discusses approaches for monitoring, managing, prioritizing, and protecting species and their populations in their natural environments.

We also need to know which species to protect. Conservation biologists and park managers do not have enough time and money to protect every species. Similarly, not all species need to be protected; many species have numerous large populations that are stable in size and cover a large area. Efforts need to be directed to identifying species in need of protection, using the most recent quantitative approaches that include both field data and modeling. The chapter ends by describing how this information is used to establish legal protection for endangered species.

Applied Population Biology

In order to effectively protect and manage a rare or endangered species, it is vital to have a firm grasp of its ecology, its distinctive characteristics (sometimes called its **natural history**), and the status of its populations, particularly the dynamic processes that affect population size and distribution (its **population biology**). With more information about a rare species, land managers can more effectively maintain it, identify factors that place it at risk of extinction, and develop alternative management options.

Scientists who study the natural history and population biology of a species have historically pursued this knowledge for its own sake, but today these fields have very real and important applications in population-level conservation efforts for rare and endangered species (see the chapter opening photo). In many cases, however, management decisions may have to be made without this information or while it is still being gathered. Several types of natural history and population biology information are important to conservation biology:

- *Environment.* What are the habitat types where the species is found, and how much area is there of each? How variable is the environment in time and space? How frequently is the environment affected by disturbances, and of what magnitude? How have human activities affected the environment?

- *Distribution.* Where is the species found in its habitat? Are individuals clustered together, distributed at random, or spaced out regularly? Do individuals of this species move and migrate among habitats or to different geographical areas over short or long time periods? How efficient is the species at colonizing new habitats? How have human activities affected the distribution of the species or its ability to move among habitats?

- *Biotic interactions.* What types of food and other resources does the species need, and how does it obtain them? What other species compete with it for these resources? What predators, parasites, or diseases affect its population size? With what mutualists (pollinators, dispersers, etc.) does it interact? Do juvenile stages disperse by themselves, or are they dispersed by other species? How have human activities altered the relationships among species in the ecosystem?

- *Morphology.* What does the species look like? What are the shape, size, color, surface texture, and function of its parts? How do the shapes of its body parts relate to their functions and help the species to survive in its environment? What are the characteristics that allow this species to be distinguished from species that are similar in appearance?

- *Physiology.* What amount of food, water, minerals, and other necessities does an individual need to survive, grow, and reproduce? How efficient is an individual at using its resources? How vulnerable is the species to extremes of climate, such as heat, cold, wind, and precipitation? When does the species reproduce, and what are its special requirements during reproduction?

- *Demography*. What is the current population size, and what was it in the past? Is the number of individuals stable, increasing, or decreasing? Does the population have a mixture of adults and juveniles, indicating that new individuals are being recruited? At what age do individuals begin to reproduce?

- *Behavior*. How do the actions of an individual allow it to survive in its environment? How do individuals in a population mate and produce offspring? Do individuals of a species interact cooperatively or competitively? How do individuals find food? At what time of day or year is the species most visible for monitoring?

- *Genetics*. How much variation occurs in morphological, physiological, and behavioral characteristics? How is the variation spread across the species range? How much of this variation is genetically controlled? What percentage of the genes is variable? How many alleles does the population have for each variable gene? Are there genetic adaptations to local sites? Is there gene flow between populations?

- *Interactions with humans*. How do human activities affect the species? What human activities are harmful or beneficial to the species? Do people harvest or use this species in any way? What do local people know about this species?

> Knowledge of the natural history and population biology of a species is crucial to its protection, but urgent management decisions often must be made before all this information is available or while it is still being gathered.

Methods for studying populations

Methods for studying populations have developed largely from the study of land plants and animals (**FIGURE 6.1**). Small organisms such as protists, bacteria, and fungi have not been investigated in comparable detail, and species that inhabit soil, freshwater, and marine habitats are particularly poorly investigated for population characteristics. In this section we will examine how conservation biologists undertake their studies of populations, while recognizing that the methods need to be modified for each species.

PUBLISHED LITERATURE   When gathering information, it is important to remember that other people may have already investigated an ecosystem or studied the same (or a related) species. Library indexes such as BIOSIS Citation Index and Biological Abstracts are accessible online and provide easy access to a variety of books, articles, and reports relating to a particular topic. This literature may contain records of previous population sizes and distributions that can be compared with the current status of the species. Sections of the library have material on similar topics shelved together, so finding one book often leads to finding others. The Internet provides ever-increasing access to databases, websites, electronic bulletin boards, journals, news articles, specialized discussion groups, and subscription

(A)

(B)

(C)

FIGURE 6.1 Monitoring populations requires specialized techniques suited to each species: (A) in rain forest research, individual trees are tagged, mapped, and measured for their girth at 5-year intervals at painted points. (B) Ornithologists catch birds using mist-nets; record their weight, size, sex, and breeding condition; and then attach a numbered metal band to a leg. (C) Plant ecologists monitor changes in plant communities by placing quadrats at fixed points in a grassland and recording the species present, their abundance, and plant height. (A,B photographs by Richard B. Primack; C, © Paul Glendell/Alamy.)

databases such as ScienceDirect and the Thomson Reuters (formerly ISI) Web of Science. Google Scholar is one of the best places to start searching on topics relating to conservation biology. (When using material online, always be sure to validate the accuracy and source of the data, especially on sites such as Wikipedia, where there is no control over what is posted.) Asking biologists and naturalists for ideas on references is another way to locate published materials. Searching indexes of newspapers, magazines, and popular journals is also an effective strategy because results of important scientific research are often covered in the popular news media.

UNPUBLISHED LITERATURE   Unpublished reports by scientists, enthusiastic citizens, government fisheries and wildlife agencies, national and regional forest and park departments, and conservation organizations contain an enormous amount of information on conservation biology. This "gray literature" is sometimes cited in published literature or mentioned by leading authorities in conversations, lectures, conference presentations, or articles. Often a report known through word of mouth can be obtained online or through direct contact with the author. Interviews with experts can also be used to gather their collective knowledge about endangered species and ecosystems (McBride et al. 2012). One example is a recent study of four threatened tree species in Fiji (Keppel et al. 2015). The researchers emailed questionnaires to both experts and locals to find previously undocumented populations, improve estimates of population sizes, and identify threats such as fire, logging, and invasive plants.

FIELDWORK   The natural history of a species usually must be learned through careful observations in the field (e.g., Beane et al. 2014). Fieldwork is necessary because only a tiny percentage of the world's species have been adequately studied, and the ecology of a species often changes from one place to another. Only in the field can the conservation status of a species be determined, as well as its relationships to the biological and physical environment. Fieldwork for species such as polar bears, humpback whales, and tropical trees can be time-consuming, expensive, and physically arduous, but it is crucial for developing conservation plans for endangered species, and it can be exhilarating and deeply satisfying as well. For example, there is a long tradition, particularly in Britain, of dedicated amateurs conducting excellent studies of species in their immediate surroundings with minimal equipment or financial support. However, many of the technical methods for investigating populations are very specialized and are best learned by studying under the supervision of an expert (**FIGURE 6.2**).

The need for fieldwork is highlighted by extensive observations of wild penguins. One study of king penguins (*Aptenodytes patagonicus*) in breeding

FIGURE 6.2   Researchers and volunteers release a collared Argali sheep ram (*Ovis ammon*) in Gun Galuut Nature Reserve, Mongolia. The Argali are the largest mountain sheep in the world and are globally endangered. Data collected from collared animals are used to understand range requirements and threats of this understudied species. This project is also an example of collaboration between a zoo and local groups. (Photo by Richard Reading.)

colonies on Possession Island, Crozet Archipelago, in the southern Indian Ocean (Viblanc et al. 2014) involved fitting male penguins with data loggers that recorded their heart rate, temperature, and overall physical activity to better understand sources of stress on the birds during breeding. Other research, on Magellanic penguins (*Spheniscus magellanicus*), used satellite tags and radiotelemetry to define the foraging area the birds use when feeding their chicks (Boersma and Rebstock 2009). It had previously been thought that the birds swim out into the ocean to forage for food only within 30 km of their nests, but this research showed that the birds actually swam up to 600 km and, during key periods of chick rearing, heavily used a fishing exclusion zone (**FIGURE 6.3**). Based on this information, the Argentinian government agreed to extend the number of months of fishing exclusion in this zone, and the survival of young penguins subsequently improved.

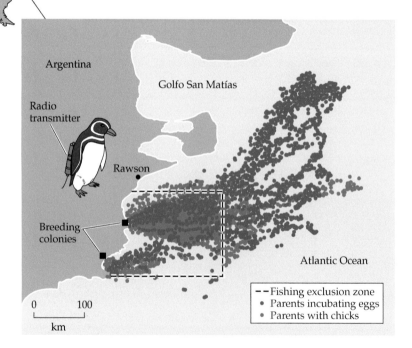

FIGURE 6.3 Satellite tracking of Magellanic penguins (*Spheniscus magellanicus*) fitted with radio transmitters shows that penguins incubating eggs forage up to 600 km from their breeding colonies. When penguins are feeding chicks, foraging takes place mainly within a seasonal fishing exclusion zone that was established to protect spawning fish. Fieldwork provided this vital information about the penguins' foraging habits, which led to the fishing zone remaining closed until the chicks left their nests. (After Boersma 2006.)

MONITORING POPULATIONS   To learn the status of a species, scientists must survey its populations in the field and monitor them over time. Monitoring plays an important role in conservation biology. Changes in population size and distribution can be determined by repeatedly surveying a population on a regular basis (Noon et al. 2012). Efforts can also be targeted at particularly sensitive species, such as butterflies, using them as indicator species of the long-term stability of ecosystems (Wikström et al. 2008). The most common types of monitoring are censuses, surveys, and population demographic studies.

The number of monitoring studies has been increasing dramatically as government agencies have become more concerned with protecting biodiversity. Some of these studies are mandated by law as part of management efforts for endangered species, and may be conducted in partnership with conservation organizations or university researchers. Long-term monitoring records can help to distinguish long-term population increases or declines (possibly caused by human disturbance) from short-term fluctuations caused by variations in weather or unpredictable natural events (Porszt et al. 2012). Monitoring records can also determine whether an endangered species is showing a positive response to conservation management or is responding negatively to factors such as the present levels of harvest or the arrival of invasive species. Observing a long-term decline in a species often provides motivation to take vigorous action to conserve it (**FIGURE 6.4**). Field work sometimes brings researchers into confrontations with people who are engaged in illegal activities or otherwise harming species and ecosystems.

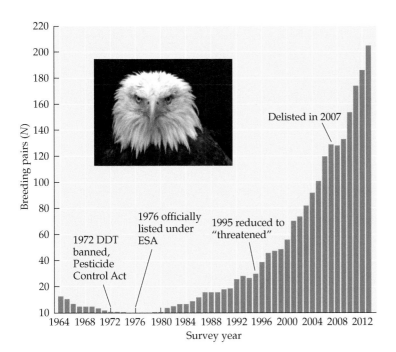

FIGURE 6.4   The American bald eagle received rigorous conservation action as a result of population monitoring. Breeding pairs observed at James Lake, Virginia, are shown here over time, as they relate to conservation activity. Alarmingly low numbers of nesting pairs observed across the US in the 1960s and 1970s spurred both research and policy action, including banning the pesticide DDT (which was harming the birds through biomagnification), increasing the penalties for hunting and trapping eagles, and listing for protection under the Endangered Species Act (ESA). Recovery after these actions meant that the species could be delisted, although it is still protected. (Inset photograph © janbugno/Shutterstock.)

The geographical range and intensity of monitoring have been greatly extended through the use of volunteers, often called "citizen scientists" (Mueller et al. 2010). Training and educating citizens not only expands the data available to scientists but often transforms these citizens into advocates for conservation (Matteson et al. 2012). Four programs that rely heavily on volunteers are the North American Amphibian Monitoring Program, Environment Canada, Project Nestwatch, and Frogwatch USA.

CENSUS   A **census** is a count of the number of individuals present in a population. It is a comparatively inexpensive and straightforward method, especially for organisms that are easy to detect or are not mobile. By repeating a census over successive time intervals, biologists can determine whether a population is stable, increasing, or decreasing in number. In one monitoring study, population censuses of the Hawaiian monk seal (*Monachus schauinslandi*) on the beaches of several islands in the Kure Atoll of the South Pacific documented a decline from almost 100 adults in the 1950s to fewer than 14 in the late 1960s (**FIGURE 6.5**). On the basis of these trends,

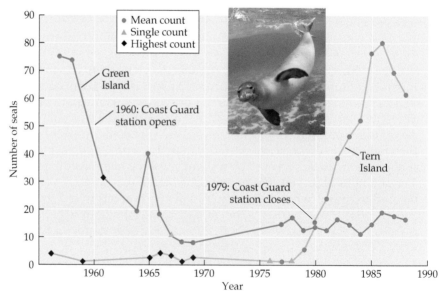

FIGURE 6.5   Censusing the populations of the Hawaiian monk seal (*Monachus schauinslandi*) on Green Island of Kure Atoll (blue line) and on Tern Island of French Frigate Shoals (green line) revealed that this species was in danger of extinction. Population counts were plotted from a single count, the mean of several counts, or the maximum of several counts. Seal populations declined when a Coast Guard station was opened on Green Island in 1960, because of disturbance by people and dogs; populations increased on Tern Island after the closing of a Coast Guard station in 1979, when there was less disturbance to seals. (After Gerrodette and Gilmartin 1990; photograph by James D. Watt, courtesy of US Department of the Interior.)

the Hawaiian monk seal was declared endangered in 1976 under the US Endangered Species Act (discussed later in this chapter; Baker and Thompson 2007). Subsequent conservation efforts reversed the trend, but only for some populations. For example, the population of the Tern Island monk seal increased after the Coast Guard station there was closed in 1979, but since the 1990s it has substantially declined because of high juvenile mortality.

Censuses of a biological community can be conducted to determine what species are currently present in a locality. Censuses conducted over a wide area can help to determine the range of a species and its areas of local abundance. As part of the North American Breeding Bird Survey (BBS; www.pwrc.usgs.gov/bbs), thousands of participants have been recording bird abundance at thousands of locations over the past 35 years along **transects**, lines often designated with measuring tape or string, along which biological data is collected (also see chapter opening photo). This information is used to determine the stability of populations of over 400 bird species over time (**FIGURE 6.6**). A comparison of current occurrences with past censuses can highlight species that have been lost and changes in species ranges. These data can also be used to determine which environmental variables (such as temperature) are important for understanding species data (Goetz et al. 2014).

SURVEY   A **survey** of a population involves using a repeatable sampling method to estimate the abundance or density of a population or species in a part of a community. Survey methods are used when a population is very large or its range is extensive, making a direct measure of all individuals impossible. An area can be divided into sampling segments and the number of individuals in certain segments can be counted. These counts can then be used to estimate the actual population size. For example, a rabbit population can be surveyed by walking transects and recording the number of rabbits observed (Dieter and Schaible 2014). Boats and planes can be used to census whales and dolphins along transects in the ocean (Hammond et al. 2013). New technologies are being employed to survey difficult-to-reach species, such as the use of drones to survey populations by taking photographs or even collecting water samples (Ore et al. 2015).

Estimating the density of organisms that are difficult to find because they are extremely rare or very small requires special methods. Surveys have expanded in recent years to include DNA analysis of scat and hair samples (Hedges et al. 2013). In some cases, specially trained dogs are used to locate scat samples of rare animal species. Such DNA studies using scats have revealed that population size is often larger than previous estimates suggested using traditional survey methods of direct observation, because some of the more elusive individuals have never been seen by observers. In an exciting new development, DNA in soil, water, and even air samples is being used to detect the presence of rare species and the first occurrence of invasive species. This genetic material in bulk samples is being referred to as environmental DNA (eDNA), and it is being heralded as a new and

(A)

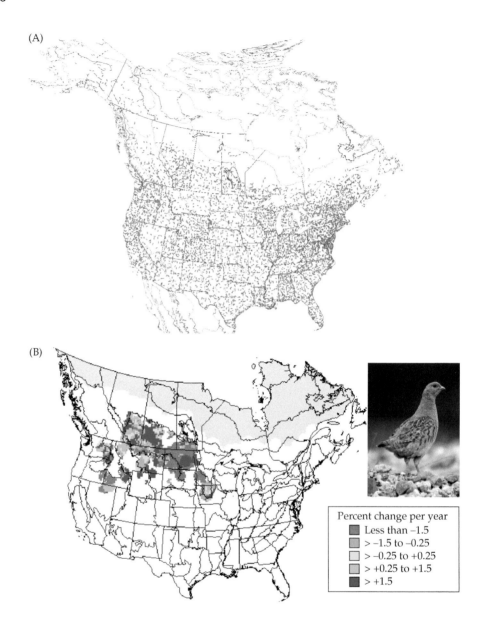

(B)

Percent change per year
- Less than –1.5
- > –1.5 to –0.25
- > –0.25 to +0.25
- > +0.25 to +1.5
- > +1.5

FIGURE 6.6 The North American Bird Survey (BBS) is one of the most comprehensive biological surveys in the world, as shown by this map of survey routes. Intensity of the shading represents the density of observer routes, each of which includes predetermined stops to record the occurrence of each bird species. (B) A map using data from the BBS of the period covering 1966–2013 for the grey partridge *Perdix perdix* (inset), for which the entire range is included in the survey (www.mbr-pwrc.usgs.gov/bbs). Locations where the bird populations are increasing (blue) can be discerned from those where it is decreasing (yellow and orange) and not changing (red); this information can be valuable for protection of this species. (A, from Sauer et al. 2013; B, after Sauer et al. 2014; inset © MikeLane45/Getty Images.)

important conservation tool with several applications (Thomsen and Willerlev 2015).

DEMOGRAPHIC STUDY   **Demographic studies** follow known individuals of different ages and sizes in a population to determine their rates of growth, reproduction, and survival (Crone et al. 2013). Either the whole population or a subsample can be followed. In a complete population study, all individuals are counted, aged if possible, sized, sexed, and marked for future identification; their position on the site is mapped; and tissue samples may be collected for genetic analysis. Techniques vary depending on the characteristics of the species and the purpose of the study. Each discipline has its own techniques for following individuals over time: ornithologists band birds' legs, mammalogists often attach tags to animals' ears, and botanists nail aluminum tags to trees; however, there is increasing concern that for animals, the stress of being captured and tagged can have harmful or even fatal effects (Jewell 2013). Information from demographic studies can be used in standard mathematical formulas (life-history formulas) to calculate the rate of population change and to identify critical stages in the life cycle (McCaffery et al. 2015).

> Demographic studies provide data on the numbers, ages, sexes, conditions, and locations of individuals within a population. These data indicate whether a population is stable, increasing, or declining and are the basis for statistical models used to predict the future of a species.

An example of the application of demographic data for conservation can be found in the work of researchers who used a combination of genetic and demographic data of a threatened Mediterranean coral species to designate and prioritize conservation units within marine protected areas (MPAs) (Arizmendi-Mejía et al. 2015). Demographic studies can also provide information on the age structure of a population. A stable population typically has an age distribution with a characteristic ratio of juveniles, young adults, and older adults. The absence or low representation of any age class, particularly juveniles, may indicate that the population is declining. Conversely, the presence of a large number of juveniles and young adults may indicate that the population is stable or even expanding.

Population viability analysis (PVA)

Monitoring data can allow us to calculate average mortality rates, average recruitment rates, the current age or size distribution of the population, and the area it occupies. This information can then be used to construct a mathematical model that estimates the ability of a population to persist in the future, a process known as **population viability analysis** (**PVA**). This is an extension of demographic analysis (e.g., Saether and Engen 2015). There are many mathematical and statistical models used for PVA, which can be thought of as risk assessment—predicting the probability that a population or a species will go extinct at some point in the future (**FIGURE 6.7**). It can also be used to identify vulnerable stages in the natural history of the species

> PVA uses mathematical and statistical methods to predict the probability that a population or species will go extinct within a certain time period. PVA is also useful in modeling the effects of habitat degradation and management efforts.

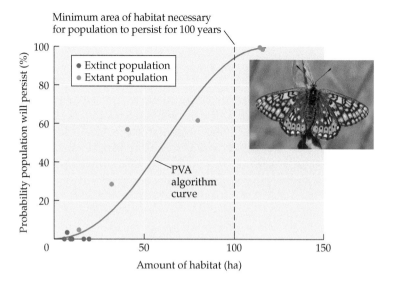

FIGURE 6.7 Population viability analyses predict that it takes 100 ha to ensure (at 95% likelihood) the persistence of a marsh fritillary butterfly population for 100 years. All of the extinct populations occupied areas much smaller than 100 ha. Four of the six extant populations occupy areas smaller than 100 ha and are predicted to go extinct in the coming decade unless their habitat is increased. (After Bulman et al. 2007; photograph © Sergey Chushkin/shutterstock.)

and to consider the effects of habitat loss, fragmentation, and deterioration (Beissinger et al. 2009).

Another important aspect of PVA is estimating how management efforts such as changing the hunting quotas or the area of protected habitat will affect the probability of extinction (Molano-Flores and Bell 2012). PVA can also model the effects of augmenting a population through the release of additional individuals caught in the wild elsewhere or raised in captivity. Here are two examples of the use of PVA:

- The Hawaiian stilt (*Himantopus mexicanus knudseni*) is an endangered, endemic bird of the Hawaiian Islands. Hunting and coastal development 70 years ago reduced the number of birds to 200, but protection has allowed recovery to the present population size of about 1600 individuals (Reed et al. 2007, 2011). The goal of government protection efforts is to allow the population to increase to 2000 birds. A PVA was made of the species' ability to have a 95% chance of persisting for the next 100 years. The model predicted that stilt numbers would increase until they occupied all available habitat but that they would show a rapid decline if nesting failure and mortality rates of first-year birds exceeded 70% or if mortality rates of adults increased above 30% per year. Keeping mortality rates below these levels would require the control of exotic predators and the restoration of wetland habitat. And most important, additional wetland would need to be protected if the goal of protecting 2000 stilts was to be achieved.

- Western lowland gorillas (*Gorilla gorilla gorilla*) are slow-reproducing primates that are critically endangered (**FIGURE 6.8**). Reintroduction of captive-bred and rehabilitated wild, orphaned gorillas appeared

FIGURE 6.8 Population viability analysis can be especially useful for long-lived and slowly reproducing species such as lowland gorillas. (© Martin Harvey/Getty Images.)

to be successful in terms of survival of the releases, but it was not known whether these reintroduced populations were likely to persist over time. Demographic studies of reintroduced populations in the Batéké Plateau region of Congo and Gabon, supplemented with data from captive populations, were used to develop a population model for PVA (King et al. 2014). The analysis revealed that for this species, annual birthrates and adult female mortality rates were parameters that particularly affected the probability of a population going extinct within 200 years. For example, increasing the birthrate from 0.18 to 0.20 decreased the probability of extinction from 29% to 9%. Given the current population structure, the PVA results also suggested that these populations had a greater than 90% likelihood of surviving for 200 years. This is important support for the reintroduction program, but it also emphasizes the necessity of monitoring over time because parameters such as birthrate and adult female survival may change in response to environmental stochasticity.

A key feature of PVA models, and a reason why they are especially used with small populations (given that small populations are particularly vulnerable to random events, as explained in Chapter 5), is that they can incorporate demographic and environmental stochasticity. For example, if the birthrate of a population has been determined to be 0.5, this means that each breeding individual has a 50% chance of producing young, which is analogous to the results of a coin flip. We know that, simply due to random chance, when we flip a coin 10 times, we will only occasionally see exactly 5 heads and 5 tails. There is even a chance (albeit small) that we may flip 10 heads in a row. Similarly, even when we are confident of the birthrate, we cannot be sure how many young will actually be born in any given year, which can have especially dramatic implications for smaller populations. Hundreds or thousands of simulations of individual populations can be run using this random variation to determine the probability of population extinction within a certain period of time, the median time to extinction, changes in population size, and changes in area occupied.

Such statistical models must be used with caution and a large dose of common sense (Jäkäläniemi et al. 2013). Generally, about 10 years of monitoring data are needed to obtain a PVA with good predictive power. The results of some models can often change dramatically with different model assumptions and slight changes in parameters. Nevertheless, PVA is widely used and has value in demonstrating the possible effectiveness of alternative management strategies (Sweka and Wainwright 2014). For this reason, attempts to utilize PVA as part of practical conservation efforts are increasingly common in management planning.

MINIMUM VIABLE POPULATION (MVP)   PVA is also used to calculate the minimum number of individuals of a given species required for a population to persist over time. Some researchers argue that PVA should be used in the establishment of **recovery criteria** (**RC**), that is, predetermined thresholds that signal that an endangered species can be removed from protection under the Endangered Species Act (Beissinger 2015). In a groundbreaking paper, Shaffer (1981) defined the number of individuals necessary to ensure the long-term survival of a species as the **minimum viable population** (**MVP**): "A minimum viable population for any given species in any given habitat is the smallest isolated population having a 99% chance of remaining extant for 1000 years despite the foreseeable effects of demographic, environmental, and genetic stochasticity, and natural catastrophes." In other words, the MVP is the smallest population size that can be predicted to have a very high chance of persisting for the foreseeable future. Shaffer emphasized the tentative nature of this definition, saying that the survival probabilities could be set at 95%, 99%, or any other percentage and that the time frame might similarly be adjusted, for example, to 100 or 500 years. The key point is that the MVP size allows a quantitative estimate to be made of how large a population must be to ensure long-term survival. In general, protecting a larger population increases the chance of the population persisting for a longer period of time (**FIGURE 6.9**).

One of the best-documented studies of MVP size tracked the persistence of 120 bighorn sheep (*Ovis canadensis*) populations (some of which have been followed for 70 years) in the deserts of the southwestern United States (Berger 1990, 1999). The striking observation is that 100% of the unmanaged populations with fewer than 50 individuals went extinct within 50 years, while within the same time period, virtually all of the populations with more than 100 individuals persisted. Thus, for bighorn sheep, the minimum population size is at least 100 individuals.

Shaffer (1981) compares MVP protection efforts to flood control. It is not sufficient to use average annual rainfall as a guideline when planning flood-control systems; instead, we must plan for extreme situations of high rainfall and severe flooding, which may occur only once every 50 or 100

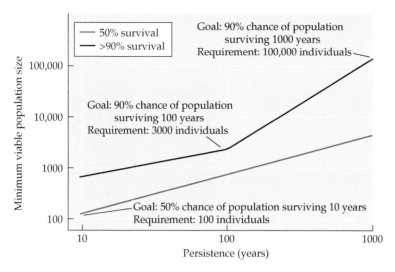

FIGURE 6.9 If the goal is persistence for a greater number of years, then a larger minimum viable population (MVP) size is needed. A greater MVP is needed to ensure a higher probability of persistence, as illustrated in this case by a 50% probability of survival and a greater than 90% probability of survival. Both axes are on log scales. The values were derived from changes in population size and persistence of 1198 species. (After Traill et al. 2010.)

years. In the same way, when attempting to protect natural systems, we understand that certain catastrophic events, such as hurricanes, earthquakes, forest fires, epidemics, and die-offs of food items, may occur at even greater intervals. To plan for the long-term protection of endangered species, we must provide for their survival, not only during average years, but also during exceptionally harsh years. Consequently, an accurate estimate of the MVP size for a species requires an analysis of its environment. This can be expensive and require months, or even years, of research. Analyses of over 200 species for which adequate data were available (mainly vertebrates) indicated that most MVP values for long time periods fall in the range of 3000–5000 individuals, with a median of 4000 (Flather et al. 2011). For species with extremely variable population sizes, such as certain invertebrates and annual plants, protecting a population of about 10,000 individuals may be the ideal strategy.

Unfortunately, many species, particularly endangered species, have population sizes smaller than these recommended minimums. For instance, half of 23 isolated elephant populations remaining in West Africa have fewer than 200 individuals, a number considered to be inadequate for long-term survival of the population (Bouché et al. 2011). Likewise, the wolf population on Isle Royale, Michigan, has been fluctuating around 20 individuals but currently has only 8 adults of breeding age and no pups (Mlot 2013).

FIGURE 6.10   Extinction rates of bird species on the Channel Islands, with an island scrub jay (*Aphelocoma insularis*) as an example of one of the species. Each dot represents the extinction percentage of all the species in that population size class; extinction rate decreases as the size of the population increases. Populations with fewer than 10 breeding pairs had an overall 39% probability of extinction over 80 years, populations of between 10 and 100 pairs averaged about 10% probability of extinction, and populations of over 100 pairs had a very low probability of extinction. (After Jones and Diamond 1976. Photograph, © Tim Zurowski/BIA/Getty Images.)

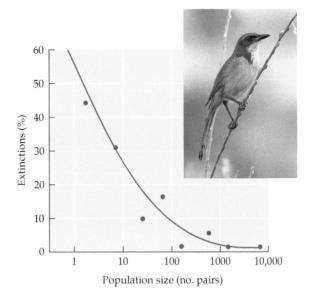

Field evidence from long-term studies of birds on the Channel Islands off the California coast supports the fact that large populations are needed to ensure population persistence; only bird populations with more than 100 breeding pairs had a greater than 90% chance of surviving for 80 years (**FIGURE 6.10**). In spite of most evidence to the contrary, however, small populations sometimes do prevail, and many populations of birds have survived for 80 years with 10 or fewer breeding pairs. Of course, birds are especially mobile and can readily recolonize areas following local extinction, whereas less mobile species lack this ability.

Once an MVP size has been established for a species, the **minimum dynamic area** (**MDA**)—the area of suitable habitat necessary for maintaining the minimum viable population—can be estimated by studying the size of the home range of individuals and colonies of endangered species (Pe'er et al. 2014; Thiollay 1989). It has been estimated that reserves in Africa of 100–1000 km^2 are needed to maintain many small mammal populations (**FIGURE 6.11**). To preserve populations of large carnivores, such as lions, reserves of 10,000 km^2 are needed.

Metapopulations

Over time, populations of a species may become extinct on a local scale while new populations may form nearby on other suitable sites. Often a species that lives in an ephemeral habitat, such as a streamside herb, is better characterized in terms of a **metapopulation** (a "population of populations") that is made up of a shifting mosaic of populations linked by some degree of migration (Nöel et al. 2013). In some species, every population in the metapopulation is short-lived and the distribution of the species

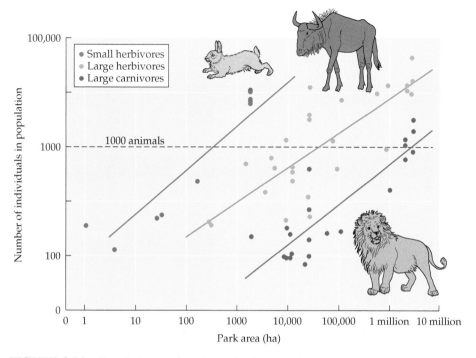

FIGURE 6.11 Population studies show that large parks and protected areas in Africa contain larger populations of each species than small parks; thus, only the largest parks may contain long-term viable populations of many vertebrate species. Each dot represents an animal population in a park. If the viable population size of a species is 1000 individuals (dashed line), parks of at least 100 ha will be needed to protect small herbivores (e.g. rabbits, squirrels), parks of more than 10,000 ha will be needed to protect large herbivores (e.g. zebras, wildebeests), and parks of at least 1 million ha will be needed to protect large carnivores (e.g., lions, hyenas). (After Schonewald-Cox 1983.)

changes dramatically with each generation. In other species, the metapopulation may be characterized by one or more **source populations** (core populations) with fairly stable numbers and several **sink populations** (satellite populations) that fluctuate in size with arrivals of immigrants. Populations in the satellite areas may become extinct in unfavorable years, but the areas may be recolonized, or rescued, by migrants from the more permanent core population when conditions become more favorable (**FIGURE 6.12**). Metapopulations may also involve relatively permanent populations between which individuals occasionally move.

Bighorn sheep (*Ovis canadensis*) also offer a well-studied example of metapopulation dynamics. These sheep have been observed dispersing between mountain ranges and occupying previously unpopu-

> Populations of a species are often connected by dispersal and can be considered a metapopulation. In such a system, the loss of one population can negatively affect other populations.

FIGURE 6.12 Possible metapopulation patterns, with the size of a population indicated by the size of the circle. The arrows indicate the direction and intensity of migration between populations. (After White 1996.)

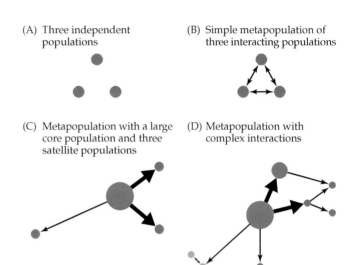

(A) Three independent populations

(B) Simple metapopulation of three interacting populations

(C) Metapopulation with a large core population and three satellite populations

(D) Metapopulation with complex interactions

lated sites, while mountains that previously had sheep populations are now unoccupied. Migration and gene flow occurs primarily between populations less than 15 km apart and is greater when the intervening countryside is hilly rather than flat (Creech et al. 2014). Maintaining dispersal routes between existing population areas and potentially suitable sites, including across international borders, is important in managing this species (Buchalski et al. 2015).

The persistence of metapopulations often depends on habitat availability. For example, destruction of the habitat of one central, core population might result in the extinction of numerous smaller populations that depend on the core population for periodic colonization. Effective management of a species often requires an understanding of these metapopulation dynamics and a restoration of lost habitat and dispersal routes.

Long-term monitoring

To understand the reasons behind population changes, monitoring of populations needs to be combined with monitoring of other environmental parameters. The long-term monitoring of ecosystem processes (e.g., temperature, rainfall, humidity, soil acidity, water quality, discharge rates of streams, and soil erosion) and community characteristics (species present, percentage of vegetative cover, amount of biomass present at each trophic level, etc.) allows scientists to determine the health of the ecosystem and the status of species of special concern (Papworth et al. 2009). The Long-Term Ecological Research (LTER) program in the United States focuses on such changes on timescales ranging from months and years to decades and centuries (**FIGURE 6.13**).

As an example of the need for long-term monitoring, certain amphibian, insect, and annual plant populations are highly variable from year to year,

	Years	Research discipline	Physical events	Biological phenomena
10^5	100 Millennia			Evolution of species
10^4	10 Millennia	Paleoecology and limnology	Continental glaciation	Bog succession Forest community migration
10^3	Millennium		Climate change	Species invasion Forest succession
10^2	Century		Forest fires CO_2-induced climate warming	Cultural eutrophication Population cycles
10^1	Decade			
10^0	Year		Sun spot cycle El Niño events Prairie fires Lake turnover Ocean upwelling	Prairie succession Annual plants Seasonal migration Plankton succession
10^{-1}	Month			
10^{-2}	Day		Storms Daily light cycle Tides	Algal blooms Daily movements
10^{-3}	Hour	Most ecology		

LTER spans from 10^2 (Century) to 10^{-1} (Month).

FIGURE 6.13 The Long-Term Ecological Research (LTER) program focuses on timescales ranging from months to centuries in order to understand changes in the structure, function, and processes of ecosystems that are not apparent from short-term observations. (From Magnuson 1990.)

so many years of data are required to determine whether a particular species is actually declining in abundance over time or merely experiencing a number of low population years that are in accord with its regular pattern of variation. In one instance, more than 50 years of observation of populations of two flamingo species (*Phoenicopterus ruber*, the greater flamingo, and *Phoeniconaias minor*, the lesser flamingo) in southern Africa revealed that large numbers of chicks fledged only in years with high rainfall, making it appear that reproduction was simply highly variable (**FIGURE 6.14**). In fact, there was a 31-year gap in which no major hatchings occurred and the population appeared to be in peril. Fortunately, the population began to increase in 2008 and 2011, when numerous chicks hatched.

The fact that environmental effects may lag for many years behind their initial causes creates a challenge to understanding change in ecosystems. For example, acid rain, nitrogen deposition, and other components of air pollution may gradually change the water chemistry, algal community, and oxygen content of forest streams, ultimately making the aquatic environment unsuitable for the larvae of certain insect species. In this case, the cause (air pollution) may have occurred years or even decades before the effect (insect decline) becomes detectable. Even habitat fragmentation can have delayed effects on losses via gradual environmental degradation and metapopulation extinction.

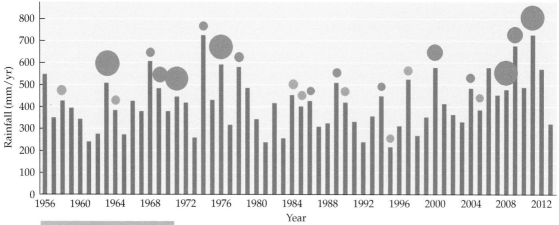

FIGURE 6.14 The bars show rainfall data from Etosha National Park in southern Africa for the years 1956–2013. The flamingo breeding events that occurred in those years are indicated by circles. Orange circles indicate failed breeding events; eggs wither laid but not chicks hatched. The small, medium, and large green circles indicate, respectively, fewer than 100 chicks hatched, hundreds of chicks hatched, and thousands of chicks hatched. There was a 31-year gap between 1976 and 2008 in which no large hatching even occurred. (After Simmons 1996, with updates from R. E. Simmons. Photograph © Kevin Schafer/DigitalVision/Photolibrary.com.)

A major purpose of monitoring programs is to gather essential data on biological communities and ecosystem functions that can be used to document changes over time. Monitoring in these studies allows managers to determine whether the goals of their projects are being achieved or whether adjustments must be made in the management plans (called adaptive management), as discussed in Chapter 10.

Conservation Categories

The IUCN uses quantitative information, including the area occupied by the species and the number of mature individuals presently alive, to assign species to conservation categories.

Once the data collected on populations and species allows us to identify those species most vulnerable to extinction, it is useful to create a system whereby extinction risk can be categorized to facilitate the prioritization of conservation efforts. To mark the status of rare and endangered species for conservation purposes, the International Union for Conservation of Nature (IUCN) has established conservation categories (**FIGURE 6.15**) (www. iucn.org). These categories have proved useful in establishing protection for threatened species at the national and interna-

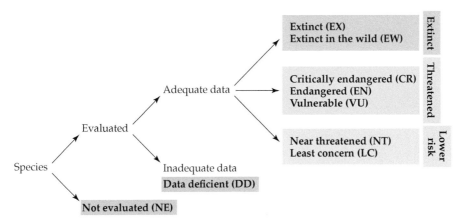

FIGURE 6.15 The IUCN categories of conservation status. This chart shows the distribution of the categories. Reading from left to right, they depend on (1) whether a species has been evaluated or not and (2) how much information is available for the species. If data are available, the species is then put into a category of lower risk, threatened, or extinct. (After IUCN 2001.)

tional levels and directing attention toward species of special concern. The conservation categories follow:

- *Extinct* (*EX*). The species (or other taxon, such as subspecies or variety) is no longer known to exist. The IUCN currently lists 709 animal species and 90 plants species as extinct.

- *Extinct in the wild* (*EW*). The species exists only in cultivation, in captivity, or as a naturalized population well outside its original range. The IUCN currently lists 32 animal species and 29 plant species as extinct in the wild.

- *Critically endangered* (*CR*). The species has an extremely high risk of going extinct in the wild, according to any of the criteria A–E (**TABLE 6.1**).

- *Endangered* (*EN*). The species has a very high risk of extinction in the wild, according to any of the criteria A–E.

- *Vulnerable* (*VU*). The species has a high risk of extinction in the wild, according to any of the criteria A–E.

- *Near threatened* (*NT*). The species is close to qualifying for a threatened category but is not currently considered threatened.

- *Least concern* (*LC*). The species is not considered near threatened or threatened. (Widespread and abundant species are included in this category.)

- *Data deficient* (*DD*). Inadequate information exists to determine the risk of extinction for the species.

TABLE 6.1 IUCN Red List Criteria for the Assignment of Conservation Categories

Red List criteria A–E	Quantification of criteria for Red List category "critically endangered"[a]
A. Observable reduction in numbers of individuals	The population has declined by 80% or more over the last 10 years or three generations (whichever is longer), either based on direct observation or inferred from factors such as levels of exploitation, threats from introduced species and disease, or habitat destruction or degradation.
B. Total geographical area occupied by the species	The species has a restricted range (<100 km² at a single location) *and* there is observed or predicted habitat loss, fragmentation, ecological imbalance, or heavy commercial exploitation.
C. Predicted decline in number of individuals	The total population size is less than 250 mature, breeding individuals and is expected to decline by 25% or more within 3 years or 1 generation.
D. Number of mature individuals currently alive	The population size is less than 50 mature individuals.
E. Probability the species will go extinct within a certain number of years or generations	Extinction probability is greater than 50% within 10 years or 3 generations.

[a]A species that meets the described quantities for *any one* of criteria A–E may be classified as critically endangered. Similar quantification for the Red List categories "endangered" and "vulnerable" can be found at www.iucnredlist.org.

- *Not evaluated* (NE). The species has not yet been evaluated against the Red List criteria.

When used on a national or other regional level, there are two additional Red List categories:

- *Regionally extinct* (RE). The species no longer exists within the country (region) but is extant in other parts of the world.
- *Not applicable* (NA). The species is not eligible for the regional Red List because, for example, it is not within its natural range in the region (it has been introduced) or because it is only a rare migrant to the region.

Species in the critically endangered, endangered, and vulnerable categories are considered **threatened** with extinction. For these three categories, the IUCN has developed quantitative measures of threat based on the probability of extinction. These **Red List criteria**, described in Table 6.1, are based on the developing methods of PVA. These criteria focus on population trends and habitat condition. The advantage of this system is that it provides a standard method of classification by which decisions can be reviewed and evaluated according to accepted quantitative criteria, using whatever information is available.

Using habitat loss as a criterion in assigning categories is particularly useful for many species that are poorly known biologically, because species can be listed as threatened if their habitat is being destroyed even if scientists know little else about

> The IUCN system has been used to identify Red Lists of threatened species and to determine whether species are responding to conservation efforts.

them. In practice, a species is most commonly assigned to an IUCN category based on the area it occupies, the number of mature individuals it has, or the rate of decline of the habitat or population; the probability of extinction is least commonly used (van Swaay et al. 2011).

Using the criteria in Table 6.1 and the categories in Figure 6.15, the IUCN has evaluated and described the threats to plant and animal species in its series of **Red Data Books** and **Red Lists** of threatened species; these detailed lists of endangered species by group and by country can be seen at www.iucn.org. Species listed as threatened include 1462 of 5499 described mammal species, 2097 of 10,052 bird species, and 2303 of 6338 amphibian species (see Table 5.1).

Most bird, amphibian, and mammal species have been evaluated using the IUCN system, but the levels of evaluation are lower for reptiles, fish, and flowering plants, resulting in low apparent levels of threat. Even though numerous species of fish (2523), reptiles (1160), molluscs (2197), insects (1259), crustaceans (1735), and higher plants (10,686) are designated as threatened with extinction, most species in these groups have still not yet been evaluated. The evaluations of insects and other invertebrates, mosses, algae, fungi, and microorganisms are even less adequate (Régnier et al. 2009). While in most cases the lack of data leads to underestimates of extinction risk, it has been argued that for some groups the risk may be overestimated when based on only presence–absence data, such as for amphibians (Cruickshank et al. 2016).

By tracking the conservation status of species over time, it is possible to determine whether species are responding to conservation efforts or are continuing to be threatened (Lacher et al. 2012). One such measure is the **Red List Index**, which demonstrates that the conservation status of certain animal groups has continued to decline since 1988, with particularly sharp declines for albatrosses, petrels, and many amphibians (Baillie et al. 2008; Quayle et al. 2007). Within Australia, the Red List Index is declining faster than the Global Red List Index, indicating the need for additional conservation action (Szabo et al. 2012). Another measure, the **Living Planet Index**, follows population sizes for 2688 vertebrate species; this index has shown an average decline of 61% for tropical species and an average 31% increase for temperate species from 1970 to 2008.

A program similar to the efforts of the IUCN is the NatureServe network of Natural Heritage programs, which covers all 50 US states, 3 Canadian provinces, and 14 Latin American countries (www.natureserve.org/explorer). This network, strongly supported by The Nature Conservancy, gathers, organizes, and manages information on the occurrence of "elements of conservation interest"—more than 64,000 species, subspecies, and biological communities, in addition to half a million precisely located populations (**FIGURE 6.16**) (De Grammont and Cuarón 2006). Elements are given status ranks based on a series of standard criteria: number of remaining populations or occurrences, number of individuals remaining (for species) or extent of area (for communities), number of protected sites,

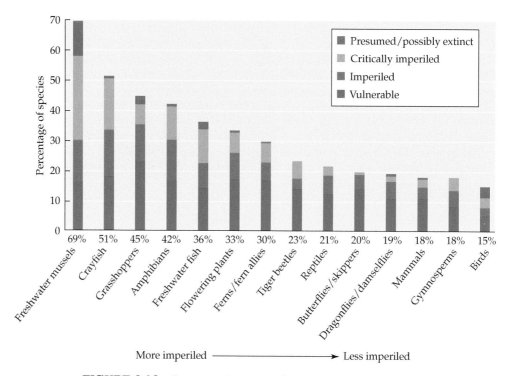

FIGURE 6.16 Some species groups from the United States ranked as presumed extinct, critically imperiled, imperiled, or vulnerable according to criteria endorsed by The Nature Conservancy and coordinated by NatureServe. The groups are arranged with those at greatest risk on the left. Freshwater species are at greater risk of extinction than terrestrial species. (After Wilcove and Master 2005.)

degree of threat, and innate vulnerability of the species or community. When methods of the Natural Heritage system and the IUCN categories are used to evaluate the same species, the resulting rankings of threat are quite similar. On the basis of these criteria, elements are assigned an imperilment rank from 1 to 5, ranging from critically imperiled (1) to demonstrably secure (5), on a global, national, and regional basis. Species are also classified as "X" (extinct), "H" (known historically with searches ongoing), and "unknown" (uninvestigated). Data on these conservation elements are available on the NatureServe website (www.natureserve.org).

Prioritization: What Should Be Protected?

Although some conservationists would argue that no species should ever be lost, the reality is that numerous species are in danger of going extinct and there are too few resources available to save them all. The challenge to conservation efforts lies in finding ways to minimize the loss of biodiversity during a period of limited financial and human resources (Watson et al. 2011).

THE SPECIES APPROACH   One approach to establishing conservation priorities involves protecting particular species—and in doing so, protecting an entire biological community and associated ecosystem processes (Branton and Richardson 2011). Protected areas are often established to protect individual species of special concern, such as rare species, endangered species, keystone species, and culturally significant species. Species such as these, which provide the impetus to protect an area and ecosystems, are known as **focal species**. One type of focal species is an **indicator species**, a species that is associated with an endangered biological community or set of unique ecosystem processes; for instance, the endangered northern spotted owl is a forest indicator species in the Pacific Northwest of the United States. Many national parks have also been created to protect **flagship species**, such as tigers and pandas, which capture public attention, have symbolic value, and are crucial to ecotourism. Flagship and indicator species are also known as **umbrella species** because protecting them automatically protects other species and aspects of biodiversity.

Another way in which the prioritization of species for conservation can be determined is by evaluating the degree to which it is DUE (*d*istinctive, *u*tilitarian, and *e*ndangered) (**FIGURE 6.17**):

1. *Distinctiveness (or irreplaceability)*. A species is often given high conservation value if it is taxonomically distinctive—that is, it is the only species in its genus or family (Faith 2008). Similarly, a population of a species having unusual genetic characteristics that distinguish it from other populations of the species might be a high priority for conservation. An ecosystem composed primarily of rare endemic species or with other unusual attributes (small area, scenic value, unique geological features) is given a high priority for conservation.

2. *Utility*. Species that have present or potential value to people, such as wild relatives of wheat, are given high conservation priority. Species with major cultural significance, such as tigers in India and the bald eagle in the United States, are given high priority.

FIGURE 6.17   The Komodo dragon (*Varanus komodoensis*) of Indonesia is an example of a species that fits all three categories: it is the world's largest lizard (distinctiveness), it has major potential as a tourist attraction in addition to being of great scientific interest (utility), and it occurs on only a few small islands of a rapidly developing nation (endangerment). (© Barry Kusuma/Getty Images.)

3. *Endangerment (or vulnerability).* Species in danger of extinction are of greater concern than species that are not; thus, the whooping crane (*Grus americana*), with only about 382 individuals, requires more protection than the sandhill crane (*Grus canadensis*), with approximately 520,000 individuals.

The species approach follows from creating survival plans for individual species, which are developed by governments and private conservation organizations. In the Americas, the Natural Heritage Programs and the NatureServe network use information on rare and endangered species to target new localities for conservation—areas where there are concentrations of endangered species or where the last populations of a declining species exist (www.natureserve.org). In the IUCN Species Survival Commission, over 100 specialist groups provide action plans for endangered animals and plants.

THE ECOSYSTEM APPROACH   A number of conservationists have argued that rather than species, ecosystems and the biological communities they contain should be targeted for conservation (Tallis and Polasky 2009). They claim that spending $1 million or so on habitat protection and management of a self-maintaining ecosystem might preserve more species and provide more value to people in the long run than spending the same amount of money on an intensive effort to save just one conspicuous species (**FIGURE 6.18**). It often is easy to demonstrate an ecosystem's economic value to policy makers and the public in terms of flood control, clean water, and recreation, whereas arguing for a particular species may be more difficult. Thus, combining species and ecosystem approaches may be a good conservation strategy in many circumstances.

When using this ecosystem-based approach, conservation planners should try to ensure that representative sites of as many types of ecosystems as possible are protected. A **representative site** includes the species and environmental conditions characteristic of the ecosystem. Although no site is perfectly representative, biologists working in the field can often identify suitable sites for protection.

Where immediate decisions must be made to determine park boundaries and which species and ecosystems to protect, biologists are being trained to make **rapid biodiversity assessments**, also known as RAPs (for *rapid assessment programs*). RAPs involve mapping vegetation, making lists of species, checking for species of special concern, estimating the total number of species, and looking out for new species and features of special interest. A bioblitz (explained in Chapter 2) is one type of RAP.

THE WILDERNESS APPROACH   Wilderness areas are a related priority, in part because they are more likely to contain species and populations that need protecting, such as old-growth trees (**FIGURE 6.19**). Large blocks of land that have been minimally affected by human activity, have a low human population density, and are not likely to be developed in the near

(A)

(B)

FIGURE 6.18 (A) The Monterey Bay National Marine Sanctuary in California was established to protect a coastal marine environment that includes marine mammals, seabirds, and ocean bottom species. (B) Feather worms and various coral species living on the ocean bottom at the sanctuary. (Photographs courtesy of NOAA.)

future are also perhaps the only places in the world where large mammals can survive in the wild. These wilderness areas can also serve as reference areas for restoration (see Chapter 10). It is worth emphasizing that even these so-called wilderness areas have had a long history of human activity and people have often affected the structure of the biological communities they contain.

THE HOTSPOT APPROACH Certain organisms can be used as **biodiversity indicators** to highlight new areas where concentrations of species can be protected. For example, a site with a high diversity of flowering plants often, but not always, will also have a high diversity of mosses, spiders, and fungi. Further, areas with high diversity often have a high percent-

FIGURE 6.19 Old-growth giant sequoias are protected in this designated wilderness area within Sequoia and Kings Canyon National Park. (© Neale Clark/robertharding/Getty Images.)

age of endemism—species occurring there and nowhere else (Joppa et al. 2011). The IUCN Plant Conservation office in England has documented about 250 global centers of plant diversity with large concentrations of species (Hoffmann et al. 2008). In a similar effort, BirdLife International has identified over 200 Important Bird Areas (IBAs) with more than 2400 restricted-range bird species; many of these localities are in urgent need of protection (www.birdlife.org).

Using a similar approach, Conservation International, World Wildlife Fund, and others have designated **hotspots** that have great biological diversity and high levels of endemism and that are under immediate threat of species extinctions and habitat destruction (www.biodiversitya-z.org/content/biodiversity-hotspots) (**FIGURE 6.20**). Using these criteria, 34 global hotspots have been targeted for new protected areas. These hotspots together encompass the entire ranges of 12,066 endemic species of terrestrial vertebrates (42% of the world's total) and at least part of the ranges of an additional 35% of the remaining terrestrial vertebrate species—all on only 2.3% of Earth's total land surface.

One major center of biodiversity is the tropical Andes, where 30,000 plant species, 1728 bird species, 569 mammal species, 610 reptile species, and 1155 amphibian species persist in tropical forests and high-altitude grasslands on about 0.3% of Earth's total land surface. The hotspot approach has generated a considerable amount of enthusiasm and funding, and it will be worth watching to see how successful it is in advancing the goals of conservation in areas of intense human pressures, some of which are sites of armed conflicts (Hanson et al. 2009).

(A)

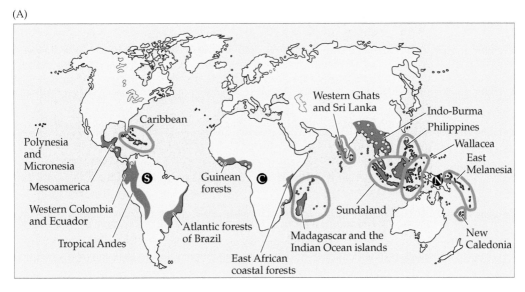

(B)

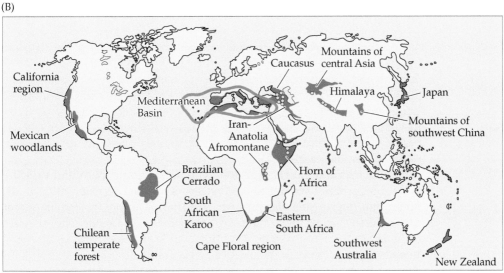

FIGURE 6.20 Hotspots are targets for protection because of their high biodiversity, endemism, and significant threat of imminent extinctions. (A) Sixteen tropical rain forest hotspots. Areas circled in green are island groups. The Polynesia/Micronesia region (far left) covers a large number of Pacific Ocean islands, including the Hawaiian Islands, Fiji, Samoa, French Polynesia, and the Marianas. Black-circled letters indicate the only three remaining undisturbed rain forest areas of any extent, in South America (S), the Congo basin of Africa (C), and the island of New Guinea (N). (B) Eighteen hotspots representing other ecosystems. Yellow dots denote areas that have experienced armed conflicts between 1950 and 2000 with over 1000 casualties. (After Hansen et al. 2009; Mittermeier et al. 2005.)

Legal Protection of Species

Once conservation biologists have identified a species as needing protection, laws can be passed and treaties can be signed to implement conservation efforts. National laws protect species within individual countries, while international agreements provide a broader framework for conservation.

> National governments protect designated endangered species within their borders, establish national parks, and enforce legislation on environmental protection.

National laws

People in many countries recognize that preserving a healthy environment and protecting species are linked to sustaining human health. National governments and national conservation organizations in such countries acknowledge this and play an important role in the protection of all levels of biological diversity. Laws are passed to establish national parks and other protected areas; to regulate activities such as fishing, logging, and grazing; and to limit air and water pollution. International treaties that restrict trade in endangered animals are implemented at the national level and enforced at the borders. The true measure of a nation's commitment to protecting biodiversity is the effectiveness with which these laws are enforced.

In European countries, endangered species conservation is accomplished through domestic enforcement of international agreements such as the Convention on International Trade in Endangered Species (CITES) and the Ramsar Convention on Wetlands (see the section "Intenational agreements" on p. 226). Species that occur on the IUCN's international Red Lists of endangered species and in national Red Data books are also protected (Fontaine et al. 2007). To indicate where these species can be found, the Fauna Europaea database provides information on the distribution of 130,000 terrestrial and freshwater species. Countries in Europe protect species and habitats through directives adopted by the European Union; these directives implement the earlier Bern Convention, which was established to protect endangered species in Europe, with a special focus on migratory species. Some countries may have additional laws, such as the Wildlife and Countryside Act of 1981 in the United Kingdom, which protects habitat occupied by endangered species.

Despite the fact that many countries have enacted legislation to preserve biodiversity, it also is true that national governments are sometimes unresponsive to requests from conservation groups to protect the environment. In some cases national governments have acted to decentralize decision making, relinquishing control of natural resources and protected areas to local governments, village councils, and conservation organizations.

THE US ENDANGERED SPECIES ACT   The United States has developed a system for designating and protecting threatened species that is separate from the IUCN system. In the United States, the principal conservation law protecting species is the **Endangered Species Act** (**ESA**), passed in 1973 and subsequently amended in 1978 and 1982. The ESA was created by the US Congress to "provide a means whereby the ecosystems upon which endangered species and threatened species depend may be conserved [and] to provide a program for the conservation of such species." Species are protected under the ESA if they are on the official list of endangered and threatened species. In addition, a recovery plan is generally required for each listed species (Himes Boor 2014).

As defined by law, "endangered species" are those likely to become extinct as a result of human activities or natural causes in all or a significant portion of their range; "threatened species" are those likely to become endangered in the near future. The secretary of the Interior Department, acting through the US Fish and Wildlife Service (FWS), and the secretary of the Commerce Department, acting through the National Marine Fisheries Service (NMFS), can add and remove species from the list based on information available to them. Since 1973, more than 1519 species in the United States have been added to the list, including many well-known species such as the whooping crane (*Grus americana*) and the manatee (*Trichechus manatus*), in addition to 625 endangered species from elsewhere in the world that face special restrictions when they are imported into the United States.

Many species are listed under the ESA only when they have fewer than 100 individuals remaining, making recovery difficult (see Chapter 5). An early listing of a declining species might allow it to recover and thus become a candidate for removal from the list sooner than if authorities wait for its status to worsen before adding it to the list. The great majority of species in the United States listed under the ESA are flowering plants (839 species) and vertebrates (403 species), despite the fact that most of the world's species are insects and other invertebrates (ecos.fws.gov/tess_public/reports/box-score-report). Only 70 insect species are currently listed; however, if the same proportion of insects as of vertebrates were protected, an estimated 29,000 species would be protected under the ESA, an awesome number to contemplate (Dunn 2005). Clearly, greater efforts must be made to study the lesser known and underappreciated invertebrate groups and extend listing to those endangered species whenever necessary (Stankey and Shindler 2006).

> The ESA mandates such strong protection for species that conservation and business groups often agree to compromises that allow some species protection along with limited development.

The protection afforded to species listed under the ESA is so strong that business interests and landowners often lobby strenuously against listing species in their area. At the extreme are landowners who destroy endangered species or species being considered for listing on their property to evade the provisions of the ESA, a practice informally known as "shoot, shovel, and shut up." Such was the fate of the threatened Preble's meadow jumping mouse (*Zapus hudsonius*), which lives in streamside habitats in Colorado and Wyoming on a quarter of the sites that contained suitable habitat (Brook et al. 2003). Many argue that landowners should be compensated in some way to provide an important incentive to conserve endangered species and their habitats. Financial incentives exist to help private landowners improve habitats for endangered species (see Chapter 11), but they may be too small to be effective in many cases.

Another important obstacle to listing is the difficulty of species recovery—rehabilitating species or reducing the threats to species to the point where they can be removed from listing under the ESA, or "delisted." So far, only about 30 of more than 1436 listed species in the United States have been delisted because of recovery, and another 25 species have shown enough recovery to be reclassified from endangered to threatened (Schwartz 2008). The most notable successes include the brown pelican (*Pelecanus occidentalis*), the American peregrine falcon (*Falco peregrinus*), and the American alligator (*Alligator mississippiensis*). In 2007, the bald eagle (*Haliaeetus leucocephalus*) was removed from the federal list of threatened and endangered species because its numbers in the lower 48 states had increased from 400 breeding pairs in the 1960s to over 9000 pairs (see Figure 6.4).

Overall, most listed species are still declining in range and abundance, and unfortunately, for around 20% of species there is insufficient data to determine whether their populations are changing over time (Leidner and Neel 2011). Due to their low numbers and consequent vulnerability, there is now recognition that even species that are candidates for delisting will still require some degree of conservation management to maintain their populations (Redford et al. 2011) (**FIGURE 6.21**).

The difficulty of implementing recovery plans for so many species is often not primarily biological but, rather, political, administrative, and

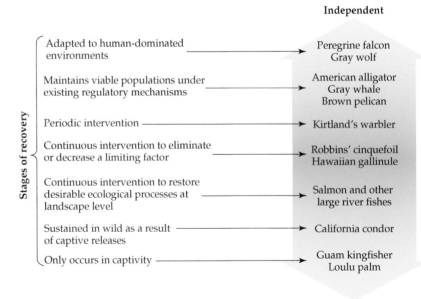

FIGURE 6.21 Endangered species often require active management and intervention as part of the recovery process. There is a continuum, with some species independent of humans and others dependent on human intervention. (After Scott et al. 2005.)

ultimately financial (Briggs 2009). For example, an endangered river clam species might need to be protected from pollution and the effects of an existing dam. Installing sewage treatment facilities and removing a dam are theoretically straightforward actions but they are expensive and difficult to carry out in practice. Total expenditures reported for fiscal year 2013 were $1.75 billion, of which $1.67 billion was reported by federal agencies and $76 million was reported by the states (USFWS 2013). However, this amount of funding is less than the amount requested for listing and recovery purposes (Gibbs and Currie 2012). The cost would be much higher if the US government granted private landowners financial compensation for ESA-imposed restrictions on the use of their property, an option that is periodically discussed in the US Congress.

Funding for the ESA has been growing steadily over the past 20 years, but the number of protected species has been growing even faster. As a result, there is less money available per species. The importance of adequate funding for species recovery is shown by a study demonstrating that species that receive a higher proportion of requested funding for their recovery plans have a higher probability of reaching a stable or improved status than species that receive a lower proportion of funding (Miller et al. 2002). The longer a species has been protected under the ESA, the higher is the probability that it is improving in status (Taylor et al. 2005) (**FIGURE 6.22**). Also, species have a higher probability of improving if critical habitat and a recovery plan have been designated for them (Gibbs and Currie 2012).

Concerns about the implications of ESA protection force business organizations, conservation groups, and goverments to develop compromises

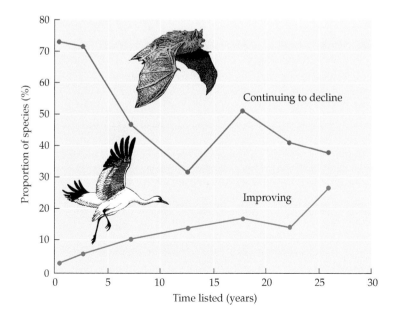

FIGURE 6.22 The longer species have been listed, protected, and managed under the Endangered Species Act, the greater is their probability of improving in status (as shown by the whooping crane) and the slower is their probability of continuing to decline in status (with the Indiana bat as an example). The numbers do not add up to 100% because some species are not changing in status and others are of unknown status. (After Taylor et al. 2005.)

that reconcile both conservation and business interests (Camacho 2007). To provide a legal mechanism to achieve this goal, Congress amended the ESA in 1982 to allow the design of **habitat conservation plans** (**HCPs**). HCPs are regional plans that allow development in designated areas but also protect remnants of biological communities or ecosystems that contain groups of actual or potentially endangered species. These plans are drawn up by the concerned parties—developers, conservation groups, citizen groups, and local governments—and given final approval by the FWS. About 650 HCPs covering about 16 million ha and over 500 species have been approved. In one case, an innovative program in Riverside County, California, allows developers to build within the historic range of the endangered Stephens' kangaroo rat (*Dipodomys stephensi*) if they contribute to a fund that will be used to buy land and establish wildlife sanctuaries to protect the species. Already, more than $42 million has been used to secure 41,000 ha, with the long-term goal of raising $100 million. As a result of the HCP and the resulting new reserves, the Stephens' kangaroo rat is being considered for removal from the endangered species list; however, the long-term effectiveness of these measures remains to be seen. While HCPs are not perfect, they are attempts to create the next generation of conservation planning. They seek to protect many species, entire ecosystems, or whole communities; extend over a wide geographical region; and include many projects, landowners, and jurisdictions.

INTERNATIONAL AGREEMENTS   International agreements to protect species and other aspects of biodiversity and ecosystems are needed for several reasons: species migrate across borders, there is international trade in biological products, the benefits of biodiversity are of international importance, and the threats to biodiversity are often international in scope. International agreements have provided a framework for countries to cooperate in protecting species, ecosystems, and genetic variation. Treaties are negotiated at international conferences and come into force when they are ratified by a certain number of countries, often under the authority of international bodies, such as the United Nations Environment Programme (UNEP), the Food and Agriculture Organization of the United Nations (FAO), and the IUCN.

The benefits of biodiversity conservation also accrue globally, as the environment can supply natural products for agriculture, medicine, and industry and genetic materials for research and breeding new or stronger varieties of species for human use. Tropical regions are also important in the global ecosystem through their influence on carbon dioxide (CO_2) levels and weather patterns. Each country is ultimately responsible for protecting its own natural environment, but until their economies become stronger, developing countries may be unable to pay for the habitat preservation, research, and management required for the task. Because biodiversity conservation is important at both the national and global levels, it is fair for the developed countries of the world (including the United

States, Canada, Japan, Australia, and many European nations) to help pay to protect biodiversity.

The protection of biodiversity must be addressed at multiple levels of government. Although the major control mechanisms that presently exist in the world are based within individual countries, international agreements among countries are increasingly used to protect species, ecosystems, and genetic variation. International cooperation is an absolute requirement for several reasons:

- *Species migrate across international borders.* Conservation efforts must protect species at all points in their ranges; efforts in one country will be ineffective if critical habitats are destroyed in a second country to which an animal migrates (Bradshaw et al. 2008) (see Chapter 8). For example, efforts to protect migratory bird species in northern Europe will not work if the birds' overwintering habitat in Africa is destroyed. Efforts to protect whales in US coastal waters will not be effective if these species are killed or harmed in international waters. Species are particularly vulnerable when they are migrating, as they may be more conspicuous, more tired, or more desperately in need of food and water. Globally, international parks, often called "peace parks," have been created to protect species living and moving through border areas, such as the Waterton–Glacier International Peace Park on the border of the United States and Canada, which protects grizzly bears (*Ursus arctos horribilis*) and lynx (*Lynx canadensis*).

- *International trade in biological products is commonplace.* A strong demand for a product in one country can result in the overexploitation of the species in another country to supply this demand. When people are willing to pay high prices for exotic pets, plants, or wildlife products such as rhino horn, poachers looking for easy profits, or poor, desperate people looking for any source of income, will take or kill even the very last animal to obtain this income. To prevent overexploitation, consumers who buy wildlife products, and the people who collect and trade them, need to be educated about the consequences of overuse of wild species. When poverty is the root of overexploitation, it is sometimes possible to provide people with economic alternatives while strictly controlling resource use (see Chapter 11). Where exploitation stems from greedy people seeking to make a profit, laws and enforcement efforts such as border checks should be strengthened.

- *Biodiversity provides internationally important benefits.* The community of nations benefits from the species and genetic variation used in agriculture, medicine, and industry; the ecosystems that help regulate climate; and the national parks and other protected areas of international scientific and tourist value. It is also widely recognized that biodiversity has intrinsic value, existence value, and option value (see Chapter 3). The developed countries of the world that use and rely on biodiversity and ecosystem services from poor tropical countries provide limited, inadequate funding to help these less-wealthy countries manage and protect globally significant resources. Funding levels need to be increased and the funds must be used more effectively.

- *Many environmental pollution problems that threaten ecosystems are international in scope.* Such threats include atmospheric pollution and acid rain; the pollution of lakes, rivers, and oceans; greenhouse gas production and global climate change; and ozone depletion (Srinivasan et al. 2008). Additionally, the environmental costs of many of these problems do not fall on countries in proportion to their role in causing them. For example, the United States and China are the world's leading producers of greenhouses gases (see Chapter 4), but many low-lying countries such as Bangladesh and the Maldives will be most affected by the rising sea levels associated with climate change. Or consider the River Danube, which carries the pollution of a vast agricultural and industrial region that spans 10 countries before it empties into the Black Sea, another international body of water bordered by 4 additional countries. Problems such as these can only be solved by countries working together.

International agreements to protect species

To address the protection of biodiversity, countries worldwide have signed a number of key international agreements. These agreements have provided a framework for countries to cooperate in protecting species, habitats, ecosystem processes, and genetic variation. Treaties are negotiated at international conferences and come into force when they are ratified by a certain number of countries, often under the authority of international bodies such as the UNEP, the FAO, and the IUCN.

As was mentioned earlier, one of the most important treaties protecting species at an international level is the **Convention on International Trade in Endangered Species** (**CITES**), established in 1973 in association with UNEP (Reeve 2014; www.cites.org). Currently there are 179 member countries. CITES, headquartered in Switzerland, establishes lists (known as Appendices) of species for which international trade is to be controlled or monitored. Member countries agree to restrict trade in these species and halt their destructive exploitation. Regulated plants include important horticultural species such as orchids, cycads, cacti, carnivorous plants, and tree ferns; timber species and wild-collected seeds are increasingly being considered for regulation as well. Closely regulated animal groups include parrots, large cats, whales, sea turtles, birds of prey, rhinos, bears, and primates. Species collected for the pet, zoo, and aquarium trades and species harvested for their fur, skin, or other commercial products are also closely monitored.

International treaties such as CITES are implemented when a country signing the treaty passes laws to enforce it. Nongovernmental organizations (NGOs) such as the IUCN, Wildlife Trade Program, the TRAFFIC network run by the World Wildlife Fund

> CITES has developed extensive lists of species for which trade is prohibited, controlled, or monitored. Many countries have established their own Red Data list of species that they protect within their borders.

(WWF) and the IUCN, and UNEP's World Conservation Monitoring Centre (WCMC) provide technical advice regarding legal and enforcement aspects of CITES to national governments. Countries may also protect species listed by national Red Data books. Once species protection laws are passed within a country, police, customs inspectors, wildlife officers, and other government agents can arrest and prosecute individuals possessing or trading in protected species and can seize the products or organisms involved (**FIGURE 6.23**). For example, in November 2013, Thai authorities at the Bangkok airport seized hundreds of endangered turtles being shipped illegally in passenger baggage. What made this story particularly unusual is that the seizure and news story took place just a day after a major CITES conference in Bangkok.

Member countries are required to establish their own management and scientific authorities to implement CITES obligations within their own borders (see Chapter 11). CITES is particularly active in encouraging cooperation among countries, in addition to fostering conservation efforts by development agencies. The CITES Secretariat periodically sends out bulletins aimed at publicizing specific illegal activities. For example, in recent years, the CITES Secretariat has recommended that its member countries

FIGURE 6.23 A fur reference collection in northern China. For some products, such as the zebra skin, the type of animal involved can be easy to identify, but for other products, such as bags, coats, rugs, and shoes, the type of animal used to make them may be hard to determine, and often requires microscopic analysis of hairs. (Photograph by Richard Primack.)

FIGURE 6.24 The burning of ivory in Kenya. To keep ivory off the international market and hopefully reduce the killing of wild elephants, wildlife authorities in Kenya burned more than 15 tons of elephant tusks seized from poachers in 2015, and made plans to burn an addtional 120 tons of both elephant tusks and rhino horns in 2016. (© Carl de Souza/AFP/Getty Images.)

halt wildlife trade with Vietnam because of that country's unwillingness to restrict the illegal export of wildlife from its territory.

CITES has been instrumental in restricting trade in certain endangered wildlife species. Its most notable success was a global ban on the ivory trade after poaching caused severe declines in African elephant populations (**FIGURE 6.24**). Recently, countries in southern Africa with increasing elephant populations have been allowed to resume limited ivory sales, resulting in an unfortunate increase in illegal harvesting.

While illegal trade in wildlife may not sound important, it is a huge business, estimated to be between $10 and $20 billion per year, excluding aquatic species (see Chapter 4). It remains a major problem and is sometimes linked to illegal trade in timber, drug smuggling, and arms trafficking. Not surprisingly, compiling accurate data on illegal wildlife trade is a challenge (Sajeva et al. 2013). One difficulty with enforcing CITES is that shipments of both living and preserved plants and animals are often mislabeled, due to either ignorance of species names or deliberate attempts to avoid the restrictions of the treaty. Also, sometimes countries fail to enforce the restrictions of the treaty because of corruption or a lack of trained staff. Finally, many restrictions are difficult to enforce because some international borders are remote, rugged, and difficult to monitor, such as that between Laos and Vietnam. As a result, the illegal wildlife trade continues to pose one of the most serious threats to biodiversity, particularly in Asia.

Another key treaty is the **Convention on the Conservation of Migratory Species of Wild Animals (CMS)**, often referred to as the **Bonn Convention**, which focuses primarily on bird species (www.cms.int). This convention, which has been signed by 119 countries, complements CITES by encouraging international efforts to conserve bird species that migrate across international borders and by emphasizing regional approaches to research, management, and hunting regulations. The convention now includes protection of bats and their habitats and cetaceans in the Baltic and North Seas. Other important international agreements that protect species include:

- Convention on the Conservation of Antarctic Marine Living Resources (www.ccamlr.org)
- International Convention for the Regulation of Whaling, which established the International Whaling Commission (www.iwc.int)
- International Convention for the Protection of Birds and the Benelux (Belgium/Netherlands/Luxembourg) Convention Concerning Hunting and the Protection of Birds
- Convention for the Conservation and Management of Highly Migratory Fish Stocks in the Western and Central Pacific Ocean (www.wcpfc.int)
- Additional agreements protecting specific groups of animals, such as prawns, lobsters, crabs, fur seals, Antarctic seals, salmon, and vicuña

A number of more broadly focused international agreements are also increasingly seeking direct protection of endangered species. For example, the Convention on Biological Diversity, described in Chapter 11, now includes recommendations for the protection of IUCN Red Listed species (www.iucnredlist.org).

A weakness of all these international treaties is that they operate through consensus, so strong measures often are not adopted if one or more countries oppose them. Also, any nation's participation is voluntary, meaning that countries can choose to ignore these conventions and pursue their own interests if they find the conditions of compliance too difficult (Carraro et al. 2006). This flaw was highlighted when several countries decided not to comply with the International Whaling Commission's 1986 ban on whale hunting, and the Japanese government announced that it would continue hunting whales under the dubious claim that further data were needed to evaluate the status of whale populations. Persuasion and public pressure are the principal means used to induce countries to enforce treaty provisions and prosecute violators, though funding through treaty organizations can also help. An additional problem is that many conventions are underfunded and are consequently ineffective in achieving their goals. Unfortunately, there are often no monitoring mechanisms in place to determine whether countries are even enforcing the treaties.

Summary

- Protecting and managing a rare or endangered species requires a firm grasp of its ecology and its distinctive characteristics (sometimes called its natural history). Long-term monitoring of a species in the field can determine if it is stable, increasing, or declining in abundance over time.

- Population viability analysis (PVA) uses demographic, genetic, and environmental data to estimate how various management actions will affect the probability that a population will persist until some future date. It can be used to calculate the minimum viable population (MVP) size: the smallest population size that can be predicted to have a high chance of persisting for the foreseeable future. The MVP for many species is at least several thousand individuals.

- A species may be best described as a metapopulation made up of a shifting mosaic of populations that are linked by some degree of migration.

- The IUCN has developed quantitative criteria for populations and ecosystems to assign species to conservation categories: extinct, extinct in the wild, critically endangered, endangered, vulnerable, near threatened, least concern, data deficient, and not evaluated.

- Priorities for protection can be determined in several ways, including the species approach, the ecosystem approach, the wilderness approach, and the hotspot approach.

- National governments protect biodiversity by establishing national parks and refuges, controlling imports and exports at their borders, and creating regulations for air and water pollution. The most effective law in the United States for protecting species is the Endangered Species Act (ESA).

- At the international level, the Convention on International Trade in Endangered Species (CITES) allows governments to regulate, monitor, and sometimes prohibit trade in individuals and products from endangered species.

For Discussion ▪

1. How might you monitor populations of a species of fish over time? Would your methods differ if the fish was a freshwater species or a marine species? Why or why not?

2. Choose a threatened species in your region. Weigh the merits and limitations of a species-centered approach, an ecosystem approach, a wilderness approach, and a hotspot approach for protecting it. Which approach will be most effective for protecting this species? Which approach is most feasible? Which approach will protect the most species in addition to your target?

3. A wide range of laws protect endangered species. Why don't species covered by such laws quickly recover?

Suggested Readings

Carroll, C., R. J. Frederickson, and R. C. Lacy. 2014. Developing metapopulation connectivity criteria from genetic and habitat data to recover the endangered Mexican wolf. *Conservation Biology* 28: 76–86. Metapopulation models demonstrate that dispersal between populations is crucial to preventing local extinction for small populations.

Douglas, L. R. and K. Alie. 2014. High-value natural resources: Linking wildlife conservation to international conflict, insecurity, and development concerns. *Biological Conservation* 171: 270–277. Social, economic, and political issues must be addressed for conservation goals to be achieved.

Duarte, A., J. S. Hatfield, T. M. Swannack, M. R. Forstner, et al. 2016. Simulating range-wide population and breeding habitat dynamics for an endangered woodland warbler in the face of uncertainty. *Ecological Modelling* 320: 52–61. Mathematical tools that incorporate stochasticity assist conservation management planning.

Hedges, S., A. Johnson, M. Ahlering, M. Tyson, and L. S. Eggert. 2013. Accuracy, precision, and cost-effectiveness of conventional dung density and fecal DNA based survey methods to estimate Asian elephant (*Elephas maximus*) population size and structure. *Biological Conservation* 159: 101–108. New DNA methods are greatly improving our ability to estimate population size.

Jäkäläniemi, A., A. H. Postila, and J. Tuomi. 2013. Accuracy of short-term demographic data in projecting long-term fate of populations. *Conservation Biology* 27: 552–559. Many years of data are needed to build a good PVA model.

Liu, P., L. Sun, J. Li, L. Wang, et al. 2015. Population viability analysis of *Gloydius shedaoensis* from northeastern China: A contribution to the assessment of the conservation and management status of an endangered species. *Asian Herpetological Research* 1(1): 48–56. PVAs have been used to evaluate current protection and management strategies for Chinese snakes.

NatureServe. 2009. http://natureserve.org. This website organizes and presents data on biodiversity surveys from North America.

Pittman, S. E., M. S. Osbourn, and R. D. Semlitsch. 2014. Movement ecology of amphibians: A missing component for understanding population declines. *Biological Conservation* 169: 44–53. Studies of juvenile and adult movement patterns are critical for the conservation management of amphibians.

Schwartz, M. W. 2008. The performance of the Endangered Species Act. *Annual Review of Ecology, Evolution, and Systematics* 39: 279–299. Many listed species are recovering, but certain goals have not been achieved.

Thomsen, P. F. and W. Willerslev. 2015. Environmental DNA—An emerging tool in conservation for monitoring past and present biodiversity. *Biological Conservation* 183: 4–18. Detecting whole organisms or just traces of organisms from bulk samples of soil, water, or air expands our capacity to survey populations and species diversity.

7

Bringing Species Back from the Brink

European bison, or
wisent (*Bison bonasus*),
at snowfall, Germany.

I n Chapters 5 and 6, we discussed the problems conservation biologists face in preserving naturally occurring populations of endangered species. This chapter discusses some exciting conservation methods used to establish new wild and semi-wild populations of rare and endangered species and increase the sizes of existing populations. These methods include breeding species in zoos, aquaria, and botanic gardens—organizations that also assist conservation through education and research programs. Captive breeding and other approaches to augment or establish new populations may allow species that have persisted only in captivity or in small, isolated populations to regain their ecological and evolutionary roles within their ecosystems. Furthermore, simply increasing the number and size of its populations generally lowers the probability a species will go extinct.

Population establishment programs are unlikely to be effective, however, unless the factors leading to the decline of the original wild populations are clearly understood and eliminated, or at least controlled (Venevsky et al. 2005). For example, endangered amphibians are increasingly being raised in captivity,

> Establishing new populations of endangered species can benefit the species itself, other species, and the ecosystem. However, such programs must identify and eliminate the factors that led to the original population's decline.

but they cannot be released back into the wild if they lack resistance to the chytrid fungal pathogen *Batrachochytrium dendrobatidis*, a nonnative and now widespread fungus that is killing them worldwide (Kolby et al. 2015). One possibility is to breed captive amphibian populations for fungal resistance before attempting to establish new populations in the wild.

Establishing and Reinforcing Populations

Three basic approaches, all involving relocation of existing captive-bred or wild-collected individuals, have been used to establish new populations of animals and plants and to enlarge existing populations. The IUCN's Reintroduction Specialist Group coordinates many of these efforts (www.iucnsscrsg.org):

- A **reintroduction program**[1] involves releasing captive-bred or wild-collected individuals at an ecologically suitable site within their historical range where the species no longer occurs (Carter et al. 2016).

- A **reinforcement program** involves releasing individuals into an existing population to increase its size and gene pool (Smyser et al. 2013); this approach can be thought of as restocking or augmentation. These released individuals may be raised in captivity or may be wild individuals collected elsewhere.

- An **introduction program** involves moving captive-bred or wild-collected animals or plants to areas suitable for the species outside their historical range.

The main objectives of a reintroduction program are to create a new population in its original environment and to help restore a damaged ecosystem. For example, a program initiated in 1995 to reintroduce gray wolves (*Canis lupus*) into Yellowstone National Park aims to restore the equilibrium of predators, herbivores, and plants that existed prior to human intervention in the region (**FIGURE 7.1A**). Wild-collected animals are also sometimes caught and then released elsewhere within the range of their species when a new protected area has been established, when an existing population is under a new threat in its present location, or when natural or artificial barriers to the normal dispersal tendencies of the species exist. If possible, individuals are released near the site where they or their ancestors were collected to ensure genetic adaptation to their environment.

---

[1]Some confusion exists about the terms denoting the establishment of populations. Reintroduction programs are sometimes called *reestablishments* or *restorations*. Another term, *translocation*, usually refers to moving individuals from a location where they are about to be destroyed to another site that, hopefully, provides a greater degree of protection.

(A)

(B)

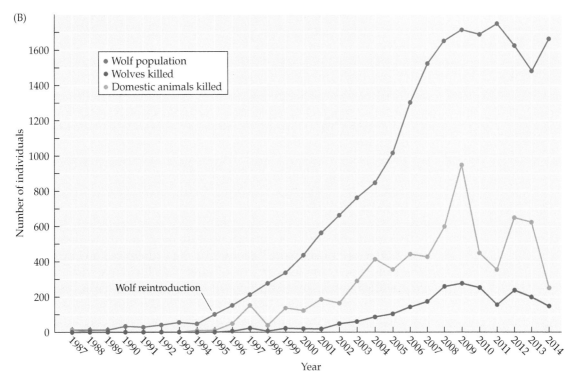

FIGURE 7.1   (A) A gray wolf in Yellowstone National Park wears a radio transmitter collar that allows researchers to follow its movements. (B) The number of wolves in Wyoming, Idaho, and Montana increased following the reintroduction of wolves to the Yellowstone area in 1995. There has also been an increase in the number of domestic animals, primarily sheep, killed by wolves, and an increase in the number of problem wolves killed by government authorities. (A, photograph courtesy of William Campbell/US Fish and Wildlife Service; B, after Musiani et al. 2003, with updates courtesy of M. Musiani, and from Clark and Johnson 2009.)

In contrast to reintroduction, *introduction* to an entirely new location outside the existing range of a species may be appropriate when the environment within the known range of a species has deteriorated to the point at which the species can no longer survive there, or when reintroduction is impossible because the factor causing the original decline is still present in the species' range. This approach is sometimes referred to as assisted colonization (Lunt et al. 2013). The New Zealand kakapo, for example, was introduced to offshore islands because nonnative predators had decimated its populations on the mainland (Moro et al. 2015). In the near future, introductions may be necessary for many species if those species can no longer survive within their current ranges because of a warming climate (Hendricks et al. 2016).

The reintroduction or introduction of a species must be carefully thought out so that the released species does not damage its new ecosystem or harm populations of any local endangered species (Olden et al. 2011). Care must be taken that released individuals have not acquired any diseases that could spread to and decimate wild populations.

One special method used in establishing new populations or reinforcing existing populations is *head-starting*, an approach in which animals are raised in captivity during their vulnerable young stages and then released into the wild. The release of sea turtle hatchlings produced from eggs collected from the wild and raised in nearby hatcheries is an example of this approach (see Chapter 1).

Considerations for animal programs

Establishing new populations is often expensive and difficult because it requires a serious, long-term commitment. The programs that capture, raise, release, and monitor sea turtles, peregrine falcons, and whooping cranes, for example, have cost millions of dollars and have required years of work. When the animals involved are long-lived, the program may have to continue for many years before its outcome is known (Grenier et al. 2007).

Population establishment programs can also become highly emotional public issues, as demonstrated by the programs for the California condor, the black-footed ferret, and the gray wolf in the United States and comparable programs in Europe. These programs are often criticized on many different fronts. They may be attacked as a waste of money ("Millions of dollars for a few ugly birds!"), unnecessary ("Why do we need wolves here when there are so many elsewhere?"), intrusive ("We just want to go about our lives without the government telling us what to do!"), poorly run ("Look at all the ferrets that died of disease in captivity!"), or unethical ("Why can't the last animals just be allowed to live out their lives in peace without being captured and put into zoos?"). The answer to all of these criticisms is straightforward. Although not appropriate for every endangered species, a well-run, well-designed captive breeding and population establishment program may be the best hope for a species' preservation.

Because of the conflicts and high emotions involved, it is crucial that population establishment programs include local people so that (ideally) the community has a stake in the program's success. (Indeed, this is true of any conservation project.) At a minimum, it is necessary to explain the need for the program and its goals and to convince local people to support it—or at least not to oppose it (Yochim and Lowry 2016). Programs are often more successful if they provide incentives to affected people rather than imposing rigid restrictions and laws. For example, direct payments are made to Wyoming residents whose domestic animals are injured by reintroduced wolves, and the few wolves that repeatedly attack livestock are either killed or moved in order to retain local support for the program (Boitani et al. 2011) **(FIGURE 7.1B)**.

Successful reintroduction programs often have considerable educational value. In Brazil, conservation and reintroduction efforts for the golden lion tamarin (*Leontopithecus rosalia*) have become a rallying point for the protection of the last remaining fragments of the Atlantic forest (see textbook cover image). In the Middle East and northern Africa, captive-bred Arabian oryx (*Oryx leucoryx*) have been successfully reintroduced into many desert areas that they formerly occupied, providing a source of national pride and opportunities for employment (Fisher 2016).

The introduction of a species to new sites needs to be carefully considered and evaluated in order to ensure that the species does not damage its new ecosystem or harm any local populations of other endangered species (Olden et al. 2011). Care must be taken to ensure that released individuals have not acquired any diseases while in captivity that could spread to and decimate wild populations. For example, captive black-footed ferrets (*Mustela nigripes*) must be carefully handled and quarantined so that they do not acquire diseases from people or dogs that they might transfer into wild populations upon their release in North American grasslands **(FIGURE 7.2A)**.

There is an important genetic component to the selection of plants or animals for reintroduction programs. Captive populations may have lost much of their genetic variation or, when raised for several generations in captive conditions, may have become genetically adapted to the benign captive environment. Such genetic changes, which lower the species' ability to survive in the wild following release, have occurred in captive populations of Pacific salmon. Individuals have to be carefully selected to guard against inbreeding depression and to produce a genetically diverse release population (Ottewell et al. 2014). Wild-caught individuals must be selected from a population living in an environment and climate that is as similar as possible to that of the release site. After release, a species may adapt genetically to its new environment such that the gene pool from the original wild population is not actually being preserved.

Some animal species may require special care and assistance immediately after release to increase their survival prospects (Harrington et al. 2013; White et al. 2012). This approach is known as **soft release**. Animals may have to be fed and sheltered at the release point until they are able to subsist on their own, or they may need to be caged temporarily at the release point and introduced gradually, once they become familiar with the sights, sounds, smells, and layout of the area (**FIGURE 7.2B**). Eighty-eight chicks of the Mauritius kestrel (*Falco punctatus*) raised in captivity by humans and then given a soft release into the wild on the Indian Ocean island of Mauritius, for example, had a survival rate that was not significantly different from that of 284 chicks born in the wild (Nicoll et al. 2004).

Animals can also be released without assistance such as food supplementation (**hard release**), although reintroductions of this type are more likely to succeed with wild-caught over captive-bred individuals (Bocci

(A)

(B)

FIGURE 7.2 (A) A black-footed ferret raised at the captive colony in Colorado. (B) Cages allow black-footed ferrets to experience the environment into which they will eventually be released. The ferrets' caretaker is wearing a mask to reduce the chance of exposing the animals to human diseases. (A, photograph by Ryan Hagerty/US Fish and Wildlife Service; B, photograph by M. R. Matchett, courtesy of US Fish and Wildlife Service.)

et al. 2016). Intervention may be necessary if animals appear unable to survive, particularly during episodes of drought or low food abundance (Blanco et al. 2011). Even when animals appear to have enough food to survive, supplemental feeding may help by increasing reproduction and allowing the population to persist and grow. Outbreaks of diseases and pests may have to be monitored and dealt with. The effects of human activities in the area, such as farming and hunting, need to be observed and possibly controlled. In every case, a decision has to be made about whether it is better to give occasional temporary help to the species or to force the individuals to survive on their own (Harrington et al. 2013).

Establishment programs for common wildlife species managed for hunting have always been widespread and have contributed a great deal of knowledge to biologists trying to establish new populations of endangered species. A number of generalizations can be made from analyses of about 200 establishment programs for such common game species (Fischer and Lindenmayer 2000; Hughes and Lee 2015):

- Success was greater for releases in excellent-quality habitat (84%) than in poor-quality habitat (38%).
- Success was greater in the core of the historical range (78%) than at the periphery of and outside the historical range (48%).
- Success was greater with wild-caught (75%) than with captive-reared animals (38%).
- Success was greater for herbivores (77%) than for carnivores (48%).
- For the bird and mammal species studied, the probability of establishing a new population increased with the number of animals being released, up to about 100 individuals. Releasing more than 100 animals did not further enhance the probability of success.

Other examinations of the published literature have found that the success rate of reintroduction projects involving endangered mammals, birds, reptiles, amphibians, and fish is generally much lower than that of projects involving wildlife managed for hunting, suggesting that improved methods need to be developed for endangered species (Griffiths and Pavajeau 2008; Harding et al. 2015). In fact, the rates of success may be even lower than these values indicate because the results of many projects, particularly those that fail, are not published and are poorly documented (Miller et al. 2014; White et al. 2012). As we'll see later in the chapter, this low rate of success emphasizes that protecting existing populations of endangered species and improving their habitat must remain our highest priorities.

Clearly, monitoring of establishment programs is crucial in determining whether the programs are achieving their stated goals (Seddon et al 2014; McCleery et al. 2014). Monitoring may need to be carried out over many years, even decades, because many reintroductions that initially appear successful eventually fail. The key elements of monitoring are determining whether released individuals survive and establish a breeding population, then following that population over time to see whether it increases in numbers of individuals and geographic range. Monitoring of important ecosystem elements is also needed to determine the broader impact of a reintroduction; for example, when a predator species is introduced, it will be crucial to determine its effect on prey species and competing species and its indirect effect on vegetation (Baker et al. 2016). In an otter reintroduction program, for instance, the returned otter populations appealed to the general public, but the otters reduced populations of fish and crustaceans, which angered commercial fishermen (Fanshawe et al. 2003).

The costs of reintroduction also need to be tracked and published so it can be determined whether reintroduction represents a cost-effective strategy. In the case of the orangutan (*Pongo* spp., **FIGURE 7.3**), it was found that reintroduction costs twelve times as much per animal as the protection of forest habitat. Reintroduction was effective in the short term, but at time scales longer than 10–20 years, habitat conservation was much more cost-effective (Wilson et al. 2014).

Behavioral ecology of released animals

To be successful, both introduction and reintroduction programs must often address the behaviors of animals that are being released (Buchholz 2007). *Behavioral ecology* is the study of an animal's behavior in the context of its environment and considers the adaptive significance of those behaviors.

When social animals, which include many mammals and some bird species, grow up in the wild, they learn from other members of their population, particularly their parents, how to interact with their environment and with other members of their species. They learn how to search for food and how to gather, capture, and consume it. When mammals and birds are raised in captivity, their environment is limited to a cage or pen, so exploration is unnecessary. Searching for food and learning about new food sources is not required because the same food items come to them day after day, on schedule. For example, when the European bison (*Bison bonasus*, see chapter opening photo), previously extinct in the wild, was reintroduced, it was necessary to determine whether the animals were successfully

FIGURE 7.3 The Sumatran orangutan (*Pongo abelii*) is a critically endangered species found only on the island of Sumatra, in Indonesia. It is rarer than the other species of orangutan, the Bornean orangutan. Although reintroduction programs for both species are common and achieve instant results, it has been argued that protecting their habitat may protect more individuals in the long term. (© Ryan Deboodt/Getty Images.)

exploring and using their new, wild environment (Schmitz et al. 2015). In a recent review of conservation projects that use behavioral ecology, it was found that while foraging and dispersal behaviors were often considered, others such as anti-predator behavior and social behaviors were not, even though these are important issues for reintroductions (Berger-Tal et al. 2016).

Social behavior may become highly distorted when animals are raised alone or in unnatural social groupings (i.e., in small groups or single-aged groups). In such cases, the animals may lack the skills to survive in their natural environment and the social skills necessary to cooperatively find food, sense danger, find mating partners, and raise young (Parlato and Armstrong 2013). The greatest threats to the survival of these animals, and the primary reasons many such establishment projects fail, are predation ("Why are these guys trying to eat me?"), starvation ("Why aren't they feeding me any more?"), and habitat quality ("My old home was way better!") (White et al. 2012).

To overcome these behavioral problems, captive-raised mammals and birds may require extensive training before and after release. In some cases, human trainers use puppets or wear costumes to mimic the appearance and behavior of wild individuals so that young animals learn to identify with their own species rather than with humans (**FIGURE 7.4**).

In other cases, wild individuals serve as "instructors" for captive individuals of the same species. For example, wild golden lion tamarins (*Leontopithecus rosalia*) are caught and held with captive-bred tamarins so that the captive-bred tamarins will learn appropriate behavior from the wild ones. After they form social groups, they are released together. Wild-caught African wild dogs (*Lycaon pictus*) that gave birth and bonded together in holding areas prior to release had a higher success rate after reintroduction in multiple sites in South Africa (Gusset et al. 2008).

> It is imperative that captive-bred mammals and birds learn predator avoidance and species-appropriate social behavior if they are to survive and reproduce after being released into the wild. They may also require some support after release.

Establishing plant populations

Methods used to establish new populations of rare and endangered plant species are fundamentally

FIGURE 7.4   California condor chicks (*Gymnogyps californianus*) raised in captivity are fed by researchers using puppets that look like adult birds. Conservation biologists hope that minimizing human contact with the birds will improve their chances of survival when they are returned to the wild. (Photograph by Ron Garrison, courtesy of US Fish and Wildlife Service.)

different from those used to establish terrestrial vertebrate species. Animals can disperse to new locations and actively seek out the most suitable microsites. The seeds of plants, however, are dispersed to new sites by agents such as wind, animals, water, or the actions of conservation biologists (e.g., Rood et al. 2015). Once a seed lands on the ground or an adult is planted at a site, it is unable to move, even if a suitable microsite exists just a few meters away. The immediate microsite is crucial for plant survival: if the environmental conditions are in any way too sunny, too shady, too wet, or too dry, the seed will not germinate, or the resulting plant will not reproduce or will die.

Plant ecologists are investigating the effectiveness of site treatments before and after planting—such as burning the leaf litter, removing competing vegetation, digging up the ground, and excluding grazing animals—as a means of enhancing population establishment (Reiter et al. 2016). They are also investigating the most effective way to establish a species: whether seeds should be sown or adult individuals or seedlings should be transplanted into the enhanced site. For example, during the introduction of Mead's milkweed (*Asclepias meadii*), a threatened perennial prairie plant in the midwestern United States, it was found that survivorship was greater for older juvenile plants than for seedlings, and that survival was higher in burned habitat than in unburned habitat. Seedling survival was also higher in 1996, which had greater rainfall than average. Researchers also use PVA (see Chapter 6) and other modeling tools to determine best practices for plant reintroduction (Halsey et al. 2015) (**FIGURE 7.5**).

> New plant populations are established by sowing seeds or transplanting seedlings or adults. Site treatments such as burning off or physically removing competing plants are often necessary for success.

Plant reintroductions frequently fail, although, as with animals, this fact is not captured in the literature because researchers do not usually publish failed experiments (Godefroid et al. 2011a). Furthermore, many apparently successful reintroductions have either failed after additional years or have never established a second generation of plants, which is an important indicator of success. In general, success increases with the numbers of individuals or populations reintroduced (Liu et al. 2015) (see Figure 7.5), but this conclusion assumes that plantings are taking place in suitable habitat. Genetic analysis of source populations can assist in determining resistance to environmental stress, and resilience to environmental change (He et al. 2016). Because of climate change, certain plant species may no longer be genetically suited to their present sites, and conservation biologists may have to look elsewhere in the range of a species for suitable genotypes to plant (Gray et al. 2011).

The status of new populations

The establishment of new populations raises some novel issues at the intersection of scientific research, conservation efforts, government regulation, and ethics. These issues need to be addressed because reintroduction, introduction, and reinforcement programs will increase in the coming years as the biodiversity crisis eliminates more species and populations from the

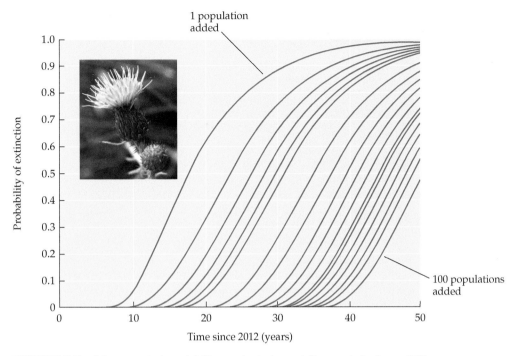

FIGURE 7.5 Metapopulation viability analysis (a modeling tool similar to PVA that considers number of populations rather than individuals) of Pitcher's thistle (*Cirsium pitcheri*; inset), a federally threatened species that grows in the Indiana Dunes. The graph shows that the probability of extinction over time was greatly reduced by increasing numbers of reestablished populations. After 40 years, the extinction probability is almost 100% with one population added, but is reduced to less than 1% with more than 100 populations added. (After Halsey et al. 2015. Photograph courtesy of the US Fish and Wildlife Service.)

wild. In addition, assisted colonization may be needed for many species if their present ranges become too hot, too dry, or otherwise unsuitable because of global climate change or some other change in the environment (Lunt et al. 2013).

Many of the reintroduction programs for endangered species are mandated by official recovery plans set up by national governments. If such plans are to be formulated and implemented, conservation biologists must be able to explain the benefits and limitations of reintroduction and introduction programs in a way that government officials and the general public can understand, and they must address the legitimate concerns of those groups (**FIGURE 7.6**).

Sometimes stakeholders' concerns can be addressed by giving varying degrees of protection to new populations. In the United States, for example, populations can be designated *experimental essential* or *experimental nonessential*. Experimental essential populations are regarded by the US Endangered Species Act as critical to the survival of the species, and they

FIGURE 7.6 Residents of southern Taiwan were surveyed to assess attitudes toward a restoration program for Formosan sika deer (*Cervus nippon taiouanus*), which had gone extinct in the wild, in Kenting National Park. After its reintroduction, the deer was associated with threats to crops (48% of respondents), and 18% of respondents had actually suffered losses to the deer. Even so, the majority of those surveyed believed the deer was a tourism resource (87%) and supported the reintroduction program (75%). (Yen et al. 2015.) (© Imagemore Co., Ltd./Getty Images.)

are as rigidly protected as naturally occurring populations. Experimental nonessential populations have less protection under the law; designating populations as nonessential often helps to overcome the fear of local landowners that having endangered species on their property will restrict how their land can be managed and developed.

Legislators and scientists alike must understand that the establishment of new populations through reintroduction programs in no way reduces the need to protect the original populations of the endangered species. Original populations are likely to have the most complete gene pool of the species and the most intact interactions with other members of the biological community. In many cases, proposals are made by developers or government departments to compensate for habitat damage or eradication of endangered populations by development projects by creating new habitat or establishing new populations. This activity is generally referred to as **mitigation**. Mitigation plans, which are often directed at legally protected species and habitats, often include (1) adjustments to the development plan to reduce the extent of damage, (2) establishment of new populations and habitat as compensation for what is being destroyed, and (3) enhancement of populations and habitat that remain after development. Given the poor success of most attempts to create new populations of rare species, mitigation plans for threatened or endangered species should be approached with great skepticism.

> The establishment of new populations through reintroduction programs in no way reduces the need to protect the original populations of endangered species.

Ex Situ Conservation Strategies

The best strategy for the long-term protection of biodiversity is the preservation of biological communities and populations (called **in situ conservation**, as discussed in Chapter 6). However, in the face of increasing threats to biodiversity, relying solely on in situ conservation is not currently a viable option for many rare and endangered species. For example, even under in situ conservation management and protection programs, species may still decline and go extinct in the wild for any of the reasons already discussed: habitat destruction, loss of genetic variation, demographic and environmental stochasticity, and so forth. Likewise, if a remnant population is too small to maintain the species, if it is still declining despite conservation efforts, or if the remaining individuals are found outside of protected areas, then in situ conservation may not be adequate. It is likely that the only way to prevent species in such circumstances from going extinct is to maintain individuals in artificial conditions under human supervision (Canessa et al. 2015). **Ex situ**, or off-site, **conservation** used in place of, or to complement, in situ conservation can mean the difference between persistence and extinction for some species. Already a number of species that have gone extinct in the wild have survived because of propagation in captive colonies. The beautiful Franklin tree (*Franklinia alatamaha*), for example, grows only in cultivation and is no longer found in the wild.

Ex situ and in situ conservation are complementary strategies (Zimmermann et al. 2007) (**FIGURE 7.7**). The long-term goal of many ex situ conservation programs is the establishment of new populations in the wild, once sufficient numbers of individuals and a suitable habitat are available. In the case of Przewalski's horse (*Equus ferus przewalskii*), which had been declared extinct in the wild, small herds descended from 14 captive-bred

> When integrated with efforts to protect existing populations and to establish new ones, ex situ conservation is an important strategy for protecting endangered species and educating the public.

FIGURE 7.7 This model shows ways in which in situ (on-site) and ex situ (off-site) conservation efforts can benefit each other and provide alternative conservation strategies. While no species conforms exactly to this idealized model, the giant panda program (see Figure 7.11) has many of its elements. (After Maxted 2001.)

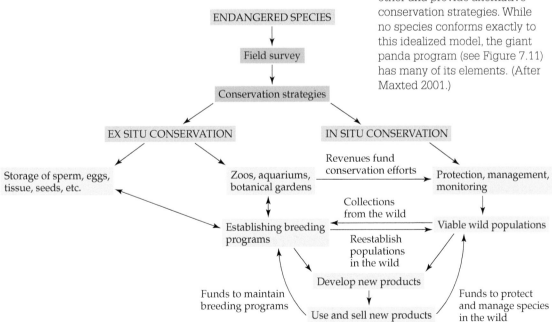

founder individuals were released in a national park in Mongolia starting in 1992 (**FIGURE 7.8A**).

An intermediate strategy that combines elements of both ex situ and in situ conservation is the monitoring and management of populations of rare and endangered species in small protected areas. Such populations are still somewhat wild, but occasional human intervention may be necessary to prevent population decline. An example is the rare Columbia Basin pygmy rabbit (*Brachylagus idahoensis*), which does not breed well in captivity (**FIGURE 7.8B**). As part of the conservation strategy for this species, groups of pygmy rabbits are released into large (2.5 ha) fenced enclosures that keep out predators, keep the rabbits in familiar surroundings, and encourage breeding. On occasion, tagged individuals from the breeding program are released back into the wild.

Research on captive populations can provide insight into the basic biology, physiology, and genetics of a species through studies that would not be possible on wild animals. The results of these studies can suggest new conservation strategies for in situ populations. Long-term, self-sustaining ex situ populations can also reduce the need to collect individuals from the wild for display or research. Ex situ facilities for animal preservation include zoos, game farms, and aquariums as well as the facilities of private breeders. Plants are maintained in botanical gardens, arboretums, and seed banks.

Ex situ conservation has several limitations. First, it is not cheap. The cost of maintaining zoos is enormous in comparison with many other conservation

(A)

(B)

FIGURE 7.8 (A) The IUCN has declared Przewalski's horse to be extinct in the wild. Several zoos around the world have maintained populations and have been successful in breeding these animals. Reintroductions of the wild horses into their natural range in Mongolia appear so far to have been successful. (B) Rare Columbia Basin pygmy rabbits breed well in large seminatural enclosures. Individuals are then released into the wild. (A, Patrick Pleul/dpa/Corbis; B, photograph by Tara Davila, courtesy of Lisa Shipley and Rod Sayler, Washington State University.)

activities, such as the budgets of national parks in developing countries; in the United States alone, zoos cost about $1 billion per year to run. Furthermore, ex situ programs protect only one species at a time. In contrast, when a species is preserved in the wild, an entire community—perhaps consisting of thousands of species—may be preserved, along with a range of ecosystem services.

There are also several ecological and evolutionary problems with ex situ conservation. Only in natural biological communities are species able to continue their process of evolutionary adaptation (Olivieri et al. 2016). Ecosystem-level interactions among species, as discussed in Chapter 2, are often crucial to a rare species' continued survival; these interactions can be quite complex and probably cannot be replicated under captive conditions. Furthermore, captive animal populations are generally not large enough to prevent the loss of genetic variation through genetic drift; the same may be true of cultivated plant species when they have special requirements for pollination that might make it difficult to ensure adequate cross-fertilization among individuals. For such species, the best solution may be in situ conservation involving careful habitat protection and management. Despite these problems, ex situ conservation remains an important approach, and many species have recovered from near extinction.

Captive individuals on display can also serve as ambassadors for their species and help to educate the public about the need to preserve the species in the wild (**FIGURE 7.9A**). Zoos, aquariums, botanical gardens, and the people who visit them regularly contribute money and expertise to in situ conservation programs (see Figure 6.2). In addition, ex situ conservation programs can be used to develop new products that can potentially generate funds from profits or licensing fees to protect species in the wild. In situ preservation of species, in turn, is vital to the survival of species that are difficult to maintain in captivity, as well as to the continued ability of zoos, aquariums, and botanical gardens to display species that do not have self-sustaining ex situ populations.

Zoos

A current goal of most major zoos is to establish viable, long-term captive breeding populations of rare and endangered animals (Zimmermann et al. 2008). The *International Zoo Yearbook* (IZY) reports births and deaths of zoo-bred animals, and a review of these data since 1972 suggests that all but five endangered mammalian species have median positive captive population growth rates, even if those rates have decreased over time (Alroy 2015). Zoos, along with affiliated universities, government wildlife departments, and conservation organizations, presently maintain over 2 million animals, including over 600,000 individual terrestrial vertebrates, representing over 7400 species and subspecies of mammals, birds, reptiles, and amphibians (**TABLE 7.1**) (www2.isis.org). While this number of captive animals may seem impressive, it is trivial in comparison to the tens of millions of domestic cats, dogs, and fish kept by people as pets. Zoos could establish breeding

TABLE 7.1 Number of Terrestrial Vertebrates Maintained in Zoos

Location	Mammals	Birds	Reptiles	Amphibians	Total
Europe	101,921	125,846	30,799	57,413	315,979
North America	50,982	62,448	31,270	50,588	195,288
Latin America	3,653	5,105	2,455	634	11,847
Asia	29,089	39,216	9,338	887	78,530
Australasia	7,674	10,312	3,890	1,875	23,751
Africa	4,185	7,939	2,435	356	14,915
Worldwide totals					
All species	197,504	250,866	80,187	111,753	640,310
Number of taxa[a]	2,238	3,753	969	544	7,486
Percentage wild-born[c]	5%	9%	15%	5%	
Rare species[b]	59,030	37,748	22,474	3,398	122,650
Number of taxa[a]	527	344	207	29	1,107
Percentage wild-born[c]	7%	9%	18%	7%	

Source: Data from ISIS, provided by Laurie Bingaman Lackey 2013.

[a]The number of taxa is not exactly equivalent to number of species because many species have more than one subspecies listed.

[b]Rare species are those covered by CITES (the Convention on International Trade in Endangered Species).

[c]The percentage of individuals born in the wild is approximate (particularly for reptiles and amphibians), since the origin of the animals is often not reported.

colonies of even more species if they directed more of their efforts toward smaller-bodied species such as insects, amphibians, and reptiles, which are less expensive to maintain in large numbers than are large-bodied mammals such as bears, elephants, and rhinoceroses. A better balance must be reached between displaying large animals that draw many visitors and displaying smaller, lesser-known animals that appeal less to the public but represent a greater proportion of the world's biodiversity.

Zoos already work together effectively to conserve some of these smaller species. For instance, seven North American zoos joined with universities and the Defenders of Wildlife to form the Panama Amphibian Rescue and Conservation Project (www.amphibianrescue.org). A major goal of this collaboration is to establish breeding populations of frogs and other amphibians that are being decimated in the wild (**FIGURE 7.9B**). Ex situ conservation efforts have been increasingly directed at saving endangered species of invertebrates as well, including butterflies, beetles, dragonflies, spiders, and molluscs. Other important targets for ex situ conservation efforts are rare breeds of domestic animals on which human societies depend for animal protein, dairy products, leather, wool, agricultural labor, transport, and recreation (Ruane 2000). Secure populations of these breeds are a potential genetic resource for the improvement and long-term health of our supplies of pigs, cattle, chickens, sheep, and other domestic animals.

(A)

(B)

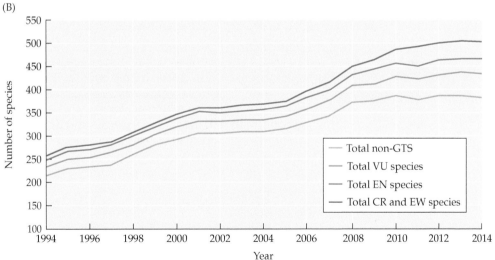

FIGURE 7.9   (A) Mother with her young daughter watch an aquarist wearing a wetsuit gently hold a Chinese giant salamander in a freshwater tank. Critically endangered in the wild, the Chinese giant salamander is the largest living species of amphibian. (B) The number of amphibian species at zoos has been steadily increasing since 1994, not only for non–globally threatened species (non-GTS), but also for vulnerable (VU), endangered (EN) and critically endangered (CR) species, and species extinct in the wild (EW). (A, photograph © Billy Hustace/Corbis; B, after Dawson et al. 2016.)

The variety of species displayed in zoos has increased in recent years, but the emphasis is still on "charismatic megafauna" because they help to attract the general public and influence them favorably toward conservation. In fact, over 90% of families enjoy seeing biodiversity at zoos and aquariums and believe they teach children about protecting species and habitat (www.aza.org). As such, ex situ animals make people aware of the threats to biodiversity (Moss et al. 2016). As part of the World Zoo Conservation Strategy, which seeks to link zoo programs with conservation efforts in the wild, the world's 2000 zoos and aquariums are increasingly incorporating ecological themes and information about the threats to endangered species into their public displays and research programs. The potential educational and financial impacts of zoos are enormous, considering that they receive approximately 600 million visitors per year (Figure 7.10A). Furthermore, the funds raised by zoos from visitor fees and other programs can directly fund in situ conservation activities.

Zoos have the needed knowledge and experience in animal care, veterinary medicine, animal behavior, reproductive biology, and genetics to establish captive animal populations of endangered species (**FIGURE 7.10B**) (Zimmermann et al. 2008). Zoos and affiliated conservation organizations have embarked on major efforts to build the facilities and develop the technology necessary to establish and house breeding colonies of endangered animals and to develop the new methods and programs needed to reintroduce species in the wild (**FIGURE 7.11**).

(A)

(B)

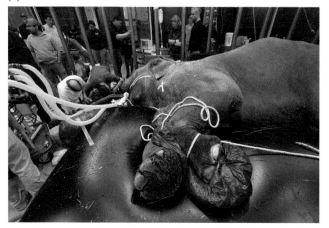

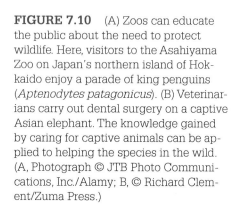

FIGURE 7.10 (A) Zoos can educate the public about the need to protect wildlife. Here, visitors to the Asahiyama Zoo on Japan's northern island of Hokkaido enjoy a parade of king penguins (*Aptenodytes patagonicus*). (B) Veterinarians carry out dental surgery on a captive Asian elephant. The knowledge gained by caring for captive animals can be applied to helping the species in the wild. (A, Photograph © JTB Photo Communications, Inc./Alamy; B, © Richard Clement/Zuma Press.)

FIGURE 7.11 China's giant panda (*Ailuropoda melanoleuca*) is one of the world's most charismatic animals and has become emblematic of the fight to save endangered species. Wolong National Nature Reserve and other facilities in China have been successful at breeding giant pandas using artificial insemination and hand rearing. The reserve has established a reintroduction program, but the loss of habitat and the fact that captive-raised individuals may lack the behavioral skills needed to survive in the wild combine to make reintroduction of pandas particularly problematic. (Photograph © LMR Group/Alamy.)

Some of these facilities are highly specialized, such as that run by the International Crane Foundation in Wisconsin, which is attempting to establish captive breeding colonies of all crane species. These collaborative efforts have paid off. Currently, fewer than 5% of the terrestrial mammals kept in zoos have been collected in the wild, and this number is declining as zoos gain more experience.

> Zoos often use the latest methods of veterinary medicine to establish healthy breeding colonies of endangered animals.

The success of captive breeding programs has been enhanced by efforts to collect and disseminate knowledge about the maintenance of rare and endangered species. The Species Survival Commission's Conservation Breeding Specialist Group, a division of the IUCN, and affiliated organizations, such as the Association of Zoos and Aquariums, the European Association of Zoos and Aquaria, and the Australasian Regional Association of Zoological Parks and Aquaria, provide zoos with the necessary information on proper care and handling of these species as well as updates on the status and behavior of the animals in the wild (www.aza.org). This information includes nutritional requirements, anesthetic techniques to immobilize animals and reduce stress during transport and medical procedures, optimal housing conditions, information about vaccinations and antibiotics to prevent the spread of disease, and breeding records. The collective effort is being aided by a central database called the Zoological Information Management System (ZIMS), maintained by the International Species Information System (ISIS), which keeps track of all relevant information on the over 2 million animals belonging to 10,000 vertebrate and invertebrate species at 825 member institutions more than 85 countries (www2.isis.org).

Many species provided with humane captive conditions reproduce with abandon—so much so that the use of contraceptives to control populations is required. However, some rare animal species, such as rhinoceroses, do not adapt or reproduce well in captivity (McCleery et al. 2014). Zoos conduct extensive research to identify management methods that can overcome these problems and promote successful reproduction of genetically appropriate mates (Wildt et al. 2009). Some of these methods come directly from human and veterinary medicine, while others are novel techniques developed at special research facilities such as the Smithsonian Conservation Biology Institute in Virginia and the San Diego Zoo Institute for Conservation Research. These techniques include **cross-fostering**, in which common species raise the offspring of rare species; **artificial incubation** of eggs under ideal hatching conditions; **artificial insemination** when adults do not show interest in mating or are living in different locations; and **embryo transfer**, which involves implanting fertilized eggs of rare species into surrogate mothers of common species (**FIGURE 7.12**). One of the most unusual techniques, known as **genome resource banking** (**GRB**), involves the freezing of purified DNA, eggs, sperm, embryos, and other tissues so that they can be used to contribute to breeding programs, to maintain genetic diversity, and for scientific research. However, many of these techniques are expensive and species specific. In any case, GRB and similar advanced methods are not substitutes for in situ and ex situ conservation programs that preserve behaviors and ecological relationships that are necessary for survival in the wild.

Raising animals in captivity places limits on the animals' conservation value. Animals living in captive breeding facilities may lose the behaviors they need to survive in the wild. Furthermore, populations raised in captivity may undergo genetic, physiological, and morphological changes that

FIGURE 7.12 A Brazilian ocelot produced by embryo transfer at the Cincinnati Zoo. (Courtesy of the Cincinnati Zoo.)

make them less able to tolerate the natural environment if they are returned to the wild. Diseases acquired in captivity may render them unsuitable for release (Minuzzi-Souza et al. 2016). Consequently, when researchers establish an ex situ program to preserve a species, they must address a series of ethical questions:

- How will establishing an ex situ population benefit the wild population?
- Is it better to let the last few individuals of a species live out their days in the wild or to breed a captive population that may be unable to adapt to wild conditions?
- Does a population of a rare species consisting of individuals that have been raised in captivity and do not know how to survive in their natural environment really represent preservation of the species?
- Are rare individuals being held in captivity primarily for their own benefit, for the benefit of their entire species, for the economic benefit of zoos, or for the pleasure of zoo visitors?
- Are the animals in captivity receiving appropriate care based on their biological needs?
- Are sufficient efforts being made to educate the public about conservation issues?

Aquariums

Approximately 600,000 individual fish, most of them obtained from the wild, are maintained in public aquariums that are open to visitors (**FIGURE 7.13**). Major efforts are being made to develop breeding techniques so that rare species can be maintained in aquariums without further collection in the wild and in the hope that some can be released back into the wild.

FIGURE 7.13 Public aquariums participate in both in situ and ex situ conservation programs and provide a valuable function by educating people about marine conservation issues. (© Jeff Greenberg/Alamy Stock Photo.)

Fish breeding programs use indoor aquarium facilities, seminatural water bodies, and fish hatcheries and farms. Many fish breeding techniques were originally developed by fisheries biologists for large-scale stocking operations involving trout, bass, salmon, and other commercial species. Other techniques were discovered in the aquarium pet trade when dealers attempted to propagate tropical fish for sale. These techniques are now being applied to endangered freshwater fauna. Currently, both public and private groups are making impressive efforts to unlock the secrets of propagating some of the more difficult species.

Aquariums have an increasingly important role to play in the conservation of endangered cetaceans, manatees, sea turtles, and other large marine animals. Aquarium personnel often respond to public requests for assistance in handling large animals stranded on beaches or disoriented in shallow waters. The aquarium community can use lessons learned from working with common species to develop programs to aid endangered species.

The ex situ preservation of aquatic biodiversity takes on additional significance due to the dramatic recent increase in aquaculture, which represents about 30% of fish and shellfish production worldwide. This aquaculture includes the extensive salmon, carp, and catfish farms in the temperate zones, the shrimp and fish farms in the tropics, and the 12 million tons of aquatic products grown in China and Japan. As fish, frogs, molluscs, and crustaceans are increasingly domesticated and raised to meet human needs, it becomes necessary to preserve the genetic stocks needed to maintain and increase commercial production of these species and to protect them against disease and unforeseeable threats. A challenge for the future will be balancing the need to increase human food production through aquaculture with the need to protect aquatic biodiversity from increasing human threats.

Botanical gardens

Botanical gardens have living collections and seed banks that provide ex situ protection and knowledge of endangered and economically important plants.

The world's 1775 botanical gardens (also known as botanic gardens) contain major collections of living plants and represent a crucial resource for plant conservation through ex situ conservation, research, and education (**FIGURE 7.14**). An **arboretum** is a specialized botanical garden focusing on trees and other woody plants. The world's botanical gardens currently contain about 4 million living plants, representing 80,000 species—approximately 30% of the world's flora (Guerrant et al. 2013; www.bgci.org). When we add in the species grown in greenhouses, subsistence gardens, and hobby gardens, the numbers are increased. One of the world's largest botanical gardens, the Royal Botanic Gardens, Kew, in England, has over 30,000 species of plants under cultivation, about 10% of the world's total; 2700 of these species are listed as threatened by the IUCN. One of the most exciting new botanical gardens is the Eden Project in southwestern England, which focuses on displaying and explaining over 5000 species of rain forest, temperate-zone, and Mediterranean plants in giant domes that constitute the world's largest greenhouse (www.edenproject.com).

(A)

(B)

(C)

(D)

FIGURE 7.14 (A) The beautiful displays at Munich Botanical Gardens in Germany play an important role in conservation by connecting the public with plants and preserving those species that no longer occur in the wild. (B) Signage at the Kirstenbosch National Botanical Garden, South Africa, educates visitors about how to promote local biodiversity in their yards. (C) Field biologists with Madagascar program of the Missouri Botanical Garden identify plants that may have medicinal benefits, work to safeguard the biodiverse areas where they grow, and raise plants for reintroductions. (D) A scientist at the Royal Botanical Gardens, Kew, United Kingdom inspects preserved seeds as a part of the Millennium Seed Project. (A, courtesy of Richard Primack; B, courtesy of Anna Sher; C, courtesy of Missouri Botanical Gardens; D, courtesy of the Royal Botanic Gardens, Kew.)

Botanical gardens increasingly focus their efforts on cultivating rare and endangered plant species, and many specialize in particular types of plants. Many botanical gardens are involved in plant conservation, especially in the reintroduction of rare and endangered plant species and the restoration of degraded ecosystems (Hardwick et al. 2011). In Yunnan, China, only 52 individuals of a rare magnolia tree, *Magnolia sinica*, are left in the wild, but the species has been successfully cultivated at Kunming Botanical Garden, providing hope that reintroductions may be possible.

Staff members at botanical gardens are often recognized authorities on plant identification, distributions, and conservation status. Botanical gardens are able to educate an estimated 200 million visitors per year about conservation issues. At an international level, Botanic Gardens Conservation International (BGCI) represents and coordinates the conservation efforts of over 700 botanical gardens (www.bgci.org). The priorities of this program include the creation of a worldwide database to support collecting activity and identification of important species that are underrepresented or absent from collections of living plants. One of its projects is the on-line PlantSearch database, which currently lists over 1.3 million records of more than 480,000 species and varieties growing in botanical gardens, of which about 3000 are rare or threatened. In addition, because most existing botanical gardens are located in the temperate zone, establishing botanical gardens in the tropics is a primary goal of the international botanical community.

Seed banks

In addition to growing plants, botanical gardens and research institutes have developed collections of seeds, sometimes known as **seed banks**, obtained from the wild and from cultivated plants. These seed banks provide a crucial backup to their living collections (**FIGURE 7.15**). The seeds of most plant species can be kept dormant in cold, dry conditions for long periods and later germinated to produce new plants. This ability of seeds to remain dormant allows the seeds of large numbers of rare species to be frozen and stored in a small space, with minimal supervision and at a low cost. Seed banks are especially important for rare and endangered species that may need to be reintroduced into the wild. At present, seeds of approximately 4000 or 10% of the world's known plant species are stored in seed banks, with the figure approaching 70% for European plants (Godefroid et al. 2011b). Efforts are made to include the full range of genetic variation found in a species by collecting seeds from populations growing across the range of the species.

More than 1000 seed banks exist worldwide, and their activities are coordinated by the Consultative Group on International Agricultural Research (CGIAR). However, if power supplies fail or equipment breaks down, an entire frozen collection could be damaged. To prevent such a loss, Norway has recently established the newest seed bank, the Svalbard Global Seed Vault, where 400,000 frozen seed samples are stored below permafrost, with millions more expected to be added in coming decades.

Seed banks have been embraced by agricultural research institutes and the agricultural industry as an effective resource for preserving and using the genetic variation that exists in agricultural crops and their wild relatives. Preserving this genetic diversity is crucial to maintaining and increasing the high productivity of modern crops and their ability to respond to changing

(A)

(B)

FIGURE 7.15 (A) At seed banks, seeds of many plant varieties are sorted, cataloged, and stored at freezing temperatures. (B) Seeds come in a wide variety of sizes and shapes. Each such seed represents a genetically unique, dormant individual. (Photographs courtesy of US Department of Agriculture.)

environmental conditions such as acid rain, global climate change, and soil erosion (Banga and Kang 2014). Researchers are in a race against time to preserve genetic variation because traditional farmers throughout the world are abandoning their diverse local crop varieties in favor of standard, high-yielding varieties (Gliessman 2015). By some estimates, 75% of global crop plant genetic diversity has been lost in this way over the last century (FAO 2007). This worldwide phenomenon is illustrated by Sri Lankan farmers, who grew two thousand varieties of rice until the late 1950s, when they switched over to just five high-yielding varieties.

A major controversy involved in the development of agricultural seed banks is who owns and controls the genetic resources of crops (Brush 2007). The genes of **landraces** of crop plants (local species that have been adapted by humans over time) and wild relatives of crop species represent the building blocks needed to develop elite, high-yielding varieties suitable for modern agriculture (Nabhan 2008). Approximately 96% of the raw genetic variation necessary for modern agriculture comes from developing countries such as India, Ethiopia, Peru, Mexico, Indonesia, and China (**FIGURE 7.16**), yet most corporate breeding programs for elite strains are located in the industrialized countries of North America and Europe. In the past, genetic material was perceived as free for the taking: the staffs of international seed banks freely collected seeds and plant tissue from developing countries and gave them to research stations and seed compa-

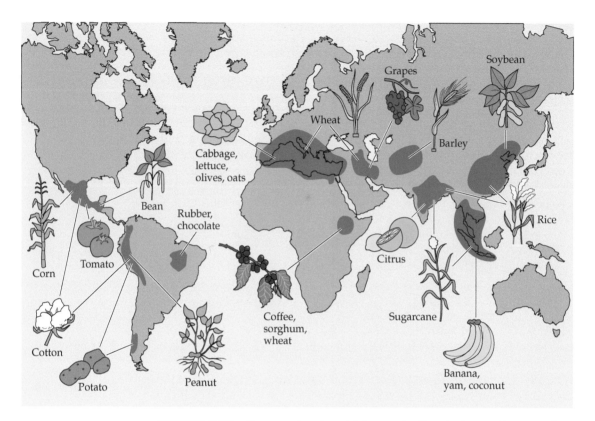

FIGURE 7.16 Crop species show high genetic diversity in certain areas of the world (shown in blue), often where the species was first domesticated or where it is still grown in traditional agricultural settings. This genetic diversity is of international importance in maintaining the productivity of agricultural crops. (Map courtesy of Garrison Wilkes.)

nies. Seed companies then developed new strains through sophisticated breeding programs and field trials. The resulting seeds were sold at high prices to maximize profits for the companies, which often totaled hundreds of millions of dollars a year, but the countries from which the original seeds were collected did not receive any profit from this activity. Developing countries now question why they should share their biological materials freely if they will have to pay for new seed varieties and cultivated plants based on those genetic resources.

One solution to this controversy involves negotiating agreements, using the framework of the Convention on Biological Diversity (see Chapter 11), in which countries agree to share their genetic resources in exchange for receiving new products and a share of the profits. A few such contracts have been negotiated, such as the one between Merck and the government of Costa Rica for the development of products based on species collected from the wild in that country (see Chapter 3 for similar examples). These agreements will be followed carefully to determine whether they are mutually satisfactory and can serve as models for future contracts.

Can Technology Bring Back Extinct Species?

The idea of using technology to bring species back from extinction is not particularly new. Medical scientists have re-created viruses and other pathogens in the past to study their behavior in hope of preventing similar diseases. In 2003, scientists even brought back an extinct mammal: the Pyrenean ibex or bucardo (*Capra pyrenaica pyrenaica*), the last of which had died in 2000. The process involved fusing frozen bucardo cells with goat egg cells from which the nuclei had been removed and implanting the resulting eggs into a domestic goat, which served as a surrogate mother. Even though an embryo did develop, the resulting cloned ibex was badly deformed and died within minutes of birth.

The process of bringing a species back—sometimes referred to as *de-extinction*—requires DNA, ideally a complete genome. This requirement limits the candidate species to those that have not been extinct so long that their genetic material has completely degraded. The genome of the extinct species, perhaps obtained from tissue preserved in a freezer, buried under a glacier, or preserved deep in an anoxic bog or oceanic mud, is then compared with that of a closely related living species—for example, the extinct passenger pigeon with the extant band-tailed pigeon—and the differences identified. At that point, the living species' DNA and cells can potentially be modified and used to create stem cells and germ cells with the DNA of the extinct species. These germ cells can then be implanted into eggs from living relatives. At the end of the gestation period, the result could be an extinct species brought back to life. This process was recently used to create early-stage embryos of an extinct gastric-brooding frog (*Rheobatrachus silus*), but they soon died. Perhaps in the future, the technique could be applied to tissue from extinct species preserved in the northern permafrost, such as the extinct mastodon, or even to bits of DNA from fossilized bones.

But is it a good idea to bring back extinct species? The possibility brings up questions of ethics, practicality, and the fundamental goals of conservation. For example, if re-created passenger pigeons were brought up by another pigeon species and lacked the flocking, feeding, and other behaviors of extinct passenger pigeons, should they really be considered passenger pigeons? Would resurrected passenger pigeons function in an ecosystem in the same ways as past passenger pigeons? Would it be acceptable to bring back a species if its habitat no longer exists?

It is also unclear how de-extinction programs would affect and interact with existing conservation efforts. Would the ability to bring species back from extinction reduce the incentive to preserve species in the wild? Will de-extinction research and efforts divert funding from research and conservation projects aimed at preserving existing species and ecosystems?

The thought-provoking questions raised by these exciting new techniques highlight the interdisciplinary nature of conservation biology—its liberal overlap with biotechnology, genetics, ecology, ethics, and economics. Just because we can do something does not necessarily mean that we should; whether or not we use de-extinction to bring back an extinct species or restore its ecosystem functions will depend on the development of new conservation technologies and the thoughtful discussions that they inspire.

Summary

- New populations of rare and endangered species can be established in the wild using either captive-raised or wild-caught individuals.

- Long-term monitoring of a species in the field can determine if it is stable, increasing, or declining in abundance over time.

- Animals sometimes require behavioral training before release as well as maintenance after release.

- Some species that are in danger of going extinct in the wild can be maintained in zoos, aquariums, botanical gardens, and seed banks; this strategy is known as ex situ conservation. These captive colonies can sometimes be used later to establish new populations in the wild.

- Technologies are being developed that could potentially revive versions of extinct organisms, however this may neither be practical nor advisable.

For Discussion ■

1. How do you judge whether a reintroduction project is successful? Develop simple and then increasingly detailed criteria to evaluate a project's success.

2. Would it be a good idea to create new wild populations of African rhinoceroses, elephants, and lions in Australia, South America, the southwestern United States, or other areas outside their current range, as described by Donlan et al. (2006)? What would be some of the legal, economic, and ecological issues involved?

3. Would biodiversity be adequately protected if every species were raised in captivity? Is this possible? Is it practical? How would freezing a tissue sample of every species help to protect biodiversity? Again, is this possible or practical?

Suggested Readings

Banga, S. S. and M. S. Kang. 2014. Developing climate-resilient crops. *Journal of Crop Improvement* 28: 57–87. Farmers are crucial to maintaining the genetic diversity of crops in the face of climate change.

Dawson, J., F. Patel, R. A. Griffiths, and R. P Young, R. P. 2016. Assessing the global zoo response to the amphibian crisis through 20-year trends in captive collections. *Conservation Biology* 30(1): 82–91. Zoos are increasing numbers of endangered amphibians in ex situ.

Dolman, P. M., N. J. Collar, K. M. Scotland, and R. J. Burnside. 2015. Ark or park: stochastic population modeling to evaluate potential effectiveness of in situ and ex situ conservation for a critically endangered bustard. *Journal of Applied Ecology* 52: 841–850. Mathematical models can inform conservation approaches with less risk to extant populations.

He, X., M. L. Johansson, and D. D. Heath. 2016. Role of genomics and transcriptomics in selection of reintroduction source populations. *Conservation Biology*. Genetic analysis of wild populations.

Liu, H., H. Ren, Q. Liu, X. Wen, M. Maunder, and J. Gao. 2015. Translocation of threatened plants as a conservation measure in China. *Conservation Biology*, 29(6): 1537–1551. Success of rare plant reintroductions in China were influenced by the number planted, the plant life form, and the source of the material.

McCleery, R., J. A. Hostetler, and M. K. Oli. 2014. Better off in the wild? Evaluating a captive breeding and release program for the recovery of an endangered rodent. *Biological Conservation* 169: 198–205. Such programs will only be successful if animals breed well in captivity and if released animals have a reasonable rate of survival and reproduction in the wild.

Miller, B., W. Conway, R. P. Reading, C. Wemmer, D. Wildt, D. Kleiman, and 4 others. 2004. Evaluating the conservation mission of zoos, aquariums, botanical gardens, and natural history museums. *Conservation Biology* 18: 86–93. Eight tough questions are asked with hopes that these institutions can become more effective.

Morrell, V. 2014. Science behind plan to ease wolf protection is flawed, panel says. *Science* 343: 719. Removing federal protection for the gray wolf in the United States is controversial, mixing together science and politics.

Sebastián-González, E., J. A. Sánchez-Zapata, F. Botella, J. Figuerola, F. Hiraldo, and B. A. Wintle. 2011. Linking cost efficiency evaluation with population viability analysis to prioritize wetland bird conservation actions. *Biological Conservation* 144: 2354–2361. Different management approaches are evaluated for their cost effectiveness on bird populations in Spain.

Tlusty, M. F., A. L. Rhyne, L. Kaufman, M. Hutchins, G. M. Reid, C. Andrews, and 4 others. 2013. Opportunities for public aquariums to increase the sustainability of the aquatic animal trade. *Zoo Biology* 32: 1–12. Aquariums should be taking the lead in conservation issues such as the ornamental fish trade and the sustainable harvesting and consumption of seafood.

Zimmer, C. 2013. Bringing them back to life: The revival of an extinct species is no longer a fantasy. But is it a good idea? *National Geographic* 223(14). A popular review of the science of de-extinction.

Zimmermann, A., M. Hatchwell, L. Dickie, and C. D. West (eds.) 2008. *Zoos in the 21st Century: Catalysts for Conservation*, pp. 1243–1248. Cambridge University Press, Cambridge. Many modern zoos see the advancement of in situ wildlife conservation as part of their mission. This often-cited work suggests that population control is the only solution to the overuse of common-property resources.

8

Protected Areas

A designated wilderness area near Gunnison, Colorado.
Wilderness areas receive some of the strongest
protections available.

One of the most used, successful, and surprisingly flexible of the tools for conservation is the creation of protected areas. A **protected area** is a clearly defined geographical space, recognized, dedicated, and managed, through legal or other effective means, to achieve the long-term conservation of nature with associated ecosystem services and cultural values (Dudley 2008). The importance of establishing protected areas is highlighted by the Aichi Biodiversity Target 11 from the United Nations Convention on Biological Diversity (see Chapter 11):

> *By 2020, at least 17 per cent of terrestrial and inland water areas and 10 per cent of coastal and marine areas, especially areas of particular importance for biodiversity and ecosystem services, are conserved through effectively and equitably managed, ecologically representative and well-connected systems of protected areas and other effective area-based conservation measures, and integrated into the wider landscape and seascape.*

Since passage of the Yosemite Grant Act and the establishment of the first formal protected area in 1864 and the first national park in 1872 (**FIGURE 8.1**), increasing numbers of protected areas have been established. However, it is estimated that to reach the goals for Target 11, an additional 2.2 million km^2 are needed (UNEP Protected Planet Report 2014).

FIGURE 8.1 American bison (*Bison bison*) graze at Yellowstone National Park, Wyoming, the first National Park established in the world. (Photograph by Richard Primack.)

Protecting areas that contain healthy, intact ecosystems is an effective way to preserve overall biodiversity, and some people argue that it is ultimately the only way to preserve many species and ecosystems, particularly those not well suited for landscapes heavily modified by humans. However, protecting restored areas can also be valuable and has been shown to be as good or better for maintaining some ecosystem services (Possingham et al. 2015). Preserving ecosystems involves:

1. establishing individual protected areas,
2. creating networks of protected areas,
3. managing those areas effectively,
4. implementing conservation measures outside the protected areas, and
5. restoring biological communities in degraded habitats (www.wri.org).

The first three of these topics are discussed here and the final two are covered in Chapters 9 and 10, respectively.

Establishment and Classification of Protected Areas

Protected areas can be established in a variety of ways, but the most common mechanisms are (roughly in decreasing order of significance):

• Government action, usually at a national level, but often on regional or local levels

- Land purchases and easements by private individuals and conservation organizations
- Actions of indigenous peoples and traditional societies
- Development of biological field stations (which combine biodiversity protection and research with conservation education) by universities and other research organizations

Although legislation and land purchases alone do not ensure habitat protection, they can lay the groundwork for it. Partnerships among governments of developing countries, international conservation organizations, multinational banks, research and educational organizations, and governments of developed countries are another way to bring together funding, training, and scientific and management expertise to establish new protected areas.

Traditional societies also have established protected areas to maintain their ways of life or simply to preserve their land (Langton et al. 2014). Many of these protected areas have been in existence for long periods of time and are linked to the religious and cultural beliefs of the inhabitants. National governments in many countries, including the United States, Canada, Colombia, Brazil, and Australia, have recognized the rights of traditional societies to own and manage the land on which they live, hunt, and farm. However, in some cases, the recognition of land rights results only after conflict in the courts, in the press, and on the land.

The International Union for Conservation of Nature (IUCN) has developed a six-category system for classifying protected areas (**TABLE 8.1**). The conservation of nature in protected areas is a primary management objective in all six categories (Dudley 2008), with lands in categories I–IV considered strictly protected. However, areas in the fifth and sixth categories are considered multiple-use or multi-management protected areas, as they are administered, not only to conserve biodiversity, but also to produce natural resources, such as timber and cattle, for human use.

> The IUCN has developed a classification system for protected areas, ranging from strict nature reserves to managed-resource protected areas, depending on the level of human impact and the needs of society for resources.

These multi-management protected areas can be particularly significant for several reasons:

- They are often much larger in area than other categories of protected areas
- They may contain many or even most of their original species
- They often adjoin or surround other protected areas
- They are more likely to benefit local people than strictly protected areas, and therefore are more likely to earn local support

A review of 171 published reports on 165 protected areas found that all types (I–VI) were more likely to have a positive impact on the local people than a negative impact and that the multi-use protected areas (V and VI) were the most likely to have positive conservation and socioeconomic outcomes (Oldekop et al 2015). This is likely because local people benefit

---

TABLE 8.1 IUCN Protected Area Designations I–VI

Category	Description
Ia. Strict nature reserves	Managed mainly for scientific research and monitoring; areas of land and/or sea possessing some outstanding or representative ecosystems, geological or physiological features, and/or species
Ib. Wilderness areas	Managed mainly for wilderness protection; large areas of unmodified or slightly modified land and/or sea retaining their natural character and influence, without permanent or significant habitation, which are protected and managed so as to preserve their natural condition
II. National parks	Managed mainly for ecosystem protection and recreation; natural areas of land and/or sea designated to (1) protect the ecological integrity of one or more ecosystems for present and future generations; (2) exclude exploitation or occupation inimical to the purposes of designation of the area; and (3) provide a foundation for spiritual, scientific, educational, recreational, and visitor opportunities, all of which must be environmentally and culturally compatible
III. Natural monuments	Managed mainly for conservation of specific natural features; areas containing one or more specific natural or natural/cultural features of outstanding or unique value because of inherent rarity, representative or aesthetic qualities, or cultural significance
IV. Habitat/species management areas	Managed mainly for conservation through management intervention; areas of land and/or sea subject to active intervention for management purposes to ensure the maintenance of habitats and/or to meet the requirements of specific species
V. Protected landscapes and seascapes	Managed mainly for conservation and recreation; areas of land, with coast and sea as appropriate, where the interaction of people and nature over time has produced an area of distinct character with significant aesthetic, ecological, and/or cultural value, and often with high biological diversity
VI. Managed-resource protected areas	Managed mainly for the sustainable use of natural ecosystems; areas containing predominantly unmodified natural systems, managed to ensure long-term protection and maintenance of biological diversity, while also providing a sustainable flow of natural products and services to meet community needs

Source: After www.iucn.org.

---

from direct use of the protected area (see Chapter 3) and are not excluded from its management.

Almost every country currently has one or more protected areas (Jenkins and Joppa 2009; www.iucn.org). Countries that protect less than 1% of their land include Syria, Iraq, Haiti, and Uruguay (UNEP Protected Planet Report 2014). Although it could be argued that virtually all countries should have at least one national park, large countries with rich biotas and a variety of ecosystem types generally benefit from having many protected areas. Brazil, for example, protects 26% of its land area and has 67 national parks.

(A)

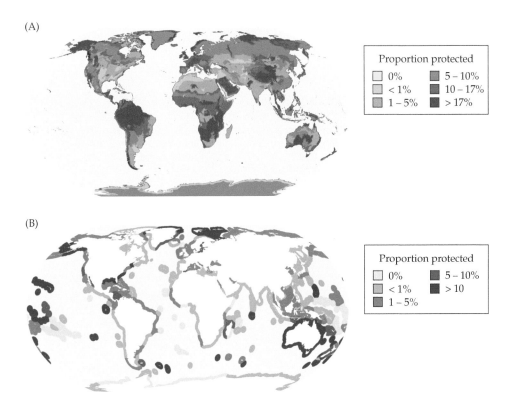

Proportion protected

☐	0%	■	5 – 10%
	< 1%	■	10 – 17%
	1 – 5%	■	> 17%

(B)

Proportion protected

☐	0%	■	5 – 10%
	< 1%	■	> 10
	1 – 5%		

FIGURE 8.2   The percentage of the world's terrestrial and marine ecoregions that are protected. The darkest color for each map indicates those that have reached the targets stipulated by the Convention of Biological Diversity. (From Watson et al. 2014.)

Around 209,000 protected areas in IUCN categories I–VI have been designated worldwide, in over 190 countries (www.wdpa.org) (**FIGURE 8.2**).[1] Nevertheless, this impressive number represents only about 15.4% of Earth's total land surface, which is about the same area as the land used to grow all of the world's crops. Much of this protected land is concentrated in areas that are at higher elevations, on steeper slopes, and in remote areas far from roads and cities, where disturbance from humans is often already minimal (Joppa and Pfaff 2009). For example, the world's largest park is in Greenland on inhospitable terrain and covers 970,000 km^2, accounting for about 3% of the global area that is protected. Far more advantageous for conservation is the establishment of protected areas guided by one of the approaches for species protection described in Chapter 6. Furthermore, only about 30% of the protected area is in categories I–IV (*strictly protected*

---

[1]Uncertainty about the number and size of protected areas stems from the different standards used throughout the world, the degree of protection actually given to a particular designated area, and variations in when the data were gathered.

in scientific reserves and national parks), with the greatest growth in protected areas that share priorities for human use (**FIGURE 8.3**).

The measurements of protected areas in individual countries and on continents are only approximate because sometimes the laws protecting national parks and wildlife sanctuaries are not strictly enforced. At the same time, there are sections of managed areas that, while not legally protected, are carefully protected in practice. Examples include the designated wilderness areas within US national forests that forbid logging, grazing, mountain bikes, and motorized vehicles.

The proportion of land that is strictly protected varies dramatically among countries: high proportions of land are protected in Germany (42%), Austria (23%), and the United Kingdom (26%), and surprisingly low proportions are protected in Bosnia and Herzegovina (1%), Ireland (2%), Hungary (5%), and Denmark (5%). Moreover, even if a country has numerous protected areas, certain unique habitats that also have high economic value may remain unprotected. In addition, protected areas may be reduced in size by the government, opened up for exploitation, or even have their protected status removed (known as **degazettement**), particularly if they are found to contain valuable natural resources (Mascia and Pallier 2011; Mascia et al. 2014).

The limited extent of protected areas highlights the biological significance of the more than 23% of the world's land that is managed for sustainable resource production, such as production forests, watersheds

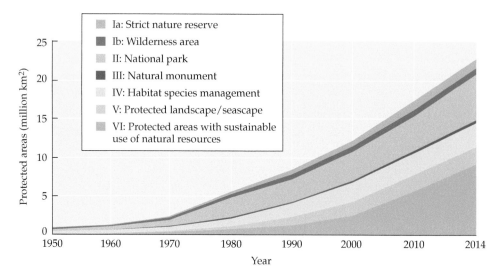

FIGURE 8.3 The increase in amount of area in both MPAs and terrestrial protected areas assigned to each of the six IUCN categories over time. Colored areas are those for which data has been provided to the World Database on Protected Areas (WDPA) and represents only 64% of total global protected area; those for which categories have not been assigned were not included. (After UNEP-WCMC Report 2014.)

around reservoirs, and grazing lands, which is described in greater detail in Chapter 9.

Marine protected areas (MPAs)

Marine conservation has lagged behind terrestrial conservation efforts; however, rapid progress is now being made in MPAs (Gaines et al. 2010). Only about 10.9% of the world's territorial seas near coastlines and about 3.4% of the total marine environment are included in protected areas (**FIGURE 8.4**). The international goal of protecting 10% of the entire marine

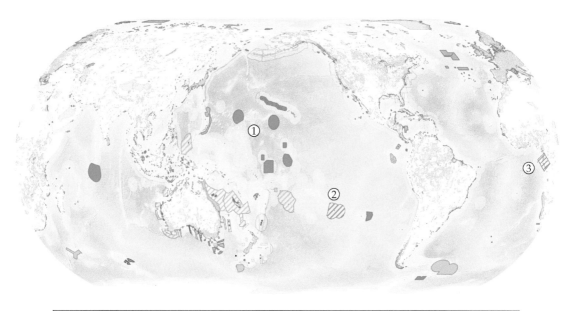

Key to marine protected areas
- ■ No-take marine reserve; officially designated and implemented
- ▨ No-take marine reserve; officially designated, not implemented
- □ Multi-zoned (with some no-take) marine proteced area; officially designated and implemented
- ▨ Multi-zoned (with some no-take) marine proteced area; officially designated, not implemented
- ■ Other MPAs

FIGURE 8.4   The world's terrestrial protected areas (shown in light brown) and Marine Protected Areas (see key). Although many small protected areas do not show up at this scale, large areas in IUCN categories I–VI are indicated, as well as many areas that are protected in some manner (e.g., privately) but do not have an official designation at the present time. Note the large protected areas in Greenland, western Europe, eastern and western North America, the Amazon Basin, northeastern Australia, and western China. Many large new MPAs have recently been designated (numbered circles), including (1) the expansion of the Pacific Remote Islands National Monument in 2014, (2) new areas around Pitcairn Islands in 2015, and (3) Gabon in 2014. (From MPAtlas.org. [Current 02/2016].)

environment by the year 2020 was motivated in part to manage declining commercial fishing stocks (Rife et al. 2013); however, even stronger measures may be required to conserve the full range of coastal and marine biodiversity (Spalding et al. 2008; www.iucn.org). Some of the current proposals for new MPAs are receiving pushback from groups that fear it will harm the fishing industry. One such example is protests of the proposal to introduce 30 new MPAs off Scotland in 2016, even though MPAs can benefit fisherman by providing opportunities for fish to reproduce and grow larger.

Over 5000 marine and coastal protected areas have been established worldwide, but most are small. A number of countries, though, are now creating very large marine protected areas (VLMPAs) covering hundreds of thousands of square kilometers of marine ecosystems. One of the largest single reserves was established in 2015 around the Pitcairn Islands in the South Pacific Ocean; 830,000 km^2 was set aside as a no-take zone, with only traditional fishing by locals permitted. This reserve is dwarfed, however, by the Pacific Remote Islands Monument, a series of areas that was expanded to eight times their original area by President Obama in 2014. It now covers a total of 2 million km^2.

Marine protected areas seek to preserve the nursery grounds of commercial species and to maintain water quality and both the physical and biological features of ecosystems (Gaines et al. 2010). In the process, high-quality protected areas can also maintain recreational activities such as swimming and diving and the economic benefits associated with tourism. Unfortunately, many MPAs exist only on maps and receive insufficient regulation, funding, and enforcement to prevent overharvesting and pollution. Even in the well-funded Phoenix Islands Protected Area, located in the central Pacific, fishing is only banned in 3% of the total area.

The effectiveness of protected areas

Studies show that protected areas generally are effective at protecting ecosystems, particularly when regulations are enforced and when areas are geographically isolated (Geldmann et al. 2013). The value of protected areas in maintaining biodiversity is abundantly clear in many tropical countries (Nolte et al. 2013) (see Figure 4.6). Inside a given park's boundaries you may see numerous trees and abundant animal life, while outside the park you may very well see cleared land and see and hear few animals (**FIGURE 8.5**). Many of these protected areas also continue to provide income and services

FIGURE 8.5   National parks and other protected areas are able to prevent damage to the natural forests in (A) the Atlantic coast forests of Brazil and (B) West Africa. In the Atlantic coast forest, there is a sharp boundary with intact forest inside the protected areas and around 50% intact forest outside the protected area. For protected areas in West Africa, there is considerable forest degradation within 16 km of the park boundary, particularly for IUCN categories V and VI, which are mainly forest reserves. See Table 8.1 for a description of IUCN categories. (After Joppa et al. 2008.)

to poor people living nearby (Andam et al. 2010). At the same time, these areas must be monitored and managed to protect them from overharvesting by both legal and illegal subsistence use and commercial production. In some cases, due to management problems and conflicts with local inhabit-

(A) Brazil

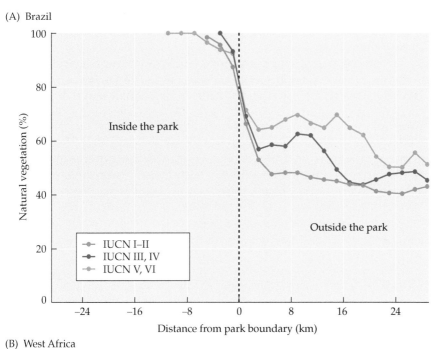

(B) West Africa

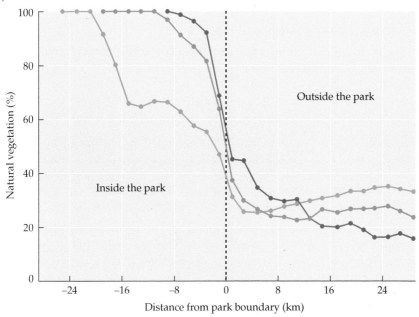

ants, national parks have become more degraded than neighboring areas (Wright et al. 2007). Consequently, the effective management of protected areas must consider activities that occur in neighboring areas. There are also cases where the protected status is violated by the very governments charged with enforcing it, such as the extensive logging in the Sochi National Park in the Western Caucasus to enable construction for the Olympic Games (Bragina et al. 2015).

If national parks can be established and maintained where concentrations of species occur, high numbers and percentages of species can be preserved (Joppa et al. 2013). A system of protected areas can include a high percentage of a country's species if it includes representatives of all major ecosystems. For example, in Britain, 88% of plant species occur in the protected area system, of which 26% are found exclusively within protected areas (Jackson et al. 2009). Likewise, China's nature reserves cover 15% of the total area and include 81% of the country's vegetation types (Wu et al. 2011). However, it is important to recognize that the long-term survival of many species in protected areas, and even of the ecosystems themselves, remains in doubt because populations of many species and the area of the ecosystems may be so reduced in size that their eventual fate is extinction.

> Although the number of species living within a protected area is an important indicator of the area's potential to protect biodiversity, protected areas need to maintain healthy ecosystems and viable populations of important species.

Measuring effectiveness: Gap analysis

Gap analysis compares biodiversity priorities with existing and proposed protected areas (see, for example, Simaika et al. 2013), but it can consider several factors. In the past, conservationists used informal means to ensure that the high-priority areas were protected, for example, by establishing national parks in different regions with distinctive ecosystems and ecological features (Shafer 1999). Now, however, conservationists are using more systematic planning processes involving gap analysis (Tognelli et al. 2009). Gap analysis generally consists of the following steps:

1. Data are compiled describing the presence and distribution of species, ecosystems, and physical features of the region, which are sometimes referred to as *conservation units*. Information on human densities and economic factors can also be included.

2. Conservation and social goals are identified, such as the amount of area to be protected for each ecosystem, the number of individuals of rare species to be protected, or the desired balance between wilderness and mixed resource management.

3. Existing conservation areas are reviewed to determine what is protected already and what is not (known as "identifying gaps in coverage").

4. Additional areas are identified to help meet the conservation goals ("filling the gaps").

5. These additional areas are reviewed in more detail and, if appropriate and practical, protected in some way (often by being directly purchased or designated as national parks). **Management plans** are then developed and implemented.

6. The new protected areas are monitored to determine whether they are meeting their stated goals. If not, the management plan can be changed or, possibly, additional areas can be acquired to meet the goals.

Gap analysis can also be used to identify holes in conservation at international scales. For example, a group of researchers from 27 institutions and organizations did an analysis of all existing protected areas and 25,380 species across the globe (Butchart et al. 2015). They calculated that protected areas currently only include 77–78% of important sites for biodiversity and there is insufficient coverage for 57% of the species they evaluated (**FIGURE 8.6**). They determined that in order to reach the Aichi Biodiversity targets of >17% and >10% marine systems protected, an additional 3.3 million km^2 outside existing protected areas were needed, plus 14.8 million km^2 to cover all threatened species. To cover these plus all documented important sites for biodiversity and individual country and biome targets, the researchers estimated that the amount of area protected needed to be roughly doubled (see Figure 11.7).

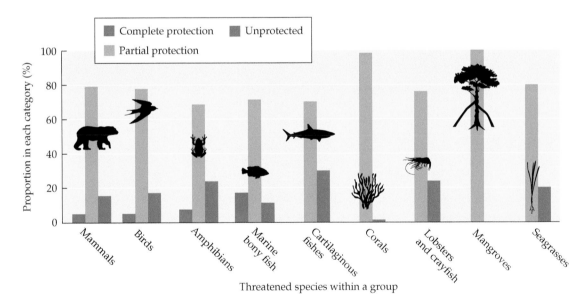

FIGURE 8.6   A gap analysis for various groups of animals and plants shows the proportion of threatened species that receive any sort of protection. Depending on the group, 10–30% of threatened species lack protection anywhere in their range; most threatened species have at least partial protection. Relatively few threatened species are completely protected. (After Butchart et al. 2015.)

GIS is an effective tool for gap analysis, which uses a wide variety of information to pinpoint critical areas and species that are priorities for protection.

Geographic information systems (GIS) are a vital tool in gap analysis; they facilitate the integration of the wealth of data on the natural environment with information on species distributions (Murray-Smith et al. 2009). The basic GIS approach involves storing, displaying, and manipulating many types of spatial data involving factors such as vegetation types, climate, soils, topography, geology, hydrology, species distributions, human settlements, and resource use (**FIGURE 8.7**). This approach can point out correlations among the abiotic, biotic, and human ele-

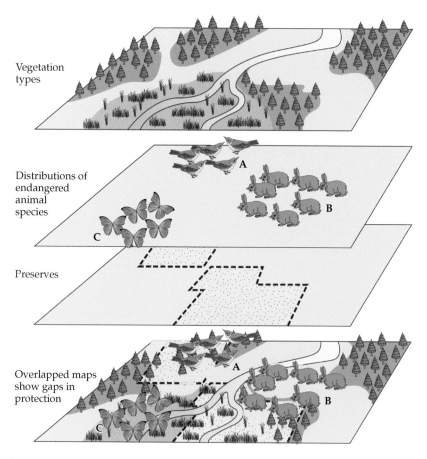

FIGURE 8.7 Geographic information systems (GIS) provide a method for integrating a wide variety of data for analysis and display on maps. In this example, vegetation types, distributions of endangered animal species, and preserved areas are overlapped to highlight areas that need additional protection. The overlapped maps show that the distribution of Species A is predominantly in a preserve, Species B is only protected to a limited extent, and Species C is found entirely outside the preserves. Establishing a new protected area to include the range of Species C would be the highest priority. (After Scott et al. 1991.)

ments of the landscape; help plan parks that include a diversity of biological communities; and even suggest sites that are likely to support rare and protected species. Aerial photographs and satellite imagery are additional sources of data for GIS analysis, and they can highlight patterns of vegetation structure and distribution over local and regional scales. These images can dramatically illustrate when current government policies are not working and need to be changed.

Designing Protected Areas

The size and placement of protected areas throughout the world are often determined by the distribution of people, potential land values, the political efforts of conservation-minded citizens, and historical factors (Armsworth et al. 2006; Mills et al. 2014). Ideally, the selection of new protected areas will follow conservation priorities, such as to protect a particular species, ecosystem, or biodiversity hotspot (see Chapter 6); however, there are many other possible scenarios. In urban and suburban areas, the fund-raising ability of private conservation groups, wealthy individuals, and government departments is often the most important factor in determining what land is acquired (Lerner et al. 2007). In other cases, certain parcels of land may be purchased to protect a critical water supply or a charismatic species, but yet others may be acquired simply because they adjoin the property of an influential citizen. Moreover, sometimes lands are set aside for conservation protection because they have no immediate commercial value—they are "the lands that nobody wants" (Scott et al. 2001).

Issues of reserve design have proved to be of great interest to governments, corporations, and private landowners, who are being urged, or even mandated, to manage their properties for both the commercial production of natural resources and the protection of biodiversity. However, conservation biologists must be cautious about using simplistic, overly general guidelines for designing protected areas, because every conservation situation requires individual consideration. Everyone benefits from increased communication between the academic scientists, who are developing theories of nature reserve design, and the managers, planners, and policymakers, who are actually creating new reserves (Braunisch et al. 2012).

Conservation biologists often start by considering the **four Rs**:

1. *Representation*. Protected areas should contain as many features of biodiversity (species, populations, habitats, etc.) as possible.
2. *Resiliency*. Protected areas must be sufficiently large to maintain all aspects of biodiversity in a healthy condition for the foreseeable future, including as climate conditions change.
3. *Redundancy*. Protected areas must include enough examples of each aspect of biodiversity to ensure its long-term existence in the face of future uncertainties.

4. *Reality.* There must be sufficient funds and political will, not only to acquire and protect lands, but also to regulate and manage the protected areas.

The following, more specific, questions about reserve establishment and design are also useful for discussing how best to construct and link protected areas:

- Given a particular amount of funding to spend on a protected area or network of areas, what is the most effective way to spend it?
- How large must a nature reserve be to effectively protect biodiversity?
- Is it better to have a single large protected area or multiple smaller reserves?
- When a network of protected areas is created, should the areas be far apart or close together, and should they be isolated from one another or connected by corridors?
- How many individuals of an endangered species must be included in a protected area to prevent the local extinction of a species?
- What is the most cost-effective way to design a protected area to achieve its conservation goals?
- What is the best shape for a nature reserve?

> Principles of design have been developed to guide land managers in establishing and maintaining networks of protected areas.

Conservationists explore many of these issues using the island biogeography model of MacArthur and Wilson (1967), which describes the relationship between the size of an area and the number of species it supports (see Chapter 5). When applied to protected areas, the model frequently assumes that parks are **habitat islands**—intact habitat surrounded by an unprotected matrix of inhospitable terrain. In fact, many species are capable of living in and dispersing through the surrounding habitat matrix (see Chapter 9). This may explain the finding that species diversity on oceanic islands increases more rapidly with size than do habitat islands, and that among habitat islands, this relationship is stronger for urban islands than for forest islands (Matthews et al. 2015). Researchers working with island biogeography models and data from protected areas have proposed various principles of reserve design, which are still being debated (**FIGURE 8.8**).

FIGURE 8.8   Principles of reserve design that are based in part on theories ▶ of island biogeography. Imagine that the reserves are "islands" of the original biological community surrounded by land that has been made uninhabitable for the original species by human activities such as farming, ranching, or industrial development. The practical application of these principles is still being studied and debated, but in general, the designs shown on the right are considered preferable to those shown on the left. (After Shafer 1997.)

	Worse	Better
(A)	Ecosystem partially protected	Ecosystem completely protected
(B)	Smaller reserve	Larger reserve
(C)	Fragmented reserve	Unfragmented reserve
(D)	Fewer reserves	More reserves
(E)	Isolated reserves	Corridors maintained
(F)	Isolated reserves	"Stepping-stones" facilitate movement
(G)	Uniform habitat protected	Diverse habitats (e.g., mountains, lakes, forests) protected
(H)	Irregular shape	Reserve shape closer to round (fewer edge effects)
(I)	Only large reserves	Mix of large and small reserves
(J)	Reserves managed individually	Reserves managed regionally
(K)	Humans excluded	Human integration; buffer zones

Also, aspects of protected area models differ depending on whether the targets are terrestrial or marine species; vertebrates, higher plants, or invertebrates; or whole ecosystems. Many aspects of marine species' life cycles and dispersal mechanisms are largely unknown (Burgess et al. 2014; Schofield et al. 2013). MPAs and complementary conservation efforts outside protected areas must be designed to accommodate the high mobility of some species and limited dispersal of other target species, thus ensuring that corridors and genetic populations are preserved.

Protected area size and characteristics

> Large reserves are generally better able to maintain many species because such reserves support larger population sizes and a greater variety of habitats. However, small reserves are important in protecting particular species and ecosystems.

An early debate in conservation biology centered on one question: Is species richness maximized in one large nature reserve or in several smaller ones of an equal total area (McCarthy et al. 2006; Soulé and Simberloff 1986)? This debate is known in the literature as the **SLOSS debate** (with "SLOSS" standing for *single large or several small*). Is it better, for example, to set aside one reserve of 10,000 ha or four reserves of 2500 ha each? Proponents argue that only large reserves can maintain sufficient numbers of large, wide-ranging, low-density species (such as large carnivores) over the long term. Large reserves also minimize the ratio of edge habitat to total habitat, encompass more species, and sometimes have greater habitat diversity than small reserves.

The advantage of large parks was effectively demonstrated by an analysis of 299 mammal populations in 14 national parks in western North America (**FIGURE 8.9**) (Newmark 1995). Extinction rates were very low or zero in parks with areas over

FIGURE 8.9 Mammals have higher extinction rates in smaller parks than in larger parks. If these same trends continue for the next 100 years, small parks such as Bryce are predicted to lose more than 30% of their mammal species, whereas large parks such as Yosemite will only lose 5%, and Kootenay-Banff-Jasper-Yoho is not expected to lose any species. Each dot represents the actual extinction rate of mammal populations (expressed as the proportion of species that have gone extinct per year) for a particular US national park, Canadian national park, or two or more adjacent parks. (After Newmark 1995.)

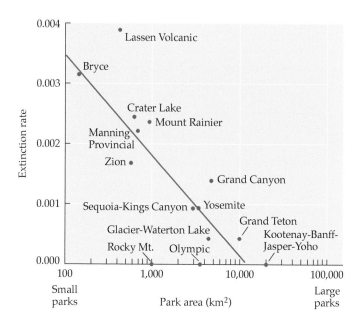

1000 km^2 but much higher in parks that are smaller than 1000 km^2. In addition to the ability to support more species because of their larger size, larger parks tend to be surrounded by lower densities of people compared to small reserves, potentially increasing their connectivity with other ecosystems and populations (Wiersma et al. 2004).

On the other hand, once a park reaches a certain size, the number of new species added with each increase in area starts to decline. At that point, creating a second large park or another park some distance away may be a more effective strategy for preserving additional species than simply adding area to the existing park.

The research on extinction rates of populations in large parks has three practical implications:

1. When a new park is being established, it should generally be made as large as possible (within the context of an overarching strategy for optimizing protected areas) in order to preserve as many species as possible, contain large populations of each species, and provide a diversity of habitats and natural resources. Keystone resources should be included, in addition to habitat features that promote biodiversity, such as elevational gradients.

2. When possible, land adjacent to protected areas should be acquired to reduce external threats to existing parks and maintain buffer zones. For example, terrestrial habitats adjacent to wetlands are often needed by semiaquatic species such as snakes and turtles. Moreover, protecting natural ecological units, such as entire watersheds or mountains, is often the best means to reduce external threats.

3. The effects of climate change, invasive species, and other threats are altering ecosystems within existing protected areas. These changes can reduce the area of habitat available for a species and lead to declines in population size and increased probability of extinction. These changes emphasize the need for preserving corridors or otherwise connecting protected areas to facilitate the dispersal of species among them.

Although research suggests that large parks may be best able to achieve conservation goals in many situations, there are some instances where several small parks or a mixture of large and small ones is better for conservation. For example, a study of wetlands in New Zealand determined that certain rare plant species were not necessarily present in the large protected areas and that several small reserves were more effective for their protection (Richardson et al. 2015).

Small reserves, even those less than a hectare in size, may effectively protect isolated populations of rare species, particularly if they contain a unique habitat type (Jarošík et al. 2011). Also, regional biodiversity may depend on small natural features (SNFs) such as temporary pools, caves, single trees, or rock outcrops (Hunter et al. 2015). These SNFs may contain keystone resources (discussed later) or ecological processes and as such have a disproportionately large conservation value, a phenomenon that has been called the *Frodo effect*.[2] For example, ephemeral desert springs should be prioritized for inclusion in protected areas because so many organisms are dependent on their cycles of wet and dry periods (Acuña and Ruhí 2016). Thus, several small protected areas that include critical SNFs will conserve more species than a single large one that does not contain those SNFs.

Even if a large reserve would protect the most species in a given system, often officials have no choice but to accept the challenge of managing species and biological communities in small reserves. The 10,000 protected areas in Britain, for instance, have an average area of only 3 km^2 (**FIGURE 8.10**) (Jackson et al. 2009). Numerous countries have many more small protected areas (less than 100 ha) than medium and large ones, yet the combined area of these small reserves is only a tiny percentage of the total area under protection. This is particularly true in places, such as Europe, China, and Southeast Asia, that have been intensively cultivated for centuries.

Singapore provides an excellent example of how small reserves provide long-term protection for numerous species. A group of reserves that have been isolated since 1860 currently protect 5% the original habitat on Singapore, yet they still contain around half of the country's original flora and fauna, including 350 of the original bird species and 26 of the mammal species (Corlett 2013). In addition,

- European areas
- International wetlands
- National areas

FIGURE 8.10 The geographic locations of protected areas in Britain managed for biodiversity conservation. Note their large number, varied sizes and shapes, and scattered distribution. At the scale of this map, most of the small protected areas cannot be seen. Many of the protected areas are covered by two or more designations. There are other areas managed for other purposes, which are not shown on this map. (Courtesy of Sarah Little.)

[2]The Frodo effect borrows its name from the J. R. R. Tolkien character in the Lord of the Rings, who is small in stature yet responsible for the future of everyone else.

small reserves located near populated areas make excellent conservation education and nature study centers that further the long-range goals of conservation biology by developing public awareness of important issues. By 2030, over 60% of the world's population will live in urban areas; thus, there is a need to develop such reserves for public use and education.

It is generally agreed that protected areas should be designed to minimize edge effects. Conservation areas that are rounded in shape minimize the edge-to-area ratio, and the center is farther from the edge than in other park shapes. Long, linear parks have the most edge—all points in the park are close to the edge (Yamaura et al. 2008). Consequently, for parks with four straight sides, a square park is a better design than an elongated rectangle of the same area. However, most parks have irregular shapes because land acquisition is typically a matter of opportunity rather than design.

As discussed in Chapter 4, the internal fragmentation of protected areas by roads, fences, farming, logging, and other human activities should be avoided as much as possible because fragmentation creates barriers to dispersal and often divides a large population into two or more smaller populations, each of which is more vulnerable to extinction than the large population. Fragmentation also provides entry points for invasive species, which may harm native species, and creates more undesirable edge effects.

Networks of Protected Areas

To overcome some of the effects of fragmentation and limits on the size of protected areas, conservationists have developed strategies to aggregate small and large protected areas into larger conservation networks (Bode et al. 2011; Van Teeffelen et al. 2012). Networks are also important to accommodate metapopulation dynamics (see Chapter 6).

Cooperation among public and private landowners is important for creating these networks, particularly in developed metropolitan areas and other areas where there are many small, isolated parks controlled by different government agencies and private organizations. An excellent example of cooperation intended to network urban protected areas is the Chicago Wilderness project, which consists of more than 300 organizations collaborating to preserve around 227,000 ha of tallgrass prairies, woodlands, rivers, streams, and other wetlands in metropolitan Chicago (www. chicagowilderness.org).

Often a protected area is embedded in a larger matrix of habitat managed for human uses such as timber forest, grazing land, and farmland (**FIGURE 8.11**). If the protection of biodiversity becomes a secondary priority of these areas and is included in their management plans, then these mixed-use areas can serve as buffers, corridors, or other key components of protected area networks (Hansen et al. 2011). Habitat managed for resource extraction can sometimes also be designated as important secondary sites for wildlife and as dispersal corridors between isolated protected areas.

FIGURE 8.11 Tina Buijs, a park guard supervisor and operations manager with The Nature Conservancy (TNC), talks with Juan Antillanca, a farmer belonging to the Huiro indigenous community that borders TNC's Valdivian Coastal Reserve in Chile. The reserve is a 61,000 ha site comprising temperate rain forest and 36 km of Pacific coastline. Keeping in close contact with their neighbors helps TNC officials realize their conservation goals. (Photograph © Mark Godfrey/The Nature Conservancy.)

Using a network approach, groups of rare species can be managed as large metapopulations to facilitate gene flow and migration among populations (Andrello et al. 2015).

Habitat corridors

Growing numbers of conservationists argue that connectivity is important, and they are taking steps to link isolated protected areas into large systems through the use of **habitat corridors**—strips of protected land running between the reserves (Beier 2011; Magrach et al. 2012). Such habitat corridors, also known as **conservation corridors** or **movement corridors**, can allow plants and animals to disperse from one reserve to another, facilitating gene flow and the colonization of suitable sites.

Corridors are clearly needed to preserve animals that must migrate seasonally among different habitats to obtain food and water, such as the large grazing mammals of the African savanna. If these animals were confined to a single reserve by fences, farms, and other anthropogenic factors, they might starve (Wilcove and Wikelski 2008). The width required for effective corridors varies depending on the species, length of the corridor, and other factors.

In many areas, roads are a primary obstruction to the creation of habitat corridors. In these cases culverts, tunnels, and overpasses can create passages under and over roads and railways that allow reptiles, amphibians, and mammals to travel between habitat fragments or protected areas (Soanes et al. 2013). An added benefit of these passageways is that they reduce collisions between animals and vehicles, which saves lives and money. For example, in Canada's Banff National Park, road collisions involving deer, elk, and other large mammals declined by 96% after fences, overpasses, and underpasses were installed along a major highway (**FIGURE 8.12A**) (Ford et al. 2009).

> Establishing habitat corridors can potentially transform a set of isolated protected areas into a linked network with populations interacting as a metapopulation.

Corridors that facilitate natural patterns of migration will probably be the most successful at protecting species. For example, large grazing animals often migrate in regular patterns across a rangeland in search of water and the best vegetation. In seasonally dry savanna habitats, animals often migrate along the riparian forests that grow along streams and rivers. In mountainous areas, many bird and mammal species regularly migrate to higher elevations during the warmer months of the year. For example, a corridor was established in Costa Rica to link two wildlife reserves, the Braulio Carrillo National Park and La Selva Biological Station, to protect migrating birds. A 7700 ha corridor of forest several kilometers wide and 18 km long, known as the Zona Protectora Las Tablas, was set aside to provide an elevational link that allows at least 75 species of birds to migrate between the two large conservation areas (Bennett 1999).

Some conservation biologists have started to plan habitat corridors on a truly huge scale. Wildlands Network has a detailed plan, called the Spine of the Continent Initiative, that would link all large protected areas in the western United States and Canada by habitat corridors, creating a system that would allow large and currently declining mammals to coexist with human society.

(A)

FIGURE 8.12 (A) An overpass above a fenced-off divided highway allows animals to migrate safely between two forested areas. (B) Individuals of a species naturally disperse between two large protected areas (areas 1 and 2, left-hand panel) by using smaller protected areas as stepping-stones. The right-hand panel shows that habitat destruction and a large edge effect zone caused by a new road have blocked a migration route. To offset the effects of the road, compensation sites (orange) have been added to the system of protected areas, and an overpass has been built over the highway to allow dispersal. (A, © Paul Zizka/Getty Images; B, after Cuperus et al. 1999.)

(B)

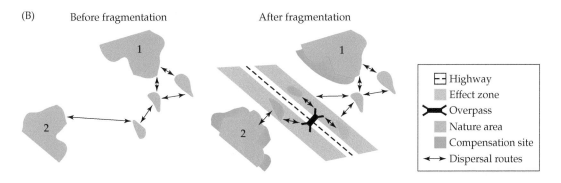

As the global climate changes, many species are moving to higher elevations and higher latitudes. Creating corridors to protect expected routes—such as north–south river valleys, ridges, and coastlines—would be a useful precaution. Extending existing protected areas in the direction of anticipated species movements could help to maintain long-term populations (Nuñez et al. 2013). Corridors that cross gradients of elevation, rainfall, and soils could also allow the local migration of species to more favorable sites.

Although the idea of corridors is intuitively appealing, there are some possible drawbacks (Ogden 2015; Orrock and Damschen 2005; Simberloff et al. 1992). In particular, corridors may facilitate the movement of pest species and disease; a single infestation could quickly spread to all of the connected nature reserves and cause the extinction of all populations of a rare species. Also, animals dispersing along corridors may be exposed to greater risks of predation because human hunters as well as animal predators tend to concentrate on routes used by wildlife. Finally, there is a risk that the corridors, which in some cases can be expensive and difficult to construct, will not be used by the intended species.

Some studies published to date support the conservation value of corridors, while other studies show no effect (Gilbert-Norton et al. 2010; Pardini et al. 2005). In general, maintaining existing corridors is probably worthwhile because many are located along watercourses that may be biologically important habitats themselves. New protected corridors are most obviously needed along known migration routes (Newmark 2008). There would also be value in leaving small clumps of original habitat between large conservation areas to facilitate movement in a stepping-stone pattern (**FIGURE 8.12B**), such as the protected areas that birds use as stopping points along the flyways of their annual migration routes.

Landscape Ecology and Park Design

The interaction of human land-use patterns, conservation theory, and park design is evident in the discipline of **landscape ecology**, which investigates patterns within the mosaic of the physical environment, ecological communities, ecosystem processes, and human–ecosystem interactions on local and regional scales (Schwenk and Donovan 2011; Wu and Hobbs 2009) (**TABLE 8.2**). It has focused on terrestrial systems, but the same approach can apply to marine and other aquatic systems (Jelinski 2015).

Landscape ecology has been most intensively studied in Europe and Asia, where long-term practices of traditional agriculture and forest management have shaped landscape patterns. In the European countryside, cultivated fields, pastures, woodlots, and hedges alternate to create a mosaic that affects the distribution of wild species. Likewise, in the traditional Japanese landscape known as *satoyama*, flooded rice fields, hay fields, vil-

In some cases, long-term traditional human use has created landscape patterns that preserve and even increase biodiversity.

TABLE 8.2	A Proposed Framework for Landscape Design That Promotes Biodiversity Conservation

1. Distinguish and delineate different patches of land covers (e.g., rice fields, hay fields, forest, villages) in the selected landscape.

2. Categorize patches as unaltered (lower human use) vs. altered (higher human use) land covers.

3. Identify the constraints on land use planning (e.g., economic, social, political).

4. Given these constraints, create a landscape plan that maximizes the total amount and diversity of unaltered land cover, especially near water (e.g., by restoring some abandoned hay fields to forest);

5. …minimize human disturbance within altered land cover, especially near water (e.g., by removing roads through forest);

6. …and aggregate altered land covers associated with high-intensity land uses, especially away from water (e.g. by planning future development to be near existing villages, rather than in currently unaltered areas).

Source: After Gagné, et al. 2015.

lages, and forests provide a rich diversity of habitat for wetland species, such as dragonflies, amphibians, and waterfowl (Kadoya et al. 2009) (**FIGURE 8.13**). These heterogeneous landscapes, which include a mix of human-created and natural features, are critical for the survival of some species. In many areas, traditional patterns of farming, grazing, and forestry are being abandoned. In some places, rural people have left the land completely and migrated to urban areas or their farming practices have become more intensive, involving more machinery and the application of fertilizer. To protect species and ecosystems in such situations, the design and management of protected areas frequently include strategies to maintain the traditional landscapes, in some cases by subsidizing traditional practices or having volunteers manage the land.

To increase the number and diversity of animals, wildlife managers sometimes create the greatest amount of landscape variation possible within the confines of some protected areas, par-

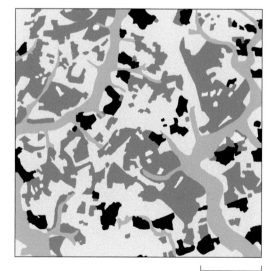

FIGURE 8.13 Schematic of a traditional rural landscape near Tokyo, Japan, with an alternating pattern of villages (black); secondary forest (dark green); paddies, or wet rice fields (light green); and hay fields (beige). Such landscapes were common in the past but are now becoming rare because of the increasing mechanization of Japanese agriculture, the movement of the population away from farms, and the urbanization of the Tokyo area. The area cover is approximately 4 km × 4 km. (After Yamaoko et al. 1977.)

1 km

ticularly refuges or other areas managed primarily for hunting and fishing. Fields and meadows are created and maintained, small thickets are encouraged, groups of fruit trees and crops are planted, patches of forests are periodically cut, small ponds and dams are developed, and numerous trails and dirt roads meander across and along all the patches. Such landscaping is often appealing to the public, who are the main visitors and financial contributors to the park. However, the species in these landscapes are likely to be principally common species that depend on human disturbance—and in some cases, invasive species. To remedy this localized approach, large animals, such as bears, mountain lions, and tigers, generally are best managed on the level of a regional landscape, in which the sizes of the landscape units more closely correlate to the natural population sizes and migration patterns of the species (Wikramanayake et al. 2011).

Managing Protected Areas

Some people believe that "nature knows best" and that humans do not need to actively manage biodiversity—that once protected areas are legally established, the work of conservation is largely complete. The reality, however, is often very different. Management is required because humans have already modified many local environments so much that the remaining species and ecosystems need human monitoring and intervention in order to survive. The world is littered with "paper parks" created by government decree but left to flounder without any management (Joppa et al. 2008). These protected areas have gradually—or sometimes rapidly—lost species as their habitat quality has degraded. In some countries, people readily farm, log, mine, hunt, and fish in protected areas because they feel that government land is owned by "everyone," so anybody can take whatever they want and nobody is willing to intervene.

Many parks are actively managed according to carefully prepared management plans designed to prevent deterioration. Important considerations in management plans include protecting biodiversity, maintaining ecosystem services and health, maintaining historical landscapes, and providing resources and experiences of value to local inhabitants and visitors (Hobbs et al. 2010). Part of the management plan involves making the public aware of which activities are encouraged (for example, wildlife photography) and which are prohibited (for example, hunting), and then enforcing the rules. The public can even be encouraged to engage in behaviors that help mitigate the negative impacts of ecotourism (**FIGURE 8.14**).

In many European, Asian, and African countries with well-established traditions of cultivation, ranching, and grazing, hundreds (and even thousands) of years of human activity have shaped habitats such as woodlands, meadows, and hedges. These habitats support high species diversity as a result of traditional land-management practices, which must be maintained if the species are to persist (Jacquemyn et al. 2011). Similarly, grasslands

FIGURE 8.14   Management of protected areas may involve advising visitors how to avoid harming wildlife, as does this sign in the parking lot of Table Mountain National Park, on the Cape of Good Hope in South Africa. (Photographs by Anna Sher.)

that have been grazed in the past by large wild animals or domesticated animals, such as cattle, still need to be grazed. If protected areas that include these types of habitats are not managed, they will undergo **ecological succession** (a predictable, gradual and progressive change in species over time) and many of their characteristic species will disappear as shrubs trees become dominant (**FIGURE 8.15**).

It is important, though, to be cautious in taking management actions. In some cases, often because of a lack of complete understanding of an ecosystem or conflicting management objectives, management practices may be ineffective or even detrimental. Some protected areas are managed to promote the abundance of a game species, such as deer, for hunting. Management has frequently involved eliminating top predators, such as wolves and cougars. However, without predators to control them, game populations (and, incidentally, rodents that feed on seeds and can spread disease) sometimes increase far beyond expectations, resulting in overgrazing, habitat degradation, and a collapse of animal and plant communities.

Overenthusiastic park managers who remove hollow or dead trees, rotting logs, and underbrush to "improve" a park's appearance may unwittingly remove a critical keystone resource needed by certain animal species for nesting and overwintering, by rare plants for seed germination, and by all species as an integral part of nutrient cycling (Keeton et al. 2007). In these instances, a "clean" park can become a biologically sterile park. Likewise, in many parks, fire is part of the natural ecology of the area (Nimmo et al. 2013). Attempts to suppress fire completely are expensive and waste scarce management resources. Suppressing the normal fire cycle may eventually lead to the loss of fire-dependent species and to massive, uncontrollable fires.

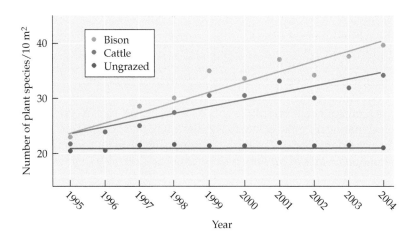

FIGURE 8.15 Large herbivores originally grazed the tallgrass prairies of the midwestern United States. The loss of these herbivores has altered the ecology of this ecosystem, with a resulting loss of plant species. Management involving grazing by cattle and bison resulted in a gradual increase in plant species in prairie research plots over a 10-year period, compared with ungrazed control plots. (After Towne et al. 2005; photograph courtesy of Jim Peaco/US National Park Service.)

> Management plans are needed that articulate conservation goals and practical methods for achieving them. Management activities can include controlled burns, enforcement of restrictions on human use, and the maintenance of keystone resources, especially water.

The most effectively managed parks are usually those whose managers have the benefit of research and monitoring programs and have funds available to implement their management plans. At these sites, the effects of different grazing methods (sheep versus cattle, light versus heavy) on populations of wildflowers, butterflies, and birds are closely followed using an experimental approach in which treated areas are compared with control areas. These experiments demonstrate that parks often must be actively managed to prevent deterioration.

Managing sites

Park managers sometimes must actively manage sites to ensure that all or particular successional stages are present so that species characteristic of each stage have a place to persist and thrive.

In some wildlife sanctuaries, grasslands and fields are maintained by live-stock grazing, burning, mowing, tree cutting, or shallow plowing in order to retain open habitat in the landscape. One common way to do this is to set localized, controlled fires periodically in grasslands, shrublands, and forests to reinitiate the successional process (Middleton 2013). Obviously, such burning must be done in a legal and carefully controlled manner to prevent damage to nearby property (**FIGURE 8.16A**). Also, prior to burning, land managers must develop a program of public education to explain to local residents the role of fire in maintaining the balance of nature. In other situations, parts of protected areas must be carefully managed to *minimize* human disturbance and fire (**FIGURE 8.16B**).

(A)

(B)

FIGURE 8.16   Conservation management: intervention versus leave-it-alone. (A) Heathland in protected areas of Cape Cod, Massachusetts, is burned on a regular basis in order to maintain the open vegetation habitat and to protect wildflowers and other rare species. (B) Sometimes management involves keeping human disturbance to an absolute minimum. Muir Woods National Monument is a forest of old-growth coast redwoods, protected in the midst of the heavily urbanized San Francisco Bay area. (A, photograph by Elise Smith, US Fish and Wildlife Service; B, photograph courtesy of US National Park Service.)

KEYSTONE RESOURCES   In many protected areas, it may be necessary to preserve, maintain, and supplement keystone resources on which many species depend. These resources include trees that supply fruit when little or no other food is available, pools of water during a dry season, exposed mineral licks, and so forth. Keystone resources and keystone species can be enhanced in managed conservation areas to increase the populations of species whose numbers have declined.

By planting areas with food plants and building an artificial pond, it may be possible to maintain vertebrate species in a smaller conservation area and at higher densities than would be predicted based on studies of species distribution in undisturbed habitat. Artificial ponds, for example, can provide habitat for insects that people enjoy watching such as dragonflies and also serve as centers of public education in urban areas (Kobori and Primack 2003). Likewise, nest boxes or drilled nesting holes in living and dead trees can provide substitute shelter resources for birds and mammals where there are few dead trees with nesting cavities (Lindenmayer et al. 2009).

In each case a balance must be struck between, at one extreme, establishing nature reserves free from human influence and, at the other extreme, creating seminatural gardens in which the plants and animals are dependent on people. Further, in some cases human replacements or augmentations of resources may be ineffective, such as placing next boxes on small trees that lack other features hollow-nesting birds need (Le Roux et al. 2015).

Rivers, lakes, swamps, estuaries, ponds, lakes, and all the other types of wetlands must receive a sufficient supply of clean water to maintain ecosystem processes. In particular, maintaining healthy wetlands is necessary for populations of waterbirds, fish, amphibians, aquatic plants, and a host of other species (Jähnig et al. 2011). Yet protected areas may end up directly competing for water resources with agricultural irrigation projects, demands for residential and industrial water supplies, flood control schemes, and hydroelectric dams. Wetlands are often interconnected, so a decision affecting water levels and quality in one place will have ramifications for other areas. In particular, the construction of dams on major rivers often completely alters the environmental conditions, eliminating or reducing the abundance of many of the native fish and other aquatic species. Consequently, one strategy for maintaining wetlands is to include entire watersheds within given protected areas.

Monitoring sites

An important aspect of managing protected areas involves monitoring components that are crucial for biodiversity, such as the quality and quantity of water in ponds and streams; the number of individuals of rare and endangered species; and the density of herbs, shrubs, and trees (Lindenmayer et al. 2011; Pocock et al. 2015). Methods for monitoring these components include recording standard observations, carrying out surveys, and taking photographs from fixed points. Monitoring an area's biodiversity is sometimes combined with monitoring social and economic aspects of surrounding communities because of the linkages between people and conservation. In particular, it is often important to monitor the amount and value of plant and animal materials that people obtain from nearby ecosystems.

Managers must continually assess the information they gain from monitoring and adjust park management practices in an adaptive manner to achieve their conservation objectives, a process sometimes referred to as **adaptive management** (**FIGURE 8.17**).

Unfortunately, monitoring programs are often discontinued after a few years from a lack of funding or interest. The lack of long-term monitoring and baseline inventories can make it difficult or impossible to design and evaluate management plans and detect the effects of relatively subtle, chronic problems, such as acid rain and climate change, which can dramatically alter ecosystems over time. To remedy this situation, many scientific research organizations and government agencies have begun to implement programs to monitor and study ecological change over the course of decades and centuries. Programs include the system of 26 Long-Term Ecological Research (LTER) sites established by the US National Science Foundation, and the World

> Parks must be monitored to determine whether their goals are being met, and management plans may need to be adjusted based on new information from monitoring.

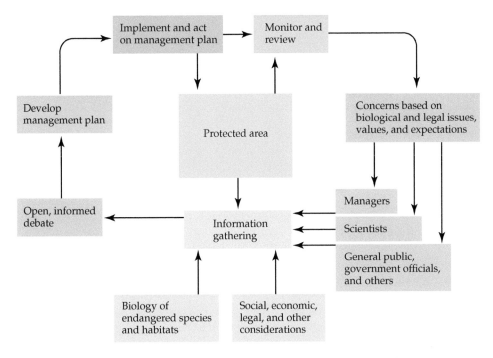

FIGURE 8.17 Model of an adaptive management process for protected areas, emphasizing the decision-making stages. Input is solicited from many sources, and then the plan is developed, implemented, and monitored. (After Cork et al. 2000.)

Network of Biosphere Reserves of the UN's Educational, Scientific and Cultural Organization (UNESCO).

The scale and methods of monitoring have to be appropriate for management needs. For large parks in remote areas, remote sensing using satellites and airplanes may be an effective method for monitoring logging, shifting cultivation, mining, and other activities, both authorized and unauthorized (Morgan et al. 2010). In some cases, the inhabitants are trained to carry out more intensive monitoring at a smaller scale. Often these local people have extensive and useful knowledge of a protected area that they are willing to share as part of the monitoring process (Anadón et al. 2009).

Management and people

> The involvement of local people is often the crucial element missing from conservation strategies. Local people need to be involved in conservation programs, as participants, employees, and leaders.

A central part of any park's management plan must be policies on the use of the park resources by different groups of people (Grumbine and Xu 2011). In developing countries especially, restricting access to protected areas can cost people—who may have traditionally used the area and its resources—access to the basic resources that they need to stay alive. Such displaced people may oppose conservation in these areas (Mascia and Claus 2009). Many parks flourish or are destroyed depending on the degree of support, neglect, or hostility they receive from the people who live in or near them. In the best-case scenario, local people become involved in park management and planning, are trained and employed by the park authority, and benefit from the protection of biodiversity and regulation of activity within the park (Ferro et al. 2011).

However, in the worst-case scenario, if there is a history of bad relations and mistrust between local people and the government or if the purpose of the park is not explained adequately, the local people may reject the park's concept and ignore its regulations. In this case, the local people may come into conflict with the park personnel, to the detriment of the conservation goals. Park personnel and even armed soldiers may have to patrol constantly to prevent illegal activity. Escalating cycles of conflict can lead to outright conflict in which park personnel and the local inhabitants may be threatened, injured, or even killed.

Unfortunately, it is sometimes necessary to exclude local people from protected areas, especially in cases when resources are being overharvested, either legally or illegally, to the point that the health of the ecosystem and the existence of endangered species are threatened (Packer et al. 2013). Such degradation and loss of biodiversity can result from overgrazing by cattle, excessive collection of fuelwood, or hunting with guns. In such cases the only solution may be a strategy of "fences and fines."

Zoning as a solution to conflicting demands

A possible way to deal with conflicting demands on a protected area is **zoning**, which considers the overall management objectives for a park and sets aside designated areas that permit or give priority to certain activities. For example, some areas of a forest may be designated for timber production, hunting, wildlife protection, nature trails, or watershed maintenance. Other zones may be established for the recovery of endangered species, restoration of degraded communities, and scientific research. The challenge in zoning is to find a compromise that people are willing to accept which provides for the long-term, sustainable use of natural resources. Managers often need to spend considerable effort informing the public about what activities are acceptable in particular areas of a park and then enforcing park regulations (Andersson et al. 2007).

> Zoning allows the separation of mutually incompatible activities. MPAs are often zoned with no-fishing areas where fish and other marine organisms can recover from harvesting.

For example, an MPA might allow fishing in certain areas and strictly prohibit it in others; certain areas might be designated for surfing, water-skiing, and recreational diving, but these sports may be prohibited elsewhere. The creation of zoned MPAs in many locations has proven to be an effective way to rebuild and maintain populations of fish and other marine organisms (**FIGURE 8.18**). In comparison with nearby unprotected sites, MPAs that restrict fishing and other activities often have greater total weight of commer-

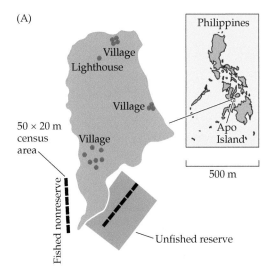

(A)

FIGURE 8.18 On Apo Island in the Philippines, large reef fish had been over-harvested to the point where they were rarely seen. (A) In response to overharvesting, a reserve was set up (blue area) on the eastern side of the island, while fishing continued as before at a specified nonreserve area on the western side. A censusing study measured the number of large reef fish at each site (six underwater census areas are shown as black rectangles for each site). (B) The resulting data show that after the marine reserve was established, the number of fish observed in the unfished reserve increased substantially. The number of fish in the unprotected area did not increase initially because the fish were still being intensively harvested; after about 8 years, however, an increase became detectable, originating from the spillover of fish from the reserve area. (After Abesamis and Russ 2005.)

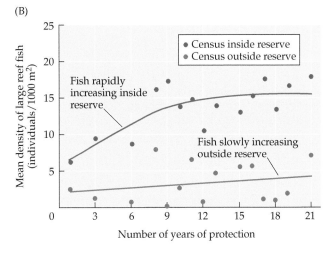

(B)

cially important fish, greater numbers of individual fish, and greater coral reef cover. Evidence shows that fish from MPAs that restrict or prohibit fishing spill over into adjacent unprotected areas, where they can help rebuild populations and also may be caught by fishermen. The enforcement of zoning is often a major challenge in MPAs because fishermen tend to fish on the edges of fishing-exclusion zones as those are the areas where the fishing is best, leading to overfishing at the margins of MPAs. A combination of local involvement, publicity, education, clear posting of warning signs, and visible enforcement significantly increases the success of any zoning plan, especially in the marine environment (Fox et al. 2012).

Biosphere reserves

UNESCO has pioneered another zoning approach, termed **biosphere reserves**, under its World Network of Biosphere Reserves Program, which integrates traditional land-use patterns (such as farming, grazing, and managing forests), research, protection of the natural environment, and sometimes tourism at a single location. These locations often have well-established human settlements and scenic landscapes. A desirable feature of the biosphere reserve program is a system in which there are zones delineating varying levels of use (**FIGURE 8.19A**). At the center is a core area in which ecosystems are strictly protected, with all human activity either prohibited or tightly regulated. This core is surrounded by a buffer zone in which traditional human activities, such as the collection of edible plants and small fuelwood, are monitored and nondestructive research is conducted. Surrounding the buffer zone is a transition zone in which some forms of sustainable development, such as small-scale farming, are allowed. In addition, some extraction of natural resources, such as selective logging, and experimental research are also permitted. This general strategy of surrounding core conservation areas with buffer and transitional zones can encourage local people to support the goals of the protected area. However, although these zones are easy to draw on paper, in practice it has been difficult to inform and gain agreement from residents who live in or near biosphere reserves about where the zones are and what uses are allowed in them.

The value of the strategy of surrounding core conservation areas with buffer and transition zones is still being debated. The approach has benefits: local people may be more willing to support park activities if they are allowed zoned access to the park, and certain desirable features of the landscape created by human use may be maintained (such as farms, gardens, and early stages of succession). Also, buffer zones may prevent parks from becoming isolated islands of nature and may create corridors that facilitate animal dispersal between highly protected core conservation areas. Yet zoning for multiple uses and resource extraction may only work if the core area is large enough to protect viable populations of all key species and if people are willing to respect the zones and their designated uses.

Respect for zones varies greatly in different parts of the world and among different social situations. In places where park management, political will, and land tenure are weak, buffer zones often are seen as a commons or as unowned and unmanaged lands that are up for grabs, which greatly reduces their effectiveness.

One instructive example of a biosphere reserve is the Kuna Yala Indigenous Reserve on the northeast coast of Panama. In this protected area comprising 60,000 ha of tropical forest and coral islands live 50,000 Kuna people, in 60 villages, who practice traditional medicine, agriculture, fishing, and forestry while documentation and research are undertaken by scientists from outside institutions (**FIGURE 8.19B**). At present, Kuna con-

(A)

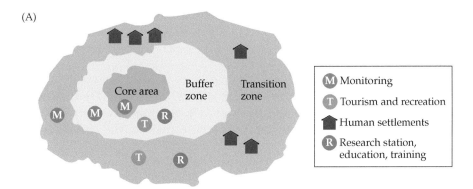

(B)

FIGURE 8.19   (A) The general pattern of a biosphere reserve: a core protected area is surrounded by a buffer zone, where human activities are monitored and managed and where research is carried out; this, in turn, is surrounded by a transition zone, where sustainable development and experimental research take place. (B) The Kuna people still practice traditional methods of catching fish in Kuna Yala Indigenous Reserve. (Photograph © Andoni Canela/AGE Fotostock.)

servation beliefs and practices are gradually changing because of outside influences, and younger Kuna often question the need to rigidly protect the reserve. Further, rising sea levels and declining marine resources are forcing village leaders to consider other options for their future (Posey and Balick 2006).

Challenges to Park Management

Managers of protected areas face many challenges. Perhaps the most basic and unavoidable of these challenges are increased pressures from growing human populations and demand for the use of natural resources. Although all threats to species discussed in Chapter 4 are relevant to protected areas, the following are some areas of particular concern.

Poaching

Human populations will continue to increase dramatically in the coming decades, while resources such as fuelwood, medicinal plants, and wild meat will become harder to find. Similarly, people who are poor and hungry will enter the nearby protected areas to take what they need to live, regardless of whether they have permission (see "Overexploitation" in Chapter 4). Within protected areas, if park rangers are underpaid, even they may be motivated to begin illegally harvesting and selling the very resources that they are charged with protecting. Addressing poverty and enforcing regulations are the most important factors to address poaching (see Chapter 12).

Trophy hunting

There is considerable debate about the role of hunting in conservation. On one side, many countries finance conservation primarily from the sale of hunting licenses. For developing countries, the focus is on large game such as the hunting of rhinos or buffalo by wealthy foreigners (see Chapter 3). Trophy hunting, when strictly regulated, has minimal effects on the overall number of individuals within a park and can even double as a management tool. One example of this is selling expensive licenses to hunt specific, individual old male rhinos that are killing young rhinos and preventing younger males from mating. However, critics argue that allowing sport hunting of any type not only sends the wrong message, that killing these animals is acceptable, but also supports a market that, in turn, promotes illegal trophy hunting. This is illustrated by the killing of Cecil the Lion (see Chapter 3 opener); because South Africa permits lion hunting, the American hunter may not have known that the lion he was shooting had been illegally taken from one of the protected areas. The financial stakes are high enough to motivate local hunting guides to steal from the parks. Making all trophy hunting illegal and removing these animal products from circulation, as was done when the Kenyan government burned tons of ivory, has helped in some cases (see Figure 6.24).

Human–animal conflict

Problems are inevitable as more people live and farm closer to high concentrations of wildlife that, when food is scarce, have nowhere to go but out of the park and into nearby agricultural fields and villages. Elephants, primates, and flocks of birds can all be significant crop raiders, while carnivores such as tigers pose a different set of challenges to nearby residents. Some nonprofit organizations and governments address these problems by creating opportunities for local people to also benefit from the animals, compensating them for their losses, and helping to build fences or other deterrents (**FIGURE 8.20**).

Degradation

Multiple-use areas can suffer from the negative effects of mining, cattle grazing, and oil exploration due to the lack of management or poor enforcement of policies. Even strictly protected areas are at risk of degradation from recreation, including wildfires, littering, fragmentation and erosion from off-road driving, and the habituation of wild animals, to name just a few (see Chapter 3). Furthermore, strictly protected areas may suffer from the same degradation as multiple-use areas, including instances when lack of oversight leads to illegal use. Generally speaking, degradation from human use is minimized in multiple-use areas where there is oversight and policies that regulate it.

Climate change

The extent to which existing protected areas will allow species and ecosystems to persist in the face of climate change is an important question

FIGURE 8.20 Elephants from protected areas trample locals farms and lower support for their conservation. In response, beehive boxes established along the perimeters of farms create anti-elephant "fences." (© Steve Taylor ARPS/ Alamy Stock Photo.)

and is being investigated (Regos et al. 2016). Species may not be able to persist in a protected area if the climate or the associated vegetation change significantly. For example, the Doñana wetlands in southern Spain are a **World Heritage site** that is already rated as under "very high threat" by UNESCO (Scheffer et al. 2015). The site contains some of the most important overwintering habitat in Europe for migrating waterbirds, yet these wetlands may dry out in coming decades due to the development of a warmer, drier climate. Similarly, in one study of the Yunnan Province in China, it was determined that as much as 65% of the region is expected to shift to a different climate zone by 2050, shifting 83% of the total protected area to a different bioclimactic stratum, to which the local species may not be able to adapt in time (Zomer et al. 2015). It may be necessary to establish new protected areas in places where a species or ecosystem may be able to disperse and survive in coming decades. In this rapidly changing environment, it is important to preserve elevational and environmental gradients, corridors, and climatic refugia so that species and ecosystems can gradually spread in response to a changing climate.

Given that species ranges are already shifting and are predicted to continue to do so in response to the rapidly changing climate, some conservationists have begun prioritizing areas for protection based on the geophysical environment rather than the species that occur there; that is, focusing on the stage (geology, soils, topography, etc.) rather than the actors (species) (Groves et al. 2012). Biodiversity often corresponds to an area's geophysical diversity, although the strength of this association varies from region to region (Anderson and Ferree 2010). The use of this strategy will likely continue because data on geophysical characteristics are available for most locations and the approach is relatively easy to integrate into existing systems for identifying high-priority conservation areas.

Funding and personnel

For park management to be effective, there must be adequate funding for a sufficient number of well-equipped, properly trained, reasonably paid, and highly motivated park personnel who are willing to carry out park policy (Becker et al. 2013; Tranquilli et al. 2012). Buildings, communications equipment, and other appropriate elements of infrastructure are necessary to manage a park. In many areas of the world, particularly in developing, but also in developed, countries, protected areas are understaffed, and the park staff lack the equipment to patrol remote areas. Without enough radios and vehicles, they may be restricted to the vicinity of headquarters and unaware of what is happening in their own park.

The importance of sufficient personnel and equipment should not be underestimated: in areas of Panama, for instance, a greater frequency of antipoaching patrols by park guards results in a greater abundance of large mammals and the seed dispersal services they provide (Brodie et al. 2009; Wright et al. 2000). International conservation organizations and governmental agencies regularly assist in providing funds for managing protected areas in developing countries, but often the funding remains insufficient. Increasing funding for the management of protected areas needs to be a priority for government agencies and conservation organizations (Watson et al. 2014).

Finally, at the end of the day, conservation biologists need to account for whether their management of protected areas achieved the stated goals and whether the funds were spent effectively. As human impact increases in most areas of the world, the importance of protected areas for the protection of biodiversity is expected only to increase (Regos et al. 2016).

Summary

- Protecting habitat is the most effective method of preserving biodiversity. About 15.4% of the Earth's land surface is included in about 209,000 protected areas, but because of the needs of human societies for natural resources, the percentage may not increase much further.

- Conservation biologists are developing guidelines for designing protected areas: the areas should be large whenever possible, they should not be fragmented, and managers should create networks of conservation areas for maximum protection.

- Habitat corridors connecting protected areas may allow species dispersal to take place and may be particularly important in maintaining known migration routes.

- Protected areas often must be actively managed in order to maintain their biodiversity. Monitoring provides information that is needed to evaluate whether management activities are achieving their intended objectives or need to be adjusted.

- Management might involve zoning to establish areas where certain activities are allowed or prohibited. Managing interactions with local people is critical to the success of protected areas and should be part of a management plan.

- Adequate staffing and funding are necessary for park management.

For Discussion

1. Obtain a map of a town, state, or nation that shows protected areas (such as nature reserves and parks) and multiple-use managed areas. Who is responsible for each parcel of land, and what is the goal in managing it? Consider the same issues for aquatic habitats (ponds, lakes, rivers, coastal zones, etc.).

2. If you could protect additional areas on the map, where would they be and why? Show their exact locations, sizes, and shapes and justify your choices.

3. Think about a national park or nature reserve you have visited. In what ways was it well run or poorly run? What were the goals of this protected area, and how could they be achieved through better management?

4. Can you think of special challenges in the management of aquatic preserves such as coastal estuaries, islands, or freshwater lakes that would not be faced by managers of terrestrial protected areas?

Suggested Readings

Burgess, S. C., K. J. Nickols, C. D. Griesmer, L. A. K. Barnett, A. G. Dedrick, E. V. Satterthwaite, and 4 others. 2014. Beyond connectivity: How empirical methods can quantify population persistence to improve marine protected-area design. *Ecological Applications* 24: 257–270. Designers of a network of MPAs should consider the dispersal ability of species.

Burrell, J. 2013. Path of the Pronghorn—Leading to New Passages: Part 3. Newswatch, National Geographic. Assessment of fencing, overpasses, and underpasses constructed to protect pronghorn migration to and from Grand Teton National Park. (newswatch.nationalgeographic.com/2013/12/06/path-of-the-pronghorn-leading-to-new-passages-part-3).

Colwell, R., S. Avery, J. Berger, G. E. Davis, H. Hamilton, T. Lovejoy, and 6 others. 2012. *Revisiting Leopold: Resource Stewardship in the National Parks.* National Park System Advisory Board, Washington, DC. Protected area management needs to consider that the environment is changing.

Danielson, F., P. M. Jensen, N. D. Burgess, R. Altamirano, P. A. Alviola, H. Andrianandrasana, and 21 others. 2014. A multicountry assessment of tropical resource monitoring by local communities. *BioScience* 64: 236–251. Monitoring by local people and by scientists produces similar results.

Edgar, G. J.R. D. Stuart-Smith, T. J. Willis, S. Kininmonth, S. C. Baker, S. Banks, and 19 others. 2014. Global conservation outcomes depend on marine protected areas with five key features. *Nature* 506: 216–220. The characteristics of successful MPAs are having no-take zones, having good enforcement, being large, being old and established, and being isolated from areas with fishing.

Hallwass, G., P. F. Lopes, A. A. Juras, and R. A. M. Silvano. 2013. Fishers' knowledge identifies environmental changes in fish abundance trends in impounded tropical rivers. *Ecological Applications* 23: 392–407. Local people can sometimes accurately describe the changes in species composition and abundance that have occurred after a dam has been built.

Hobbs, R. J., D. N. Cole, L. Yung, E. S. Zavaleta, G. H. Aplet, F. S. Chapin III, and 10 others. 2010. Guiding concepts for park and wilderness stewardship in an era of global environmental change. *Frontiers in Ecology and the Environment* 8: 483–490. Excellent statement of the need for guiding principles in park management.

Joppa, L. N., P. Visconti, C. N. Jenkins, and S. L. Pimm. 2013. Achieving the Convention on Biological Diversity's goals for plant conservation. *Science* 341: 1100–1103. It is possible but will be difficult in practice to achieve the goals of formally protecting 17% of the terrestrial world and 60% of plant species.

Maron, M., J. R. Rhodes, and P. Gibbons. 2013. Calculating the benefit of conservation actions. *Conservation Letters* 6: 359–367. Most of the time, conservation benefits are calculated wrong, and this article offers improved methods.

Mascia, M. B., S. Pallier, R. Krithivasan, V. Roshchanka, D. Burns, M. J. Mlotha, and 2 others. 2014. Protected area downgrading, downsizing, and degazettement (PADDD) in Africa, Asia, and Latin America and the Caribbean, 1900–2010. *Biological Conservation* 169: 355–361. PADD represents a widespread and generally underappreciated threat to biodiversity.

Packer, C., A. Loveridge, S. Canney, T. Caro, S. T. Garnett, M. Pfeifer, and 52 others. 2013. Conserving large carnivores: Dollars and fence. *Conservation Letters* 16: 635–641. In a study of lions, sometimes fencing off a protected area is the best and least expensive strategy to maintain populations.

Regos, A., M. D'Amen, N. Titeux, S. Herrando, A. Guisan, and L. Brotons. 2016. Predicting the future effectiveness of protected areas for bird conservation in Mediterranean ecosystems under climate change and novel fire regime scenarios. *Diversity and Distributions*, 22(1): 83–96. Models can be used to manage future risk.

Skelly, D. K., S. R. Bolden, and L. K. Freidenburg. 2014. Experimental canopy removal enhances diversity of vernal pool amphibians. *Ecological Applications* 24: 340–345. Limited tree removal is an effective strategy for encouraging certain amphibian populations.

Vidal, O., J. López-García, and E. Rendón-Salinas. 2014. Trends in deforestation and forest degradation after a decade of monitoring in the Monarch Butterfly Biosphere Reserve in Mexico. *Conservation Biology* 28: 177–186. Management actions have been effective at reducing deforestation and protecting butterfly habitat.

Wuerthner, G., E. Crist, and T. Butler (eds.). 2015. *Protecting the Wild: Parks and Wilderness*. The Foundation For Conservation. Island Press. Researchers from around the world make a strong case for the importance of protected areas for conservation.

9

Conservation Outside Protected Areas

The black rhino (*Diceros bicornis*) has benefited from conservation efforts outside the boundaries of national parks. At Imire Game Farm in Zimbabwe, rhinos are protected from poaching by 24-hour armed guards.

Establishing protected areas with intact eco-systems is essential for species conservation. It is, however, shortsighted to rely solely on protected areas to preserve biodiversity. That kind of reliance can create a paradoxical situation in which species and ecosystems inside the protected areas are preserved while the same species and ecosystems outside are allowed to be damaged, which in turn results in the decline of biodiversity within the protected areas (Kinnaird and O'Brien 2013; Newmark 2008). This decline is due to at least three reasons:

- Most protected areas are too small to protect viable, long-term populations of many species, especially large animals and migratory species.

- Many species are attracted to resources available outside protected areas. For example, it is not uncommon for primate species to raid crops in villages that are adjacent to protected areas in which they live.

- Many species migrate between protected areas seasonally to avoid freezing temperatures or other climate extremes or to access mates and other resources.

In general, the smaller a protected area, the more dependent it is on neighboring unprotected lands for the long-term maintenance of biodiversity. A crucial component of conservation strategies must be the

FIGURE 9.1 Planting a variety of species in urban parks and gardens is one way in which landscapes dominated by humans can support biodiversity. Many insects, birds, and other taxa are dependent on these habitat stepping-stones to travel between protected areas. (© johandersonphoto/Getty Images.)

protection of biodiversity inside and *outside*—both immediately adjacent to and away from—protected areas (Troupin and Carmel 2014). According to even the most optimistic predictions, in the future more than 80% of the world's land will remain outside protected areas (PAs). Given this reality, the UN Convention on Biological Diversity (see Chapter 11) established global conservation targets that include both PAs and what they call "other effective area-based conservation measures" (OECM's; Aichi Target 11). Conservation biologists must help define "effective conservation" and determine to what degree OECM's contribute to global targets (Watson et al. 2016).

Some conservation biologists argue that, given how much land is now occupied by people, we must find ways to promote biodiversity in human-dominated landscapes (**FIGURE 9.1**). This is an idea ecologist Michael Rosenzweig (2003) named **reconciliation ecology**. In many cases, this involves creating habitat in urban settings, such as enhancing parks or planting *green roofs* (see chapter opening photo for Chapter 11), and it generally involves increasing the environmental complexity (Loke et al. 2015). An innovative example of reconciliation ecology is an underwater restaurant in Eilat, Israel, which has frames built outside the windows to support a coral nursery. The creation of new reef habitat on the frames helps mitigate damage by snorkeling tourists on the main reef nearby, while creating a beautiful view for diners at the restaurant. Other examples of how nature is promoted in urban areas, and also some of the problems of this approach, are discussed in more detail later in the chapter.

Jeff McNeely (1989), an International Union for Conservation of Nature (IUCN) expert in protected areas, suggested that the park boundary

> is too often also a psychological boundary, suggesting that since nature is taken care of by the national park, we can abuse the surrounding lands, isolating the national park as an 'island' of habitat which is subject to the usual increased threats that go with insularity.

Many endangered species and unique ecosystems are found partly or entirely on unprotected lands. Consequently, the conservation of biodiversity in these places must be considered.

In the worst case, a devastated landscape polluting the air and water will strangle the protected area it surrounds and block the movement of dispersing animals and plants. In the best case, however, unprotected areas surrounding protected areas will provide additional space for ecosystem processes and new populations. In many ways, conservation outside protected areas should strive to blur the distinctions between protected and unprotected ecosystems as much as possible by maintaining unprotected areas in a state of reasonable ecological health (Radeloff et al. 2010).

In this chapter, we explore strategies to include biodiversity protection as a management objective for unprotected areas as a way to complement conservation in protected areas.

The Value of Unprotected Habitat

The human use of ecosystems varies greatly in unprotected lands, but significant portions are not used intensively by humans and still harbor some of their original biota (**FIGURE 9.2**). In almost every country, numerous

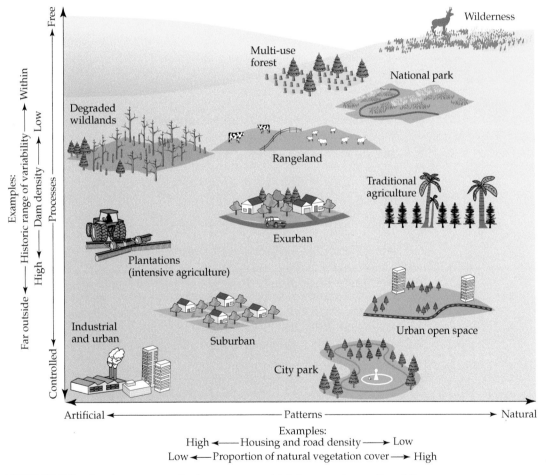

FIGURE 9.2 Landscapes vary in the extent to which humans have altered the patterns of species composition and natural vegetation through activities such as agriculture, road construction, and housing; ecosystem processes (water flow, nutrient cycling, etc.) also vary because of fire control activities, dam construction, and other activities that alter plant cover. Wilderness areas retain most of their original patterns and processes, urban areas retain the least, and other landscapes retain various intermediate amounts. (After Theobald 2004.)

rare species and ecosystems exist primarily or exclusively on unprotected public lands or on lands that are privately owned (Deguise and Kerr 2006). In the United States, 60% of species that are globally rare or listed under the US Endangered Species Act live on private forested lands (Robles et al. 2008). Even when endangered species are found on a country's public land, it is often not land managed for biodiversity but rather land managed primarily for timber harvesting, grazing, mining, or other economic uses. For example, 75% of the remaining orangutans in Indonesia live outside protected forests, often in logged forests and tree plantations (Meijaard et al. 2010). Clearly, strategies for reconciling human needs and conservation interests in unprotected areas are critical to the success of conservation plans (Cox and Underwood 2011; Koh et al. 2009).

The situation of the Florida panther (*Felis* [*Puma*] *concolor coryi*) provides an excellent example of the importance of unprotected habitat. This endangered subspecies of mountain lion lives in South Florida and has a population of only 100–120 individuals (Thatcher et al. 2009). About 31% of the land in the present range of the panther is privately owned, and animals tracked with radio collars have all spent at least some of their time on private lands (**FIGURE 9.3**). Private lands typically have better soils, which support more prey species than the public lands. Thus, panthers that spend most of their time on private lands have a better diet and are in better condition. Although panther habitat was protected on 870,362 ha of public lands as of 2013, much more habitat must be protected in order to ensure the panther's continued survival. The acquisition of additional habitat has been proposed (Kautz et al. 2006), but until that can be achieved, even slowing down the pace of land development may prove impractical. Two viable possibilities are educating private landowners on the value of conservation and paying willing landowners to practice management options that allow for the panthers' continued existence—specifically, minimizing habitat fragmentation and maintaining their preferred habitats of hardwood forest and cypress swamp. In addition, special road underpasses have been built in the hopes of reducing panther deaths from collisions with motor vehicles. Other species of big cats also frequently live outside protected areas; for example, leopards in India often adapt well to rural landscapes of farms and villages, hunting wild prey at night, when they are rarely or never seen by villagers (Athreya et al. 2013).

Next, we will discuss the importance of several types of land that are not contained within traditional protected areas but nonetheless are important for biodiversity.

Military land

Native species can often continue to live in unprotected areas, especially when those areas are set aside or managed for some other purpose that is not harmful to the ecosystem, such as security zones surrounding government installations and military reservations. For example, the US Department of Defense manages more than 11 million ha, much of it undevel-

(A)

FIGURE 9.3 (A) The Florida panther is found on both public and private lands in South Florida. (B) The green dots represent 55,000 radio telemetry records of 79 collared panthers. Public lands are outlined in red (A, photograph courtesy of Larry Richardson, USFWS; B, from Kautz et al. 2006, with updates from R. Kautz.)

(B)

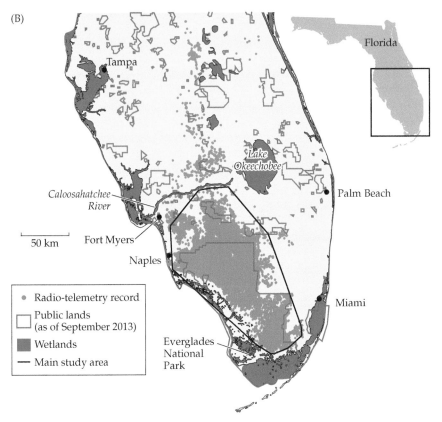

Florida

Tampa

Lake Okeechobee

Palm Beach

Caloosahatchee River

50 km

Fort Myers

Naples

Miami

- Radio-telemetry record
- Public lands (as of September 2013)
- Wetlands
- Main study area

Everglades National Park

oped, containing about 420 threatened and endangered species of plants and animals. The White Sands Missile Range in New Mexico alone is almost 1 million ha in area, about the same size as Yellowstone National Park. While certain sections of military reservations may be damaged by military activities, much of the habitat remains as an undeveloped buffer zone with restricted access. After these areas are no longer needed for military purposes, they also make excellent candidates for protected areas; nearly 31,000 hectares of former military bases are being converted to nature reserves by the German Federal Agency for Nature Conservation (*Huffington Post* 2015).

The impact of military training itself, including accidental fires, tank exercises, and artillery practice, provides the open habitat required by certain species, such as the Karner blue butterfly and its host plants near Fort McCoy in Wisconsin. As a result, many military bases have become de facto refuges for about 420 federally listed species of plants and animals, many of which have their largest populations on military bases (Stein et al. 2008). Rare and endangered desert tortoises (*Gopherus agassizii*), manatees (*Trichechus manatus*), red-cockaded woodpeckers (*Picoides borealis*), bald eagles, Atlantic white cedars (*Chamaecyparis thyoides*), and the least Bell's vireo (*Vireo bellii pusillus*) all have found a safe haven on military lands. Personnel at the Barksdale Air Force Base in Shreveport, Louisiana, have re-flooded wetlands along the Red River, restoring wetlands for wading birds. The US Department of Defense's emphasis on conservation has increased dramatically; spending on threatened and endangered species jumped 45%, from about $50 million in 2003 to $73 million in 2012 (Watson 2013).

On the other hand, many military bases contain toxic waste dumps and high levels of chemical pollutants. In addition, severe disturbance in the form of bomb explosions, artillery practice, and the use of heavy vehicles can have significant negative effects on the resident wildlife.

Unprotected forests

Forests that are either selectively logged on a long cutting cycle or are cut down for farming using traditional shifting cultivation methods may still contain a considerable percentage of their original biota and maintain most of their ecosystem services (Adum et al. 2013; MacKay et al. 2014) (**FIGURE 9.4**). This is particularly true if fires and erosion have not irreversibly damaged the soil and if native species can migrate from nearby undisturbed lands, such as steep hillsides, swamps, and river forests, and colonize the sites. For example, in Malaysia, most forest bird species are still found in rain forests 30 years after selective logging was carried out, and undisturbed forest is available nearby to act as a source of colonists (Peh et al. 2005). Likewise, in African tropical forests, gorillas, chimpanzees, and elephants can tolerate selective logging and other land uses that involve low levels of disturbance, though only when hunting levels are controlled by active antipoaching patrols (Stokes et al. 2010).

Unprotected grasslands

The mown edges of roadsides often provide an open grassland community that is a critical resource for many species, such as butterflies (Saarinen et al. 2005). A similar habitat is provided by the surprisingly large amount of mown fields occupied by power lines. In the United States, corridors for

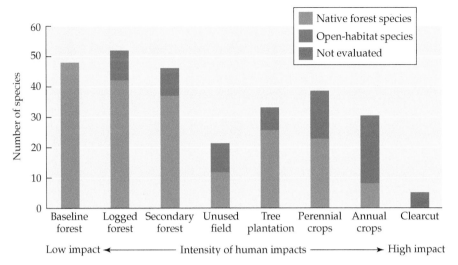

FIGURE 9.4 For a range of land uses in West Africa, when the intensity of human impacts increases, the average number of vertebrate native forest species declines and the number of open-habitat species increases. Some native forest species are still present even along with intensive land use, such as tree plantations, but the overall number of species and the proportion of native species are much lower. (After Norris et al. 2010.)

power line rights-of-way occupy over 2 million ha. Power line corridors managed with infrequent mowing and without herbicides maintain high densities of birds, insects, and other animals (King et al. 2009). If such management practices could be extended over a greater proportion of power line rights-of-way, these areas could become additional habitat for insects and a wide range of other species. Remnant prairies in the United States also represent an important habitat for many grassland species, especially where the prairies can be managed with grazing or burning.

Unprotected waters

Many heavily altered aquatic ecosystems can also have value for conservation. For example, in estuaries and seas managed for commercial fisheries, many of the native species remain because commercial and noncommercial species alike require an undamaged chemical and physical environment. It has been determined that most marine plant diversity (such as sea grass and mangroves) occurs outside existing MPAs (Daru and le Roux 2016). Also, many marine animals such as salmon, whales, and sea turtles migrate great distances, including across areas that are not protected.

Even though dams, reservoirs, canals, dredging operations, port facilities, and coastal development harm native aquatic communities, some bird, fish, and other aquatic species are capable of adapting to the altered conditions, particularly if the water is not polluted. However, there is abundant

> Even ecosystems that are managed primarily for the production of natural resources can retain considerable biodiversity, and they are important to the success of conservation efforts.

research that suggests that even when there is no obvious pollution, many species are likely to have higher abundance within MPAs than outside of them (e.g., Pikesley et al. 2016).

Land that is undesirable to humans

Other areas that are not protected by law may retain species because the human population density and degree of use are typically very low. Border areas, such as the demilitarized zone between North Korea and South Korea, often have an abundance of wildlife because they remain undeveloped and generally unoccupied by people. Governments frequently manage mountain areas, which are often too steep and inaccessible for development, as valuable watersheds that produce a steady supply of water and prevent flash flooding and erosion. They also harbor important natural communities. Likewise, desert and tundra species and ecosystems may be at less risk than other unprotected communities because such regions are marginal for human habitation and use (MEA 2005). However, it is also true that in areas like the Arctic region, the current warming of the climate will result in further development of the transportation infrastructure and a greater interest in mining deposits of oil, gas, and minerals.

Private land

In many parts of the world, wealthy individuals have acquired large tracts of land for their personal estates and for private hunting. These estates are frequently used at very low intensity, often in a deliberate attempt by the landowners to maintain large wildlife populations. In particular, some estates in Europe preserve unique old-growth forests that have been owned and protected for hundreds of years by royal families. Such privately owned lands, whether owned by individuals, families, corporations, or tribal groups, often contain important aspects of biodiversity.

Management for biodiversity can vary a great deal between landowners, of course; a study of private landowners enrolled in the Indiana Classified Forest and Wildlands Program found that size, environmental motives, and those landowners who had seen improvements occur on their land were more likely to be good stewards (Farmer et al. 2016). Strategies that encourage private landowners and government land managers to protect rare species and ecosystems are obviously essential to the long-term conservation of biodiversity. This chapter and Chapter 12 describe these strategies.

Even small yards and home gardens can be useful for supporting biodiversity, particularly of insects (Ribeiro et al. 2016). The National Wildlife Federation has a backyard "wildlife certification program" in which homeowners can receive a certificate and a sign once they ensure that their property contains all the elements of wildlife habitat, including a food source, water, and sheltering plant cover for protection and reproduction. Homeowners' associations may require the use of native plants or

(A)

(B)

FIGURE 9.5   (A) This housing development in Lake Worth, Florida has wetland conservation areas that are managed largely by the local homeowners' association. The creation of areas such as this are one way that developers can legally mitigate their negative impact on the habitat that was displaced, while also increasing the value of the houses they build. (B) The importance of these conservation areas is apparent by the diversity of plants and animals that live there, often appearing in people's yards, like this sandhill crane (*Grus canadensis*). Many sandhill cranes migrate from protected areas in the north each winter, and are dependent on such remnants of habitat, while others are year-round residents. (Photographs by Anna Sher.)

a minimum number of trees in landscaping to support biodiversity, and developments may even have homeowners'-association-supervised natural areas that increase the value of their properties (**FIGURE 9.5**).

Conservation in Urban and Other Human-Dominated Areas

Many native species can persist even in urban areas, in public parks, streams, ponds, and other, less altered, habitats (Meffert and Dziock 2012). Preserving these remnants of biodiversity within a human-dominated matrix not only presents special challenges but also provides unique opportunities to educate the public about biodiversity conservation. For example, the discovery of a new species of salamander at a popular swimming site in Austin, Texas, required a change in how the site is managed to allow the people and the salamanders to coexist. Likewise, in Europe, storks (*Ciconia ciconia*) often nest in chimneys and towers, and endangered raptors such as the peregrine falcon (*Falco peregrinus*) and bald eagle (*Haliaeetus leucocephalus*) make nests and raise their young in the skyscrapers of downtown New York, where numerous small animals (including the ubiquitous pigeons

FIGURE 9.6 "Pale Male" (on the left) is a famous red-tailed hawk that has lived on a Fifth Avenue residence in New York city since 1991. The unusually light-colored hawk has had eight mates thus far and several broods. He has inspired a website (www.palemale.com), at least three children's books, and even a movie. He is one of the first hawks known to have built a nest on a building in this city, and when the homeowners attempted to displace the hawks by removing the support structures for the nest, local bird lovers successfully protested to keep it. Pale Male is seen here with his then-current mate, Lola. (© D. Bruce Yolton.)

and rats common to urban centers) provide abundant food sources (**FIGURE 9.6**). Even ponds at golf courses in urban areas and gravel pits dug for construction materials may be suitable habitats for certain newts, dragonflies, and other wetland species provided the water is not polluted (Colding et al. 2009). In one study of 27 artificial water bodies in Australia, researchers found that greater than 70% of the regional diversity of fish species could be found in these human constructions (Davis and Moore 2015). Whether intentional or not, these are examples of reconciliation ecology because they demonstrate ways in which humans and other species can coexist.

As exciting as such examples of urban adaptations might be, we cannot assume that all species have the potential to live within human-dominated landscapes. For example, the value of urban parks for biodiversity found in some developed countries may not apply to rapidly growing megacities; in South America, they were found to be dominated by European weeds (Fischer et al. 2016). We have a lot to learn about just what habitat and disturbance features are important for various species and how to integrate those features into our urban and suburban landscapes. In general, increasing the intensity of land use will decrease the number of native species found in a location, and adaptable, generalist species (often nonnative invasives) will tend to do best. The size and configuration of landscape features will determine which species and ecosystem processes are maintained. For example, abandoned industrial sites in Germany of at least 5 ha in area are necessary to provide habitat for many bird species of special conservation concern (Meffert and Dziock 2012). More work is needed to evaluate how general conservation principles apply in specific locations.

Increasing the presence of wild animals in the urban landscape comes with fairly serious consequences for both animals and humans. For example, as woodland areas and mountain canyons become urbanized or suburbanized, people tend to create yards and gardens that attract deer. Deer bring with them a host of problems: they can carry ticks that transmit illnesses to humans, such as Lyme disease and Rocky Mountain spotted fever; they are a significant potential road hazard; and the bucks can become aggressive toward humans during mating season. In some areas, deer that live within developments also attract predators including cougars, thus increasing the potential for human–wildlife conflicts for a scarce and ecologically important top carnivore.

Understanding the ecology, the ecosystem processes, and the characteristics of the human use of a location is critical for implementing policies to promote conservation in unprotected urban areas. Deciding on the proper tools, though, requires good information on ecology and complex urban human–natural systems and knowledge of how best to motivate people to behave in conservation-friendly ways. These areas of research are growing and beginning to provide insights that are improving urban conservation.

Other human-dominated landscapes

Most of the world's landscapes have been affected in some way by human activity, but fortunately, considerable biodiversity can be maintained in well-managed and low-intensity traditional agricultural systems, grazing lands, hunting preserves, forest plantations, and recreational lands (Carrière et al. 2013; Wright et al. 2012). Birds, insects, and other animal and plant species are often abundant in traditional agricultural landscapes, with their mixture of small fields, hedges, and woodlands. Some species are found almost exclusively in these traditional human-dominated habitats. In comparison with more intensive, so-called modern, agricultural practices, which emphasize high yields of crops for sale in the market, mechanization, and external inputs, these traditional landscapes experience less exposure to herbicides, fertilizers, and pesticides and have more heterogeneity of habitat. Similarly, farmlands worked using organic methods support more birds than farmlands worked using nonorganic methods, in part because organic farms have more insects for the birds to eat. In many areas of the world, however, the best agricultural lands are being more intensively used while less optimal lands are abandoned as people leave for urban areas (Phelps et al. 2013).

Conservation biologists are increasingly discussing the value of the strategy of **land sharing**, in which low-intensity human activities, such as traditional or organic agriculture, can coexist with some elements of biodiversity. The alternative is **land sparing**, in which intensive human activities, such as modern agriculture, are practiced on some of the lands while allowing the rest to remain in their natural state. The best strategy for any given location will depend on the local circumstances, the price of land and crops, and potential financial incentives (Baudron and Giller 2014; Tscharntke et al. 2012).

FIGURE 9.7 Two types of coffee management systems. (A) Shade coffee is grown under a diverse canopy of trees, providing a forest structure in which birds, insects, and other animals can live. (B) Sun coffee is grown as a monoculture, without shade trees. In a monoculture system, animal life is greatly reduced. (A, © Aurora Photos/Alamy; B, © PisitBurana/istock.)

(A)

(B)

One notable example of preserving biodiversity in an agricultural setting comes from tropical countries and their traditional shade coffee plantations, in which coffee is grown under a wide variety of shade trees, with often as many as 40 tree species per farm (Philpott et al. 2007) **(FIGURE 9.7)**. In northern Latin America alone, shade coffee plantations cover 2.7 million ha. These plantations have structural complexity created by multiple vegetation layers and a diversity of birds and insects comparable to the adjacent natural forest, and they represent a rich repository of biodiversity (Vandermeer et al. 2010). The presence of such coffee plantations can also potentially slow the pace of deforestation (Hylander et al. 2013). Therefore, programs are being developed to encourage and subsidize farmers to maintain their shade-grown coffee plantations and to market the product at a premium price as "environmentally friendly" shade-grown coffee. But let the buyer beware; there are currently no uniform standards for shade coffee. Thus, some coffee marketed as "environmentally friendly, shade coffee" may actually be grown as sun coffee with only a few small, interspersed trees. Shade-grown chocolate and other tropical tree crops are similarly unregulated (Waldron et al. 2012).

In developing countries, conservation biologists have recently started innovative programs in which local people living in rural areas are paid directly for protecting individuals and populations of flagship species, including rhinos, tigers, gorillas, and other species of conservation interest (Dinerstein et al. 2013). When the animals do well, the people are paid directly or receive money for village improvements (see the case studies that follow).

In many countries, large parcels of government-owned land are designated as **multiple-use habitat**; that is, they are managed to provide a variety of goods and services. An emerging and important research area involves the development of innovative ways to reconcile competing claims on land use, such as logging, mining, species conservation, and tourism. This will require careful analyses and consideration of the trade-offs of pursuing alternative development options in regard to both environmental and socioeconomic priorities (Koh et al. 2010). A different approach is to use regulations, the legal system, and political pressure to prevent government-approved activities on public lands if these activities threaten the survival of endangered species.

In the United States, the Bureau of Land Management oversees more than 110 million ha, including 83% of the state of Nevada and large amounts of Utah, Wyoming, Oregon, and Idaho. National forests cover over 83 million ha, including much of the Rocky Mountains, the Cascade Range, the Sierra Nevada, the Appalachian Mountains, and the southern coast of Alaska. In the past, these lands have been managed for logging, mining, grazing, wildlife, and recreation. The challenge is that often, each one of these activities is managed by itself but their cumulative effects threaten biodiversity. Increasingly, multiple-use lands also are being valued and managed for their ability to protect species, biological communities, and ecosystem services (Kemp et al. 2013). The US Endangered Species Act of 1973 and other similar laws, such as the 1976 National Forest Management Act, require landowners, including government agencies, to avoid activities that threaten listed species. One such activity is overgrazing by cattle; when cattle grazing is reduced or eliminated on overgrazed rangelands, some of these ecosystems can recover in a few years or decades (Earnst et al. 2012).

Another approach to protecting biodiversity in human-dominated landscapes has been to define standards of best practices so that the use of resources does not harm biodiversity. The Forest Stewardship Council has been one such organization by working to promote the certification of timber produced from sustainably managed forests. For the Forest Stewardship Council and similar organizations to grant certification, the forests need to be managed and monitored in the interests of their long-term environmental health, and the rights and well-being of local people and workers need to be protected. The certification of forests is increasing rapidly in many areas of the world, especially in response to buyers in

Europe, who often request certified wood products. At the same time, major industrial organizations representing such industries as logging, mining, and agriculture are lobbying for their own alternative certification programs, which generally have lower requirements for monitoring and weaker standards for judging whether practices are sustainable.

Ecosystem Management

Ecosystem management links private and public landowners, businesses, and conservation organizations in a planning framework that facilitates acting together on a large scale.

Resource managers around the world are increasingly being urged by their governments and conservation organizations to think on larger geographic scales, particularly given climate change–driven shifts in the distributions of species and makeup of ecosystems. Traditionally, these managers may have focused on the production of goods and services that could be managed on the local scale, such as volume of timber or number of park visitors. But today, these managers are being asked to expand their emphasis to a broader perspective that includes the conservation of biodiversity and the protection of ecosystem processes (Altman et al. 2011). That is, they are shifting to **ecosystem management**, a system of large-scale management involving multiple stakeholders, the primary goal of which is preserving ecosystem components and processes for the long term while still satisfying the current needs of society (**FIGURE 9.8**). Rather than having each government agency, private conservation organization, business, or landowner act in isolation and in its own interests, ecosystem management envisions them cooperating to achieve common objectives (Redpath et al. 2013). For example, in a large forested watershed along a coast, ecosystem management would link all owners and users located from the tops of the hills to the seashore, including foresters, farmers, business groups, townspeople, and the fishing industry.

Important themes in ecosystem management include the following:

- Using the best science available to develop a coordinated plan for the area that is sustainable; includes biological, economic, and social components; and is shared by all levels of government as well as business interests, conservation organizations, and private citizens

- Ensuring viable populations of all species, representative examples of all biological communities and successional stages, and healthy ecosystem functions

- Seeking and understanding connections between all levels and scales in the ecosystem hierarchy—from the individual organism to the species, community, ecosystem, and even regional and global scales

- Monitoring significant components of the ecosystem (numbers of individuals of significant species, vegetation cover, water quality, etc.), gathering the needed data, and then using the results to adjust management in an adaptive manner—a process sometimes referred to as adaptive management (see Figure 8.17)

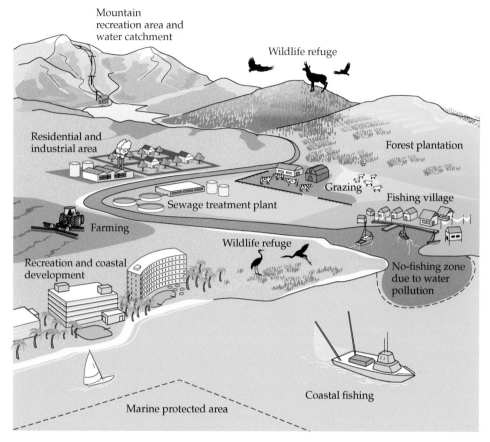

Mountain
recreation area and
water catchment

Wildlife refuge

Residential and
industrial area

Forest plantation

Grazing

Fishing village

Sewage treatment plant

Farming

Wildlife refuge

Recreation and coastal
development

No-fishing zone
due to water
pollution

Coastal fishing

Marine protected area

FIGURE 9.8 Ecosystem management involves bringing together all the stakeholders that affect a large ecosystem and receive benefits from it. In this case, a watershed needs to be managed for a wide variety of purposes, many of which influence one another. (After Miller 1996.)

One successful example of ecosystem management is the work of the Malpai Borderlands Group, a nonprofit cooperative enterprise formed by ranchers and other local landowners who promote collaboration among conservation organizations such as The Nature Conservancy, private landowners, scientists, and government agencies (www.malpaiborderlands-group.org). The group is developing a network of cooperation across the Malpai planning area, which comprises nearly 400,000 ha of unique, rugged mountain and desert habitat along the Arizona and New Mexico border. The Malpai Borderlands Group uses controlled burning as a range management tool, reintroduces native grasses, applies innovative approaches to cattle grazing, incorporates scientific research into management plans, and takes action to avoid habitat fragmentation by using **conservation easements** (agreements not to develop land) to prevent residential development. Their goal is to create "a healthy, unfragmented landscape to support a diverse, flourishing community of human, plant and animal life in the Borderlands Region" (Allen 2006).

A logical extension of ecosystem management is **bioregional management**, which integrates protection with human use and often focuses on a single large ecosystem, such as the Caribbean Sea or the Great Barrier Reef of Australia, or on a series of linked ecosystems, such as the protected areas of Central America. A bioregional approach is particularly appropriate where there is a single, continuous, large ecosystem that crosses international boundaries or when activity in one country or region will directly affect an ecosystem in another country. For the European Union and the 21 individual countries that participate in the Mediterranean Action Plan (MAP), for example, bioregional cooperation is absolutely necessary because the enclosed Mediterranean Sea has large human populations along the coasts, heavy oil tanker traffic, and weak tides that cannot quickly remove pollution resulting from cities, agriculture, and industry (**FIGURE 9.9**). This combination of problems threatens the health of the entire Medi-

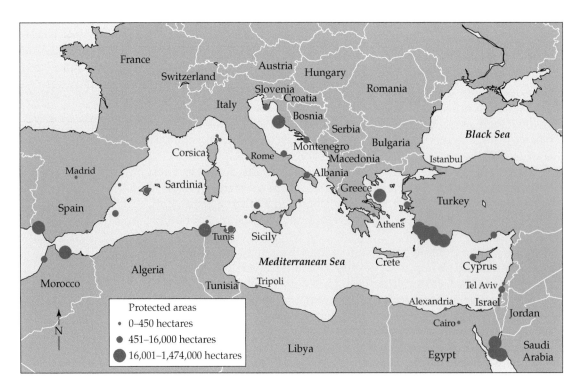

FIGURE 9.9 The countries participating in MAP cooperate in monitoring and controlling pollution and coordinating the management of their protected areas. Major protected areas along the coast are shown as dots. Note that there are no major protected areas on the coasts of France, Libya, or Egypt due to private land ownership and the promotion of economic development. (After Miller 1996.)

terranean ecosystem, including the sea, its surrounding lands, and its associated tourist and fishing industries. Cross-boundary management is also necessary because pollution from one country can significantly damage the natural resources of neighboring countries. At a conference celebrating MAP's fortieth year, a representative of the United Nations Environmental Program was quoted as saying, "While 2015 will be remembered as a major milestone in terms of international agreements, 2016 has been called the year for implementation and delivery. We must seize that opportunity, and ride that momentum." (UNEPMAP 2016.)

Working with Local People

Even remote regions that are considered "wilderness" by governments and the general public often have small, sparse human populations. Societies that practice a traditional way of life in rural areas with relatively little outside influence in terms of modern technology are variously referred to as "tribal people," "indigenous people," "native people," or more generally, "traditional people" (Timmer and Juma 2005; www.iwgia.org). These people regard themselves as the original inhabitants or long-standing residents of a region and are often organized at the community or village level.

> In many parts of the world, areas with high biodiversity are inhabited by indigenous people with long-standing systems for resource protection and use. These people are important, and may be essential, to conservation efforts in those areas.

It is necessary to distinguish these established traditional peoples from more recent settlers, who may not be as concerned with the health of surrounding biological communities or as knowledgeable about the species living there and the land's ecological limits. In many countries, such as India and Mexico, there is a striking correspondence between areas occupied by traditional people and the areas of high conservation value and intact forest (Toledo 2001). Such local people often have established systems of rights to natural resources, which sometimes are recognized by their governments, and they are potentially important partners in conservation efforts (Rai and Bawa 2013).

Worldwide, there are approximately 400 million traditional people living in more than 70 countries, which occupy about 20% of the Earth's land surface (Scariot 2013; indigenouspeople.net). Rather than being a threat to a pristine environment, in some cases traditional peoples have been an integral part of these environments for thousands of years (Middleton 2013). The present mixture and relative densities of plants and animals in many biological communities may reflect the historic activities—such as fishing, selective hunting of game animals, and planting or encouraging of useful plant species in fallow agricultural plots—of people in the area. These activities often do not degrade the environment as long as human population density remains low and there are abundant land and resources.

Many traditional societies do have strong conservation ethics. These ethics are often subtler and less clearly stated than Western conservation beliefs, but they tend to influence people's actions in their day-to-day lives, perhaps more than Western beliefs (Ban et al. 2013). In such societies, people use their traditional ecological knowledge to create management practices that are linked to belief systems and enforced by village consent and the authority of leaders. One well-documented example of such a conservation perspective is that of the Tukano Indians in northwest Brazil who have strong religious and cultural prohibitions against cutting the forest along the Upper Río Negro, which they recognize as important to the maintenance of fish populations (Andrew-Essien and Bisong 2009).

Local people who support conservation as an integral part of their livelihood and traditional values are often inspired to take the lead in protecting biodiversity. Empowering them by helping them to obtain **legal title**—a right to ownership that is recognized by the government—to their traditionally owned lands is often an important component of efforts to establish locally managed protected areas in developing countries (Rai and Bawa 2013). Today, indigenous communities own 97% of the land in Papua New Guinea. Reserves for indigenous people in the Amazon Basin of Brazil occupy over 100 million ha (22%) of its incredibly diverse habitats. The Inuit people govern one-fifth of Canada. In Australia, tribal people control 90 million ha, including many of the most important areas for conservation. Together these regions encompass a substantial percentage of the world's biodiversity (**FIGURE 9.10**).

The challenge, then, is to develop strategies for including these local peoples in conservation programs and policy development both outside and inside protected areas (Gavin et al. 2015). The partnership of traditional people, government agencies, and conservation organizations working together has been termed **co-management** (Borrini-Feyerabend et al. 2004). Co-management involves sharing management decisions and their consequences. The new strategies have been developed in an effort to avoid **ecocolonialism**, the practice by some governments and conservation organizations of disregarding the traditional rights and practices of local people in order to establish new conservation areas. The practice is called ecocolonialism because of its similarity to the historical abuses of native rights by colonial powers of past eras (Cox and Elmqvist 1997). The involvement of these people in the conservation of their lands is an issue of social justice issue; this aspect will be discussed in more detail in Chapter 11.

In many new conservation projects, the economic needs of local people are included in conservation management plans, to the benefit of both the people and the reserves (Roe et al. 2013). Such projects, known as **integrated conservation development projects** (**ICDPs**), are now regarded as worthy of serious consideration, though in practice they are often problematic to implement, as described later in the chapter. There are many possible strategies that could be classified as ICDPs, ranging from wildlife management projects to ecotourism, and may or may not include formally protected areas. These projects normally attempt to combine the protection of bio-

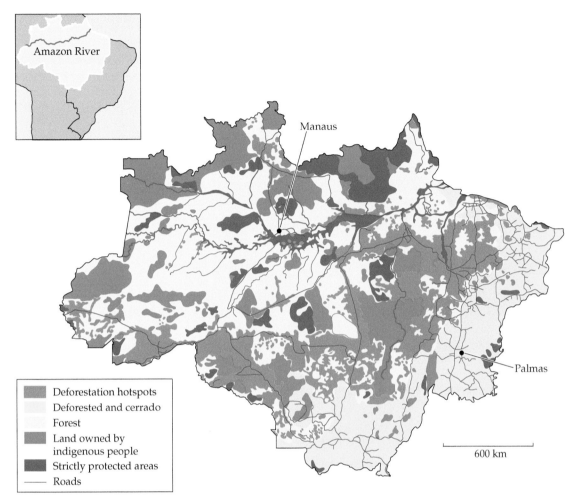

FIGURE 9.10   Large blocks of indigenous lands are important in the overall conservation strategy for the Brazilian Amazon. Many national parks and other protected area have been established since 2002. Human activities of logging, farming, and ranching over the past decade have created an "arc of deforestation." Development and deforestation are also associated with the expanding network of paved and unpaved roads. (From Soares-Filho et al. 2010.)

diversity and the customs of traditional societies with aspects of economic development, including reducing poverty, creating jobs, improving health, and ensuring food security. A large number of such programs have been initiated over the last 25 years, and they have provided opportunities for evaluation and improvement. Involving local people in ongoing monitoring efforts may increase information and also help to determine how the people themselves perceive the benefits and problems of the project (Braschler 2009). The hope of such projects is that the local people will decide that sustainable use of their local resources is more valuable than destructive use of those resources and that these people will become involved in biodiversity conservation. The following are some examples of the types of ICDPs currently in practice.

Biosphere reserves

In UNESCO's World Network of Biosphere Reserves, traditional people are allowed to use resources from designated buffer zones around strictly protected core areas (see Figure 8.19). The program is a successful example of the ICDP approach, at least in terms of its widespread adoption of as a model of conservation; there are 621 Biosphere Reserves in 117 countries, covering over 260 million ha. The Biosphere Reserve Program recognizes the role of people in shaping the natural landscape as well as the need to find ways in which people can sustainably use natural resources without degrading the environment.

One instructive example of a biosphere reserve is the Kuna Yala Indigenous Reserve on the northeast coast of Panama. In this protected area, which comprises 60,000 ha of tropical forest and coral islands, 50,000 Kuna people in 60 villages practice traditional medicine, fishing, agriculture, and forestry (**FIGURE 9.11**). Scientists from outside institutions carry out management research, and in the process they train and hire local people as guides and research assistants. The Kuna local government attempts to control the type and rate of economic development in the reserve. However, a change appears to be occurring among the Kuna: traditional conservation beliefs are eroding in the face of outside influences, often in tandem with the growing tourism industry, and younger Kuna are beginning to question the need to rigidly protect the reserve (Posey and Balick 2006). Also, the Kuna people have had difficulties establishing a stable organization that can manage the reserve and work with external conservation and donor groups. Furthermore, rising sea levels and declining marine resources are forcing village leaders to consider other options for their future. This example illustrates how empowering traditional people is no guarantee that biodiversity will be preserved. This is particularly true when traditions change or disappear, economic pressures for exploitation increase, or programs are mismanaged. The challenge will be to de-

FIGURE 9.11 Kuna boys fishing. (© Alvaro Leiva/AGE Fotostock.)

termine a way to integrate conservation into the cultural evolution of Kuna society, which cannot—and from an ethical standpoint, should not—be prevented.

> ICDPs involve local people in sustainable activities that combine biodiversity conservation and economic development.

In situ agricultural conservation

The long-term health of modern agriculture depends on the preservation of the genetic variability maintained in local varieties of crops cultivated by traditional farmers (Bisht et al. 2007). One innovative suggestion has been for an international agricultural body, such as the Consultative Group on International Agricultural Research, to subsidize villages as in situ (in-place) custodians of traditional varieties of crop species (Brush 2004). Along these lines, in China, a government program that involves paying farmers to interplant high-quality traditional and high-yielding hybrid rice varieties maintains the crop's genetic variability (Zhu et al. 2003). Villages that participate in such programs have an opportunity to maintain their culture in the face of a rapidly changing world.

A different approach linking traditional agriculture and genetic conservation is being used in arid regions of the American Southwest, with a focus on dryland crops with drought tolerance (www.nativeseeds.org). A private organization, Native Seeds/SEARCH, collects the seeds of 1800 traditional crop cultivars for long-term preservation. The organization also encourages a network of 4600 farmers and other members to grow traditional crops, provides them with the seeds of traditional cultivars, and buys their unsold production. The value of this and related genetic conservation programs is being increasingly understood (Jarvis et al. 2016).

Countries have also established special reserves to conserve areas containing wild relatives and ancient landraces of commercial crops (Barazani et al. 2008). Species reserves protect the wild relatives of wheat, oats, and barley in Israel and of citrus in India.

Extractive reserves

In many areas of the world, traditional people have extracted products from natural communities for decades and even centuries. The use, sale, and barter of these natural products are a major part of people's livelihood. Understandably, local people are very concerned about retaining their rights to continue collecting natural products from the surrounding countryside. In areas where such collection represents an integral part of traditional society, the establishment of a national park that excludes the traditional collection of products will meet with as much resistance from the local community as will a landgrab that involves the exploitation of the natural resources and their conversion to other uses. A type of protected area known as an **extractive reserve** may present a sustainable solution to this problem.

One such example is found in the Brazilian Amazon, where the government is trying to address the legitimate demands of local citizens by establishing extractive reserves from which settled people collect natural materials, such as medicinal plants, edible seeds, rubber, resins, and Brazil nuts, in ways that minimize damage to the forest ecosystem (Duchelle et al. 2012) (**FIGURE 9.12**). These extractive reserves, which comprise about 3 million ha, guarantee the ability of local people to continue their way of life and guard against the possible conversion of the land to cattle ranching and farming. However, populations of large animals in extractive reserves are often substantially reduced by subsistence hunting by local people, and the density of Brazil nut seedlings is reduced by the intense collection of mature nuts.

Many countries in East and southern Africa have started aggressively applying community development and sustainable harvesting strategies in their efforts to preserve wildlife populations. Governments are attempting to develop programs to generate income from trophy hunting and wildlife tourism that can be operated at the village level and provide clear benefits to local people (Naidoo et al. 2016). One example is the Community Based Natural Resource Management program, in which local communities working with the government sell sport-hunting rights of high-value trophy species, such as lions and elephants, to safari companies (see Chapter 3). Revenue is also generated through operating tourist facilities. To maintain the needed densities of wildlife, the village community must work together with government officials to prevent illegal hunting (see the chapter opening photo).

There has been vocal support from some conservation biologists for selling hunting licences as a means of conserving species, especially when

FIGURE 9.12 Extractive reserves established in Brazil provide a reason to maintain forests. The trunks of wild rubber trees are cut for their latex, which flows down the grooves into the cup. Later the latex will be processed and used to make natural rubber products. (Photograph © Edward Parker/Alamy.)

local people are involved (e.g., Di Minin et al. 2016). However, trophy hunting is considered ethically questionable both those who believe that killing purely for sport (rather than for food) is morally wrong and point to the faulty reasoning behind consequentialism, that is, that the ends justify the means (Nelson et al. 2016). Furthermore, they argue that revenues from trophy hunting for conservation are insufficient, usually do not reach the local community, and are decreased via corruption (Lindsey et al. 2016). Finally, the market for hunting licenses creates pressure to "produce an animal" that inevitably leads to poaching, as was seen in the Cecil story (see Chapter 3; Richard Reading, pers. comm). Thus, some question whether sport hunting belongs in the same category as extractive reserves that support locals with food, firewood, or other resources.

Community-based initiatives

In many cases, local people already protect natural areas and resources such as forests, wildlife, rivers, and coastal waters in the vicinity of their homes. Protection of such areas, sometimes called **community conserved areas**, or **community-based conservation (CBC)**, is often enforced by village elders because of the clear benefit to the local people (see Chapter 11). These benefits include maintaining natural resources (e.g., food supplies and drinking water) and the use of the land for religious and traditional practices. The protection of biodiversity may even be an intrinsic aspect of local beliefs (Borrini-Feyerabend et al. 2004). In this way, the goal of a CBC is to align ecological, economic, and social goals. A review of 136 CBC projects across the globe found that degree of local participation, environmental education and skills-training programs all significantly contributed to win–win outcomes for the people and to biodiversity (Brooks 2016). The most important feature, however, was institutional capacity building: efforts to improve infrastructure and communication and decision-making processes. Governments and conservation organizations can assist local conservation initiatives by providing access to scientific expertise, training programs, and financial assistance to develop needed infrastructure, in addition to simply offering legal title to traditional lands.

One example of a local initiative is the Community Baboon Sanctuary in eastern Belize, which was created by a collective agreement among a group of villages to maintain the forest habitat required by the local population of black howler monkeys (known locally as baboons). Ecotourists visiting the sanctuary pay a fee to the village organization, and additional payments are made if they stay overnight and eat meals with a local family. Conservation biologists working at the site have provided training for local nature guides, a body of scientific information on the local wildlife, funds for a local natural history museum, and business training for the village leaders.

In the Pacific islands of Samoa, much of the rain forest land and marine area is under "customary ownership": it is owned by communities of indigenous people (Boydell and Holzknecht 2003). Villagers are under increasing pressure to sell logs from their forests to pay for schools and other necessities. Despite this situation, the local people have a strong desire to preserve the land because of the forest's religious and cultural significance, as well as its value for medicinal plants and other products. A variety of solutions are being developed to meet these conflicting needs. In 1988, in American (or Eastern) Samoa, where about 90% of the land is under customary ownership, the US government leased forest and coastal land from the villages to establish a new national park (americansamoa.noaa.gov). Under this agreement, the villages gained needed income yet retained ownership of the land and their traditional hunting and collecting rights (www.nps.gov).

Payments for ecosystem services

A creative strategy involves making direct payments to individual landowners and local communities that protect critical ecosystems and the services they provide, in effect paying the community to be a good steward of the land (Wunder 2013; Wünscher and Engel 2012). These types of programs are sometimes referred to as **payments for ecosystem services** (**PES**), and they are becoming increasingly popular (**FIGURE 9.13**). Governments, nongovernmental conservation organizations, and businesses develop markets in which local villagers and landowners can participate through protecting and restoring ecosystems. For example, owners of a forest may receive direct payments from a city government for the ecosystem services provided by the forest, such as controlling floods and providing drinking water. Local landowners and farmers can be paid for allowing large predators such as wolves, bears, tigers, and mountain lions to be on their land, with additional payments given as compensation if their livestock is attacked (Dickman et al. 2011).

Rural people can be drawn into newly developing international markets for eco-

(A)

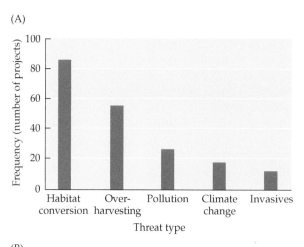

(B)

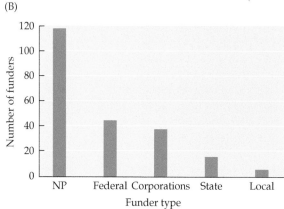

FIGURE 9.13 Patterns of payments for 103 ecosystem services (PES) projects from 37 countries. (A) Number of projects addressing different types of threat. Most projects address issues of habitat conversion (from forest to agricultural land) and overharvesting of trees. (B) Funding sources for the projects are primarily nonprofit (NP) conservation organizations but also include government agencies at national, state, and local levels, as well as corporations. (After Tallis et al. 2009.)

system services (www.ecosystemmarketplace.com), especially programs that trade in carbon credits for reducing atmospheric levels of greenhouse gases. In the Kasigu Corridor Reducing Emissions from Deforestation and Forest Degradation (REDD) project in southeastern Kenya, villagers protect a natural migration corridor for elephants between Tsavo East National Park and Tsavo West National Park (**FIGURE 9.14**). The villagers earn carbon credits and funding from international programs for protecting the forest and maintaining wildlife populations, as determined by an independent outside evaluation. Funds from the program have been used to pay for wildlife patrols, build schools and other infrastructure, and start local businesses, while in the process creating 350 new jobs and generating a total of $1.2 million (Dinerstein et al. 2012). Programs addressing carbon sequestration and climate change are likely to expand greatly in coming years, and they may provide substantial funds for land protection. However, at present such programs are sometimes unable to pay enough money to prevent landowners from converting their land to other uses (Banerjee et al. 2013).

New markets are being developed in which local people and landowners are paid for providing ecosystem services, such as protecting forests to maintain water supplies and planting trees to absorb carbon dioxide. Programs that address climate change issues are predicted to become more common in the coming years.

PES have been effective for improving habitat for the giant panda (*Ailuropoda melanoleuca*) in China, as a part of the Natural Forest Conservation Program (Tuanmu et al. 2016). In fact, an evaluation of the changes in vegetation revealed that often more improvement occurred in areas managed by locals than those managed by the government. However, this success only took place when the financial incentives were very high.

FIGURE 9.14 Elephants (*Loxodonta africana*) in Kenya depend on this wildlife corridor for their annual migration between two national parks. The corridor is maintained by locals who benefit financially. (© Morkel Erasmus/Getty Images.)

Evaluating conservation initiatives that involve traditional societies

Unfortunately, when external funding ends, and if the projected income stream fails to develop, many of these integrated conservation and development projects (ICDPs) end abruptly. Even for projects that appear successful, there is often no monitoring of ecological and social parameters to determine whether the project goals are being achieved. It is essential that any conservation program design include mechanisms for evaluating the progress and success of measures being taken.

A key element in the success of many of the projects discussed in the preceding sections is the opportunity for conservation biologists to complement and work with stable, flexible, local communities with effective leaders and competent government agencies (Baker et al 2012). Certain projects appear to be successful at combining the protection of biodiversity with sustainable development and poverty reduction. The Equator Initiative of the United Nations is cosponsored by many leading conservation organizations, businesses, and governments and is helping to fund and publicize such efforts.

However, in many cases a local community may have internal conflicts and poor leadership, making it incapable of administering a successful conservation program. Moreover, conservation initiatives involving recent immigrants or impoverished, disorganized local people may be difficult to carry out, and government agencies working on the project may be ineffective or even corrupt. Consequently, while working with local people may be a desirable goal, in some cases it simply is not possible.

Case Studies: Namibia and Kenya

Throughout the world, the protection of biodiversity is being included as an important objective of land management. We conclude the chapter by examining two case studies of successful community-based programs in Namibia and Kenya that illustrate some of the challenges and successes of managing biodiversity outside protected areas.

Community-Based Natural Resource Management (CBNRM) programs in Africa represent an approach in which local landowners and communal groups are given the authority to manage and profit from the wildlife on their own property. In many African countries, wildlife both inside and outside national parks is often managed by government officials, often with no input from the local people, who gain little or no economic benefit from the wildlife on their own land and have no incentive to protect the wildlife. By changing the management system to CBNRM, government officials and conservation organizations hope to counterbalance pressures threatening local wildlife while simultaneously contributing to rural economic development. There is a long history of CBNRM programs in Africa, but it has been difficult to develop stable programs that are economically viable and effectively managed.

One of the most ambitious new programs for local communities managing wildlife is found in Namibia in southern Africa (Riehl et al 2015). Namibia has over 2.3 million people, with 62% of them living in rural areas and farming or raising livestock (NACSO 2014). Namibia includes an impressive six different biomes, from the desert to the subtropical (**FIGURE 9.15**). These biomes support high levels of endemism, including 700 plant species, 91 bird species, and 26 mammal species (UNCBD report 2010).

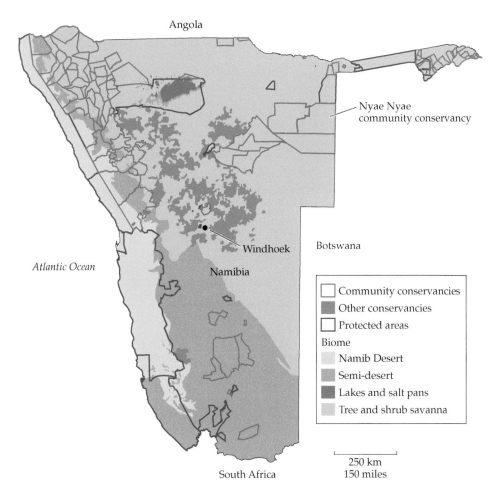

FIGURE 9.15 The distribution of current and emerging community conservancies in Namibia, in which communal groups agree to protect biodiversity. State-protected lands are also shown. The "Other conservancies" category denotes areas in which communal groups have not yet committed to forming conservancies to protect wildlife. It can be seen here that several biomes are more represented in these community conservancies than in formal protected areas (After NACSO 2008, with updates from Riehl et al. 2015.)

Beginning in 1996, the Namibian government granted traditional communal groups the right to use and manage the wildlife on their own lands. To obtain these rights, a group needs to form a management committee and determine the boundaries of its land. The government then designates the group as a "community conservancy." The benefits of forming a conservancy and participating in wildlife management are fourfold:

1. The conservancy can form joint ventures with tour operators, with about 5–10% of the gross earnings paid to the conservancy. A certain number of the employees in the tourist operation are hired from among the communal group. Revenues from the joint ventures are used to train and pay game guards, again hired from the communal group, who monitor the wildlife populations and prevent poaching.

2. Using funds from the joint ventures, the conservancy members can build and operate campsites for tourist groups, providing direct revenue, employment, and experience for the communal group.

3. The conservancy can apply to the government for a trophy-hunting quota, which will be granted if the wildlife populations are large enough, as indicated by monitoring. Hunting licenses can then be sold or auctioned off to professional hunters, who bring in wealthy foreign tourists willing to pay a high price for an African hunting experience. Payments to the conservancy for high-value animals such as lions and elephants can be as large as $11,000 per animal. Meat from the hunted animals is distributed to the group members as an added benefit. This approach to funding is not without controversy, however (see Chapter 3 and "Extractive reserves" in this chapter). Some economic analysis suggests that tourism alone (i.e., without hunting) is not sufficient to cover operating costs (Naidoo et al. 2016).

4. Once the conservancy has formed a wildlife management plan, four species of wildlife—gemsbok, springbok, kudu, and warthog—can be hunted for subsistence. In practice, the hunting is often done by game guards and professional hunters and the meat is distributed to everyone in the community.

Over the last 18 years, 79 conservancies in Namibia have been established, covering 19.4% of Namibia's land surface (Riehl et al. 2015) (see Figure 9.15). Help in the initial establishment of the conservancies has come from external funding agencies, such as the US Agency for International Development. Conservancy members have received further training in tourism, finance, and marketing, along with effective advocacy to gain support from the government and the private sector. Although the financial gain by the CBNRM is well documented and social benefits such as improved health in these areas relative to adjacent areas have been documented, whether other expected social benefits, such as education, have improved is less clear (Riehl et al. 2015). This is due in part to the large degree of variability between conservancies with a great deal of tourism and those that have none. Some analyses also suggest that although hunting and tourism in conservancies could earn more per hectare than livestock rearing (Lindsey et al. 2013), conservancies may not be economically viable in the long term (Humavindu and Stage 2015). PES and other revenue-sharing systems have been proposed to address the problem of viability (Lapeyre 2015).

So far, the communal management system seems to be having positive effects on conservation. Namibia currently claims to host the world's largest populations of free-ranging cheetah and black rhino, both international species of concern (see chapter opening photo). A report from a consortium of conservation nongovernmental organizations (NGOs), including WWF Namibia, working with the Ministry of Environment, documented dramatic increases in many large mammals since the CBNRM programs were initiated (NASCO 2008). Many species, especially ungulates, have been observed as having greater diversity and higher numbers within the conservancies than in adjacent, unprotected land (Lindsey et al. 2015).

Other African countries have programs that are similar to Namibia's. In Kenya, for example, about two-thirds of the country's 650,000 large animals—including giraffes, elephants, zebras, and ostriches—live outside park boundaries in rangelands used by commercial ranches and as traditional grazing lands by local people (Western et al. 2009; Young et al. 2005). The rangelands outside the parks are increasingly unavailable to wildlife, though, because of fences, poaching, and agricultural development, which have led to a gradual decline in wildlife numbers.

A combination of regulations, community involvement, and economic incentives is contributing to the persistence of substantial populations of wildlife in certain of these unprotected areas of Kenya in spite of the challenges (Kinnaird and O'Brien 2013). In some places, private ranching in which wildlife and livestock are managed together for both meat and ecotourism is more profitable than managing livestock alone because the livestock and the wildlife use different food resources. As in Namibia, many ranches have also developed facilities for foreign tourists who want to view wildlife, which creates an additional source of revenue and an incentive for protecting these species.

Although these community-based management programs have been successful in many cases, their dependence on tourism and subsidies from outside donor governments and conservation and development organizations can make them vulnerable. When these outside subsidies cease, the wildlife programs often end as well, suggesting that the programs are often not really profitable on their own. The ineffectiveness and corruption of some local government agencies are additional factors that can cause such programs to fail. These community wildlife programs will be judged successful when they can demonstrate that they can both protect wildlife and provide a stable income source for the local people.

Summary

- Considerable biodiversity exists outside protected areas, particularly in habitat managed for multiple-use resource extraction. Such unprotected habitats are vital for conservation because in almost all countries, protected areas account for only a small percentage of total area. Animals and plants living in protected areas often disperse to unprotected land, where they are vulnerable to hunting/harvest, habitat loss, and other threats from humans.

- Governments are increasingly encouraging the protection of biodiversity as a priority on multiple-use land, including forests, grazing lands, agricultural areas, military reservations, and urban areas. All of these can be managed for conservation, keeping in mind that there are species that are too sensitive to ever exist outside of strictly protected areas.

- Government agencies, private conservation organizations, businesses, and private landowners can cooperate in large-scale ecosystem management projects to achieve conservation objectives and use natural resources sustainably. Bioregional management involves cooperation across large regions to manage large ecosystems, which frequently cross international borders.

- In Africa, many of the characteristic large animals are found predominantly in rangeland outside the parks. Local people and landowners often maintain wildlife on their land for a variety of purposes. Local communities are now generating income by combining wildlife management and ecotourism, sometimes including trophy hunting.

For Discussion ■

1. Consider a national forest that has been used for decades for logging, hunting, and mining. If endangered plant species are discovered in this forest, should these activities be stopped? Can logging, hunting, and mining coexist with endangered species, and if so, how? If logging has to be stopped or scaled back, do the logging companies or their employees deserve any compensation? Explain your answer.

2. Do you think that trophy hunting on private reserves is a good means by which to preserve species? Why or why not? On what basis should this decision be made: economic, ethical, past success, or future potential for conservation?

3. Choose a large aquatic ecosystem that includes more than one country, such as the Black Sea, the Rhine River, the Caribbean, the St. Lawrence River, or the South China Sea. What agencies or organizations have responsibility for ensuring the long-term health of the ecosystem? In what ways do they, or could they, cooperate in managing the area?

Suggested Readings

Athreya, V., M. Odden, J. D. Linnell, J. Krishnaswamy, and U. Karanth. 2013. Big cats in our backyards: Persistence of large carnivores in a human dominated landscape in India. *PLoS ONE* 8(3): e57872. Leopards live near farmlands and villages but do not affect people or livestock because they hunt at night.

Baudron, F. and K. E. Giller. 2014. Agriculture and nature: Trouble and strife? *Biological Conservation* 170: 232–245. The authors consider the alternatives of land sharing and land sparing.

Brooks, J. S. 2016. Design features and project age contribute to joint success in social, ecological, and economic outcomes of community-based conservation projects. *Conservation Letters*. Capacity building, particularly when it improves infrastructure, is the most important predictor of success in CBC projects.

Davis, A. M. and A. R. Moore. 2015. Conservation potential of artificial water bodies for fish communities on a heavily modified agricultural floodplain. *Aquatic Conservation: Marine and Freshwater Ecosystems*. Many species of fish can be found in gravel pits and other haphazard water bodies, but the greatest diversity was found in constructed wetlands.

Farmer, J. R., Z. Ma, M. Drescher, E. G. Knackmuhs, and S. L. Dickinson. 2016. Private landowners, voluntary conservation programs, and implementation of conservation friendly land management practices. *Conservation Letters*. Attitudes and motivations of private landowners affect how well they manage for biodiversity.

Gavin, M. C., J. McCarter, A. Mead, F. Berkes, J. R. Stepp, D. Peterson, and R. Tang. 2015. Defining biocultural approaches to conservation. *Trends in Ecology and Evolution* 30(3): 140–145. For conservation to work, local people must be involved.

Hylander, K., S. Nemomissa, J. Delrue, and W. Enkosa. 2013. Effects of coffee management on deforestation rates and forest integrity. *Conservation Biology* 27: 1011–1019. Traditional forms of coffee plantations can maintain forest cover and some level of biodiversity.

Lindsey, P. A., C. P. Havemann, R. M. Lines, A. E. Price, T. A. Retief, T. Rhebergen, and 2 others. 2013. Benefits of wildlife-based land uses on private lands in Namibia and limitations affecting their development. *Oryx* 47(1): 41–53. CBNRM is effective for protecting large mammals, especially ungulates, in part due to their management for sport hunting.

Loke, L. H., R. J. Ladle, T. J. Bouma, and P. A. Todd. 2015. Creating complex habitats for restoration and reconciliation. *Ecological Engineering*, 77: 307–313. Sometimes that which benefits people can also benefit biodiversity.

Naidoo, R., L. C. Weaver, R. W. Diggle, G. Matongo, G. Stuart-Hill, and C. Thouless. 2016. Complementary benefits of tourism and hunting to communal conservancies in Namibia. *Conservation Biology*. Tourism and hunting are distinct funding sources and become profitable at different periods during a conservancy's development.

Riehl, B., H. Zerriffi, and R. Naidoo. 2015. Effects of community-based natural resource management on household welfare in Namibia. *PLoS ONE* 10(5). The benefits of this approach can be seen on multiple levels of society but not always in ways one would expect.

Restoration Ecology

Volunteers and fishermen plant mangrove trees in Kondang Merak, located on the eastern coast of Java, Indonesia, in an effort to restore the local marine ecosystem. This area has been heavily damaged by illegal fishing methods that have included the use of cyanide and dynamite.

Ecosystems that have been damaged or destroyed by intensive human activities such as mining, ranching, and logging may lose much of their **ecological resilience**, or natural ability to recover. In some cases, recovery would require centuries or even millennia without human assistance. The restoration of ecosystems can be motivated by the protection of species, improving aesthetics, recreation, strengthening ecosystem function or connectivity, or the reestablishment of ecosystem services (Keenelyside 2012). Rebuilding damaged ecosystems also can be used to enlarge, enhance, and connect protected areas, as well as to create buffers around them (see Chapter 8).

Degraded ecosystems provide important opportunities to apply research findings by helping to restore historical species and communities (Clewell and Aronson 2008). **Ecological restoration** is the practice of restoring the species and ecosystems that occupied a site at some point in the past but were damaged or destroyed (www.ser.org) (**FIGURE 10.1**). **Restoration ecology** is the science of restoration—the research and scientific study of restored populations, communities,

FIGURE 10.1 (A) Trout stream habitat that has been degraded by human activities. (B) Trout stream habitat that has been restored by installing fencing to exclude cattle, planting native species, and reinforcing stream banks with rocks. (From Hobbs et al. 2010; photographs courtesy of K. Matthews.)

(A)

(B)

Some ecosystems have been so degraded by human activity that their resilience, or ability to recover on their own, is severely limited. Ecological restoration reestablishes functioning ecosystems, with some or all of the original species or, sometimes, a different group of species.

and ecosystems (Falk et al. 2006). These are overlapping disciplines: the process of ecological restoration provides useful scientific data, while restoration ecology interprets and evaluates restoration projects in a way that can lead to improved methods.

Restoration ecology will play an increasingly important role in the conservation of biodiversity as degraded lands and aquatic communities are partially or completely restored to their original species compositions and are integrated into existing conservation reserve networks. Because many degraded areas are unproductive and of little economic value, governments may be willing to restore them to increase their economic productivity and conservation value. For example, degraded areas may be subject to soil erosion and increased risk of flooding; restoration in these cases may be motivated by a desire to mitigate threats to human life or property. Restoration efforts also can be part of **compensatory mitigation** or **biodiversity offsets**, in which a new site is created or rehabilitated in compensation for a site that has been destroyed elsewhere by development (Maron et al. 2012). This is particularly true for wetlands, for which a "no net loss policy" has been adopted by many jurisdictions. At other times, ecological processes rather than ecosystems need to be restored. For example, annual floods disrupted by the construction of dams and levees or natural fires stopped by efforts at fire suppression may need to be reintroduced if the absence of these processes proves harmful to species and ecosystems.

Ecological restoration has its origins in older, applied technologies that attempted to restore ecosystem functions or species of known economic value, such as wetland creation (to prevent flooding), mine site reclamation (the final stage of closing a mine, usually involving plantings to prevent soil erosion), range management of overgrazed lands (to increase the production of grasses), and technologies to facilitate tree planting on cleared land (for timber, recreational, and ecosystem values). However, these approaches often produce biological communities that are overly simplified or cannot maintain themselves. As concern for biodiversity has grown, restoration plans have included as a major goal the reestablishment of original or historical species assemblages and processes. For example, one of the objectives of the European Union Biodiversity Convention was to restore at least 15% of degraded ecosystems by 2020 (European Commission 2011). The input of conservation biologists is needed to achieve these goals.

Where to Start?

To be successful in the long term, restoration projects must first establish clear goals, followed by an assessment of site conditions to determine if those goals can be met, and if so, by what means (**FIGURE 10.2**). Often, the goal of restoration efforts is to create ecosystems that are comparable in function or species composition to existing **reference sites** (Humphries and Winemiller 2009). Reference sites are central to the very concept of restoration; they act as comparison sites, providing explicit restoration goals and allowing for quantitative measures of the project's success (Higgs et al. 2014). Unrestored areas can act as "negative" reference sites or controls to further determine the impact of the restoration actions.

If practical, the successfully restored ecosystem should be dominated by native species, contain representatives of all key functional groups of species, have a physical environment suitable for native species and ecosystem processes, and be secure from detrimental outside disturbances. In some cases, such as at arid and cold sites, achieving such recovery may take decades or even centuries.

But is the reestablishment of native species always the goal of restoration? Site conditions or limitations of resources may make this undesirable or impossible. There are four main approaches that define outcomes when

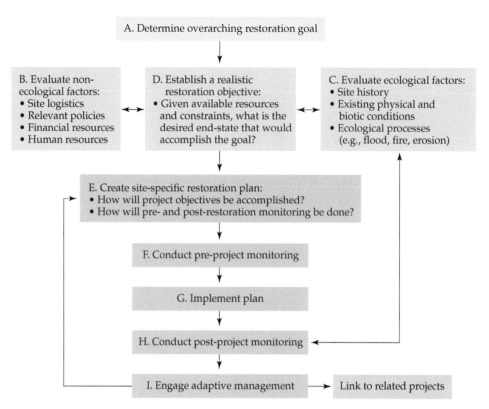

FIGURE 10.2 A flow diagram of a scientific approach to restoration. The first step is to (A) clarify the overarching goal of the restoration project, such as to increase biodiversity at a site. Then, to determine how this will be achieved, it is necessary to evaluate both (B) nonecological factors, such as how much money is available for the project, and (C) ecological factors, such as the condition of the soil and what plant species currently are found there. These factors will inform and be influenced by (D) the specific objectives for restoration, such as establishing plant species that will promote diversity at higher levels. Only after these steps have been taken can (E) a specific, realistic plan for how to implement restoration be created, such as removing weeds, improving the soil, or planting seeds. The implementation of the plan should accompany both pre- and post-monitoring, (F–H). Monitoring progress of the restoration objectives will then facilitate (I) improvement over time at both that site and future projects. (After Shafroth et al. 2008.)

considering the restoration of biological communities and ecosystems (**FIGURE 10.3**) (Bradshaw 1990):

1. *No action*. Restoration is deemed too expensive, previous attempts have failed, or experience has shown that the ecosystem will recover on its own. Letting the ecosystem recover on its own, also known as **passive restoration**, is typical for old agricultural fields in eastern North Amer-

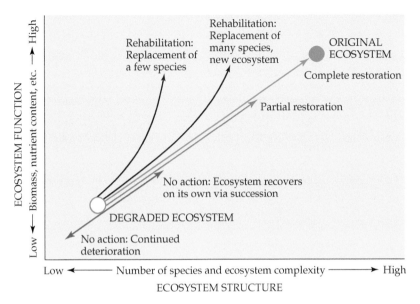

FIGURE 10.3 Decisions must be made about whether the best course of action is to completely restore a degraded site (green arrow), partially restore it (blue arrow), rehabilitate it by introducing different species (black arrow), or take no action (red arrow). (After Bradshaw 1990.)

ica, which sometimes return to forest within a few decades after being abandoned. However it should be noted that even in this case, the species composition may be quite different, especially for the understory (Flinn and Marks 2007).

2. *Rehabilitation.* A degraded ecosystem is replaced with a different but productive ecosystem type. For example, a degraded forest might be replaced with a productive pasture or a tree plantation. Just a few species may be replaced, or a larger-scale replacement of many species may be attempted. As the ultimate goal is not to restore the original ecosystem, some authors consider the term *restoration* inappropriate to refer to rehabilitated ecosystems (Perring et al. 2014).

3. *Partial restoration.* At least some of the ecosystem functions and some of the original, dominant species are restored. An example is replanting a degraded grassland with a few species that can survive. Partial restoration typically focuses on dominant species or particularly resilient species that are critical to ecosystem function, delaying action on the rare and less common species that would be part of a complete restoration program.

4. *Complete restoration.* The area is completely restored to its original species composition and structure by an active program of site modification and reintroduction of the original species. For complete restoration, the first step is to determine and then mitigate the source of ecological degradation. For example, the source of pollution of a lake ecosystem

must be identified and controlled before the ecosystem can be restored. Natural ecological processes must be reestablished because they help the system recover and contribute to long-term resilience.

In practice, ecological restoration must also consider the speed of restoration, the cost, the reliability of results, and the ability of the target community to persist with little or no further maintenance. Practitioners of ecological restoration must have a clear grasp of how natural systems work and what methods of restoration are feasible (Falk et al. 2006). Considerations of the cost and availability of seeds, when to water plants, how much fertilizer to add, how to remove invasive species, and how to prepare the surface soil may become paramount in determining a project's outcome. Dealing with such practical details generally has not been the focus of academic biologists in the past, but they must be considered in ecological restoration. Fortunately, restoration ecology often involves professionals from other fields, who can lend their expertise, ultimately enriching the restoration process. However, these practitioners sometimes have different goals than conservation biologists. For instance, civil engineers involved in major projects seek economical ways to permanently stabilize land surfaces, prevent soil erosion, make the site look better to the general public, and if possible, restore the productive value of the land.

> Restoration projects require monitoring to determine whether goals such as costs and speed of recovery are being met. Such projects may also provide new insights into ecological processes.

The degree of alteration will generally dictate what type of action is required; the most degraded systems will likely require both biological and physical alterations to the habitat (**FIGURE 10.4**). If the damage has been caused by abiotic factors such as soil erosion or lowered water tables, then the source of the problem should be addressed or at least considered before making any attempt to reestablish species (Hobbs and Harris 2001).

One particular quandary in the restoration of highly degraded ecosystems is ecological assembly order; that is, in what order and when should the species components of the system be put back together? When trophic order is apparent (e.g., predators require an established prey species), the decision seems obvious, but the overlap in functional relationships among many species complicates such decisions. For example, degraded parks in Africa may have contained a dozen or more ungulate species with similar ecological roles. In what order should these be reintroduced? Knowledge about the specific ecology of individual species can help guide such complicated decisions (Temperton and Hobbs 2004).

Another important issue is the genotypes of plant species that are being reintroduced to a restoration site. In general, it is advisable to follow the local-is-best (LIB) approach, which prioritizes locally adapted genotypes. This is important for at least two reasons. First, there may be local adaptations within a species that will make it either more or less likely to thrive at the restoration sites. Second, there is a risk of outbreeding depression or genetic swamping of local strains by introducing non-local genotypes (see Chapter 5). There are also problems with this approach, including

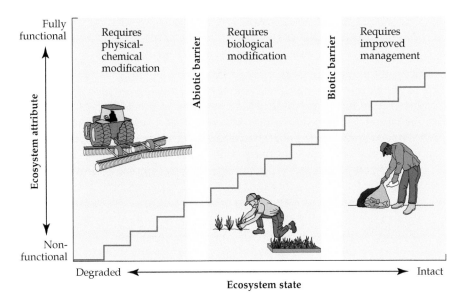

FIGURE 10.4 A conceptual model that considers thresholds for the restoration of ecosystem function. Generally, the most degraded, and therefore nonfunctional, sites will require overcoming abiotic constraints that contribute to the problem, such as removing levees from a river, building structures for coral to grow on, or amending soil that is too acidic. Biotic barriers can be overcome by planting, reintroducing missing trophic levels, or providing a food source. Of course, biotic changes can lead to abiotic ones, as in the case of an invasive plant species increasing the frequency or intensity of fires. If an ecosystem is mostly intact and functional, improved management rather than restoration is needed. (Parks Canada and the Canadian Parks Council, 2008; after Whisenant 1999, and Hobbs and Harris 2001.)

that such genetic matching may be unfeasible or impossible (Smith et al. 2007). In other cases, harvesting wild native plants risks hurting the source populations (Meissen et al. 2015), or the local genotypes have such low genetic variability as to risk inbreeding depression at the restoration site. For this reason, restoration practitioners may need to resort to less related plant stock or even nonnative species (Jones 2003). In at least some cases, the use of cheaper, commercial seed rather than seed that was hand collected or specially reared has no adverse ecological effects (Reiker et al. 2015).

Once restoration has begun, to determine whether these goals are being achieved, both the restoration and the reference sites must be monitored over time (González et al. 2015). Because restoration efforts need to be customized for individual sites, it is often advisable to experimentally test different restoration methods (Lloyd et al. 2013). These restored sites then need to be monitored for years or even decades to determine how well management goals are being achieved and whether further intervention is required, an approach called adaptive management or **adaptive resto-**

ration (Wagner et al. 2008). In particular, native species may have to be reintroduced if they have not survived, and invasive species may have to be removed if they are still abundant or there has been a **secondary invasion** by a different invader.

Restoration ecology is valuable to the broader science of ecology because it provides an acid test of how well we understand a biological community; the extent to which we can successfully reassemble a functioning ecosystem from its component parts demonstrates the depth of our knowledge and points out deficiencies (Bradshaw 1990). For example, Grman and colleagues (2014) defined rules for the assembly of plant communities on prairies after examining 29 restoration projects in southwestern Michigan.

Efforts to restore degraded terrestrial communities generally have emphasized the establishment of the original plant community, as it typically contains the majority of the biomass and provides structure for animals and other elements of the community. However, some researchers argue that in the future, restoration ecology should devote more attention to the other major components of the community (Fraser et al. 2015). Fungi and bacteria (see Chapter 3) play vital roles in soil decomposition and nutrient cycling; soil invertebrates are important in creating soil structure; herbivorous animals are important in reducing plant competition and maintaining species diversity; birds and insects are essential pollinators; and many birds and mammals have vital functions as insect predators, soil diggers, and seed dispersers (Morandin and Kremen 2013). Many birds, insects, and other animals may be able to recolonize the site on their own, but other large animals and above-ground invertebrates may have to be reintroduced from existing populations or captive breeding populations (see Chapter 7) if they are unable to disperse to the site on their own (Audino et al. 2014). Restoration efforts may also succeed by focusing on reestablishing ecological processes that support native communities rather than just planting specific plant taxa (Moreno-Mateos et al. 2015b).

Another practical consideration in restoration projects is gaining the support and participation of local stakeholders. Many restoration efforts are supported and even initiated by local conservation groups because they recognize the direct connection between a healthy environment and people's personal and economic well-being (Felson et al. 2013; Higgs 2003) (see Chapter 3).

Restoration in Urban Areas

Highly visible restoration efforts are taking place in many urban areas. These efforts seek to reduce the intense human impact on ecosystems and enhance the quality of life for city dwellers (Felson et al. 2013; Shwartz et al. 2014). Local citizen groups often welcome the opportunity to work with government agencies and conservation groups to restore degraded urban areas. Unattractive drainage canals in concrete culverts can be replaced with winding streams bordered with large rocks and planted with native

FIGURE 10.5 The Cheonggyecheon Stream in the center of Seoul, Korea, is an example of an urban restoration project that has provided many ecosystem services and improved the quality of life for residents. (© Alex Barlow/Getty Images.)

wetland species. Vacant lots and neglected lands can be replanted with native shrubs, trees, and wildflowers. Gravel pits can be packed with soil and restored as ponds. Establishing native plant species in these urban areas often leads to increases in populations of native birds and insects (Burghardt et al. 2009). These efforts have the additional benefits of fostering neighborhood pride, creating a sense of community, and enhancing property value (**FIGURE 10.5**). However, such restorations are often only partially successful because of their small size and the fact that they are embedded in the highly modified urban environment. Developing urban places where people and biodiversity can coexist has been termed **reconciliation ecology** (Rosenzweig 2003).

An example of the value of restoration projects to people in urban settings is evident in Japan, where parents, teachers, and children in Tokyo and Yokohama have built over 500 small ponds next to schools and in public parks to provide habitat for dragonflies and other native aquatic species (Kobori 2009). The ponds are planted with aquatic plants; many dragonflies colonize them on their own, and some species are carried in as nymphs from other ponds. Dragonflies are an important symbol in Japanese culture, and dragonfly ponds are useful for teaching zoology, ecology, chemistry, and principles of conservation. The schoolchildren are responsible for the regular weeding and maintenance of these "living laboratories," which helps them to feel an ownership of the project and to develop environmental awareness.

Highly visible restoration efforts are taking place in many urban areas to reduce the intense human impact on ecosystems and enhance the quality of life for city dwellers.

Restoring native communities on huge landfills presents one of the most unusual opportunities for urban restoration. In the United States, 150 million tons of trash are buried in over 5000 active landfills each year. When the landfills have reached their maximum capacity, they are usually capped with sheets of plastic and layers of clay to prevent toxic chemicals and pollutants from seeping out. If these sites are left alone, they are often colonized by weedy, exotic species. However, these eyesores can instead be the focus of conservation efforts; planting native shrubs and trees attracts birds and mammals that will bring in and disperse the seeds of a wide range of native species. The Fresh Kills restored landfill site on Staten Island in New York City is a good example of such restoration practices (www.nycgovparks.org/park-features/freshkills-park). It occupies almost 1000 ha and has garbage mounds as tall as the Statue of Liberty, with a volume 25 times that of the Great Pyramid of Giza. The landfill was closed in 2001 and is now undergoing restoration to create a huge public park with many elements of a native ecosystem, a project that is being implemented in six phases to be completed by the year 2036. The eventual goal is to create a large public parkland area (almost three times the size of New York City's Central Park) with abundant wildlife and many recreational, cultural, and educational amenities.

Restoration Using Organisms

Restoration is often limited in geographic scope due to the associated cost, but there is a growing movement to consider ways to repair ecosystems at grand scales, made possible in some cases by reestablishing certain ecosystem dynamics. Introducing animals (and other types of organisms such as bacteria or fungi) can accomplish what would not be logistically or financially possible otherwise. **Rewilding** is a term first introduced by Michael Soulé in the mid-1990s to describe the reintroduction of top carnivores in order to regulate the system from the top down. Since that time, the term rewilding has been used in a variety of contexts, especially attempts to restore aspects of ecosystems that last existed in the Pleistocene era, more than 11,000 years ago. The most famous of these is the reintroduction of wolves in Yellowstone National Park (see Chapter 7). The idea that the restoration of an entire ecosystem can be facilitated by the reintroduction of one or more missing functional groups of animals is an approach that has also been referred to as *trophic rewilding* (Svenning et al. 2015).

> Animals and other organisms can facilitate restoration at scales that might otherwise not be possible.

One place where trophic rewilding has been successful is the Oost-vaardersplassen ("eastward-sailing wetland"), a nature reserve covering about 56 km^2 in a densely populated area just 52 km outside Amsterdam, the capital of the Netherlands (www.staatsbosbeheer.nl/English). The European landscape has arguably been more altered by human activities than any other in the world; no forest, river, grassland, wetland, or almost any other ecosystem of Europe has escaped human influence. But Frans Vera, a

FIGURE 10.6   Reintroduced horses and other large herbivores in the Oost-vaardersplassen helped decrease the dominance of trees through grazing. (Photograph by Richard Primack.)

Dutch government scientist, believed that the reintroduction of large herbivores that had been absent from the landscape for hundreds of years could return the Oostvaardersplassen to a former, more functional state. Because many of these species are now extinct, he introduced modern mammals as ecological surrogates; beginning in the 1980s, he brought in Heck cattle in place of extinct aurochs (wild cattle) and Konik ponies in place of tarpans, the last of Europe's wild horses (**FIGURE 10.6**). Vera also reintroduced red deer, which were among the original herbivores in the area.

The rewilding effort had remarkable effects on the landscape. Populations of horses and deer exploded, grasslands and marshes began to thrive as woody vegetation retreated, and many endangered birds took up residence in the newly opened habitat. In 2013, the carcass of a wolf was found in the Netherlands, marking their appearance for the first time since the nineteenth century. However, the lack of significant predation has led to booming herbivore populations that have no opportunity to expand beyond the isolated reserve to other areas in search of food. Photos and videos of starving animals were shown on television and in other media, and people objected to such cruel "treatment" of animals, even though it was a natural process (*Economist* 2013), leading to new management policies of shooting suffering animals. There has also been public debate over why managers are waiting for top predators to arrive on their own rather than reintroducing them. Future plans for the reserve include developing corridors to connect it to other nature reserves, as a part of a

European network of protected areas, to facilitate the natural expansion of predators and allow for migration of the herbivores. Although public opinion has sometimes been critical, the proximity of the reserve to a metropolitan area with 2.5 million people has enormous educational value, allowing visitors to witness an ecosystem beginning to function as it did thousands of years ago. More modest urban and suburban restoration projects can serve similar purposes, but by European standards, this is a major accomplishment.

Another type of restoration by animals is the release of biological control and bioremediation organisms. Unlike rewilding, these biological introductions do not help restore ecological balance by mimicking historical conditions, but rather are used to remove unwanted elements that were introduced by humans. **Bioremediation** is the use of an organism to clean up pollutants, such as prokaryotes that break down the oil in an oil spill or wetland plants that take up agricultural runoff to clean the water (see the section "Ecosystem services" in Chapter 3), whereas biological control (also known as **biocontrol**) is the use of one type of organism, such as an insect, to manage another, undesirable, species, such as an invasive plant. Historically, a focus on human needs has led to problems in some cases, when the released organism itself became a pest (see Chapter 4). However, these experiences have informed a broader view that considers the whole ecosystem, facilitating the use of both bioremediation and biocontrol in restoration contexts with conservation-oriented goals (Seastedt 2014).

One example of the use of biocontrol for ecological restoration is the release of the tamarisk leaf beetle (*Diorhabda* spp.) along rivers in the western United States. Rivers and their riparian plant communities are frequently the object of restoration efforts, but the geographic scale of the problems often limits what can be done (Gonzalez et al. 2015). The focus of restoration efforts in Texas, New Mexico, Arizona, and other western states has frequently been the removal of exotic tamarisk (*Tamarix* spp.) trees, which now dominate many riparian zones. When it behaves invasively, this species is associated with a host of problems that negatively affect both plants and animals (Sher 2013). Efforts at removing the tree with bulldozers or herbicides risk harm to native species, are difficult to use in remote regions, and are too expensive to implement on a large scale. In response to these problems, more than a decade of research on biological control of the tamarisk eventually led government scientists to release the beetle in the wild in 2003 (Bean et al. 2013). By 2015, the beetle had spread to cover hundreds of miles of rivers, feeding on tamarisk leaves and turning acres of the invasive tree brown. The goal was to facilitate the recovery of native trees and other species, and there is evidence that this is occurring in some locations (**FIGURE 10.7**). However, it may be too early to determine the long-term response of the ecosystem, especially in the context of climate change (Hultine et al. 2015).

Just as the rewilding projects are not without problems, in this case there have been criticisms about unintended effects of the biological control on wildlife, particularly herpetofauna (reptiles and amphibians)

(A) 2006

(B) 2013

(C) 2015

FIGURE 10.7   (A) Tamarisk (salt cedar, *Tamarix*) was introduced from Eurasia in the 1800s and has since formed monoculture thickets along rivers, as shown here in a 2007 photograph of the Colorado River, from Fossil Point near Moab, Utah. The restoration of such inaccessible places is being accomplished with the assistance of a biological control insect, the tamarisk leaf-beetle (*Diorhabda* spp.), which feeds on tamarisk leaves. (B) Stretches of tamarisk trees turned brown after attack by the beetle, as shown in a photograph taken in 2013. (C) This defoliation of the leaves allows the restoration of native species, such as the green willows shown in this photograph taken in 2015, which are colonizing along the riverbank and growing up through the dead branches of the tamarisk tree. (A, Photograph by Anna Sher; B, Photograph by Wayne Ranney; inset, courtesy of Eric Coombs, Oregon Department of Agriculture, Bugwood.org; C, Photograph by Wright Robinson.)

(Bateman et al. 2014) and some species of birds (Sogge et al. 2013). Concern for a federally listed bird that nests in the tamarisk even resulted in litigation by an environmental group against the agency that released the beetle. Several scientists have argued that the overall ecological benefits of reducing the tamarisk are worth such problems (Tamarisk Coalition

> In some cases, restoration may be inadvisable due to economic costs or possible negative impacts on the ecosystem.

2016). This case illustrates the point that the benefits of any restoration effort must always be weighed against perceived, potential, and actual costs (Hinz et al. 2014).

A frequent scientific critique of both rewilding and biocontrol restoration projects is that the practitioners spend too much of their resources in active conservation and not enough on monitoring, researching, or publishing findings. This problem is caused in large part by the limited funding these projects receive; they are often chronically underfunded and rely on volunteers and nonprofit support. The lack of scientific publications generated by the world's handful of rewilding projects in particular may be a reason why they have not gained wider publicity and acceptance. Even though the pace is slow, long-term efforts like these will provide important lessons for restoration efforts elsewhere.

Moving Targets of Restoration

Since ecosystems change over time in response to climate change, plant succession, the varying abundance of common species, and other factors, the goals of restoration may have to be changed over time as well or modified to include temporal dynamics to remain realistic. In many situations in which human activities have drastically altered the environment, some biologists say we will have to accept **novel ecosystems**, in which there is a mixture of native and nonnative species coexisting in a community unlike the original or reference site (Hobbs et al. 2013; Kueffer and Kaiser-Bunbury 2014). These novel ecosystems may differ in species composition and function compared to historical conditions, reflecting the shifting nature of species, the alteration of the environment, and even human values (Harris et al. 2006).

Restoration ecology is increasingly addressing the issue of the moving target, especially as it becomes evident that so many ecosystems simply cannot be restored in the traditional sense (Arthington et al. 2014). The soil chemistry or water availability may be too different or the elimination of an introduced species may be impractical or even undesirable if that species can perform an ecological role similar to that of a missing native species. For example, research on native versus novel forests in Hawaii found that species richness was greater in the novel forest that included both native and exotic species and that total plant biomass and nutrient cycling were either the same or greater there than in the native forest (Mascaro et al. 2012). Even though the native plant species in this system are declining, proponents of novel ecosystems may not consider the novel forests for restoration efforts. In response to such perspectives, other researchers argue that the concept of novel ecosystems may be mistakenly applied when the barriers are social or political rather than ecological, and that projects may use novelty as an excuse to not attempt complete restoration when it would have otherwise been possible (Murcia et al. 2014; Simberloff et al. 2015).

Restoration of Some Major Communities

In addition to rivers, many efforts to restore ecosystems have focused on wetlands, lakes, prairies, and forests. These environments have been severely altered by human activities and are good candidates for restoration work.

Wetlands

Some of the most extensive restoration work has been done on wetlands, including swamps and marshes (Halpern et al. 2007) (**FIGURE 10.8**). Because of wetland protection under the Clean Water Act and the US government policy of no net loss of wetlands, large development projects that damage wetlands must repair them or create new wetlands to compensate for those damaged beyond repair (Robertson 2006). The focus of these efforts has been on re-creating the natural hydrology of the area and then planting native species (Brinson and Eckles 2011). Many successful restoration projects have resulted from this legislation; however, it has fallen short of expectations due to a lack of monitoring and oversight (Clare and Creed 2014). Strategies to restore the biodiversity of rivers include the complete removal of dams and other structures and controlled releases of water from dams (Helfield et al. 2007). For peatlands degraded by harvesting for horticultural peat, a well-recognized restoration approach is the moss layer transfer technique, which consists of spreading native plant material collected from the top 10 cm of natural peatlands over the restored area to facilitate reestablishment (Rochefort and Lode 2006).

Wetland restoration is motivated by more than just a concern for biodiversity, however. The 2005 destruction of New Orleans and other Gulf Coast cities by Hurricane Katrina, and to a lesser extent by Hurricane Rita soon after Katrina, was in part a result of the loss due to the development of the region's wetlands, which had protected the coast from the force of hurricanes. The ensuing natural disaster has become a classic example of the importance of ecosystem services to biological and human communities alike (see Chapter 3). Ironically, the damage that followed these hurricanes had been predicted seven years earlier by the Louisiana Coastal Wetlands Conservation and Restoration Task Force (1998), which had stressed the urgent need for immediate action to restore lost wetlands. Restoration projects have begun, but if they are not adequately funded and large enough in scope, New Orleans will remain vulnerable to another destructive flood.

Experience has shown that efforts to restore wetlands often fail to closely match the species composition or hydrologic characteristics of reference sites. The subtleties of species composition, water movement, and soils, as well as the site history, can be too difficult to match. Often the restored wetlands are dominated by exotic, invasive species. However, the restored wetlands usually do have some of the wetland plant species, or at least similar ones, and can provide some of the functions of the reference sites (Meyer et al. 2010). The restored wetlands also have some of the beneficial

(A) **1973** Before drainage

(B) **2000** After drainage

(C) **2005** After partial restoration

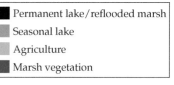

■	Permanent lake/reflooded marsh
▦	Seasonal lake
□	Agriculture
■	Marsh vegetation

Iraq

50 km

FIGURE 10.8 Marsh restoration in Iraq as shown by Landsat images. The images have false color, with marshland in red, agriculture in pink, and wetlands in black and blue. (A) In 1973, marshes covered extensive areas of southern Iraq and were home to about 400,000 Marsh Arabs. Three main marshes are labeled 1, 2, and 3. (B) As shown in the 2000 image, the marshes were drained by the government for political reasons. (C) The reflooding of lakes and wetlands in recent years has resulted in the restoration of some of the marsh vegetation, with major restored areas indicated as A, B, and F. Other letters indicate sampling sites. The new canals are still visible. (From Richardson and Hussain 2006; courtesy of Curtis J. Richardson, Duke University Wetland Center.)

ecosystem characteristics, such as flood control and pollution reduction, and they are often valuable for wildlife habitat. Additional research into restoration methods may result in further improvement.

Aquatic systems

Both freshwater and marine systems of all types are subject to degradation by pollution, overexploitation of resources, global warming, and other factors (see Chapter 4), making them candidates for restoration. Aquatic restoration may deal with problems regarding water chemistry, trophic relationships with exotic species, and physical conditions of the shore or bank. Although aquatic systems are often considered more resilient than terrestrial systems, once damage has become severe, restoration can be more complex.

One of the most common types of damage to lakes and ponds is **cultural eutrophication**, or the accumulation of excess nutrients in the water caused by human activity. Signs of eutrophication include an increased prevalence of algal species (particularly surface scums of blue-green algae), decreased water clarity and oxygen content, fish kills, and an eventual increase in the growth of floating plants and other water weeds (see Figure 4.23). In many lakes, the eutrophication process can be reversed by reducing the amounts of mineral nutrients entering the water through better sewage treatment or by diverting polluted water. One of the most dramatic and expensive examples of lake restoration has been the effort to restore Lake Erie (Sponberg 2009). Lake Erie was the most polluted of the Great Lakes in the 1950s and 1960s, suffering from deteriorating water quality, extensive algal blooms, oxygen depletion in deeper waters, declining indigenous fish populations, and collapsed commercial fisheries. To address this problem, the governments of the United States and Canada have invested billions of dollars since 1972 in wastewater treatment facilities, reducing the annual discharge of phosphorus into the lake from 15,000 tons in the early 1970s to around 2000 tons today (International Joint Commission 2014).

> Lake restorations help to improve water quality and restore the original species composition and community structure.

Many of the issues associated with lake restoration apply equally well to marine ecosystems. These include shorelines, coral reefs, saltmarshes, and mangroves. A number of large-scale projects are restoring estuaries and bays damaged by human activities, including the Chesapeake Bay in the eastern United States (**FIGURE 10.9**). Chesapeake Bay is one of the most important fishing grounds and recreational areas in the United States. However, pollution from residential, agricultural, and industrial lands bordering the bay has caused a dramatic decline in the water quality, which affects all aspects of biodiversity. The economic consequences of this pollution have also been apparent: harvests of fish and shellfish have declined and the water has become unsafe for swimming. This type of general pollution from an entire landscape is referred to as **nonpoint source pollution**, and it requires a comprehensive restoration approach as no single source of the pollution can be readily identified and contained.

(A)

(B)

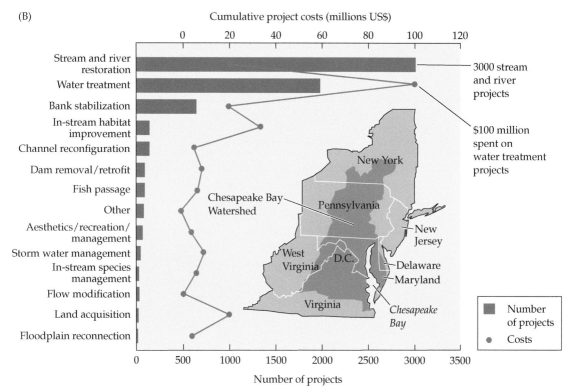

FIGURE 10.9 (A) A variety of measures have been taken to restore the health of the Chesapeake Bay ecosystem. (B) This graph shows the cumulative costs for each type of project and the number of projects. Stream and river restoration is the most common type of project, and the most money has been spent on water treatment projects. The map shows the watersheds that drain into the bay. (A, © Mary F. Calvert/MCT/Getty Images; B, after Hassett et al. 2005.)

In 1987 the federal, state, and local government bodies responsible for the bay signed an agreement to reduce nutrient and sediment loads coming into the bay by 40%, to be achieved mainly through improving the health of streams and watersheds feeding water in. Since that time, over 4700 individual restoration projects have been implemented at a total cost of over $400 million (Stokstad 2009). The largest number of projects involve stream and river restoration, which includes regrading slopes and planting native vegetation. However, the most money has been spent on water treatment projects.

A review of 235 studies of marine restoration projects found that although millions of dollars are being spent on these projects, success has been related more to the type of ecosystem (salt marshes and coral reefs had greatest organism survival rates), site selection, and techniques rather than to the amount of money spent (Bayraktarov et al. 2015). It also found that most projects were short-lived and poorly monitored. For example, a major weakness of the Chesapeake Bay restoration project was that only 5% of these projects have been monitored, and mainly just for vegetation structure, and even fewer have been monitored for water quality to determine whether they have been achieving the original goal of reducing nutrient and sediment loads. This and other projects demonstrate that, while society has accepted the need to restore large aquatic ecosystems, scientists need to do a better job of ensuring that projects deliver the services as promised.

Prairies and farmlands

Because they are species-rich, have many beautiful wildflowers, and can be established within a few years, prairies represent ideal subjects for restoration work (Foster et al. 2009). Many techniques have been used in attempts at prairie restoration, but the basic method involves site preparation by shallow plowing, burning, and raking if prairie species are present, or by eliminating all vegetation by plowing or applying herbicides if only exotics are present. Native plant species are then established by transplanting them in prairie sods obtained elsewhere, planting individuals grown from seed, or scattering prairie seeds collected from the wild or from cultivated plants. The simplest method is to gather hay from a native prairie and spread it on the prepared site. Native species are more likely to become established in the absence of fertilizer, which tends to favor nonnative species. (Of course, reestablishing the full range of plant species, soil structure, and invertebrates could take centuries or might never occur.) The Chicago metropolitan area is particularly well known for such projects; some involve creating prairie grasslands with native prairie species, rather than lawns, in suburban neighborhoods, while others involve converting forests back to prairies (**FIGURE 10.10**).

A conundrum for restoration ecology, illustrated by work in the prairies, is to determine the target state of the ecosystem; humans have had an

FIGURE 10.10 (A) In the late 1930s, members of the Civilian Conservation Corps (one of the organizations created by President Franklin Roosevelt in order to boost employment during the Great Depression) participated in a University of Wisconsin project to restore the wild species of a midwestern prairie. (B) The prairie as it looks today. (A, photograph courtesy of the University of Wisconsin Arboretum and Archives; B, photograph courtesy of Molly Field Murray.)

(A)

(B)

impact on many ecosystems for centuries or even millennia. For example, early humans hunted many North American mammals to extinction more than 12,000 years ago. Should North American grasslands be restored to a state resembling those that existed either before European colonization, a few hundred years ago, or those that existed before human colonization, more than 12,000 years ago? One of the most ambitious proposed restorations involves re-creating a short-grass prairie ecosystem, or "buffalo commons," on about 380,000 km² of the Great Plains states, from the Dakotas to Texas and from Wyoming to Nebraska (Adams 2006). Some of this land is currently used for environmentally damaging and often unprofitable agriculture and grazing supported by government subsidies. The human population of this region is declining as farmers and townspeople go out of business and young people move away. From the ecological, sociological, and economic perspectives, the best long-term use of much of this region might be as a restored prairie ecosystem. The human population of the region could stabilize around nondamaging core industries such as tourism, wildlife management, and low-level grazing by cattle and bison, leaving only the best lands in agriculture. As mentioned in "Restoration Using Organisms" earlier in this chapter, some have argued for a process of North American rewilding whereby large game animals from Africa and Asia, such as elephants, cheetahs, camels, and even lions, would be released in an attempt to re-create the types of ecological interactions that occurred in North America before humans arrived on the continent (Hayward 2009). Both these proposed projects are controversial because many of the farmers and ranchers in the region want to continue their present way of life without alteration, and they tend to be highly resentful of unwanted advice and interference from scientists and the government. The projects are also controversial because of the proposed release of nonnative mammals in North American ecosystems.

Tropical dry forest in Costa Rica

An exciting restoration process has been ongoing since 1985 in northwestern Costa Rica. The tropical dry forests of Central America have long suffered from large-scale conversion to cattle ranches and farms. This destruction has gone largely unnoticed, as international scientific and public attention has focused on more glamorous rain forests elsewhere in South and Central America (e.g., see Chapters 6 and 11). The American ecologists Daniel Janzen and Winnie Hallwachs have been working with Costa Rica's Ministry of the Environment, the resident staff, and the Guanacaste Dry Forest Conservation Fund to restore the biology and cultural connectivity of 130,000 ha of land and 43,000 ha of overfished marine habitat in Area de Conservación Guanacaste (ACG) (Ehrlich and Pringle 2008; www.acguanacaste.ac.cr) (**FIGURE 10.11**).

The restoration of these marginal ranchlands, low-quality farms, and forest fragments includes eliminating brush fires started by people, banning logging and hunting, and occasionally planting both native and exotic trees to shade out introduced African grasses. Light livestock grazing initially reduced grass volume and then was phased out as the forest returned through natural animal- and wind-borne seed dispersal. In 29 years, this process has converted tens of thousands of hectares of pastures and old fields to a species-rich, dense young forest, with abundant and growing populations of native animals. Nonetheless, the area will require an estimated 200–500 years to regain the original forest structure.

A key element in the restoration plan is what has been termed **biocultural restoration**, meaning that ACG staff members teach basic biology and ecology on-site to 2500 students in grades 4 through 6 from 53

(A)

(B)

(C)

FIGURE 10.11 The Area de Conservación Guanacaste (ACG) is an experiment in restoration ecology—an attempt to restore the devastated and fragmented tropical dry forest of Costa Rica. (A) A barren grassland with scattered forest fragments was heavily grazed by cattle and frequently burned. (B) Native trees and other species became established once again in this young forest after 17 years without cattle and fire. Note the person in the lower left for scale. (C) Daniel Janzen, an ecologist from the United States, is a driving force behind the restoration project in Guanacaste. Here he explains a land purchase deal to the board of directors of the Guanacaste Dry Forest Conservation Fund. (Photographs courtesy of Brad Zlotnick.)

neighboring schools and also give presentations to citizen groups, all as part of the ACG core mission. This effort, combined with the fact that all 95 members of ACG's staff and administration are resident Costa Ricans, has resulted in residents viewing ACG as if it were a large ranch producing "wildland resources" for the community rather than an exclusionary "national park." Janzen (quoted in Allen 1988) believes that, in rural areas such as Guanacaste Province, providing an opportunity for learning about and from nature can be one of the most valuable functions of conserved wildlands:

> *The goal of biocultural restoration is to give back to people the understanding of the natural history around them that their grandparents had. These people are now just as culturally deprived as if they could no longer read, hear music, or see color.*

This restoration effort has accomplished so much and has attracted widespread attention in part because a highly articulate individual—Janzen—commits most of his time and resources to a cause in which he passionately believes (Laurance 2008a). Janzen's enthusiasm and vision have inspired many other people to join his cause, and he is a classic example of how one individual can be a potent force for conservation.

The Future of Restoration Ecology

E.O. Wilson predicted in 1992, "Here is the means to end the great extinction spasm. The next century will, I believe be the era of restoration in ecology." He may be right. Restoration ecology is an evolving and rapidly growing discipline, with its own scientific society (the Society for Ecological Restoration) and increasing numbers of journals: *Restoration Ecology, Ecological Restoration, The Journal of Ecosystem Restoration, Ecological Management and Restoration*, and others. Ecosystems are being restored using methods developed by the discipline, books are being written about the subject, and new courses are being taught at more universities. Scientists are increasingly able to make use of the growing range of published studies and suggest improvements in restoration techniques. At its best, restored land can provide new opportunities for protecting biodiversity and generating an appreciation of nature.

Conservation biologists in this field must take care to ensure that restoration efforts are not simply public relations endeavors taken by environmentally damaging corporations that are only interested in continuing business as usual (Maron et al. 2012). A 5 ha "demonstration" or "best practices" project in a highly visible location does not compensate for thousands or tens of thousands of hectares damaged elsewhere and should not be accepted as adequate by conservation biologists. Attempts to mitigate or offset the destruction of an intact biological community by the building of a similar species assemblage at a new location is almost certainly not going to provide a home for the same species or provide similar ecosystem functions (Moreno-Mateos et al. 2015a).

> Ecological restoration is an important and growing tool for conservation, but the protection of existing biodiversity remains the first priority.

The best long-term strategy remains protecting and managing biological communities where they are found naturally; many researchers argue that the requirements for the long-term survival of all species are most likely to be found there, and the protection of intact systems is generally cheaper and easier than repairing systems that have been degraded. In addition, we need to consider restoring ecosystems in anticipation of the impacts of climate change (see Chapter 8). There are many technical, scientific, logistical, and economic challenges for future restoration ecologists and managers to address in our pursuit to repair damage we have done as a species.

Summary

- Ecological restoration is the practice of reestablishing populations, ecosystems, and landscapes that include degraded, damaged, or even destroyed habitat. Restoration ecology provides methods for reestablishing species, whole biological communities, and ecosystem functions in degraded habitat.

- The establishment of new communities on degraded or abandoned sites provides an opportunity to enhance biodiversity and can improve the quality of life for the people living in the area. Restoration ecology can also provide insight into community ecology by testing our ability to reassemble a biological community from its native species.

- Restoration projects begin by eliminating or neutralizing factors that prevent the system from recovering. Then some combination of site preparation, habitat management, and reintroduction of original species gradually allows the community to regain the species and ecosystem characteristics of designated reference sites. Attempts to restore habitat need to be monitored to determine whether they are reestablishing the composition of historical species and the functions of the ecosystem.

- Biological control and bioremediation are tools whereby organisms such as insects or protists can be used to remove invasive species and pollutants (respectively).

- In some cases, restoration to a former state is impractical or impossible due to the nature or extent of the degradation, the presence of invasive species, or climate change. In such cases, a novel ecosystem that may have some of the same functionality of the original one may be considered, but it should not be valued over the native ecosystem.

- Creating new habitat to replace lost habitat elsewhere, which is known as compensatory mitigation or biological offsetting, has value but should be regarded as only part of an overall conservation strategy that includes the protection of species and ecosystems where they naturally occur.

For Discussion ▪

1. Restoration ecologists are improving their ability to restore biological communities. Does this mean that biological communities can be moved around the landscape and positioned in convenient places that do not inhibit the further expansion of human activities?

2. What methods and techniques could you use to monitor and evaluate the success of a restoration project? What timescale would you suggest using?

3. What do you think are some of the easiest ecosystems to restore? The most difficult? Why?

4. Aldo Leopold encouraged humans to "keep every cog and wheel" in order to maintain healthy ecosystems. Is it necessary, or even possible, to return every missing species back into a restored ecosystem?

Suggested Readings

Cole, I. A., S. M. Prober, I. D. Lunt, and T. B. Koen. 2016. A plant traits approach to managing legacy species during restoration transitions in temperate eucalypt woodlands. *Restoration Ecology*. doi: 10.1111/rec.12334. It is important to preserve desirable species while removing undesirable ones.

Corlett, R. T. 2016. Restoration, reintroduction and rewilding in a changing world. *Trends in Ecology and Evolution*. doi: 10.1016/j.tree.2016.02.017.

Dodds, W. K., K. C. Wilson, R. L. Rehmeier, G. L. Knight, S. Wiggam, J. A. Falke, and 2 others. 2008. Comparing ecosystem goods and services provided by restored and native lands. *BioScience* 58: 837–845. Within 10 years of restoration, restored ecosystems provide 31%–93% of the benefits of native lands.

Felson, A. J., M. A. Bradford, and T. M. Terway. 2013. Promoting Earth stewardship through urban design experiments. *Frontiers in Ecology and the Environment* 11: 362–367. Cities can be restored to reduce environmental impacts and create a healthier environment for people.

Fraser, L. H., W. L. Harrower, H. W. Garris, S. Davidson, P. D. N. Hebert, R. Howie, and 8 others. 2015. A call for applying trophic structure in ecological restoration. *Restoration Ecology* 23: 503–507. The traditional focus of restoration ecology on vegetation must be challenged.

Galatowitsch, S. M. *Ecological Restoration*. 2012. Sinauer Associates, Sunderland, MA. A comprehensive overview of the strategies being used around the world to reverse human impacts to landscapes, ecosystems, and species.

González E., A. A. Sher, E. Tabacchi, A. Masip, and M. Poulin. 2015. Restoration of riparian vegetation: A global review of implementation and evaluation approaches in the international, peer-reviewed literature. *Journal of Environmental Management* 158: 85–94. The restoration of riverbank communities can have a wide range of goals and utilize a variety of methodological approaches.

Handel, S. N. 2016. Greens and greening: Agriculture and restoration ecology in the city. *Ecological Restoration* 34(1): 1–2. Urban restoration provides many benefits to both wildlife and people.

Keenelyside, K., N. Dudley, S. Cairns, C. Hall, and S. Stolton. 2012. *Ecological restoration for protected areas:principles, guidelines and best practices* (Vol. 18). IUCN. A best practices guide based on input from restoration ecologists from more than a dozen countries.

Kueffer, C. and C. N. K. Kaiser-Bunbury. 2014. Reconciling conflicting perspectives for biodiversity conservation in the Anthropocene. *Frontiers in Ecology and the Environment* 12: 131–137. Restored landscapes and novel ecosystems with mixtures of native and nonnative species might be best suited to survive increasing human impacts.

Oppenheimer, J. D., S. K. Beaugh, J. A. Knudson, P. Mueller, N. Grant-Hoffman, A. Clements, and M. Wight. 2015. A collaborative model for large-scale riparian restoration in the western United States. *Restoration Ecology* 23(2): 143–148. Federal agencies and nonprofit organizations can work together to facilitate restoration.

11

The Challenges of Sustainable Development

Growing plants on rooftops (green roofs) helps pollinators, absorbs CO_2, and can even provide habitat in what is otherwise an urban desert. This rooftop garden in Japan provides vegetables and fruits for local residents.

Efforts to preserve biological diversity sometimes conflict with both real and perceived human needs (**FIGURE 11.1**). Increasingly, many conservation biologists, policymakers, and land managers are recognizing the need for sustainable development—economic development that satisfies both present and future needs for resources and employment while minimizing the impact on biodiversity and functioning ecosystems (Selomane et al. 2015). Sustainable development, a term sometimes used interchangeably with *sustainability*, can be contrasted with more typical development that is *unsustainable*. Unsustainable development cannot continue indefinitely because it destroys or uses up the resources on which it depends. As defined by some environmental economists, **economic development** implies improvements in the efficiency, organization, and distribution of resource use or other economic activity but not necessarily increases in resource consumption. Economic development is clearly distinguished from **economic growth**, which is defined as material increases in the amount of resources used. Sustainable development is a useful and important concept in conservation biology because it emphasizes improving current economic development and limiting unsustainable economic growth.

FIGURE 11.1 Sustainable development seeks to address the conflict that exists between development to meet human needs and the preservation of the natural world. (Top photograph © FloridaStock/Shutterstock; bottom photograph © kavram/Shutterstock.)

> The goal of sustainable economic development is to provide for the current and future needs of human society while at the same time protecting species, ecosystems, and other aspects of biodiversity.

Investment in national park infrastructure that improves the protection of biodiversity and provides revenue opportunities for local communities is an example of movement toward sustainable development, as is the implementation of less destructive logging and fishing practices. Especially in developing countries, sustainability is tightly intertwined with issues of *social justice*; one cannot have sustainable development without ensuring that all parties benefit. This is true in part because environmental degradation disproportionately affects the poor. It has been argued that the combination of environmental, economic, and social considerations explicit in sustainable development will inevitably lead to improved quality of life, corporate profits, and human health in addition to environmental benefits (Dernbach and Cheever 2015).

Unfortunately, the term *sustainable development* has become overused and is often misappropriated. Few politicians or businesses are willing to proclaim themselves to be against sustainable development. Thus, many large corporations, and the policy organizations that they fund, misuse the notion of sustainable development to "greenwash" their industrial activities, with only limited change in actual practice. In these cases, both ecosystems and people will likely suffer.

For instance, a plan to establish a huge mining complex in the middle of a forest wilderness cannot justifiably be called sustainable development simply because a small percentage of the land area is set aside as a park. Not only will precious habitat be lost, but local inhabitants who depend on resources from that forest will be impacted as well. Waste from the mine can poison fish and people alike. Similarly, building huge houses filled with "energy-efficient" appliances and cars that boast the latest energy-saving technology but are routinely driven long distances cannot really be called sustainable development or "green technology" when the net result is increased energy use. Alternatively, some people champion the opposite extreme, claiming that sustainable development means that vast areas must be kept off-limits to all development and should remain as, or be allowed to return to, wilderness. This may not be the best option for either conservation goals or people: some places require active restoration (see Chapter 10) or active management (see Chapters 8 and 9) to best protect biodiversity, and barring all people will inevitably harm local interests.

The primary conflict of sustainable development is often not between people and nature so much as it is between the powerful and the vulnerable. As with all such disputes, informed scientists and citizens must study the issues carefully, identify which groups are advocating which positions and why, and then make careful decisions that best meet both the needs of human society and the protection of biodiversity and ecosystems. In many cases this involves compromise, and in most cases compromises form the basis of government policy and laws, with conflicts resolved by government agencies and the courts.

Sustainable Development at the Local Level

Most efforts to find approaches that promote both the preservation of species and habitats and the needs of society rely on initiatives from concerned citizens, conservation organizations, and government officials. These efforts may take many forms, but they begin with individual and group decisions to prevent the destruction of habitats and species in order to preserve things of perceived economic, cultural, biological, scientific, or recreational value. The results of these initiatives often end up codified into environmental regulations or laws.

Local and regional conservation regulations

In modern societies, local (city and town) and regional (county, state, and provincial) governments pass laws to provide protection for species and habitats while at the same time allowing development for the continued needs of society. Often, but not always, these local and regional laws are comparable to, or stricter than, national laws, particularly for protections of clean water and air, and less often for endangered species. Such laws are passed because citizens and political leaders feel that they represent the will of the majority and provide long-term benefits to society. The most prominent of these laws govern when and where hunting and fishing can occur; the size, number, and species of animals that can be taken; and the types of weapons, traps, and other equipment that can be used. Restrictions are enforced through licensing requirements and patrols by game wardens and police. In some settled and protected areas, hunting and fishing are banned entirely. Similar laws affect the harvesting of plants, seaweed, and shellfish. Certification of the origin of biological products may be required to ensure that wild populations are not depleted by illegal collection or harvest. These restrictions have long applied to certain animals such as trout and deer and to plants of horticultural interest such as orchids, azaleas, and cacti. More recently, there are certification programs for the origin of ornamental fish, timber, and other products.

Laws that control the ways in which land is used are another means of protecting biodiversity (Reed et al. 2014). For example, on a more local scale, vehicles and even people on foot may be restricted from habitats and resources that are sensitive to damage, such as birds' nesting areas, bogs, sand dunes, wildflower patches, and sources of drinking water. Uncontrolled fires may severely damage habitats, so practices, such as campfires, that contribute to accidental fires are often rigidly controlled. Zoning laws, among the strongest and most widely used restrictions, sometimes prevent construction in sensitive areas such as barrier beaches and floodplains. Even where development is permitted, building permits are often reviewed carefully to ensure that damage is not done to endangered species or ecosystems, particularly wetlands. For major regional and national projects, such as dams, canals, mining and smelting operations, oil extraction, and highway construction, environmental impact statements must be prepared that describe the damage that such projects will or could possibly cause so that these projects can be conducted in a more environmentally sensitive manner.

One of the most powerful strategies in protecting biodiversity at the local and regional levels is the designation of intact biological communities as nature reserves, conservation land, and state and provincial parks and forests (see Chapter 8). Government bodies buy land and establish protected areas for various uses—local parks for recreation, conservation areas to maintain biodiversity, forests for timber production and other uses, and watersheds to protect water supplies.

The passage and enforcement of conservation-related laws on a local level can become an emotional experience that divides a community and can even lead to violence. To avoid such counterproductive outcomes, conservationists must be able to convince the public that using resources in a thoughtful and sustainable manner creates the greatest long-term benefit for the community. The general public must be made to look beyond the immediate benefits that can come with the rapid and destructive exploitation of resources. For example, towns often need to restrict development in watershed areas to protect water supplies; this may mean that houses and businesses are not built in these sensitive areas and landowners may have to be compensated for these lost opportunities. It is essential that conservation biologists clearly communicate the reasons for restrictions that protect biodiversity and ecosystems. Those affected by the restrictions can become allies in the protection of resources if they understand the importance and long-term benefits of reduced access. These people must be kept informed and consulted throughout the decision-making process. Conservation biologists must develop the necessary skills, combined with the best science, to negotiate and compromise; to encourage conservation actions in others and understand their perspectives; and to explain positions, regulations, and restrictions. These skills are part of the growing field of conservation psychology (Clayton and Myers 2015; Van Vugt 2009). Having a fervent belief in one's cause is no longer enough.

Land trusts and related strategies

In many countries, nonprofit, private conservation organizations are among the leaders in acquiring land for conservation (Bode et al. 2011). In the Netherlands, about half the protected areas are privately owned. In the United States, over 15 million ha of land are protected at the local level by about 1700 **land trusts**, which are private, nonprofit corporations established to protect land and natural resources (www.landtrustalliance.org). At a national level, major organizations such as The Nature Conservancy and the National Audubon Society have protected an additional 10 million ha in the United States (**FIGURE 11.2**). While the purchase of land may seem the most straightforward approach to conservation, in practice, property law can be quite complex and may differ considerably between countries.

> Land trusts are private conservation organizations that purchase and protect land. Conservation easements and limited development agreements are also used by land trusts to increase the amount of land under protection.

Land trusts are common in Europe. In Britain, the National Trust has more than 3.8 million members and 62,000 volunteers and owns about 250,000 ha of land, much of it farmland, including 57 National Nature Reserves, 466 Sites of Special Scientific Interest, 355 Areas of Outstanding Natural Beauty, and 40,000 archaeological sites. The Royal Society for the Protection of Birds has more than 1 million members and manages 200 reserves with an area of almost 130,000 ha. A major emphasis of many of these reserves is nature conservation and education, often linked to school programs. These private reserve networks are collectively referred to as Conservation, Amenity, and Recreation Trusts (CARTs), a name that reflects their varied objectives.

In addition to purchasing land outright, both governments and conservation organizations protect land through conservation easements (CEs),

FIGURE 11.2 Land trusts may own and manage land or may give it to local or national governments in special agreements. Here, the chief operating officer of The Nature Conservancy signs over 10 acres to establish the Everglades Headwaters National Wildlife Refuge. (Photograph by Tom MacKenzie, USFWS, Jan. 18, 2012, at the FFA training facility in Haines City, FL, about 50 miles south of Orlando.)

also called *conservation covenants*, in which landowners give up the right to develop, build on, or subdivide their property, typically in exchange for a sum of money, lower real estate taxes, or some other tax benefit (Farmer et al. 2011; see Chapter 9). CEs can have a variety of goals that may or may not explicitly include protection of species and/or habitats, which have important implications for how they are managed and the resulting effectiveness for conservation (**FIGURE 11.3**). For many landowners, accepting a conservation easement is an attractive option: they receive a financial gain while still owning their land and are able to feel that they are assisting conservation objectives. In general, landowners are most willing to consider CEs when they are well paid; the agreement is short term, lasting just a few years rather than permanent; and there is no legal contract involved (Sorice et al. 2013). Of course, the offer of lower taxes or money is not always necessary; many landowners will voluntarily accept conservation restrictions without compensation.

Another strategy that land trusts and local and regional governments use is **limited development**, also known as **conservation development** (Milder et al. 2008). In these situations, a landowner, a property developer, and a government agency and/or conservation organization reach

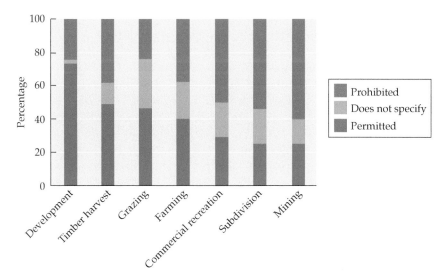

FIGURE 11.3 Conservation easements are intended to protect land that might otherwise be developed; however, a large proportion of them do allow some forms of development and other uses. These figures reflect a review of 269 CE documents that spanned six US states. When activities were not specified, they were likely to be allowed by default. (After Rissman et al. 2015.)

a compromise that allows part of the land to be commercially developed while the remainder is protected by a conservation easement. Limited development allows the construction of necessary buildings and other infrastructure for an expanding human society; the projects are often successful precisely because being located adjacent to conservation land enhances the value of the developed land.

Governments and conservation organizations can further encourage conservation on private lands through other mechanisms, including compensating private landowners for desisting from some activity that damages the environment and for implementing conservation activity (Knoot et al. 2015). **Conservation leasing** involves providing payments to private landowners who actively manage their land for biodiversity protection. Tax deductions and payments can also be obtained for any costs of restoration or management, including weeding, controlled burning, establishing nest holes, and planting native species. In some cases, private landowners may still be allowed to develop their land later, even if endangered species come to live there.

A related idea is **conservation banking**, in which a landowner deliberately preserves an endangered species or a protected habitat type such as wetlands, or even restores degraded habitat and creates new habitat. A conservation bank is like a financial bank in that it is intended to be a stable protector, but instead of money, it protects natural resource values. The US Fish and Wildlife Service (USFWS) evaluates proposals for new conservation banks "in the context of unavoidable impacts of proposed projects to listed species" (USFWS 2003). Once approved, the USFWS awards a landowner habitat or species credits in exchange for permanently protecting the land and managing it for these species. These credits can then be purchased by developers in compensation or as a biodiversity offset for a similar habitat that is being destroyed elsewhere by a construction project (Pilgrim et al. 2013). For example, the Muddy Boggy Conservation Bank in eastern Oklahoma was expanded to nearly 230 ha in exchange for 522 credits for the endangered American burying beetle (*Nicrophorus americanus*) (BusinessWire 2015). This conservation bank was specifically established to assist Oklahoma industries such as energy development, pipeline construction, and transportation projects that might have otherwise been shut down if the endangered species was found on the land. The industry can be relieved of the responsibility for how their actions may affect the beetle by purchasing credits that are then used to support the Muddy Boggy Conservation Bank's ability to protect the beetle there.

A related approach is the payments for ecosystem services (PES) program (see Chapter 9), in which a landowner is paid for providing specific conservation services (**FIGURE 11.4**) (Naeem et al. 2015). Utility companies may also gain carbon credits by paying for habitat protection (e.g., paying a landowner for not cutting down a forest) and restoration (e.g., paying a landowner for planting trees and establishing a new forest); these carbon credits are then used to offset the carbon emissions produced through the burning of fossil fuels. On a larger scale, carbon offset payments by govern-

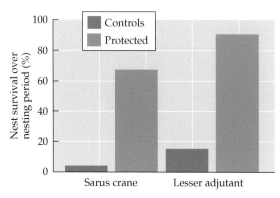

FIGURE 11.4 Fledgling success for nests of the Sarus crane (*Grus antigone*) and lesser adjutant (*Leptoptilos javanicus*) that are protected by villagers in the northern plains of Cambodia versus those that were not. Villagers participating in the program were paid by the Wildlife Conservation Society, an international conservation organization based in the United States. The local inhabitants were able to significantly supplement their incomes with payments that provided extra incentives for achieving successful nests. (After Clements et al. 2013. Crane photograph © ScratchArt/Shutterstock; adjutant photograph © Arco Images GmbH/Alamy.)

ments and international corporations can be used to compensate for emissions of greenhouse gases (Kiesecker et al. 2009).

Conservation concessions are an innovative approach in which conservation organizations outbid extractive industries such as logging companies, not for ownership of the land, but for the rights to use and protect it. The government or large landowner receives the same annual income from a conservation organization that would have been paid by a logging company, and the animals and plants of the area are protected rather than destroyed.

> Conservation lands require ongoing management and continuous vigilance, but they often provide extensive benefits to the local society and economy.

Enforcement and public benefits

The conservation measures described in this section and elsewhere in this book must be continuously monitored to make sure that regulations and laws are enforced and that agreements are being carried out, particularly in cases where destruction cannot be easily reversed (Rissman and Butsic 2011). In one scenario, a developer may agree to limit the amount of development and conserve an area of forest but then obtain construction permits, ignore the agreement, and clear all the trees. By the time action can be taken to stop the developer, the trees and the habitat they provided are gone and cannot be easily replaced. Even if sanctions such as fines or forfeiting of bonds are imposed, the developer may feel that the potential profits outweigh such considerations, and managers and officials usually take a "what's done is done" approach and allow the cleared land to be developed. Conservation workers need to raise awareness so that the public and the judicial system view "breach of promise" against the environment with the same seriousness as similar crimes against personal property.

The most powerful tool for enforcement has been fining violators, especially when these funds are directly applied to conservation, such as the restoration of damaged land. When the terms of a conservation easement are broken or other environmental harm is done, there may be a *restoration remedy*; that is, the violator may be required to pay to return the degraded property to as much of its original state as possible. For example, if a land trust is obligated to preserve a stand of trees under the terms of a conservation easement but then allows those trees to be cut down, the land trust may be obligated to replant trees and nurture them to maturity at a cost that well exceeds that of the land itself.

Public perception can also be a source of problems. Local efforts by land trusts to protect land are sometimes criticized as being elitist because they provide tax breaks only to individuals wealthy enough to take advantage of them while decreasing the revenue collected from land and property taxes. Other analysts argue that land used in other ways, such as for agriculture or commercial activity, is more productive. Although land in trust may initially yield lower tax revenues, the loss is often offset by the increased value and consequent increased property taxes of houses and land adjacent to the conservation area. In addition, the employment, recreational activities, tourist spending, and research projects associated with nature reserves and other protected areas generate revenue throughout the local economy, which benefits local residents (Di Minin et al. 2013). Finally, by preserving important features of the landscape and natural communities, local nature reserves also protect and enhance the cultural heritage of the local society, a consideration that must be valued if sustainable development is to be achieved.

Conservation at the National Level

Throughout the modern world, national governments play a leading role in conservation activities. The level of a government's conservation actions can substantially affect the conservation outcomes within its borders. Unintentionally triggering conflict with local government during conservation implementation and lack of government funding are two significant problems faced by parties actively working to reestablish endangered species (Crees et al. 2016). Conservation biologists contribute to these efforts by providing government officials with key information on threats to biodiversity, with the hope and expectation that resulting laws and regulations will be used to protect biodiversity (**FIGURE 11.5**).

Similar to local and regional governments, national governments can use their revenues and authority to buy land for conservation. In the United States, special funding mechanisms exist at the national level, such as the Lands Legacy Initiative and the Land and Water Conservation Fund, to purchase land for conservation purposes.

The establishment of national parks is a particularly important conservation strategy. National parks are the single largest source of protected lands in many countries. For example, Costa Rica's national parks protect about 62,000 ha, or about 12% of the nation's land area (www. costarica-nationalparks.com). Outside the protected areas, deforestation is occurring rapidly, and soon national parks may represent the only undisturbed habitat and source of natural products, such as timber, in the whole country. The US National Park Service protects about 34 million ha with 410 sites. The United States government also protects biodiversity in more than 598 National Wildlife Refuges covering 61 million ha, and the U.S. Bureau of Land Management's National Landscape Conservation System has 873 sites covering 13 million ha, including National Conservation Areas, National Monuments, and Wilderness Areas.

FIGURE 11.5 A field biologist takes a US congressman into a woodland to explain how habitat fragmentation and the loss of biodiversity contribute to increasing rodent populations and the rising incidence of Lyme disease. (Photograph by J. Halpern.)

National legislatures and governing agencies are the principal bodies for developing policies that regulate environmental pollution. Laws are passed by legislatures and then implemented in the form of regulations imposed by government agencies. Laws and regulations affecting air emissions, sewage treatment, waste dumping, and the development of wetlands are often enacted to protect human health and property and resources such as drinking water, forests, and commercial and sport fisheries. The level of enforcement of these laws demonstrates a nation's determination to protect the health of its citizens and the integrity of its natural resources. At the same time, these laws protect biological communities that would otherwise be destroyed by pollution and other human activities. The air pollution that exacerbates human respiratory disease, for instance, also damages commercial forests and biological communities, and pollution that ruins drinking water also kills terrestrial and aquatic species such as turtles, fish, and aquatic plants.

National governments can also have a substantial effect on the protection of biodiversity through the control of their borders, ports, industry, and commerce. To protect forests and regulate their use, governments can ban logging, as was done in Thailand following disastrous flooding; they can restrict the export of logs, as was done in Indonesia; and they can penalize timber companies that damage the environment. Certain kinds of environmentally destructive mining can be banned. Methods of shipping oil and toxic chemicals can be regulated. Conservation biologists can provide government officials with key information for developing the needed policy framework, and resource managers and others can then use the resulting laws and regulations to protect biodiversity.

> National governments protect designated endangered species within their borders, establish national parks, and enforce legislation on environmental protection.

Despite the fact that many countries have enacted legislation to protect endangered species, forests, wetlands, and other aspects of biodiversity, it is also true that national governments are sometimes unresponsive to requests from conservation groups to protect the environment. Governments have even acted to remove the legal protected status of national parks, sacred forests, and other conservation areas (degazettement; see Chapter 8), in order to facilitate the extraction of natural resources and economic development (Hardy et al. 2016). Governments sometimes do this because they feel that the needs of the broader regional and national society for natural resources and economic development are more important than the needs of the local people, the environment, or the ecosystem services they may forgo. There are also cases where the downgrading, downsizing, and degazettement of protected areas are linked to certain government officials who profit personally from their actions. In some cases, national governments recognize that local people are best able to protect ecosystems close to where they live and have relinquished control of these resources to local governments, village councils, and conservation organizations.

International Approaches to Sustainable Development

International cooperation and agreements to protect biodiversity are needed for migratory species and for occasions when threats occur across countries.

The biological diversity needed for humanity's future well-being is concentrated in the tropical countries of the developing world, most of which are relatively poor and experiencing rapid rates of population growth, development, and habitat destruction (see Figure 6.20). Developing countries may be willing to preserve biodiversity, but they are often unable to pay for the habitat preservation, research, and management required for the task. The developed countries of the world (including the United States, Canada, Japan, Australia, and many European nations) must work together with tropical countries to preserve the biodiversity needed by the world as a whole.

While the major legal and policing mechanisms that presently exist in the world are based within individual countries, international cooperation to protect biodiversity is an absolute requirement for several reasons:

- *Species migrate across international borders.* Conservation efforts must protect species at all points in their ranges; efforts in one country will be ineffective if critical habitats are destroyed in a second country to which an animal migrates (Ripple et al. 2014). For example, efforts to protect migratory bird species in northern Europe will not work if the birds' overwintering habitat in Africa is destroyed.

- *International trade in biological products is commonplace.* A strong demand for a product in one country can result in the overexploitation of the species in another country to supply this demand.

- *Biodiversity provides internationally important benefits.* The community of nations benefits from the species and genetic variation used in agriculture, medicine, and industry; the ecosystems that help regulate climate; and the national parks and other protected areas of international scientific and tourist value. These benefits have been estimated to be in the trillions of dollars (McCarthy et al. 2012).

- *Many environmental pollution problems that threaten ecosystems are international in scope.* Such threats include atmospheric pollution and acid rain; the pollution of lakes, rivers, and oceans; greenhouse gas production and global climate change; and ozone depletion (Lin et al. 2014).

International Earth summits

Given these realities, there is motivation to make progress on conservation issues by bringing together leaders at international meetings. There have been six of these meetings that consider environmental issues in the context of sustainable development:

- UN Conference on the Human Environment (1972)
- World Commission on Environment and Development (1987)

- UN Conference on Environment and Development (1992)
- General Assembly Special Session on the Environment (1997)
- World Summit on Sustainable Development (2002)
- UN Conference on Sustainable Development (2012)
- Paris Climate Conference (2015)

Several of these conferences have resulted in significant international agreements, often termed "conventions," that provide frameworks for countries to cooperate in protecting species, habitats, ecosystem processes, and genetic variation. Treaties that are negotiated at international conferences come into force when they are ratified by a certain number of countries and then implemented and enforced at the national level (**FIGURE 11.6**). Those that specifically protect species, such as the Convention on International Trade in Endangered Species (CITES), were discussed in Chapter 6. Next we elaborate on the most important conferences and other international agreements that impact habitat use more generally, especially in the context of sustainable development.

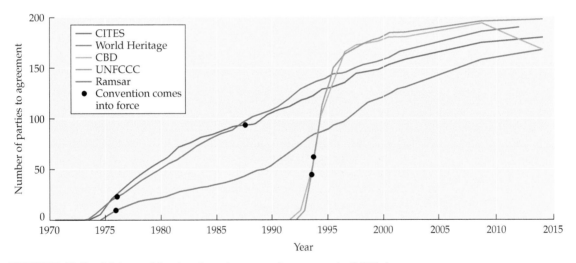

FIGURE 11.6 Major multinational environmental agreements (MEAs) are negotiated and then ratified by the governments of individual countries, which become "parties," or participants, in the provisions of the agreements or treaties. A treaty comes into force (i.e., countries begin to follow the provisions of the treaty) when it has been signed by a certain number of countries (indicated by a dot). The plot lines show the numbers of countries that have ratified various treaties that provide for biodiversity protection by protecting habitat (the Ramsar Convention on Wetlands of International Importance, the World Heritage Convention/WHC concerning the Protection of the World Cultural and Natural Heritage), species (the Convention on International Trade in Endangered Species/CITES, the Convention on Biological Diversity/CBD), and the environment (the United Nations Framework Convention on Climate Change/UNFCCC). (After WRI 2003, with updates from MEA websites.)

The United Nations Conference on Environment and Development (UNCED), held for 12 days in June 1992 in Rio de Janeiro, Brazil, was one of the most significant steps toward adopting a global approach to sound environmental management. Known unofficially as the **Earth Summit** or the **Rio Summit**, the conference brought together representatives from 178 countries, including heads of state, leaders of the United Nations, and individuals from major conservation organizations and other groups representing religions and indigenous peoples. Their purpose was to discuss ways of combining increased protection of the environment with sustainable economic development in less-wealthy countries (United Nations 1993a,b).

In addition to initiating many new projects, conference participants discussed, and most countries eventually signed, four major documents:

1. *The Rio Declaration*. This nonbinding declaration provides general principles to guide the actions of both wealthy and poor nations on issues of the environment and development. The right of nations to use their own resources for economic and social development is recognized, as long as the environments of other nations are not harmed in the process. The declaration affirms the "polluter pays" principle, in which companies and governments take financial responsibility for the environmental damage that they cause.

2. *The United Nations Framework Convention on Climate Change (UNFCCC)*. Almost universally ratified (194 signatories), this agreement requires industrialized countries to reduce their emissions of carbon dioxide (CO_2) and other greenhouse gases and to make regular reports to the United Nations on their progress. While specific emission limits were not decided on, the convention states that greenhouse gases should be stabilized at levels that will not interfere with the Earth's climate.

3. *The Convention on Biological Diversity (CBD)*. This convention has three objectives: (1) protecting the various components of biodiversity, (2) using the components sustainably, and (3) sharing the benefits of new products that are made with the genetic resources of wild and domestic species (www.cbd.int). Developing international laws for intellectual property rights that fairly share the financial benefits of biodiversity among countries, biotechnology companies, and local people is proving to be a major challenge to the convention. Because of concerns about the use or misuse of biological materials, certain developing countries have established highly restrictive procedures for granting permits to scientists who want to collect biological samples for their research (Watanabe 2015). In other cases, new research facilities have been built in developing countries and local people have been trained in scientific procedures so that biological samples do not have to be exported. In 2010, the participants in the CBD developed a list of goals to achieve sustainability, called the **Aichi Biodiversity Targets** (www.cbd.int/sp/targets). The main goal was to slow or stop the loss of biodiversity by reducing the impact of human activities. This was to be achieved in

part by changing governmental policies and increasing the percentage of land and ocean under protection; targets are far from being met (Butchart et al. 2015) (**FIGURE 11.7**).

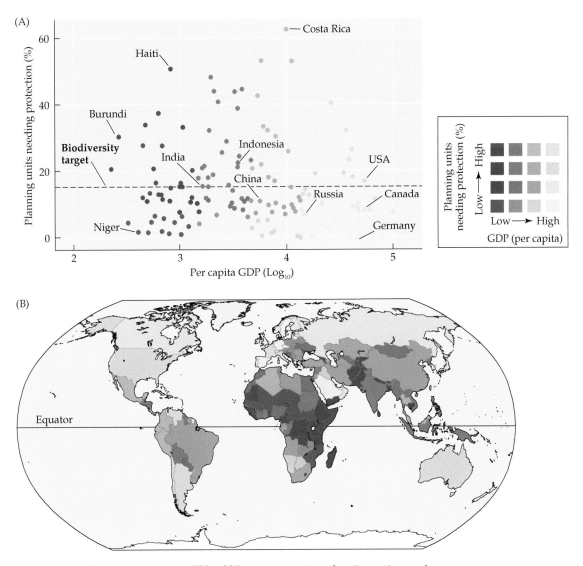

FIGURE 11.7   The proportion of 30 × 30 km conservation-planning units needing protection (in addition to existing protected areas). (A) The dotted line on the graph indicates the Aichi Biodiversity Target of 17%. Wealthier countries (i.e., those with a higher per capita gross domestic product [GDP] such as Canada and Germany) were more likely to have reached this target than developing countries (such as Haiti and Costa Rica). (B) The greatest number of units needing additional protection (as indicated in red and pink) are concentrated in tropical areas. Colors in the graph correspond to those in the map. (After Butchart et al. 2015.)

4. *Agenda 21*. This 800-page document is an innovative attempt to comprehensively describe the policies governments need to implement for environmentally sound development. Agenda 21 links the environment with other development issues, which are most often considered separately, such as child welfare, poverty, gender issues, technology transfer, and the unequal division of wealth. Plans of action address the problems of atmospheric, terrestrial, and aquatic pollution; land degradation and desertification; mountain development; unsustainable agriculture and rural development; and deforestation. Financial, institutional, technological, legal, and educational mechanisms that governments can use to implement the action plans are also described.

After the UNFCC, international agreements to reduce global greenhouse gas emissions to below-1990 levels resumed with the Kyoto Protocol in December 1997. The agreement was ratified in 2004 under the UNFCCC (see item 2), and many countries have established policies that have reduced their emissions of greenhouse gases, primarily CO_2. There were additional talks in Bali in December 2007, in Copenhagen in December 2009, in Warsaw in November 2013, and in Paris in December 2015. The **Paris Accord** is the strongest international agreement on climate change yet, requiring action from all 195 signatory nations. Previous agreements did not require reductions from China, which has become the greatest contributor of greenhouse gasses. Goals of the Paris Accord include taking actions to prevent global temperature from increasing more than 2°C above pre-industrial levels. The National Climate Action Plans already shared by 186 nations alone are expected to cut global emissions in half (Davenport 2015).

The **World Summit on Sustainable Development**, held in Johannesburg, South Africa, in 2002, emphasized achieving the social and economic goals of sustainability. This shift in focus from the Rio Summit highlights a significant, ongoing debate over whether the emphasis in conservation should be to promote sustainable use of natural resources for the benefit of poor people or to protect natural areas and biodiversity (Roe et al. 2013).

The **UN Conference on Sustainable Development** (unofficially the Rio+20), held in 2012, linked biodiversity conservation to sustainable development and climate change and emphasized the need for market-based solutions (Carrière et al. 2013). It resulted in what is referred to as the 2030 Agenda, which includes 169 targets. A working group from that conference then developed a set of 17 sustainable development agenda items, presented at the UN Sustainable Development Summit in 2015 (**FIGURE 11.8**). On January 1, 2016, the 17 goals became officially in force from that meeting, with the ambitious aims to eliminate poverty, ensure equitability (social justice), and stop climate change.

UNITED NATIONS
SUSTAINABLE
DEVELOPMENT
SUMMIT 2015
25-27 SEPTEMBER

FIGURE 11.8 A poster advertising the World Summit on Sustainable Development. The logo of colored arrows moving forward reflects the focus on social and economic goals over biodiversity goals.

International agreements that protect habitat

Seven international conventions protect biodiversity either directly or through the protection of habitat (**TABLE 11.1**) and each seeks to balance conservation concerns with development needs. Those that include provisions for land use and sustainable development include: (1) the **Ramsar Convention on Wetlands**, (2) the **World Heritage Convention** (**WHC**) (or the **Convention Concerning the Protection of the World Cultural and Natural Heritage**), and (3) the **Convention on Biological Diversity** (**CBD**).

The Ramsar Convention on Wetlands (www.ramsar.org) was established in 1971 to halt the continued destruction of wetlands, particularly those that support migratory waterfowl, and to recognize the ecological, scientific, economic, cultural, and recreational values of wetlands (**FIGURE 11.9**). The Ramsar Convention covers freshwater, estuarine, and coastal marine habitats and includes 2170 sites with a total area of more than 207 million ha. The 168 countries that have signed the Ramsar

> Countries can gain international recognition for protected areas through the Ramsar Convention, the WHC, and the Biosphere Reserves Program. Transfrontier parks in border areas provide opportunities for both conservation and international cooperation.

TABLE 11.1	The Seven Biodiversity-related International Agreements, with Their Primary Objectives
	Convention name and primary objective
CBD	**Convention on Biological Diversity (CBD)**: Promotes the conservation of biological diversity, sustainable use, and equitable sharing of benefits.
	Convention on International Trade in Endangered Species (CITES): Ensures that trade in animals and plants does not threaten their survival.
CMS	**Convention on the Conservation of Migratory Species of Wild Animals**: Provides guidelines for the conservation and sustainable use of migrating animals throughout their ranges.
Ramsar	**Convention on Wetlands (aka the Ramsar Convention)**: Promotes the conservation and sustainable use of wetlands and their resources that contribute to both biodiversity and human well-being.
MaB	**UNESCO Man and the Biosphere Program**: Biosphere Reserves are internationally recognized areas established to balance human and biodiversity needs.
	International Treaty on Plant Genetic Resources for Food and Agriculture: Promotes the conservation of plant genetic resources and the equitable sharing of the benefits that arise from them.
	World Heritage Convention (WHC): Mandates the identification and conservation of the world's cultural and natural heritage by protecting a specific list of sites.

Source: After Convention on Biological Diversity, www.cbd.int/brc.

FIGURE 11.9 The Djoudj National Bird Sanctuary is a Ramsar-listed site in Senegal noted for millions of waterbirds, such as these great white pelicans (*Pelecanus onocrotalus*), which stop here during their annual migration south from Europe in autumn. (© BSIP/UIG/Getty Images.)

Convention agreed to conserve and protect their wetland resources and designate at least one wetland site of international significance for conservation purposes. The fourth Ramsar Strategic Plan was launched August 21, 2015, to cover the period from 2016 to 2024.

The goal of the World Heritage Convention is to protect cultural areas and natural areas of international significance through its World Heritage Site program (whc.unesco.org). The convention is unique because it emphasizes the cultural as well as the biological significance of natural areas and recognizes that the world community has an obligation to financially support the sites. Limited funding for World Heritage Sites comes from the United Nations Foundation, which also supplies technical assistance. As with the Ramsar Convention, the World Heritage Convention seeks to give international recognition and support to protected areas that are established initially by national legislation. The 981 World Heritage Sites protecting natural areas cover about 142 million ha and include some of the world's premier conservation areas: Serengeti National Park in Tanzania, Sinharaja Forest Reserve in Sri Lanka, Iguaçu Falls in Brazil (**FIGURE 11.10**), Manu National Park in Peru, the Queensland Rain Forest of Australia, Komodo National Park in Indonesia, and Great Smoky Mountains National Park in the United States, to name a few.

UNESCO's World Network of Biosphere Reserves was founded in 1971. Biosphere reserves are designed to be models that demonstrate the compatibility of conservation efforts and sustainable development for the benefit of local people, as described in Chapter 8 (see Figure 8.19). A total of 621 biosphere reserves have been created in 117 countries, covering more than 263 million ha and including 47 reserves in the United States, 41 in Mexico,

FIGURE 11.10   World Heritage Sites include some of the most revered and well-known conservation areas in the world. (© Joris Van Ostaeyen/istock.)

41 in Russia, 32 in China, 16 in Bulgaria, and 15 in Germany. The largest biosphere reserve, located in Greenland, is more than 97 million ha in area.

Along with provisions of the CBD (see the next section, "Funding for Conservation"), these three conventions establish an overarching consensus regarding the appropriate conservation of protected areas and certain habitat types. More limited international agreements protect unique ecosystems and habitats in various regions, including the Western Hemisphere, the Antarctic, the South Pacific, Africa, the Caribbean, and the European Union. Other international agreements have been ratified to prevent or limit pollution that poses regional and international threats to the environment. For example, the Convention on Long-Range Transboundary Air Pollution in the European region recognizes the role that the long-range movement of air pollution plays in acid rain, lake acidification, and forest dieback; the Convention for the Protection of the Ozone Layer was signed in 1985 to regulate and phase out the use of chlorofluorocarbons; and the Convention on the Law of the Sea promotes the peaceful use and conservation of the world's oceans.

Conservation measures can also potentially contribute to promoting cooperation between governments. Such is often the case when countries need to manage areas collectively. In many areas of the world, largely uninhabited mountain ranges mark the boundaries between countries. These areas often are designed as national parks, with each country managing its own wildlife and ecosystems. As an alternative, countries can establish transfrontier parks on both sides of boundaries to cooperatively manage

FIGURE 11.11 The Greater Limpopo Transfrontier Park, which includes Gonarezhou National Park, Kruger National Park, Limpopo National Park, and several smaller conservation areas, has the potential to unite wildlife management activities in South Africa, Mozambique, and Zimbabwe. A larger transfrontier conservation area (hashed area) will include Zinave National Park, Banhine National Park, private game reserves, and private farms and ranches. (After www.peaceparks.org.)

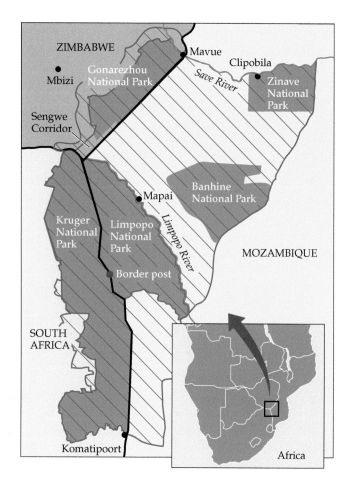

whole ecosystems and promote conservation on a large scale (Thondhlana et al. 2015). An early example of this collaboration was the decision to manage Glacier National Park in the United States and Waterton Lakes National Park in Canada as the Waterton–Glacier International Peace Park. Today, intensive efforts are being made to link national parks and protected areas in Zimbabwe, Mozambique, and South Africa into the Greater Limpopo Transfrontier Park and other, larger management units (**FIGURE 11.11**). An added advantage of this joint management is that the seasonal migratory routes of large animals will be protected. As another example, the establishment of the Red Sea Marine Peace Park between Israel and Jordan is important, not only for conservation, but also for its potential for building trust in a war-ravaged region.

Marine pollution is another issue of vital concern because of the extensive areas of international waters not under national control and the ease with which pollutants released in one area can spread to another area. Agreements covering marine pollution include the Convention on the Law of the Sea, the Regional Seas Program of the United Nations Environmental Programme (UNEP), and the Convention on the Prevention of Marine Pollution by Dumping of Wastes and Other Matter. Regional agreements cover the northeastern Atlantic Ocean, the Baltic Sea, and other specific locations, particularly in the North Atlantic region. The pelagic zone of the open ocean (the area of the ocean far from the shore) is still largely unexplored and unregulated at this point and is in urgent need of protection.

Funding for Conservation

In general, there has been movement away from funding conservation efforts solely with local taxpayer dollars in favor of a variety of other approaches. Nongovernmental organizations and policies that require industry and development to pay for mitigation have been important sources of funding within developed countries. However, funding of conservation in developing countries has been less straightforward. One of the most contentious issues for international conferences and treaties has been deciding how to fund the proposals, particularly the CBD and other programs related to sustainable development, conservation, and climate change. Over the past 20 years, international funding for conservation by developed countries, foundations, and private donors has increased, though not as much as developing countries and conservation biologists had hoped. In anticipation of the UN Sustainable Development Summit in September 2015, a group of officials from various nations convened in Ethiopia's capital for the third International Conference on Financing for Development (FFD3). They produced the Addis Ababa Agenda, intended as the global plan of how to implement and support the post 2015 agenda (Bhattacharya et al. 2015). However ambitious and well-intentioned, like many agreements of this type it was immediately attacked for lacking any concrete commitments or "teeth" to increase funding (Barcia 2015).

The World Bank and international nongovernmental organizations (NGOs)

At the time of the Earth Summit, the cost of conservation programs was estimated to be about $600 billion per year, of which $125 billion was to come from developed countries as part of their overseas development assistance (ODA). Because the level of ODA from all countries in the early 1990s totaled approximately $60 billion per year, implementing these conventions would have required a tenfold increase in the aid commitment at that time. The developed countries did not agree to this increase in funding, but they offered an alternative: each country would increase its level of foreign assistance to 0.7% of its gross national product (GNP) by the year 2000, which would have roughly doubled the ODA from developed nations. In the past 15 years, the ODA has been increasing steadily, now 66% more than in 2000. However, no schedule was set to meet the target date, and as of 2014, only a few wealthy northern European countries had met the GNP target percentage: Luxembourg (1.06%), Sweden (1.09%), Norway (1.00%), Denmark (0.86%), and the United Kingdom (0.70%). The United States is the largest contributor by volume at $32.7 billion in 2014, but as this represents only 0.19% GNP, it is still well below the 0.70% target (www.oecd.org). In the United States government, funding for international conservation programs is spread across many departments, including the Agency for International Development, the National Science Foundation, the Smithsonian Institution, and the USFWS.

Through the 1980s, international funding for conservation projects was approximately $200 million per year, but starting in the early 1990s, it increased to $1 billion per year (**FIGURE 11.12**). Though this funding increase was dramatic, it was still not as much as originally promised. Much of the increase in conservation funding by developed countries has been channeled through the **World Bank** (www.worldbank.org) and the associated **Global Environment Facility** (**GEF**) (www.thegef.org/gef). The World Bank is a multilateral development bank established to promote international trade and economic activity. It is governed mainly by developed countries, and only a small portion of its activities are related to conservation. The related organization, the GEF, was established specifically to channel money from developed countries to conservation and environmental projects in

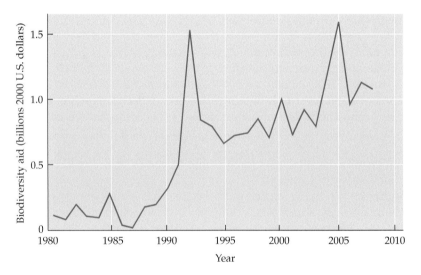

FIGURE 11.12 International funding for biodiversity projects on an annual basis for the period 1980–2008. Annual fluctuations in funding are caused by the tendency to bunch large projects together and by the momentum created by the Rio Earth Summit in 1992. The total funding over the entire time period is $18.6 billion, of which 31% is from the World Bank and 28% is from the Global Environment Facility (GEF). Another 27% of the funding is provided by donor countries, including the United States, the Netherlands, and Germany. (After Miller et al. 2013.)

developing countries, with much of its funding distributed by the World Bank, other multilateral development banks with a regional focus, and the United Nations Development Programme.

Over the past 30 years, a total of approximately $18 billion in international funding has been provided for almost 10,000 biodiversity projects in 171 countries (Miller et al. 2013). After the World Bank ($6.8 billion) and the GEF ($5 billion), the top donors were the United States ($1.3 billion), the Inter-American Development Bank ($1.1 billion), and the Netherlands ($783 million). Other large sources of project funding come from the United Nations Development Programme and UNEP. These organizations provide funds to national governments and conservation organizations for projects including establishing protected areas, protecting endangered species, restoring degraded habitats, training conservation staff, developing conservation infrastructure, addressing global climate change, and managing forest, freshwater, and marine resources (Hickey and Pimm 2011).

The enormous impact of such international funding is illustrated by the joining of the World Bank, a source of funding, with the World Wildlife Fund (WWF), an international conservation organization, to initiate a new program, the Forest Alliance, to protect and mange over 100 million ha of forest in countries around the world (worldwildlife.org) (Bowler et al. 2012). The World Bank is also one of the leaders in efforts to reduce CO_2 emissions caused by deforestation in tropical countries including Indonesia. Through its Forest Carbon Partnership Facility, companies and developed countries can offset their present production of greenhouse gases by purchasing carbon credits for maintaining these tropical forests. The World Bank has partnered with the WWF and other large NGOs in implementing such programs.

International conservation NGOs (also called INGOs) (e.g., the WWF, Conservation International, BirdLife International, The Nature Conservancy, and the Wildlife Conservation Society) implement conservation activities directly, often through a carefully articulated set of priorities and programs (Robinson 2012). These NGOs have also emerged as leading sources of conservation funding, raising funds from membership dues, donations from wealthy individuals, sponsorship from corporations, and grants from foundations and international development banks (**FIGURE 11.13**). Although foundations and conservation organizations provide only a small fraction of the funding for biodiversity projects, they can sometimes be more flexible and can fund innovative small projects and provide more intensive management than is typical from projects with government funding. The growing importance of private funding throughout the world is illustrated by a recent $260 million donation given by the Gordon and Betty Moore Foundation in the United States to the NGO Conservation International.

International NGOs are often active in establishing, strengthening, and funding both local NGOs and government agencies in the developing world that run conservation programs (Cohen-Shacham et al. 2015).

FIGURE 11.13 Over the past four decades, there has been a dramatic increase in the annual contributions to many conservation organizations, as illustrated by the revenues of four large NGOs from the United States: The Nature Conservancy, World Wildlife Fund (WWF), Environmental Defense Fund, and Sierra Club. Note that the revenue for The Nature Conservancy should be multiplied by 100; for 2013, contributions were approximately $5 billion. (After Zaradic et al. 2009, with updates from P. Zaradic.)

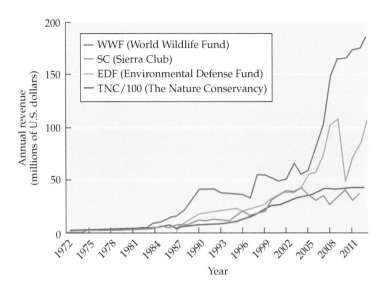

From the perspective of an INGO such as the WWF, working with local organizations in developing countries is an effective strategy because it relies on local knowledge and it trains and supports groups of citizens within the country, who can then be conservation advocates for years to come. NGOs are often perceived to be more effective at carrying out conservation projects than government departments, but programs initiated by NGOs may end after a few years when funding runs out, and often they fail to achieve a lasting effect. Moreover, the income of NGOs can be quite variable, depending on the state of the economy.

Environmental trust funds

In addition to direct grants and loans for projects provided by the World Bank and other institutions, another important mechanism used to provide secure, long-term support for conservation activities in developing countries is the **national environmental fund (NEF)**. NEFs are typically set up as conservation trust funds or foundations in which a board of trustees—composed of representatives of the host government, conservation organizations, and donor agencies—allocates the annual income from an endowment to support inadequately funded government departments and nongovernment conservation organizations and activities. NEFs have been established in over 50 developing countries with funds contributed by developed countries and by major organizations such as the World Bank, the GEF, and the WWF.

One important early example of an NEF, the Bhutan Trust Fund for Environmental Conservation (BTF), was established in 1991 by the government of Bhutan in cooperation with the World Bank and the WWF. The BTF has already received about $44 million (exceeding its goal of $20 million). The fund provides more than $1 million per year for surveying the rich biological resources of this eastern Himalayan country. NEFs have proliferated in recent years, with the Latin American and Caribbean Network of Environmental Funds (RedLAC) alone comprising 13 countries and over 3000 projects supported by an annual budget of over $70 million (www.redlac.org).

Debt-for-nature swaps

Many countries in the developing world have accumulated huge international debts that they are unable to repay. As a result, some developing countries have rescheduled their loan payments, unilaterally reduced them, or stopped making them altogether. Because of the low expectation of repayment, the commercial banks that hold these debts have sometimes sold them at a steep discount on the international secondary debt market. For example, Costa Rican debt has traded for only 14%–18% of its face value.

> Government, World Bank, and NGO funding for conservation projects has increased in recent decades. Environmental trust funds and debt-for-nature swaps provide additional mechanisms to support conservation activities.

In a creative approach, debt from the developing world is used as a vehicle for financing projects to protect biodiversity in so-called **debt-for-nature swaps** (Cassimon et al. 2014). In one common type of debt-for-nature swap, an NGO in the developed world (such as Conservation International) buys up the debts of a developing country; the NGO agrees to forgive the debt in exchange for the country carrying out a conservation activity. This activity could involve land acquisition for conservation purposes, park management, development of park facilities, conservation education, or sustainable development projects. In another type of swap, a government of a developed country that is owed money directly by a developing country may decide to cancel a certain percentage of the debt if the developing country will agree to contribute to an NEF or some other conservation project.

Debt-for-nature swaps have converted debt valued at $1.5 billion into conservation and sustainable development activities in Colombia, Poland, the Philippines, Madagascar, and a dozen other countries. However, spending money on conservation programs may divert funds from other necessary domestic programs such as medical care, schools, and agricultural development. Furthermore, these programs have become less common because external debt of countries is not generally available at such steep discounts.

How effective is conservation funding?

Conservation organizations have developed a number of tools to evaluate the effectiveness of funded projects (Mascia et al. 2014). Projects funded by the GEF and World Bank so far have received mixed evaluations (www.thegef.org/gef). On the positive side, increased funding was available for conservation and biodiversity projects, planning national biodiversity strategies and legislation, identifying and protecting important ecosystems and habitats, and enhancing staff training. However, the lack of participation by community groups, local scientists, and government leaders; an overreliance on foreign consultants; and an elaborate and time-consuming application procedure were identified as major problems. An additional problem was the mismatch of short-term funding with the long-term needs of poor countries.

It must be recognized that many environmental projects supported by international aid do not provide lasting solutions to the problems because of failure to deal with the "4 Cs"—concern, contracts, capacity, and causes. Environmental aid will be effective only when applied to situations in which both donors and recipients have a genuine *concern* to solve the problems (Do key people really want the project to be successful, or do they just want the money?); when mutually satisfactory and enforceable *contracts* for the project can be agreed on (Will the work actually be done once the money is given out? Will money be siphoned off into private hands?); where there is the *capacity* to undertake the project in terms of institutions, personnel, and infrastructure (Do people have the skills to do the work, and do they have the necessary resources, such as vehicles, research equipment, buildings, and access to information, to carry out the work?); and when the *causes* of the problem are addressed (Will the project treat the underlying causes of the problem or just provide temporary relief of the symptoms?). Despite these problems, international funding of conservation projects continues. Past experiences are informing new projects, which are more effective, but with the result that the application and accounting processes can be extremely cumbersome and time-consuming.

> While the recent increased funding for the protection of biodiversity is welcome, further funding is needed to accomplish the task.

The need for increased funding for biodiversity remains great at the local, national, and international levels. It is estimated that it will take \$3.4–\$4.8 billion annually to reduce the extinction risk of all globally threatened species by just one International Union for Conservation of Nature (IUCN) category (e.g., from GS1 to GS2), and that adequately managing and protecting all areas with such species will cost \$76 billion annually (McCarthy et al. 2012). While \$76 billion is an enormous amount of money, it is dwarfed by the more than \$750 billion spent on the military defense of the United States in 2014. Similarly, while the conservation funds provided by the United States government and the World Bank seem large, they are small compared with the other activities that these and other major international organizations support. Certainly the world's priorities could be modestly adjusted to give more resources to the protection of biodiversity (Rands et al. 2010).

There is also a role to be played by conservation organizations and businesses working together to market "green products." Already the Forest Stewardship Council and similar organizations are certifying wood products from sustainably managed forests and coffee companies are marketing shade-grown coffee. If consumers are educated to buy these products at a somewhat higher price, this could be a strong force in international conservation efforts.

Finally, a potentially huge new funding source to protect tropical forests is being adopted as part of international efforts to address climate change. Because about 20% of global greenhouse gas emissions result from destruction of the tropical forest, a funding mechanism called **Reducing Emissions from Deforestation and Forest Degradation** (**REDD**) could pay to protect tropical forests (Munawar et al. 2015). REDD rewards poorer nations for preserving forests by paying them for the carbon that is stored in their forests. There are huge concerns about whether this money will be well spent in protecting forests and reducing poverty in developing countries or whether it will be diverted to other purposes or cause even worse deforestation in other places (Phelps et al. 2013). Organizations at all levels will be involved in designing, implementing, and monitoring what happens as REDD becomes a reality. These types of new and existing funding mechanisms will all be needed in coming years as part of international efforts to protect biodiversity.

Summary

- Sustainable development is economic development that satisfies the present and future needs of human society while minimizing its impact on biodiversity. Achieving sustainable development is a challenge for conservation biology and society.

- Legal efforts to protect biodiversity occur at the local, regional, and national levels and regulate activities affecting both private and public lands. Governments and private land trusts buy land for conservation purposes or acquire CEs and development rights for future protection. Associated laws limit pollution, regulate or ban certain types of development, and set rules for hunting and other activities—all with the aim of preserving biodiversity and protecting human health.

- International agreements and conventions that protect biological diversity are needed because species migrate across borders, there is an international trade in biological products, the benefits of biological diversity are of international importance, and the threats to diversity are often international in scope and require international cooperation.

- The 1992 Rio Summit (also known as the Earth Summit) resulted in four major documents that are the foundation of many programs and subsequent international sustainability conferences.

- Conservation groups, governments in developed countries, and the World Bank are increasing funding to protect biodiversity, especially in developing countries. NEFs and debt-for-nature swaps are also used to fund conservation activities. However, the amount of money is still inadequate to deal with the problems.

For Discussion ■————————————————————————

1. What are the roles of government agencies, private conservation organizations, businesses, community groups, and individuals in the conservation of biodiversity? Can they work together, or are their interests necessarily opposed to each other?

2. How can conservation biologists provide links between basic science and a public environmental movement? What suggestions can you make for ways in which conservation biologists and environmental activists can energize and enrich each other in working toward an economically and environmentally stable world?

3. Do you believe that we will be successful in reaching the global targets established by the Earth Summit (for biodiversity) or the Paris Accord (for global warming)? Why or why not? If they are not reached, was there a value to setting them?

Suggested Readings ————————————————————————

Ban, N. C., M. Mills, J. Tam, C. C. Hicks, S. Klain, N. Stoeckl, and 5 others. 2013. A social-ecological approach to conservation planning: Embedding social considerations. *Frontiers in Ecology and the Environment* 11: 194–202. Concern for the rights and welfare of local people must be included when designing systems of protected areas.

Brundtland, G., M. Khalid, S. Agnelli, S. Al-Athel, B. Chidzero, L. Fadika, and M. M. de Botero. 1987. *Our Common Future: Report of the World Commission on Environment and Development* (the "Brundtland Report"). This product of the United Nations World Commission on Environment and Development (WCED) laid the groundwork for the important international summits and agreements to follow, including the 1992 Earth Summit.

Colglazier, W. 2015. Sustainable development agenda 2030. *Science* 349: 1048–1050. The Global Sustainable Development Report can be used as a bridge between the United Nations Sustainable Development Goals and scientific communities.

Dernbach, J. C. and F. Cheever. 2015. Sustainable development and its discontents. *Transnational Environmental Law* 4(2): 247–287. Provides answers to many of the criticisms of sustainable development and a critique of the so-called alternatives to sustainability.

Fehr-Duda, H. and E. Fehr. 2016. Sustainability: Game human nature. *Nature* 530. Sustainable development can be promoted by utilizing particular aspects of human nature.

Minteer, B. A. and T. R. Miller. 2011. The new conservation debate: Ethical foundations, strategic trade-offs, and policy opportunities. *Biological Conservation* 144: 945–947. Part of a special issue of the journal devoted to the balance between protecting biodiversity and providing opportunities for rural people.

Muradian, R., M. Arsel, L. Pellegrini, F. Adaman, B. Aguilar, B. Agarwal, and 21 others. 2014. Payments for ecosystem services and the fatal attraction of win–win solutions. *Conservation Letters* 6: 274–279. Programs involving PES are being presented as an outstanding opportunity, but they are unlikely to be as successful as promised.

Naeem, S, A. V. Ingram, T. Agardy, P. Barten, G. Bennett, E. Bloomgarden, and 39 others. 2015. Get the science right when paying for nature's services. *Science* 347(6227): 1206–1207. Science needs to play a larger role in determining PES.

Negrón-Ortiz, N. 2014. Pattern of expenditures for plant conservation under the Endangered Species Act. *Biological Conservation* 171: 36–43. Plants receive far less funding than animals.

Nieto-Romero, M., A. Milcu, J. Leventon, F. Mikulcak, and J. Fischer. 2016. The role of scenarios in fostering collective action for sustainable development: Lessons from central Romania. *Land Use Policy* 50: 156–168. Barriers to (and solutions for) movement toward sustainable development in rural communities are explored through interviews with Transylvanians.

Nolte, C., A. Agrawal, K. M. Silvus, and B. S. Soares-Filho. 2013. Governance regime and location influence avoided deforestation success of protected areas in the Brazilian Amazon. *Proceedings of the National Academy of Sciences USA* 110: 4956–5961. Once a protected area is established, a variety of factors can affect its ability to protect biodiversity.

Reed, S. E., J. A. Hilty, and D. M. Theobald. 2014. Guidelines and incentives for conservation development in local land-use regulations. *Conservation Biology* 28: 258–268. Conservation development involves building housing in a dense cluster, with the remaining land protected for conservation purposes.

Selomane, O., B. Reyers, R. Biggs, H. Tallis, and S. Polasky. 2015. Towards integrated social-ecological sustainability indicators: Exploring the contribution and gaps in existing global data. *Ecological Economics* 118: 140–146. The contribution of nature to development is largest for poor rural communities.

Sorice, M. G., C-O. Oh, T. Gartner, M. Snieckus, R. Johnson, and C. J. Donlan. 2013. Increasing participation in incentive programs for biodiversity conservation. *Ecological Applications* 23: 1146–1155. Innovative programs are being developed to encourage local participation in conservation projects.

Wunder, S. 2013. When payments for environmental services will work for conservation. *Conservation Letters* 6: 230–237. PES are best suited to stable societies with good legal and financial institutions and may not be suited to many developing countries.

12

An Agenda for the Future

The future of species like this rare Penland's beardtongue flower (*Penstemon penlandii*) depends on the work of conservation scientists collaborating with others.

As we have seen throughout this book, the causes of the rapid, worldwide decline in biodiversity are no mystery. Ecosystems are destroyed and species are driven to extinction by human resource use, which is propelled by the need of many rural people to survive, by the excessive consumption of resources by affluent people and countries, and by the desire to make money (Fischer et al. 2012). The destruction may be caused by local people in the region, people recently arrived from outside the region, local business interests, large businesses in urban centers, suburban sprawl into rural areas, multinational corporations in other countries, military conflicts, governments, or others. Moreover, people may also be unaware of, or apathetic toward, the impact of human activities on the natural world. Furthermore, although there have never before been so many people involved in documenting new species, hundreds of thousands of species are yet to be discovered and may be lost before they are even named (Costello et al. 2015).

For conservation policies to work, people at all levels of society must believe that the conservation of biodiversity is in their own interest (Manfredo et al. 2016). If conservationists can demonstrate that protecting biodiversity has more value than destroying it, people and their governments may become more willing to preserve biodiversity. This assessment should include not only immediate monetary value but also less tangible aspects, including existence value, option value, and intrinsic value (see Chapter 3).

Ongoing Problems and Possible Solutions

There is a consensus among conservation biologists that our efforts to preserve biodiversity face several major problems and that certain changes must be made to policies and practices (Gustafsson et al. 2015; Sutherland et al. 2011). We list some of these problems and suggest some solutions next. Note that for the purposes of this text we have simplified the solutions, leaving out many of the intricacies that must be addressed to provide comprehensive, real-world answers.

Problem: Protecting biodiversity is difficult when most of the world's species remain undescribed by scientists and are not known by the general public. Furthermore, most ecosystems lack monitoring to determine how they are changing over time.

Solution: We must train more scientists and enthusiastic amateurs to identify, classify, and monitor species and ecosystems, and we must increase funding in these areas (Costello et al. 2015). In particular, we need to train more scientists and establish research institutes in developing countries. Citizen scientists often can play an important role in protecting and monitoring biodiversity if they are given some training and guidance by conservation biologists (Danielson et al. 2014) (**FIGURE 12.1**). People who are interested in conservation biology should be taught basic skills, such as species identification and environmental-monitoring techniques. For example, following the *Deepwater Horizon* oil spill in the Gulf of Mexico, the US Fish and Wildlife Service, Cornell Lab of Ornithology, and National Audubon Society quickly mobilized thousands of volunteers and adapted the online bird-monitoring program eBird to allow volunteers to collect and submit field observations that enabled scientists and managers to assess damage to birds and other wildlife and to target cleanup efforts. Other examples are Project Budburst, which recruits the public to submit phonological data as a way of tracking the impacts of climate change, and iNaturalist, an international program with which amateurs can record observations of plants and wildlife. iNaturalist also connects these enthusiastic amateurs with scientists who can use their skills to collect data. Such citizen-science programs are becoming increasingly common and valuable to conservation (Sullivan et al. 2014).

Other types of citizen-science programs are those based in conservation education. These programs provide specific information to particular audiences, such as schoolchildren, senior citizens, or rural people living

> Conservation biologists can spread the message of conservation and do better science by educating members of the public and including them in their projects. These citizens then often become advocates for protecting biodiversity.

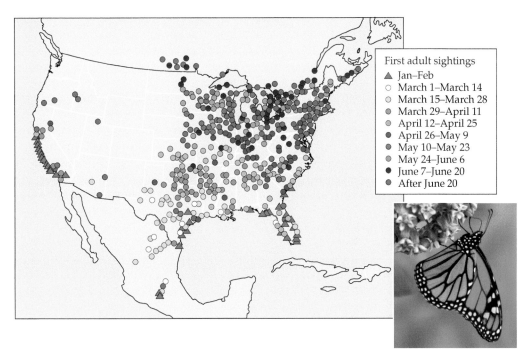

FIGURE 12.1 Each spring, thousands of citizen observers, especially students, contribute their observations to the Journey North website, which tracks the migration of the monarch butterfly. This map presents the observations of the first appearance of monarch butterflies across North America in spring 2009. The butterflies primarily overwinter in central Mexico and start migrating into the southern United States in March, arriving in the northern United States and southern Canada in late May and June. There are secondary overwintering sites in California, Florida, and Texas. Efforts by citizen observers can be used to detect the earlier migration of butterflies in response to a warming climate and a decline of butterfly numbers as a result of habitat damage in the Mexican overwintering grounds. (After www.learner.org/jnorth; inset by David McIntyre.)

near national parks, and can help promote conservation-oriented behaviors (Şekercioğlu 2012). In these types of projects, volunteers not only observe how to conduct science and conservation, but actually perform science and conservation. In this way they can experience a sense of service and ownership of the project and become advocates. In some cases, this work translates into lasting advances in conservation and can inspire students to pursue careers in the field. Many conservation biologists and ecologists attribute their passion for the field to a transformative outdoor experience such as those provided by these programs.

Other education approaches have been developed that improve people's understanding of conservation, including the creative use of smart phones, web resources, digital photography, and other advanced technology. Information on biodiversity must be made more accessible; this may be accomplished in part through two new sites, the Encyclopedia of Life (eol. org) and the Tree of Life Web Project (tolweb.org), which serve as central clearinghouses for data. The question of how to improve the science, conservation, and education outcomes of citizen science and other educational techniques is an active area of research.

Problem: Many conservation issues are global in scope and involve many countries, making it important but difficult to coordinate conservation actions. Furthermore, international agreements are extremely difficult to enforce, making noncompliance an issue.

Solution: We must focus on and build upon successful international efforts. Countries are increasingly willing to discuss international conservation issues, as shown by the 2015 Paris Accord (see Chapter 11). Nations are also becoming more willing to sign and implement treaties such as the Convention on Biological Diversity, the United Nations Framework Convention on Climate Change, and the Convention on International Trade in Endangered Species. International conservation efforts are expanding, and further participation in these activities by conservation biologists and the general public should be encouraged. Another positive development is the trend toward establishing transfrontier parks that straddle national borders; these parks benefit wildlife and encourage cooperation between countries.

Citizens and governments of developed countries must also become aware that they bear a direct responsibility for destroying biodiversity through their overconsumption of the world's resources and the specific products they purchase (Sayre et al. 2013) (**FIGURE 12.2A**). To encourage the enactment of legislation, conservation professionals must demonstrate how changes in the actions and lifestyles of individuals on the local level, and those of cities and nations on a larger scale, can exert positive influences far beyond the immediate community. Improvements in understanding conservation psychology will help in the design of programs to encourage conservation-minded behaviors.

Problem: Developing countries often want to protect their biodiversity but face great pressure to develop their natural resources to generate needed income.

Solution: Conservation organizations, zoos, aquariums, botanical gardens, natural history museums, governments, and even businesses in developed countries, as well as international organizations such as the United Nations and the World Bank, should continue to provide technical and financial support to developing countries for conservation activities, in particular establishing and maintaining protected areas. It is important for these organizations to support the training of conservation biologists in developing countries so local people can become advocates for biodiversity within their own country (Paknia et al. 2015). Support from developed nations should continue until countries can protect biodiversity with their own re-

Conservation biologists need to support approaches that provide benefits for people and protect biodiversity. One approach is to compensate landowners and local people for the ecosystem services that their land provides.

(A)

(B)

FIGURE 12.2 (A) A farmer in Nicaragua grows and processes sustainably produced coffee berries in hope of receiving a higher price for the crop than for conventionally produced coffee berries. (B) Sustainably produced and certified coffee is available for purchase in an increasing number of stores and has a recognizable Fair Trade logo on the package. (A, © Janet Jarman/Corbis; B, courtesy of David McIntyre.)

sources and personnel. Requiring support from developed nations until that time is fair and reasonable since they have the funds to support these parks, drive much of the degradation of biodiversity through consumption, benefit from the exploitation of natural resources in these countries, and make use of biological resources in their agriculture, industry, research programs, zoos, aquariums, botanical gardens, and educational systems. Solutions for economic and social problems in developing countries must be addressed simultaneously, particularly those relating to reducing poverty and ending armed conflicts (see Chapter 11). A variety of financial mechanisms exist to achieve these goals, including direct grants, payments for ecosystem services, debt-for-nature swaps, and trust funds. Individual citizens in developed countries can donate money and participate in organizations and programs that further advance these conservation goals.

Problem: Economic analyses often paint a falsely encouraging picture of development projects that damage the environment. Economic decision making often fails to include and evaluate ecosystem services and intrinsic values.

Solution: Cost–benefit analyses must evaluate development projects comprehensively by comparing potential project benefits with the full range of costs, including environmental and human costs such as soil erosion, air and water pollution, greenhouse gas production, declines in the availability of natural products and other ecosystem services, and the loss of places for people to live (Maron et al. 2013; Pilgrim et al. 2013). Local communities and the general public should be presented with all available information and asked to provide input into the decision process. We must adopt the "polluter pays" principle, in which industries, governments, and individual citizens pay for cleaning up the environmental damage their activities have caused (Szlávik and Füle 2009; Zhu and Zhao 2015). And financial subsidies to industries that damage the environment—such as the pesticide, fertilizer, transportation, petrochemical, logging, fishing, and tobacco industries—should end, particularly subsidies to industries that damage human health as well (see Chapter 3). Funds from these so-called perverse subsidies should be redirected to activities that enhance the environment and human well-being, especially to people whose lands provide ecosystem services to the public.

Problem: Disadvantaged people simply trying to survive are frequently blamed for the destruction of the world's biodiversity.

Solution: In places where local actions lead to biodiversity losses, conservationists can help recruit development and humanitarian organizations that have the skills to assist local people in organizing and developing sustainable economic activities that do not damage biodiversity (Kahler et al. 2013). Conservation biologists and conservation organizations increasingly participate in programs for poor, rural areas that promote smaller families, establish reliable food supplies, provide training in economically useful skills (Moro et al. 2013), and promote the benefits and pleasures of positive experiences as opposed to physical possessions. These programs should be closely linked to efforts aimed at recognizing basic human rights, especially ownership of the land and the right to a healthy environment, that is, social justice movements (see Chapter 11).

Conservation organizations and businesses should work together to market green products produced by rural communities, with some of the profits shared with those communities. Already the Forest Stewardship Council and similar organizations certify wood products derived from sustainably managed forests, coffee companies market shade-grown coffee (Edwards and Laurance 2012), and aquariums and ocean conservation organizations have developed lists of seafood that is harvested sustainably. Products meeting environmental, labor, and developmental standards can be Fair Trade Certified (**FIGURE 12.2B**). As of 2016, over 1210 organizations in 74 developing countries and representing over 1.5 million farmers and workers were certified to sell fair trade products, including bananas, tea, cotton, and fresh fruit (www.fairtrade.net). If consumers choose to buy these certified products instead of noncertified products, despite the fact that they are slightly more expensive, their purchases can strengthen local and international conservation efforts and provide tangible benefits to disadvantaged people in rural areas. However, the source of these products must be carefully investigated to ensure that their manufacturers have used truly sustainable production and are not simply using the certification label for marketing purposes (Christian et al. 2013). In some cases, such as some wine from Argentina, investigators found no meaningful difference between the practices of the fair trade and conventional wine sectors (Staricco and Ponte 2015).

Problem: Central governments often make decisions about establishing and managing protected areas with little input from the people and organizations in the affected region. Consequently, local people sometimes feel alienated from conservation projects and do not support them.

Solution: For a conservation project to succeed, the local inhabitants must know they will benefit from it and that their involvement is important (see Chapter 11). To achieve this goal, environmental impact statements, economic forecasts, and other project information should be made publicly available to encourage open discussion at all stages of the project's development and implementation. Local people should be provided with the assistance they need to understand and evaluate the implications of the project. Often they will want to protect biodiversity and associated ecosystem services because they understand that their livelihood, quality of life, and sometimes, even their survival depend on protecting the natural environment (Gupta et al. 2015).

Mechanisms should be established to ensure that government agencies, conservation organizations, and local communities and businesses share input into the management process. Conservation biologists working in national parks and other protected areas should engage in meaningful dialogue to explain the purpose and results of their work to nearby communities and school groups and listen to what local people have to say. In some cases, regional strategies, such as habitat conservation plans or community conservation plans, may have to be developed to reconcile the need for some development (and resulting loss of habitat) with the need to protect species and ecosystems.

Problem: Revenues, business activities, and scientific research associated with national parks and other protected areas may not directly benefit the people living in the surrounding communities.

Solution: People living in local communities often bear the costs, but do not receive the benefits, of nearby protected areas. Addressing this problem requires developing mechanisms to benefit the local communities (**FIGURE 12.3**). For example, when possible, local people should be trained and employed in protected areas as a way of using local knowledge and providing income. Governments or organizations can also assist local people in developing businesses related to tourism and other park activities A portion of park revenues should fund local community projects such as schools, clinics, roads, cultural activities, sports programs and facilities, and community businesses—infrastructure that benefits an entire village, town, or region; this establishes a link between conservation programs and the improvement of local lives.

Problem: National parks and conservation areas often have inadequate budgets to pay for conservation activities.

Solution: It is often possible to increase funds for park management by raising rates for admission, lodging, or meals so that they reflect the actual cost of maintaining the area. Concessions selling goods and services may be required to contribute a percentage of their income to the park's operation. Also, zoos and conservation organizations in the de-

FIGURE 12.3 A women's cooperative in Puno, Peru, makes and sells hats featuring the endangered Lake Titicaca frog to sell to tourists. The cooperative was started by conservation biologists with the Denver Zoo seeking to create income streams for local people that did not involve harvesting the animals. (Photograph by Richard Reading.)

veloped world should continue to make direct financial contributions to conservation efforts in developing countries. For example, members of the American Zoo and Aquarium Association and their partners participate in over 3700 in situ conservation projects in 100 countries worldwide (see Chapter 7).

Problem: Many endangered species and ecosystems occur on private land or on government land managed for timber production, grazing, mining, and other activities. Timber companies that lease forests and ranchers who rent rangeland from the government often damage biodiversity and reduce the productive capacity of the land in the pursuit of short-term profits. Private landowners often regard endangered species on their land as a burden because of restrictions on the use of their habitats.

Solution: Laws should be changed so people are only permitted to obtain leases to harvest trees and use rangelands if they maintain the health of the ecosystem (Stocks 2005). Governments should also eliminate perverse tax subsidies that encourage exploiting natural resources and establish payments for land management, especially on private land, that enhance conservation. Additionally, landowners should be educated to protect endangered species and praised publicly for their efforts. It is vital to develop connections among farmers, ranchers, conservation biologists, and hunting and fishing groups because biodiversity, wildlife, and the rural way of life are all intertwined.

We must also seek ways to integrate biodiversity considerations into plans, policies, and practices that depend on, or in some way relate to, the environment, on both the local and the international scales (Redford et al. 2015). To be effective, this requires communication between those who want the ecosystem services and those who are more concerned with the ecosystem itself (QUINTESSENCE Consortium 2016). This approach has been called *mainstreaming biodiversity*, and it is frequently used in the United Nations and other international treaty bodies. Article 10(a) of the Convention on Biological Diversity (CBD) calls for integrating "as far as possible and as appropriate, the conservation and sustainable use of biological resources into national decision-making," and Decision VII/14 adopted the CBD Addis Ababa Principles and Guidelines for the Sustainable Use of Biodiversity (AAPG) in 2004. Spurred by such agreements and other, more local, motivators, governments are increasingly incorporating biodiversity into formal laws and standard practices.

Problem: In many countries, governments are inefficient and bound by excessive regulation. Consequently, they are often slow and ineffective at protecting endangered species and ecosystems.

Solution: Local nongovernmental organizations (NGOs) and citizen groups often represent the most efficacious agents for promoting conservation, especially if they work in concert with government policies, such as economic incentives for conservation (G. Wright et al. 2015). Accordingly, these citizen groups should be encouraged and supported politically, scientifically, and financially. Conservation biologists must educate citizens about local environmental issues and encourage them to take action when necessary. Building the capacity of universities, the national media, and NGOs to evaluate, propose, and implement conservation-related policies also effectively encourages national-level action. Individuals, organizations, and businesses should start new foundations to financially support conservation efforts. One of the most important trends in conservation funding and policy is the increased strength of international NGOs, such as the World Wide Fund for Nature (with about 5 million members) and the Royal Society for the Protection of Birds. The number of NGOs has risen dramatically in past decades, and they can substantially influence local conservation programs and environmental policy at the local, national, and international levels (WRI 2003).

Problem: Many businesses, banks, and governments remain uninterested in, and unresponsive to, conservation issues.

Solution: Leaders may become more willing to support conservation efforts if they receive additional information about the benefits of more sustainable practices or perceive strong public support for conservation initiatives (Robinson 2012). In countries with fairly open societies, lobbying and similar efforts may be effective in changing the policies of unresponsive institutions because politicians and other officials generally want to

Conservation biologists often work with environmental activists and nongovernmental organizations to protect biological diversity. This often means finding a balance between doing the best possible science and political advocacy.

avoid bad publicity. Petitions, rallies, letter-writing campaigns, and economic boycotts all have their place if requests for change are ignored. In many situations, radical environmental groups such as Greenpeace and Earth First! dominate media attention with dramatic, publicity-grabbing actions, while mainstream conservation organizations follow after them to negotiate a compromise. In closed societies, focusing on identifying and educating key leaders is usually a better strategy because of the dangers faced by any opponents of the government. A better understanding of the diverse values that different cultures attribute to biodiversity also can help in promoting sustainable practices.

The Role of Conservation Biologists

The problems and solutions outlined here underscore the importance of conservation biologists, who are among the primary participants in this arena. Conservation biology differs from many other scientific disciplines in that it plays an active role in preserving biodiversity in all its forms: species, genetic variability, biological communities, and ecosystem functions. As a result, conservation biologists must balance political advocacy with scientific credibility, maintaining the greatest possible objectivity in their scientific research (Horton et al. 2016). Members of the diverse disciplines that contribute to conservation biology share the common goal of protecting biodiversity in practice, rather than simply investigating and talking about it (Laurance et al. 2013). However, conservationists must work together to provide practical solutions to address real-world situations (Redpath et al. 2013).

Working together is easier said than done, and there is concern among some conservation biologists that disagreements between saving biodiversity for its intrinsic value and biodiversity's instrumental value (i.e., use values; see Chapter 3) pose a danger for the future of conservation (Tallis et al. 2014). In a 2014 issue of *Nature*, Heather Tallis, Jane Lubchenco, and 238 cosignatories petitioned for cooperation between these camps and suggested that the inclusion of women conservation biologists plays a critical role in achieving cooperation. Certainly, the inclusion of all bright minds and good ideas is necessary for the continued success of the discipline.

Challenges for conservation biologists

Decisions about park management and species protection increasingly incorporate the ideas and theories of conservation biology (e.g., Natural Resource Stewardship and Science 2015). The need for maintaining large parks and protecting large populations of endangered species has received widespread attention in both academic and popular literature. The vulnerability of small populations to local extinction, even when they are being carefully protected and managed, and the alarming rates of species extinction and destruction of unique ecosystems worldwide have also been highly publicized. As a result of this publicity, the need to protect biodiversity has entered the political debate and been targeted as a priority for government conservation programs. At the same time, botanical gardens, museums, nature centers, zoos, national parks, and aquariums are reorienting their programs to meet the challenges of protecting biodiversity and embracing the concepts of Earth stewardship (Sayre et al. 2013). The sense of urgency is heightened by the recognition that many endangered species living in cold climates, such as polar bears and penguins, face immediate threats due to a warming climate and the melting of sea ice. What is ultimately required, however, is including the principles of conservation biology into the broader domestic policy arena and economic-planning process (Halpern et al. 2013). Incorporating conservation biology into economic policy and reprioritizing domestic policy goals will require substantial public education and political effort.

Achieving the agenda

Successfully meeting conservation challenges requires that conservation biologists take on several active roles. Conservationists must become more effective *educators, leaders,* and *motivators* in the public forum as well as in the classroom (**FIGURE 12.4**). Conservation biologists need to educate as broad a range of people as possible about the problems that stem from a loss of biodiversity, convey a positive message about what needs to be done, and then give examples of some successes (Bickford et al. 2012; Laurance et al. 2013). The Society for Conservation Biology has made disseminating knowledge the first item in its new code of ethics. Conservation biologists must provide actionable research and leadership that counters the pessimism and passivity so frequently encountered in modern society

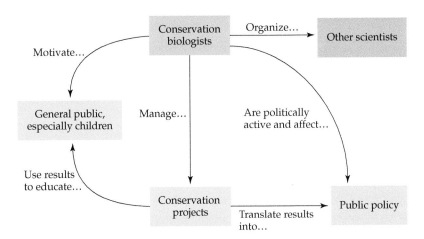

FIGURE 12.4   Conservation biologists must be active in various ways to achieve the goals of conservation biology and to protect biodiversity. Not every conservation biologist can be active in each role, but all of the roles are important.

(Carlson 2013). The International Union for the Conservation of Nature demonstrates this approach with their campaign emphasizing "Love. Not Loss," which attempts to capture people's positive feelings toward nature rather than simply convey dry facts about nature's demise (www.iucn.org). Many groups, including anglers, hunters, bird-watchers, hikers, religious groups, and artists, may help conservation efforts once they become aware of the issues or recognize that their self-interest or emotional well-being depends on conservation (Granek et al. 2008). Some evidence suggests that such awareness is increasing worldwide, even in some of the developing countries, such as India (Varma et al. 2015).

Conservation biologists often teach college students and write technical papers addressing conservation issues, but these actions only reach limited audiences: remember that only a few dozen or at most few thousand people read most scientific papers. In contrast, millions of adults watch nature programs on television, especially those produced by the Public Broadcasting Service and the British Broadcasting Corporation, and tens of millions of children watch television programs such as *Go, Diego, Go* and *Wild Kratts,* and movies such as *Avatar, Wall-E,* and *The Lord of the Rings*, which have powerful conservation themes. An especially powerful example of the role of media was the release of *Blackfish* in 2013, a movie that highlighted the treatment of captive killer whales (*Orcinus orca*), including the questionable practice of how they are captured in the wild. The filmmakers worked with scientists and rigorously researched all the information presented. The huge following of the film resulted in legislation proposed in the New York and California legislatures in 2014 to protect the whales (A. J. Wright et al. 2015), and Congress unanimously passed an amendment to the US Agriculture Appropriations Act to update regulations concerning captive marine mammals (Charky 2014). Moreover, on March 16, 2016, SeaWorld announced that the killer whales currently in its care will be their last (CNN 2016).

Conservation biologists must reach a wider range of people by speaking in villages, towns, cities, elementary and secondary schools, parks, neighborhood gatherings, and religious organizations (Swanson et al. 2008). Conservationists should also work to more widely incorporate the themes of conservation into public discussions. Conservation biologists must spend more of their time writing articles and editorials for newspapers, magazines, and blogs, as well as effectively speaking on radio, television, and other mass media, in ways the public can understand (Jacobson 2009). They must make a special effort to talk to children's groups and to write versions of their work that children can read. Hundreds of millions of people visit zoos, aquariums, natural history museums, and botanical gardens, making these places prime venues for communicating conservation messages to the public. Conservation biologists must continue to seek out creative ways to reach wider audiences and avoid repeatedly "preaching to the converted." Potential constituents are becoming more technologically savvy, more diverse, speaking different languages, and viewing wildlife through different cultural lenses. These changes require that conservation biologists use new approaches and social media, such as YouTube, Facebook, Twitter, and blogs, for effective outreach and communication (Jacobson et al. 2014). Data from these approaches can even be used in conjunction with surveys to evaluate how effective scientists are in conveying what they intend to convey and whether they are reaching their intended audiences (Bombaci et al. 2016) (**FIGURE 12.5**).

The efforts of Merlin Tuttle and Bat Conservation International (BCI) illustrate how public attitudes toward even unpopular species can change. BCI has campaigned throughout the United States and the world to educate

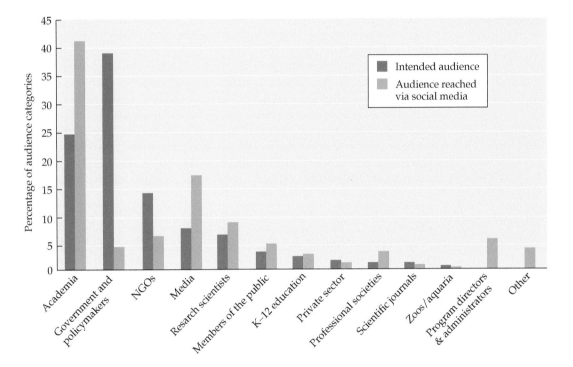

FIGURE 12.5   During the 2013 International Congress for Conservation Biology, efficacy of communication was evaluated by comparing the audience intended by the presenters (as determined via survey) with the audience reached via Twitter, a social media platform. This graph shows the percentage for each of 13 audience categories. "Audience reached" was determined by counting the senders and recipients of 700 live "tweets" during the conference and 1711 "retweets" (forwarding of a Twitter message to another set of followers). The mean number of followers for tweeters and retweeters was 2404, with 40% having over 10,000 followers across 40 countries; thus, considerably more than the 1500 attendees to the conference were reached. Although academics were an intended audience and were, in fact, the greatest proportion of people who were reached, the remainder of the presenters' intended audience differed from those who were reached via tweets; the media and program directors were a greater proportion of their audience than government officials, despite the opposite intent of the presenters. However, these results suggest that using social media to create media interest may be a good way for scientists to reach the general public. (After Bombaci et al. 2016.)

people on the importance of bats in ecosystem health, emphasizing their roles as insect eaters, pollinators, and seed dispersers. A valuable part of this effort has involved producing bat photographs and films of exceptional beauty. In Austin, Texas, Tuttle intervened when citizens petitioned the city government to exterminate the hundreds of thousands of Brazilian free-tailed bats (*Tadarida brasiliensis*) that live under a downtown bridge. He and his colleagues convinced people that the bats are both fun to watch and critical in controlling insect populations that feed on important crop plants and bite people. The situation has changed so drastically that the government now protects these bats as a matter of civic pride and practical pest control, and hundreds of citizens and tourists gather every night to watch the swarm of bats emerge from under the bridge on their nightly foraging forays (**FIGURE 12.6**).

Conservation biologists must also become *politically active leaders* to influence public policy and new laws (Sale 2015). Scientists can, and should, participate in politics without sacrificing professional credibility, in part by practicing transparency and being honest to themselves and others about the impact of personal values and funding sources (Garrard et al. 2015). Involvement in the political process allows conservation biologists to influence the passage of new laws that support the preservation of biodiversity and to argue against legislation that would prove harmful to species or ecosystems.

FIGURE 12.6 Citizens and tourists gather in the evening to watch Brazilian free-tailed bats emerge from their roosts beneath the Congress Avenue bridge in Austin, Texas. (Photograph © Merlin D. Tuttle, Bat Conservation International, www.batcon.org)

An important first step in this process is joining conservation organizations or mainstream political parties to strengthen them by working within these groups and learning more about the issues. It is important to note that there is also room for people who prefer to work by themselves. Difficulties in getting the US Congress to reauthorize the Endangered Species Act and to ratify the CBD and the United Nations Framework Convention on Climate Change illustrate the need for greater political activism on the part of scientists who understand the implications of failing to take immediate action. Though much of the political process is time-consuming and tedious, it often represents the only way to accomplish major conservation goals, such as acquiring new land for reserves or preventing the overexploitation of old-growth forests. Conservation biologists must master the languages and methods of legal and political processes and form effective alliances with environmental lawyers, citizen groups, and politicians.

An example of the effectiveness of alliances between conservation biologists and other groups is the recent expansion of the Ikh Nart Nature Reserve in the Dornogobi province of Mongolia. Populations of endangered cinerous vultures (*Aegypius monachus*), argali sheep (*Ovis ammon*), Siberian ibex (*Capra sibirica*), and many other species of animals and plants have been monitored since 2001 by a partnership of conservation biologists at the Denver Zoo, Mongolian National University, Mongolian National Education University, University of Vermont, and Mongolian Academy of Sciences (IkhNart.com). The scientists communicated the findings of their research to the local government, which responded by doubling the size of the park to about 130,000 ha and to restore some areas that had been badly degraded (Reading 2015).

A key role for conservation biologists to assume is that of *translators*; that is, scientists who can take the data and results of conservation science and translate them into legislation, public policy, and management actions (Courchamp et al. 2015). To be effective, conservation biologists must communicate the relevance of their research and demonstrate that their findings are unbiased, while respecting the values and concerns of most or all stakeholders. Conservation scientists have to be aware of the range of issues that may affect their programs and that their programs might affect, and they should be able to speak to general audiences in terms that they can understand. Conservation biologists must take the lead, as their expertise is needed.

Conservation biologists should become *motivators*, convincing a broad range of people to support conservation efforts and even assisting in formal marketing campaigns to increase awareness (A. J. Wright et al. 2015). Psychological research suggests that successful campaigns to motivate people to take conservation actions must include efforts to provide: (1) information, (2) a sense of belonging, (3) trust, and (4) incentives (Van Vugt 2009). In general, conservationists must demonstrate to local people that protecting the environment, not only saves species and ecosystems, but also improves the long-term health of their families and communi-

> The goal of conservation biology is not just to reveal new knowledge, but also to use that knowledge to protect biodiversity. Conservation biologists must learn to demonstrate the practical application of their work.

ties, their own economic well-being, and their quality of life (Morrison 2016). Public discussions, education, and publicity must be a major part of any such program, and they should be delivered in ways that instill trust and a sense of belonging and community. In particular, conservationists should devote careful attention to convincing business leaders and politicians to support conservation efforts. National leaders may be among the most difficult people to convince, since they must respond to diverse interests. However, whether their conversion occurs due to reason, sentiment, or professional self-interest, once they join the cause of conservation, these leaders may be in a position to make major contributions.

Finally, and most importantly, conservation biologists must become effective *managers* and *practitioners* of conservation projects (Blickley et al. 2013). They must be willing to walk on the land and go out on the water to find out what is really happening, to get dirty, to knock on doors and talk with local people, and to take risks. Conservation biologists must learn everything they can about the species and ecosystems they are trying to protect, and then make that knowledge available to others in a form that can be readily understood and can affect decision making.

If conservation biologists are willing to put their ideas into practice and to work with park managers, land-use planners, politicians, and local people, progress will follow. Getting the right mix of models, new theories, innovative approaches, and practical examples is necessary for the discipline to succeed. Once this balance is found, conservation biologists, working with energized citizens and government officials, will be in a position to protect the world's biodiversity during this unprecedented era of change.

Summary

- Major problems face us in protecting biodiversity. To address these problems, many policies and practices must change. Changes must occur at the local, national, and international levels and will require action on the part of individuals, conservation organizations, and governments.

- Conservation biologists must demonstrate the practical value of the theo-

ries and approaches of their discipline and actively work with all components of society to protect biodiversity and restore the degraded elements of the environment.

- To achieve the long-term goals of conservation biology, practitioners need to become involved in conservation education and the political process.

For Discussion ▪——————————————

1. Consider a current conflict over conservation of biodiversity in the news. Does it fit into one or more of the problem/solution pairs provided in this chapter? What are some possible solutions to the conflict, given what you now know?

2. As a result of studying conservation biology, have you decided to change your lifestyle or your level of political activity?

Do you think you can make a difference in the world, and if so, in what way?

3. Sutherland and colleagues (2009) posed 100 questions for conservation biology (see Suggested Readings). Provide answers for the questions you consider to be the most urgent and important.

Suggested Readings ——————————

Blickley, J. L., K. Deiner, K. Garbach, I. Lacher, M. H. Meed, L. M. Porensky, and 3 others. 2013. Graduate student's guide to necessary skills for nonacademic conservation careers. *Conservation Biology* 27: 24–34. Interpersonal and project-management skills are needed to conduct practical conservation work.

Cafaro, P. 2015. Three ways to think about the sixth mass extinction. *Biological Conservation* 192: 387–393. How we view the biodiversity crisis will impact how we respond to it in the future.

Colón-Rivera, R. J., K. Marshall, F. J. Soto-Santiago, D. Ortiz-Torres, and C. E. Flower. 2013. Moving forward: Fostering the next generation of Earth stewards in the STEM disciplines. *Frontiers in Ecology and the Environment* 11: 383–391. Conservation must be made a part of the educational process.

Costello, M. J., B. Vanhoorne, and W. Appeltans. 2015. Conservation of biodiversity through taxonomy, data publication, and collaborative infrastructures. *Conservation Biology* 29(4): 1094–1099. Collaboration between scientists is essential for protecting biodiversity and should be facilitated.

Danielson, F., K. Pirhofer-Walzi, T. P. Adrian, D. R. Kapijimpanga, N. D. Burgess, P. M. Jensen, and 5 others. 2014. Linking public participation in scientific research to the indicators and needs of international environmental agreements. *Conservation Letters* 7: 12–24. Most indicators of environmental and conservation status can be monitored by trained citizens.

Granek, E. F., E. M. P. Madin, M. A. Brown, W. Figueira, D. S. Cameron, and Z. Hogan. 2008. Engaging recreational fishers in management and conservation: Global case studies. *Conservation Biology* 22: 1125–1134. Recreational fishers can become strong advocates for conservation.

Jacobson, S. K. 2009. *Communication Skills for Conservation Professionals*. 2nd ed. Island Press, Washington, D.C. Researchers have identified practical ways to increase public support for conservation.

Morrison, S. A. 2016. Designing virtuous socio-ecological cycles for biodiversity conservation. *Biological Conservation* 195: 9–16. To accomplish successful conservation, the public must be educated about the relevance of biodiversity to their lives and be inspired to act.

Redford, K. H., B. J. Huntley, D. Roe, T. Hammond, M. Zimsky, T. E. Lovejoy, and 3 others. 2015. Mainstreaming biodiversity: Conservation for the twenty-first century. *Frontiers in Ecology and Evolution* 3: 137. Considerations of biodiversity should be incorporated in policies, practices, and principles for any human activity that affects biodiversity.

Sullivan, B. L., J. L. Aycrigg, J. H. Barry, R. E. Bonney, N. Bruns, C. Cooper and 26 others. 2014. The eBird enterprise: An integrated approach to development and application of citizen science. *Biological Science* 169: 31–40. The user-friendly approach of the eBird program is engaging tens of thousands of bird-watchers and thus providing a valuable data source for conservation biologists.

Sutherland, W. J., W. M. Adams, R. B. Aronson, R. Aveling, T. M. Blackburn, S. Broad, and 38 others. 2009. One hundred questions of importance to the conservation of global biological diversity. *Conservation Biology* 23: 557–567. Presents 100 scientific questions whose answers would have the greatest impact on conservation practice and policy.

Tallis, H., Lubchenco, J., Adams, V. M., Adams-Hosking, C., Agostini, V. N., and Kovács-Hostyánszki, A. 2014. Working together: A call for inclusive conservation. *Nature* 515(7525): 27–28. To be effective, conservation planning must include women as well as men.

Appendix

Selected Environmental Organizations and Sources of Information

The following are useful printed resources:

Conservation Directory 2017: The Guide to Worldwide Environmental Organizations (2016), published by Carrel Press. This directory lists over 4000 local, national, and international conservation organizations; conservation publications; and more than 18,000 leaders and officials in the field of conservation.

Pursuing Sustainability: A Guide to the Science and Practice (2016), published by Princeton University Press. A framework for connecting science with practice that draws upon the current literature.

The ECO Guide to Careers that Make a Difference: Environmental Work for a Sustainable World (2004), published by Island Press.

Online searches, especially using Google, provide a powerful way to search for information concerning people, organizations, places, and topics. The following are more specialized searchable databases on species and countries:

Encyclopedia of Life
www.eol.org
Developing resource for species biology.

Global Biodiversity Information Facility
www.gbif.org
Free and open access to biodiversity data.

IUCN Red List
www.iucnredlist.org
Information on all plants, fungi and animals that have been globally evaluated for conservation status.

USDA Plants List
plants.usda.gov
Data on plants, including distribution, nativity, and wetland status.

Below is a list of some major conservation organizations and resources:

Association of Zoos and Aquariums (AZA)
8403 Colesville Road, Suite 710
Silver Spring, MD 20910 USA
www.aza.org
Preservation and propagation of captive wildlife.

BirdLife International
The David Attenborough Building, 1st floor
Pembroke Street
Cambridge, CB2 3QZ, UK
www.birdlife.org
Determines status, priorities, and conservation plans for birds throughout the world.

Center for Plant Conservation
15600 San Pasqual Valley Rd.
Escondido, CA 92027, USA
www.centerforplantconservation.org, www.
mobot.org
Major center for worldwide plant conservation
activities.

Convention on Biological Diversity Secretariat
413 Rue Saint-Jacques, Suite 800
Montreal, Quebec, H2Y 1N9, Canada
www.cbd.int
Promotes the goals of the CBD: sustainable
development, biodiversity conservation, and
equitable sharing of genetic resources.

CITES Secretariat of Wild Fauna and Flora
International Environment House
11 Chemin des Anémones
CH-1219 Châtelaine, Geneva, Switzerland
www.cites.org
Regulates trade in endangered species.

Conservation International (CI)
2011 Crystal Drive, Suite 500
Arlington, VA 22202 USA
www.conservation.org
Active in international conservation efforts and
developing conservation strategies; home of the
Center for Applied Biodiversity Science.

Earthwatch Institute
114 Western Avenue
Boston, MA 02134 USA
www.earthwatch.org
Clearinghouse for international conservation
projects in which volunteers can work with
scientists.

Environmental Defense Fund (EDF)
1875 Connecticut Ave, NW, Suite 600
Washington, DC 20009 USA
www.edf.org
Involved in scientific, legal, and economic
issues.

Employment Opportunities
Various organizations have websites with
environmental and conservation opportunities
and internships throughout the world: www.
webdirectory.com/employment, www.ecojobs.
com, to name a few. A publication of interest is
Careers in the Environment by Mike Fasulo and
Paul Walker, published by McGraw-Hill.

European Center for Nature Conservation (ECNC)
P. O. Box 90154
5000 LG Tilburg, the Netherlands
www.ecnc.org
Provides the scientific expertise that is required
for formulating conservation policy.

Fauna & Flora International
Fauna & Flora International
The David Attenborough Building
Pembroke Street
Cambridge, CB2 3QZ UK
www.fauna-flora.org
Long-established international conservation
body acting to protect species and ecosystems.

Food and Agriculture Organization of the United Nations (FAO)
Viale delle Terme di Caracalla
00513 Rome, Italy
www.fao.org
A UN agency supporting sustainable
agriculture, rural development, and resource
management.

Friends of the Earth
1101 15th Street NW, 11th floor
Washington, DC 20036 USA
www.foe.org
Attention-grabbing organization working to
improve and expand environmental policy.

Global Environment Facility (GEF) Secretariat
1818 H Street NW, MSN N8-800
Washington, DC 20433 USA
www.thegef.org/gef
Funds international biodiversity and
environmental projects.

Greenpeace International
Ottho Heldringstraat 5
1006 AZ Amsterdam, the Netherlands
www.greenpeace.org/international
Activist organization known for grassroots
efforts and dramatic protests against
environmental damage.

National Audubon Society
225 Varick Street, 7th floor
New York, NY 10014 USA
www.audubon.org
Involved in wildlife conservation, public
education, research, and political lobbying,
with emphasis on birds.

National Council for Science and the Environment (NCSE)

1101 17th Street NW, Suite 250
Washington, DC 20036 USA
www.ncseonline.org
Works to improve the scientific basis for environmental decision making; their website provides extensive environmental information.

National Wildlife Federation (NWF)

11100 Wildlife Center Drive
Reston, VA 20190 USA
www.nwf.org
Advocates for wildlife conservation. Publishes the *Conservation Directory 2005–2006*, as well as the children's publications *Ranger Rick* and *Your Big Backyard*.

Natural Resources Defense Council (NRDC)

40 West 20th Street
New York, NY 10011 USA
www.nrdc.org
Uses legal and scientific methods to monitor and influence government actions and legislation.

The Nature Conservancy (TNC)

4245 North Fairfax Drive, Suite 100
Arlington, VA 22203 USA
www.nature.org
Emphasizes land preservation.

NatureServe

4600 North Fairfax Drive, 7th floor
Arlington, VA 22203 USA
www.natureserve.org
Maintains databases of endangered species for North America.

The New York Botanical Garden (NYBG) Institute of Economic Botany (IEB)

International Plant Science Center, the New York Botanical Garden
2900 Southern Boulevard
Bronx, NY 10458 USA
www.nybg.org
Conducts research and conservation programs involving plants that are useful to people.

Rainforest Action Network

425 Bush Street, Suite 300
San Francisco, CA 94108 USA
www.ran.org
Works for rain forest conservation and human rights.

Ocean Conservancy

1300 19th Street NW, 8th floor
Washington, DC 20036 USA
www.oceanconservancy.org
Focuses on marine wildlife and ocean and coastal habitats.

Royal Botanic Gardens, Kew

Richmond, Surrey, TW9 3AB, UK
www.kew.org
The famous Kew Gardens are home to a leading botanical research institute and an enormous plant collection.

Sierra Club

85 Second Street, 2nd floor
San Francisco, CA 94105 USA
www.sierraclub.org
Leading advocate for the preservation of wilderness and open space.

Smithsonian National Zoological Park

3001 Connecticut Avenue NW
Washington, DC 20008 USA
www.nationalzoo.si.edu
The National Zoo and the nearby U.S. National Museum of Natural History represent a vast resource of literature, biological materials, and skilled professionals.

Society for Conservation Biology (SCB)

1133 15th St. NW, Suite 300
Washington, DC 20005 USA
www.conbio.org
Leading scientific society for the field. Develops and publicizes new ideas and scientific results through the journal *Conservation Biology* and annual meetings.

Student Conservation Association (SCA)

689 River Road
P.O. Box 550
Charlestown, NH 03603 USA
www.thesca.org
Places volunteers and interns with conservation organizations and public agencies.

United Nations Development Programme (UNDP)

1 United Nations Plaza
New York, NY 10017 USA
www.undp.org
Funds and coordinates international economic development activities.

United Nations Environment Programme (UNEP)

United Nations Avenue, Gigiri
P.O. Box 30552, 00100
Nairobi, Kenya
www.unep.org
International program of environmental research and management.

United Nations Environment Programme World Conservation Monitoring Centre (UNEP-WCMC)

219 Huntingdon Road
Cambridge, CB3 0DL, UK
www.unep-wcmc.org
Monitors global wildlife trade, the status of endangered species, natural resource use, and protected areas.

United States Fish and Wildlife Service (USFWS)

Department of the Interior
1849 C Street NW
Washington, DC 20240 USA
www.fws.gov
The leading U.S. government agency concerned with conservation research and management; with connections to state governments and other government units, including the National Marine Fisheries Service, the U.S. Forest Service, and the Agency for International Development, which is active in developing nations. The *Conservation Directory 2005–2006*, mentioned previously, shows how these units are organized.

Wetlands International

P.O. Box 471
6700 AL Wageningen, the Netherlands
www.wetlands.org
Focus on the conservation and management of wetlands.

The Wilderness Society

1615 M Street NW
Washington, DC 20036 USA
www.wilderness.org
Devoted to preserving wilderness and wildlife.

Wildlife Conservation Society (WCS)

2300 Southern Boulevard
Bronx, NY 10460 USA
www.wcs.org
Leaders in wildlife conservation and research.

The World Bank

1818 H Street NW
Washington, DC 20433 USA
www.worldbank.org
Multinational bank involved in economic development; increasingly concerned with environmental issues.

World Conservation Union (IUCN)

Rue Mauverney 28
1196, Gland, Switzerland
www.iucn.org
Coordinating body for international conservation efforts. Produces directories of specialists and the Red List of endangered species.

World Resources Institute (WRI)

10 G Street NE, Suite 800
Washington, DC 20002 USA
www.wri.org
Produces environmental, conservation, and development reports.

World Wildlife Fund (WWF)

1250 24th Street NW
Washington, DC 20037 USA
www.worldwildlife.org, www.wwf.org
Major conservation organization, with branches throughout the world. Active in national park management.

The Xerces Society

628 NE Broadway, Suite 200
Portland, OR 97232 USA
www.xerces.org
Focuses on the conservation of insects and other invertebrates.

Zoological Society of London (ZSL)

Outer Circle, Regent's Park
London, NW1 4RY UK
www.zsl.org
Center for worldwide activities to preserve nature.

Glossary

Numbers in brackets indicate the chapter in which the term is defined.

A

acid rain Rainwater that has become acidic due to air pollution. [4]

adaptive management Implementing a management plan and monitoring how well it works, then using the results to adjust the management plan. [8]

adaptive restoration Using monitoring data to adjust management plans to achieve restoration goals. [10]

Aichi Biodiversity Targets A list of goals to achieve sustainability and the protection of biodiversity developed by the Convention on Biological Diversity (CBD). [11]

Allee effect Inability of a species' social structure to function once a population of that species falls below a certain number or density of individuals. [5]

alleles Different forms of the same gene (e.g., different alleles of the genes for certain blood proteins produce the different blood types found among humans). [2]

alpha diversity The number of different species in a community or specific location; species richness. [2]

amenity value Recreational value of biodiversity, including ecotourism. [3]

arboretum Specialized botanical garden focusing on trees and other woody plants. [7]

artificial incubation Conservation strategy that involves humans taking care of eggs or newborn animals. [7]

artificial insemination Introduction of sperm into a receptive female animal by humans; used to increase the reproductive output particularly of endangered species. [7]

B

background extinction rates The rate of species loss expected to occur in the absence of human impact. [5]

beneficiary value *See* bequest value. [3]

bequest value The benefit people receive by preserving a resource or species for their children and descendants or future generations, and quantified as the amount people are willing to pay for this goal. Also known as beneficiary value. [3]

beta diversity Rate of change of species composition along a gradient or transect. [2]

binomial The unique two-part Latin name taxonomists bestow on a species, such as *Canis lupus* (gray wolf) or *Homo sapiens* (humans). [2]

bioblitz A one-day event in which scientists and citizen scientists perform an intensive biological survey of a designated area in a short time with the goal of documenting all living species in that area. [2]

biocontrol The use of one type of organism, such as an insect, to manage another, undesirable, species, such as an invasive plant. [10]

biocultural restoration Restoring lost ecological knowledge to people to give them an appreciation of the natural world. [10]

biodiversity indicators Species or groups of species that provide an estimate of the biodiversity in an area when data on the whole community is unavailable. Also known as surrogate species. [6]

biodiversity offsets *See* compensatory mitigation. [10]

biodiversity The complete range of species, biological communities, and their ecosystem interactions and genetic variation within species. Also known as biological diversity. [1]

biological community A group of species that occupies a particular locality. [2]

biological definition of species Among biologists, the most generally used of several definitions of "species." A group of individuals that can potentially breed among themselves in the wild and that do not breed with individuals of other groups. *Compare with* morphological definition of species. [2]

biological diversity *See* biodiversity. [1]

biomagnification Process whereby toxins become more concentrated in animals at higher levels in the food chain. [4]

biophilia The postulated predisposition in humans to feel an affinity for the diversity of the living world. [1]

biopiracy Collecting and using biological materials for commercial, scientific, or personal use without obtaining the necessary permits. [3]

bioprospecting Collecting biological materials as part of a search for new products. [3]

bioregional management Management system that focuses on a single large ecosystem or a series of linked ecosystems, particularly where they cross political boundaries. [9]

bioremediation The use of an organism to clean up pollutants, such as bacteria that break down the oil in an oil spill or wetland plants that take up agricultural runoff to clean the water [10]

biosphere reserves Protected areas established as part of a United Nations program to demonstrate the compatibility of biodiversity conservation and sustainable development to benefit local people. [8]

biota A region's plants and animals. [2]

Bonn Convention Treaty to protect European species, particularly migratory species. Also called the Convention on the Conservation of Migratory Species of Wild Animals. [6]

bushmeat Meat from any wild animal. [3]

C

carnivores An animal species that consumes other animals to survive. Also called a secondary consumer or predator. *Compare with* primary consumer. [2]

carrying capacity The number of individuals or biomass of a species that an ecosystem can support. [2]

census A count of the number of individuals in a population. [6]

co-management Local people working as partners with government agencies and conservation organizations in protected areas. [9]

commodity values *See* direct use values. [3]

common-property resources Natural resources that are not controlled by individuals but collectively owned by society. Also known as open-access resources or common-pool resources. [3]

community conserved areas Protected area managed and sometimes established by local people. [9]

community-based conservation (CBC) Protection of natural areas and resources that is controlled, owned, and/or managed by the local people; an alternative to government-based conservation. [9]

compensatory mitigation When a new site is created or rehabilitated in compensation for a site damaged or destroyed elsewhere. Also known as biodiversity offset. [10]

competition A contest between individuals or groups of animals for resources. Occurs when individuals or a species use a limiting resource in a way that prevents others from using it. [2]

conservation banking A system involving developers paying landowners for the preservation of an endangered species or protected habitat type (or even restoration of a degraded habitat) to compensate for a species or habitat that is destroyed elsewhere. [11]

conservation biology Scientific discipline that draws on diverse fields to carry out research on biodiversity, identify threats to biodiversity, and play an active role in the preservation of biodiversity. [1]

conservation concessions Methods of protecting land whereby a conservation organization pays a government or other landowner to preserve habitat rather than allow an extractive industry to damage the habitat. [11]

conservation corridors Connections between protected areas that allow for dispersal and migration. Also known as habitat corridors, or movement corridors. [8]

conservation development *See* limited development. [11]

conservation easements (CEs) Method of protecting land in which landowners give up the right to develop or build on their property, often in exchange for financial or tax benefit. [9]

conservation leasing Providing payments to private landowners who actively manage their land for biodiversity protection. [11]

consumptive use value Value assigned to goods that are collected and consumed locally. [3]

Convention Concerning the Protection of the World Cultural and Natural Heritage *See* World Heritage Convention. [11]

Convention on Biological Diversity (CBD) A treaty that obligates countries to protect the biodiversity within their borders, and gives them the right to receive economic benefits from the use of that biodiversity. [11]

Convention on International Trade in Endangered Species (CITES) The international treaty that establishes lists (known as Appendices) of species for which international trade is to be prohibited, regulated, or monitored. [6]

Convention on the Conservation of Migratory Species of Wild Animals (CMS) *See* Bonn Convention. [6]

cost–benefit analysis Comprehensive analysis that compares values gained against the costs of a project or resource use. [3]

cost–effectiveness analysis An alternative to cost–benefit analysis that compares the impact (financial and otherwise) and costs of alternative means of accomplishing an objective, such as the protection of a species. [3]

cross-fostering Conservation strategy in which individuals from a common species raise the offspring of a rare, related species. [7]

cultural eutrophication Algal blooms and associated impacts caused by excess mineral nutrients released into the water from human activity. [10]

D

debt-for-nature swaps Agreements in which a developing country agrees to fund additional conservation activities in exchange for a conservation organization canceling some of its discounted debt. [11]

decomposers A species that feeds or grows on dead plant and animal material. Also called a detritivore. [2]

deep ecology Philosophy emphasizing biodiversity protection, personal lifestyle changes, and working towards political change. [3]

degazettement Government actions taken to remove the legal status of protected areas. [8]

demographic stochasticity Random variation in birth, death, and reproductive rates in small populations, sometimes causing further decline in population size. Also called demographic variation. [5]

demographic studies Studies in which individuals and populations are monitored over time to determine rates of growth, reproduction, and survival. [6]

demographic variation *See* demographic stochasticity. [5]

detritivores *See* decomposer. [2]

direct use values Value assigned to products, such as timber and animals, that are harvested and directly used by the people who harvest them. Also known as commodity value or private goods. [3]

E

Earth Summit An international conference held in 1992 in Rio de Janeiro that resulted in new environmental agreements. Also known as the Rio Summit. [11]

ecocolonialism Practice of governments and conservation organizations disregarding the land rights and traditions of local people in order to establish new conservation areas. [9]

ecological economics Discipline that includes valuations of biodiversity in economic analyses. [3]

ecological footprint The influence a group of people has on both the surrounding environment and locations across the globe as measured by global hectares per person. [4]

ecological resilience A natural ability to recover after disturbance. [10]

ecological restoration Altering a site to reestablish an indigenous ecosystem. [10]

ecological succession A predictable, gradual and progressive change in species over time [8]

ecologically extinct A species that has been so reduced in numbers that it no longer has a significant ecological impact on the biological community. *See* functionally extinct. [5]

economic development Economic activity focused on improvements in efficiency and organization but not necessarily on increases in resource consumption. [11]

economic growth Economic activity characterized by increases in the amount of resources used and in the amount of goods and services produced. [11]

economics The study of factors affecting the production, distribution, and consumption of goods and services. [3]

ecosystem diversity The variety of ecosystems present in a place or geographic area. [2]

ecosystem engineers Species that modify the physical structure of an ecosystem. [2]

ecosystem management Large-scale management that often involves multiple stakeholders, the primary goal of which is the preservation of ecosystem components and processes. [9]

ecosystem management Large-scale management that often involves multiple stakeholders, the primary goal of which is the preservation of ecosystem components and processes. [1]

ecosystem services Range of benefits provided to people from ecosystems, including flood control, clean water, and reduction of pollution.

ecosystem A biological community together with its associated physical and chemical environment. [2]

ecotourism Tourism, especially in developing countries, focused on viewing unusual and/or especially charismatic biological communities and species that are unique to a country or region. [3]

edge effects Altered environmental and biological conditions at the edges of a fragmented habitat. [4]

effective population size (N_e) The number of breeding individuals in a population. [5]

embryo transfer The surgical implantation of embryos into a surrogate mother; used to increase the number of individuals of a rare species, with a common species used as the surrogate mother. [7]

Endangered Species Act (ESA) An important US law passed to protect endangered species and the ecosystems in which they live. [6]

endemic species Species found in one place and nowhere else (e.g., the many lemur species found only on the island of Madagascar). [2]

endemic Occurring in a place naturally, without the influence of people (e.g., gray wolves are endemic to Canada). [5]

environmental economics Discipline that examines the economic impacts of environmental policies and decisions. [3]

environmental ethics Discipline of philosophy that articulates the intrinsic value of the natural world and people's responsibility to protect the environment. [3]

environmental impact assessments Evaluation of a project that considers its possible present and future impacts on the environment. [3]

environmental justice Movement that seeks to empower and assist poor and politically weak people in protecting their own environments; their well-being and the protection of biological diversity are enhanced in the process. [3]

environmental stochasticity Random variation in the biological and physical environment. Can increase the risk of extinction in small populations. [5]

environmentalism A widespread movement, characterized by political activism, with the goal of protecting the natural environment. [1]

eutrophication Process of degradation in aquatic environments caused by nitrogen and phosphorus pollution and characterized by algal blooms and oxygen depletion. [4]

evolutionary definition of species A group of individuals that share unique similarities of their DNA and hence their evolutionary past. [2]

ex situ conservation Preservation of species under artificial conditions, such as in zoos, aquariums, and botanical gardens. [7]

existence value The benefit people receive from knowing that a habitat or species exists and quantified as the amount that people are willing to pay to prevent species from being harmed or going extinct, habitats from being destroyed, and genetic variation from being lost. [3]

extant Presently alive; not extinct. [5]

externalities Hidden costs or benefits that result from an economic activity to individuals or a society not directly involved in that activity. [3]

extinct in the wild A species no longer found in the wild, but individuals may remain alive in zoos, botanical gardens, or other artificial environments. [5]

extinct The condition in which no members of a species are currently living. [5]

extinction cascade A series of linked extinctions whereby the extinction of one species leads to the extinction of one or more other species. [2]

extinction vortex Tendency of small populations to decline toward extinction. [5]

extirpated Local extinction of a population, even though the species may still exist elsewhere. [5]

extractive reserve Protected area in which sustainable extraction of certain natural products is allowed. [9]

F

flagship species A species that captures public attention, aids in conservation efforts, such as establishing a protected area, and may be crucial to ecotourism. [6]

focal species A species that provides a reason for establishing a protected area. [6]

food chains Specific feeding relationships between species at different trophic levels. [2]

food web A network of feeding relationships among species. [2]

founder effect Reduced genetic variability that occurs when a new population is established ("founded") by a small number of individuals. [5]

four Rs Guidelines used by conservation biologists when designing nature reserves: representation, resiliency, redundancy, and reality. [8]

frontier forest Intact blocks of undisturbed forest large enough to support all aspects of biodiversity. [4]

functional diversity The diversity of organisms categorized by their ecological roles or traits rather than their taxonomy. [2]

functionally extinct The state in which a species persists at such reduced numbers that its effects on the other species in its community are negligible. *See* ecologically extinct. [5]

G

gamma diversity The number of species in a large geographic area. [2]

gap analysis Comparing the distribution of endangered species and biological communities with existing and proposed protected areas to determine gaps in protection. [8]

gene pool The total array of genes and alleles in a population. [2]

genes Units (DNA sequences) on a chromosome that code for specific proteins. Also called loci. [2]

genetic diversity The range of genetic variation found within a species. [2]

genetic drift Loss of genetic variation and change in allele frequencies that occur by chance in small populations. [5]

genetically modified organisms (GMOs) Organisms whose genetic code has been altered by scientists using recombinant DNA technology. [4]

genome resource banking (GRB) Collecting DNA, eggs, sperm, embryos, and other tissues from species that can be used in breeding programs and scientific research. [7]

genotype Particular combination of alleles that an individual possesses. [2]

geographic information systems (GIS) Computer analyses that integrate and display spatial data; relating in particular to the natural environment, ecosystems, species, protected areas, and human activities. [8]

Global Environment Facility (GEF) A large international program involved in funding conservation activities in developing countries. [11]

globalization The increasing interconnectedness of the world's economy. [4]

globally extinct No individuals are presently alive anywhere. [5]

greenhouse effect Warming of the Earth caused by carbon dioxide and other "greenhouse gases" in the atmosphere that allow the sun's radiation to penetrate and warm the Earth but prevent the heat generated by sunlight from re-radiating. Heat is thus trapped near the surface, raising the planet's temperature. [4]

greenhouse gases Gases in the atmosphere, primarily carbon dioxide, that are transparent to sunlight but that trap heat near the Earth's surface. [4]

guild A group of species at the same trophic level that use approximately the same environmental resources. [2]

H

habitat conservation plans (HCPs) Regional plans that allow development in designated areas while protecting biodiversity in other areas. [6]

habitat corridors *See* conservation corridor. [8]

habitat fragmentation The process whereby a large, continuous area of habitat is both reduced in area and divided into two or more fragments. [4]

habitat islands Intact habitat surrounded by an unprotected matrix of inhospitable terrain. [8]

habitat The location or type of environment in which a specific animal or plant species lives. [2]

hard release In the establishment of a new population, when individuals from an outside source are released in a new location without assistance. *Compare with* soft release. [7]

healthy ecosystem Ecosystem in which processes are functioning normally, whether or not there are human influences. [2]

herbivores A species that eats plants or other photosynthetic organisms. Also called a primary consumer. [2]

herbivory Predation on plants. [2]

heterozygous Condition of an individual having two different allele forms of the same gene. [2]

homozygous Condition of an individual having two identical allele forms of the same gene. [2]

hotspots Regions with numerous species, many of which are endemic, that are also under immediate threat from human activity. [6]

hybridize Interbreeding between different species. [2]

hybrids Intermediate offspring resulting from mating between individuals of two different species. [2]

I

in situ conservation Preservation of natural communities and populations of endangered species in the wild. [7]

inbreeding depression Lowered reproduction or production of weak offspring following mating among close relatives or self-fertilization. [5]

indicator species Species used in a conservation plan to identify and often protect a biological community or set of ecosystem processes. [6]

indirect use values Values provided by biodiversity that do not involve harvesting or destroying the resource (such as water quality, soil protection, recreation, and education). Also known as public goods. [3]

integrated conservation development projects (ICDPs) Conservation projects that also provide for the economic needs and welfare of local people. [9]

integrated pest management An approach to controlling undesirable plants or animals that has the goal of minimizing harm to the ecosystem and people, while being cost-effective. [4]

International Union for Conservation of Nature (IUCN) *See* IUCN. [4]

intrinsic value Value of a species and other aspects of biodiversity for their own sake, unrelated to human needs. [3]

introduction program Moving individuals to areas outside their historical range in order to create a new population of an endangered species. [7]

invasive An introduced species that increases in abundance at the expense of native species. [4]

island biogeography model Formula for the relationship between island size and the number of species living on the island; the model can be used to predict the impact of habitat destruction on species extinctions, viewing remaining habitat as an "island" in the "sea" of a degraded ecosystem. [5]

IUCN International Union for the Conservation of Nature is a major international conservation organization; previously known as The World Conservation Union. [4]

K

keystone resources Any resource in an ecosystem that is crucial to the survival of many species; for example, a watering hole. [2]

keystone species A species that has a disproportionate impact (relative to its numbers or biomass) on the organization of a biological community. Loss of a keystone species may have far-reaching consequences for the community. [2]

L

land ethic Aldo Leopold's philosophy advocating human use of natural resources that is compatible with or even enhances ecosystem health. [1]

land sharing Land use which combines resource use and conservation. [9]

land sparing Land which is protected when other lands are used more intensively. [9]

land trusts Conservation organizations that protect and manage land. [11]

landraces A variety of crop that has unique genetic characteristics; local species that have been adapted by humans over time. [7]

landscape ecology Discipline that investigates patterns of habitat types and their influence on species distribution and ecosystem processes. [8]

legal title The right of ownership of land, recognized by a government and/or judicial system; traditional people often struggle to achieve this recognition. [9]

limited development Compromise involving a landowner, a property developer, and a conservation organization that combines some development with protection of the remaining land. [11]

limiting resource Any requirement for existence whose presence or absence limits a population's size. In the desert, for example, water is a limiting resource. [2]

Living Planet Index A measure of the conservation status of species, based on the IUCN categories. [6]

locally extinct A species that no longer exists in a place where it used to occur, but still exists elsewhere. [5]

loci (singular, locus) *See* genes. [2]

M

management plans A statement of how to protect biodiversity in an area, along with methods for implementation. [8]

market failure Misallocation of resources in which certain individuals or businesses benefit from using a common resource, such as water, the atmosphere, or a forest, but other individuals, businesses or the society at large bears the cost. [3]

metapopulation Shifting mosaic of populations of the same species linked by some degree of migration; a "population of populations." [6]

minimum dynamic area (MDA) Area needed for a population to have a high probability of surviving into the future. [6]

minimum viable population (MVP) Number of individuals necessary to ensure a high probability that a population will survive a certain number of years into the future. [6]

mitigation Process by which a new population or habitat is created to compensate for a habitat damaged or destroyed elsewhere. [7]

morphological definition of species A group of individuals, recognized as a species, that is morphologically, physiologically, or biochemically distinct from other groups. *Compare with* biological definition of species. [2]

morphospecies Individuals that are probably a distinct species based on their appearance but that do not currently have a scientific name. [2]

movement corridors *See* conservation corridors. [8]

multiple-use habitat An area managed to provide a variety of goods and services. [9]

mutations Changes that occur in genes and chromosomes, sometimes resulting in new allele forms and genetic variation. [2]

mutualism When two species benefit each other by their relationship. [2]

N

national environmental fund (NEF) A trust fund or foundation that uses its annual income to support conservation activities. [11]

natural history The ecology and distinctive characteristics of a species. [6]

non-use values Values of something that is not presently used; for example, existence value. [3]

nonconsumptive use value Value assigned to benefits provided by some aspect of biodiversity that does not involve harvesting or destroying the resource (such as water quality, soil protection, recreation, and education). [3]

nonpoint source pollution Pollution coming from a general area rather than a specific site. [10]

normative discipline A discipline that embraces ethical commitment rather than ethical neutrality. [1]

novel ecosystems Ecosystems in which there is a mixture of native and nonnative species coexisting in a community unlike the original or reference site. [10]

O

omnivores Species that eat both plants and animals. [2]

open-access resources Natural resources that are not controlled by individuals but are collectively owned by society. [3]

option value Value of biodiversity in providing possible future benefits for human society (such as new medicines). [3]

P

parasites Organisms that live on or in another organism (host), receiving nutritive benefit while decreasing the fitness of the host, which remains alive. [2]

Paris Accord An agreement made in Paris in 2015 by 195 nations to lower greenhouse gas emissions with the goal of preventing atmospheric temperatures from increasing more than 2°C. [11]

passive restoration Letting an ecosystem recover on its own. [10]

pathogens Disease-causing organisms. [2]

payments for ecosystem services (PES) Direct payments to individual landowners and local communities that protect species or critical ecosystem characteristics. [9]

perverse subsidies Government payments or other financial incentives to industries that result in environmentally destructive activities. [3]

phenotype The morphological, physiological, anatomical, and biochemical characteristics of an individual that result from the expression of its genotype in a particular environment. [2]

polymorphic genes Within a population, genes that have more than one form or allele. [2]

population biology Study of the ecology and genetics of populations, often with a focus on population numbers. [6]

population bottleneck A radical reduction in population size (e.g., following an outbreak of infectious disease), sometimes leading to the loss of genetic variation. [5]

population viability analysis (PVA) Demographic analysis that predicts the probability of a population persisting in an environment for a certain period of time; sometimes linked to various management scenarios. [6]

population A geographically defined group of individuals of the same species that mate and otherwise interact with one another. *Compare with* metapopulation. [2]

precautionary principle Principle stating that it may be better to avoid taking a particular action due to the possibility of causing unexpected harm. [3]

predation Act of killing and consuming another organism for food. [2]

predator release hypothesis An hypothesis which attributes the success of invasive species to the absence of specialized natural predators and parasites in their new range. [4]

predators *See* carnivores. [2]

preservationist ethic A belief in the need to preserve wilderness areas for their intrinsic value. [1]

prey An animal that is eaten as food by another species. [2]

primary consumers *See* herbivores. [2]

primary producers Organisms such as green plants, algae, and seaweeds that obtain their energy directly from the sun via photosynthesis. Also known as autotrophs. [2]

private goods *See* direct use values. [3]

productive use value Value assigned to products that are sold in markets. [3]

protected area A habitat managed primarily or in large part for biodiversity. [8]

public goods Nonconsumptive benefits that belong to society in general, without private ownership. Also known as indirect use values. [3]

R

Ramsar Convention on Wetlands A treaty that promotes the protection of wetlands of international importance. [11]

rapid biodiversity assessments Species inventories and vegetation maps made by teams of biologists when urgent decisions must be made on where to establish new protected areas. Also known as *rapid assessment plans* (RAPs). [6]

recombination Mixing of the genes on the two copies of a chromosome that occurs during meiosis (i.e., in the formation of egg and sperm, which contain only one copy of each chromosome). Recombination is an important source of genetic variation. [2]

reconciliation ecology The science of developing urban places in which people and biodiversity can coexist. [9]

reconciliation ecology The science of developing urban places in which people and biodiversity can coexist. [10]

recovery criteria Predetermined thresholds (such as numbers of individuals alive in the wild) that signal that an endangered species can be removed from protection under the Endangered Species Act. [6]

Red Data Books Compilations of lists ("Red Lists") of endangered species prepared by the IUCN and other conservation organizations. [6]

Red List criteria Quantitative measures of threats to species based on the probability of extinction. [6]

Red List Index Measure of the conservation status of species based on the IUCN categories. [6]

Red Lists Lists of endangered species prepared by the IUCN. [6]

Reducing Emissions from Deforestation and Forest Degradation (REDD) Program using financial incentives to reduce the emissions of greenhouse gases from deforestation. [11]

reference sites Control site that provides goals for restoration in terms of species composition, community structure, and ecosystem processes. [10]

reinforcement program Releasing new individuals into an existing population to increase population size and genetic variability. [7]

reintroduction program The release of captive bred or wild-collected individuals at a site within their historical range where the species does not presently occur. [7]

replacement cost approach How much people would have to pay for an equivalent product if what they normally use is unavailable. [3]

representative site Protected area that includes species and ecosystem properties characteristic of a larger area. [6]

resilience The ability of an ecosystem to return to its original state following disturbance. [2]

resistance The ability of an ecosystem to remain in the same state even with ongoing disturbance. [2]

resource conservation ethic Natural resources should be used for the greatest good of the largest number of people for the longest time. [1]

restoration ecology The scientific study of restored populations, communities, and ecosystems. [10]

rewilding Returning species, in particular large mammals to landscape, to approximate their natural condition prior to human impact. [10]

Rio Summit *See* Earth Summit. [11]

S

secondary consumers *See* carnivores. [2]

secondary invasion When the removal of an invasive species is followed by an invasion by a different species. [10]

seed banks Collections of stored seeds, collected from wild and cultivated plants; used in conservation and agricultural programs. [7]

Shannon diversity index A species diversity index that takes into account the numbers of different species and their relative abundance. [2]

shifting cultivation Farming method in which farmers cut down trees, burn them, plant crops for a few years, and then abandon the site when soil fertility declines. Also called "slash-and-burn" agriculture. [4]

sink populations Populations that receive an influx of new individuals from a source population. [6]

sixth extinction episode The present mass extinction event which is just beginning. [5]

SLOSS debate Controversy concerning the relative advantages of a single large or several small conservation areas. "SLOSS" stands for *single large or several small*. [8]

soft release In the establishment of a new population, when individuals are given assistance during or after the release to increase the chance of success. *Compare with* hard release. [7]

source populations Established populations from which individuals disperse to new locations. [6]

species diversity The entire range of species found in a particular place. [2]

species richness The number of species found in a community. [2]

species–area relationship The number of species found in an area increases with the size of the area; i.e., more species are found on large islands than on small islands. (5)

stable ecosystems Ecosystems that are able to remain in roughly the same compositional state despite human intervention or stochastic events such as unseasonable weather. [2]

stochasticity Random variation; variation happening by chance. [5]

survey Repeatable sampling method to estimate population size or density, or some other aspect of biodiversity. [6]

sustainable development Economic development that meets present and future human needs without damaging the environment and biodiversity. [1]

symbiotic A mutualistic relationship in which neither of the two species involved can survive without the other. [2]

T

taxonomists Scientists involved in the identification and classification of species. [2]

tertiary consumers The fourth trophic level, in which predators eat other predators. [2]

threatened Species that fall into the endangered or vulnerable to extinction categories in the IUCN system. Under the US Endangered Species Act, refers to species at risk of extinction, but at a lower risk than endangered species. [6]

tragedy of the commons   The unregulated use of a public resource that results in its degradation. [3]

transects   Lines often designated with measuring tape or permanent markers, along which biological data is collected. [6]

trophic cascade   Major changes in vegetation and biodiversity resulting from the loss of a keystone species. [2]

trophic levels   Levels of biological communities representing ways in which energy is captured and moved through the ecosystem by the various types of species. *See* primary producer; herbivore; predator; detritivore. [2]

U

umbrella species   Protecting an umbrella species results in the protection of other species. [6]

UN Conference on Sustainable Development
Held in 2012, this conference linked biodiversity conservation to sustainable development and controlling climate change, and emphasized the need for market-based solutions. Also unofficially called the Rio+20. [11]

use values   The direct and indirect values provided by some aspect of biodiversity. [3]

W

World Bank   International bank established to support economic development in developing countries. [11]

World Heritage Convention (WHC)   A treaty that protects cultural and natural areas of international significance. [11]

World Heritage site   A cultural or natural area officially recognized as having international significance. [8]

World Summit on Sustainable Development
Held in Johannesburg, South Africa, in 2002, this gathering emphasized achieving the social and economic goals of sustainability. [11]

Z

zoning   A method of managing protected areas that allows or prohibits certain activities in designated places. [8]

Chapter Opening Photograph Credits

Bibliography

Numbers in parentheses indicate the chapter(s) in which the reference is cited.

Abesamis, R. A. and G. R. Russ. 2005. Density-dependent spillover from a marine reserve: Long-term evidence. *Ecological Applications* 15: 1798–1812. (8)

Abson, D. J. and M. Termansen. 2011. Valuing eco-system services in terms of ecological risks and returns. *Conservation Biology* 25: 250–258. (3)

Acuña, V. and A. Ruhí. 2016. Temporary streams: current management challenges and promising solutions. *Restoration Ecology.* In press. (8)

Adams, J. S. 2006. *The Future of the Wild: Radical Conservation for a Crowded World.* Beacon Press, Boston. (10)

Adum, G. B., M. P. Eichhorn, W. Oduro, C. Ofori-Boateng, et al. 2013. Two-stage recovery of amphibian assemblages following selective logging of tropical forests. *Conservation Biology* 27: 354–363. (9)

Aguirre-Acosta, N., E. Kowaljow, and R. Aguilar. 2014. Reproductive performance of the invasive tree *Ligustrum lucidum* in a subtropical dry forest: Does habitat fragmentation boost or limit invasion?. *Biological Invasions* 16(7): 1397–1410. (4)

Albert, A., K. McKonkey, T. Savini, and M. C. Huynen. 2014. The value of disturbance-tolerant cercopithecine monkeys as seed dispersers in degraded habitats. *Biological Conservation* 170: 300–310. (2)

Alexander, S. (ed.). 2009. *Voluntary Simplicity: The Poetic Alternative to Consumer Culture.* Stead and Daughters, Whanganui, New Zealand. (3)

Allee, W. C. 1931. Animal aggregations: a study in general sociology. *The University of Chicago Press, Chicago.* (5)

Allen, C., R. S. Lutz, and S. Demarais. 1995. Red imported fire ant impacts on northern bobwhite populations. *Ecological Applications* 5: 632–638. (4)

Allen, L. S. 2006. Collaboration in the borderlands: the Malpai Borderlands Group. *Rangelands* 28(3): 17–21. (9)

Allen, W. H. 1988. Biocultural restoration of a tropical forest: architects of Costa Rica's emerging Guanacaste National Park plan to make it an integral part of local culture. *BioScience* 38: 156–161. (10)

Allendorf, F. W. and G. Luikart. 2007. *Conservation and the Genetics of Populations.* Blackwell Publishing, Oxford, UK. (5)

Allendorf, T. D. and K. Allendorf. 2012. What every conservation biologist should know about human population. *Conservation Biology* 26(6): 1523–1739. (4)

Alroy, J. 2015. Limits to captive breeding of mammals in zoos. *Conservation Biology* 29(3): 926–931. (7)

Alter, S. E., E. Rynes, and S. R. Palumbi. 2007. DNA evidence for historic population size and past ecosystem impacts of gray whales. *Proceedings of the National Academy of Sciences USA* 104: 15162–15167. (4)

Altman, I., A. M. H. Blakeslee, G. C. Osio, C. B. Rillahan, et al. 2011. A practical approach to implementation of ecosystem-based management: A case study using the Gulf of Maine marine ecosystem. *Frontiers in Ecology and the Environment* 9: 183–189. (9)

Alva-Basurto, J. C. and J. Arias-Gonzalez. 2014. Modelling the effects of climate change on a Caribbean coral reef food web. *Ecological Modelling* 289: 1–14. (2)

Anadón, J. D., A. Gimenez, R. Ballestar, and I. Pérez. 2009. Evaluation of local ecological knowledge as a method for collecting extensive data on animal abundance. *Conservation Biology* 23: 617–625. (8)

Andam, K. S., P. J. Ferraro, K. R. E. Sims, A. Healy, and M. B. Holland. 2010. Protected areas reduced poverty in Costa Rica and Thailand. *Proceedings of the National Academy of Sciences USA* 107: 9996–10001. (8)

Anderson, M. G. and C. E. Ferree. 2010. Conserving the stage: climate change and the geophysical underpinnings of species diversity. *PLoS ONE* 5: e11554. (8)

Andersson, E., S. Barthel, and K. Ahrne. 2007. Measuring social-ecological dynamics behind the generation of ecosystem services. *Ecological Applications.* 17: 1267–1287. (8)

Andrello, M., M. N. Jacobi, S. Manel, W. Thuiller, and D. Mouillot. 2015. Extending networks of protected areas to optimize connectivity and population growth rate. *Ecography* 38(3): 273–282. (8)

Andrew-Essien, E. and F. Bisong. 2009. Conflicts, conservation and natural resource use in protected area systems: An analysis of recurrent issues. *European Journal of Scientific Research* 25: 118–129. (9)

Angelsen, A., P. Jagger, R. Babigumira, B. Belcher, et al. 2014. Environmental income and rural livelihoods: A global-comparative analysis. *World Development* 64: S12–S28. (3)

Área de Conservación Guanacaste (ACG). *www.acguanacaste.ac.cr.* (10)

Arima, E. Y., R. T. Walker, S. Perz, and C. Souza, Jr. 2015. Explaining the fragmentation in the Brazilian Amazonian forest. *Journal of Land Use Science*. doi: 10.1080/1747423X.2015.1027797. (5)

Arizmendi-Mejía, R., C. Linares, J. Garrabou, A. Antunes, et al. 2015. Combining Genetic and Demographic Data for the Conservation of a Mediterranean Marine Habitat-Forming Species. *PLoS ONE*, 10(3), e0119585. (6)

Armstrong, D. P. and P. J. Seddon. 2008. Directions in reintroduction biology. *Trends in Ecology and Evolution* 23: 20–25. (7)

Armsworth, P. R., G. C. Daily, P. Kareiva, and J. N. Sanchirico. 2006. Land market feedbacks can undermine biodiversity conservation. *Proceedings of the National Academy of Sciences USA* 103: 5403–5408. (8)

Arthington, A. H., J. M. Bernardo, and M. Ilhéu. 2014. Temporary rivers: Linking ecohydrology, ecological quality and reconciliation ecology. *River research and Applications* 30(10): 1209–1215. (10)

Association of Zoos and Aquariums. 2009. *www.aza.org* (7)

Athreya, V., M. Odden, J. D. Linnell, J. Krishnaswamy, and U. Karanth. 2013. Big cats in our backyards: Persistence of large carnivores in a human dominated landscape in India. *PLoS ONE* 8(3): e57872. (9)

Audino, L. D., J. Louzada, and L. Comita. 2014. Dung beetles as indicators of tropical forest restoration: is it possible to recover species and functional diversity? *Biological Conservation* 169: 248–257. (10)

Azam, F. and A. Z. Worden. 2004. Oceanography: Microbes, molecules, and marine ecosystems. *Science* 303: 1622–1624. (2)

Bagla, P. 2010. Hardy cotton-munching pests are latest blow to GM crops. *Science* 327: 1439. (4)

Baillie, J. E., B. Collen, R. Amin, H. R. Akcakaya, H. R., et al. 2008. Toward monitoring global biodiversity. *Conservation Letters* 1(1): 18–26. (6)

Baker, C. M., M. Bode, and M. A. McCarthy. 2016. Models that predict ecosystem impacts of reintroductions should consider uncertainty and distinguish between direct and indirect effects. *Biological Conservation*. doi: 10.1016/j.biocon.2016.01.023. (7)

Baker, J., E. J. Milner-Gulland, and N. Leader-Williams. 2012. Park gazettement and integrated conservation and development as factors in community conflict at Bwindi Impenetrable Forest, Uganda. *Conservation Biology* 26: 160–170. (9)

Baker, J. D. and P. M. Thompson. 2007. Temporal and spatial variation in age-specific survival rates of a long-lived mammal, the Hawaiian monk seal. *Proceedings of the Royal Society B* 274: 407–415. (6)

Balmford, A. 1996. Extinction filters and current resilience: the significance of past selection pressures for conservation biology. *TREE* 11: 193–196. (5)

Balmford, A., J. Beresford, J. Green, R. Naidoo, et al. 2009. A global perspective on trends in nature-based tourism. *PLoS Biology* 7: e1000144. (3)

Balmford, A., A. Bruner, P. Cooper, R. Costanza, et al. 2002. Economic reasons for conserving wild nature. *Science* 297: 950–953. (3)

Ban, N. C., M. Mills, J. Tam, C. C. Hicks, et al. 2013. A social-ecological approach to conservation planning: Embedding social considerations. *Frontiers in Ecology and the Environment* 11: 194–202. (9)

Banerjee, S., S. Secchi, J. Fargione, S. Polasky, et al. 2013. How to sell ecosystem services: a guide for designing new markets. *Frontiers in Ecology and the Environment* 11: 297–304. (9)

Banga, S. S. and M. S. Kang. 2014. Developing climate-resilient crops. *Journal of Crop Improvement* 28: 57–87. (7)

Barazani, O., A. Perevolotsky, and R. Hadas. 2008. A problem of the rich: prioritizing local plant genetic resources for ex situ conservation in Israel. *Biological Conservation* 141: 596–600. (9)

Barber-Meyer, S. M. 2010. Dealing with the clandestine nature of wildlife-trade market surveys. *Conservation Biology* 24: 918–923. (3)

Barbier, E. B., J. C. Burgess, J. T. Bishop, and B. A. Aylward. 1994. The economics of the tropical timber trade. Earthscan Publications, London. (3)

Barcia, L. 2015. Addis Ababa: Financing the future or financing failure? *devex. www.devex.com/news/addis-ababa-financing-the-future-or-financing-failure-86561.* (11)

Barnosky, A. D., N. Matzke, S. Tomiya, G. O. U. Wogan, et al. 2011. Has Earth's sixth mass extinction already arrived? *Nature* 471: 51–57. (1)

Baskin, Y. 1997. *The Work of Nature: How the Diversity of Life Sustains Us.* Island Press, Washington, DC. (3)

Basset, Y., H. Barrios, S. Segar, R. B. Srygley, et al. 2015. The Butterflies of Barro Colorado Island, Panama: Local Extinction since the 1930s. *PLoS ONE* 10(8): e0136623. (5)

Bateman, H. L., D. M. Meritt, E. P. Glenn, P. L. Nagler. 2014. Indirect effects of biocontrol of an invasive riparian plant (*Tamarix*) alters habitat and reduces herpetofauna abundance. *Biological Invasions* 17: 87–97. (10)

Bateman, I. J., A. R. Harwood, G. M. Mace, R. T. Watson, et al. 2013. Bringing ecosystem services into economic decision-making: land use in the United Kingdom. *Science* 341: 45–50. (3, 4)

Baudron, F. and K. E. Giller. 2014. Agriculture and nature: Trouble and strife? *Biological Conservation* 170: 232–245. (9)

Bayraktarov, E., M. I. Saunders, S. Abdullah, M. Mills, et al. 2015. The cost and feasibility of marine coastal restoration. *Ecological Applications*. (10)

Bean, D., T. Dudley, and K. Hultine. 2013. Bring on the beetles! The history and impact of tamarisk biological control. *In* A. Sher and M. F. Quigley (eds.), *Tamarix: A Case Study of Ecological Change in the American West,* pp. 377–403. Oxford University Press. (10)

Beane, J. C., S. P. Graham, T. J. Thorp, and L. T. Pusser. 2014. Natural history of the southern hognose snake

(*Heterodon simus*) in North Carolina, USA. *Copeia* 1: 168–175. (6)

Beans, C. M. and D. A. Roach. 2015. An invasive plant alters pollinator-mediated phenotypic selection on a native congener. *American Journal of Botany* 102: 50–57. (4)

Bearzi, G. 2009. When swordfish conservation biologists eat swordfish. *Conservation Biology* 23: 1–2. (3)

Beattie, A. and P. Ehrlich. 2010. The missing link in biodiversity conservation. *Science* 328: 307–308. (3)

Becker, C. G., C. R. Fonseca, C. F. B. Haddad, and P. I. Prado. 2010. Habitat split as a cause of local population declines of amphibians with aquatic larvae. *Conservation Biology* 24: 287–294. (4)

Becker, M., R. McRobb, F. Watson, E. Droge, et al. 2013. Evaluating wire-snare poaching trends and the impacts of by-catch on elephants and large carnivores. *Biological Conservation* 158: 26–36. (8)

Beebee, T. J. C. 2013. Effects of road mortality and mitigation measures on amphibian populations. *Conservation Biology* 27: 657–668. doi: 10.1111/cobi.12063. (4)

Beier, P., W. Spencer, R. F. Baldwin, and B. H. McRae. 2011. Towards best practices for developing regional connectivity maps. *Conservation Biology* 25: 879–892. (8)

Beissinger, S. R. 2015. Endangered species recovery criteria: reconciling conflicting views. *BioScience* 65(2): 121–122. (6)

Beissinger, S. R., E. Nicholson, and H. P. Possingham. 2009. Application of population viability analysis to landscape conservation planning. *In* J. J. Millspaugh and F. R. Thompson, III (eds.), *Models for Planning Wildlife Conservation in Large Landscapes*, pp. 33–50. Academic Press, San Diego, CA. (6)

Bell, C. D., J. M. Blumenthal, A. C. Broderick, and B. J. Godley. 2010. Investigating potential for depensation in marine turtles: How low can you go? *Conservation Biology* 24: 226–235. (5)

Bennett, A. F. 1999. *Linkages in the Landscape: The Role of Corridors and Connectivity in Wildlife Conservation*. IUCN, Gland, Switzerland. (8)

Berger, J. 1990. Persistence of different-sized populations: An empirical assessment of rapid extinctions in bighorn sheep. *Conservation Biology* 4: 91–98. (6)

Berger, J. 1999. Intervention and persistence in small populations of bighorn sheep. *Conservation Biology* 13: 432–435. (6)

Berger, J., B. Buuveibaatar, and C. Mishra. 2013. Globalization of the cashmere market and the decline of large mammals in Central Asia. *Conservation Biology*, 27: 679–689. (4)

Berger-Tal, O., D. T. Blumstein, S. Carroll, R. N. Fischer, et al. 2016. A systematic survey of the integration of behavior into wildlife conservation and management. *Conservation Biology*. doi: 10.1111/cobi.12654. (7)

Bevan, E., T. Wibbels, B. M. Z. Najera, M. A. C. Martinez, et al. 2014. In situ nest and hatchling survival at Rancho Nuevo, the primary nesting beach of the Kemp's ridley sea turtle, *Lepidochelys kempii. Herpetological Conservation and Biology* 9: 563–577. (1)

Beyer, H. L., E. H. Merrill, N. Varley, and M. S. Boyce. 2007. Willow on Yellowstone's northern range: Evidence for a trophic cascade? *Ecological Applications* 17: 1563–1571. (2)

Bhagwat, S. A., N. Dudley, and S. R. Harrop. 2011. Religious following in biodiversity hotspots: Challenges and opportunities for conservation and development. *Conservation Letters* 4: 234–240. (3)

Bhagwat, S. A., S. Nogué, and K. J. Willis, K. J. 2012. Resilience of an ancient tropical forest landscape to 7500 years of environmental change. *Biological Conservation* 153: 108–117. (2)

Bhattacharya, A., J. Oppenheim, and N. Stern. 2015. Driving sustainable development through better infrastructure: Key elements of a transformation program. *Global Economy and Development Working Paper* 91. (11)

Bhatti, S., S. Carrizosa, P. McGuire, and T. Young (eds.). 2009. *Contracting for ABS: The Legal and Scientific Implications of Bioprospecting Contracts*. IUCN, Gland, Switzerland. (3)

Bickford, D., M. R. C. Posa, L. Qie, A. Campos-Arceiz, et al. 2012. Science communication for biodiversity conservation. *Biological Conservation* 151: 74–76. (12)

Bisht, I. S., P. S. Mehta, and D. C. Bhandari. 2007. Traditional crop diversity and its conservation on-farm for sustainable agricultural production in Kumaon Himalaya of Uttaranchal state: A case study. *Genetic Resources and Crop Evolution* 54: 345–357. (9)

Blackwell, M. 2011. The Fungi: 1, 2, 3… 5.1 million species? *American Journal of Botany* 98(3): 426–438. (2)

Blanco, G., J. A. Lemus, and M. Garía-Montijano. 2011. When conservation management becomes contraindicated: Impact of food supplementation on health of endangered wildlife. *Ecological Applications* 21: 2469–2477. (7)

Blickley, J. L., K. Deiner, K. Garbach, I. Lacher, et al. 2013. Graduate student's guide to necessary skills for nonacademic conservation careers. *Conservation Biology* 27: 24–34. (12)

Bobbink, R., K. Hicks, J. Galloway, T. Spranger, et al. 2010. Global assessment of nitrogen deposition effects on terrestrial plant diversity. *Ecological Applications* 20: 30–59. (4)

Bocci, A., S. Menapace, S. Alemanno, and S. Lovari. 2016. Conservation introduction of the threatened Apennine chamois *Rupicapra pyrenaica ornata*: post-release dispersal differs between wild-caught and captive founders. *Oryx* 50(01): 128–133. (7)

Bode, M., W. Probert, W. R. Turner, K. A. Wilson, and O. Venter. 2011. Conservation planning with multiple organizations and objectives. *Conservation Biology* 25: 295–304. (8, 11)

Boersma, P. D. 2006. Landscape-level conservation for the sea. *In* M. J. Groom, G. K. Meffe, and C. R. Carroll (eds.), *Principles of Conservation Biology*, 3rd Ed., pp. 447–448. Sinauer Associates, Sunderland, MA. (6)

Boersma, P. D. and G. A. Rebstock. 2009. Foraging distance affects reproductive success in Magellanic penguins. *Marine Ecology Progress Series* 375: 263–275. (6)

Boitani, L., P. Ciucci, and E. Raganella-Pelliccioni. 2011. Ex-post compensation payments for wolf predation on livestock in Italy: a tool for conservation?.*Wildlife Research* 37(8): 722–730. (7)

Bombaci, S. P., C. M. Farr, H. T. Gallo, A. M. Mangan, et al. 2016. Using Twitter to communicate conservation science from a professional conference. *Conservation Biology* 30(1): 216–225. (12)

Bonsall, M. B., C. A. Dooley, A. Kasparson, T. Brereton, et al. 2014. Allee effects and the spatial dynamics of a locally endangered butterfly, the high brown fritillary. *Ecological Applications* 24: 108–120. (5)

Borrini-Feyerabend, G., M. Pimbert, T. Farvar, A. Kothari, and Y. Renard. 2004. *Sharing Power: Learning by Doing in Co-Management of Natural Resources throughout the World.* IIED and IUCN/CEESP/CMWG, Cenesta, Tehran. (9)

Botanic Gardens Conservation International (BGCI). 2005. *www.bgci.org* (7)

Botts, E. A., B. F. N. Erasmus, and G. J Alexander. 2013. Small range size and narrow niche breadth predict range contractions in South African frogs. *Global Ecology and Biogeography* 22: 567–576. (5)

Bouché, P., I. Douglas-Hamilton, G. Wittemyer, A. J. Nianogo, et al. 2011. Will elephants soon disappear from West African savannahs? *PLoS ONE* 6(6): e20619. (6)

Bouwman, H., R. Bornman, C. Van Dyk, and I. Barnhoorn. 2015. First report of the concentrations and implications of DDT residues in chicken eggs from a malaria-controlled area. *Chemosphere* 137: 174–177. (4)

Bowen, B. W., L. A. Rocha, R. J. Toonen, S. A. Karl, and the ToBo Laboratory. 2013. The origins of tropical marine biodiversity. *Trends in Ecology and Evolution* 28: 359–366. Patterns and processes of speciation in the marine environment have both differences and similarities to the terrestrial environment. (2)

Bowler, D. E., L. M. Buyung-Ali, J. R. Healey, J. P. G. Jones, et al. 2012. Does community forest management provide global environmental benefits and improve local welfare? *Frontiers in Ecology and the Environment* 10: 29–36. (11)

Boydell, S. and H. Holzknecht. 2003. Land—caught in the conflict between custom and commercialism. *Land Use Policy* 20: 203–207. (9)

Boyles, J. G., P. M. Cryan, G. F. McKracken, and T. H. Kunz. 2011. Economic importance of bats in agriculture. *Science* 332: 41–42. (3)

Bradley, B. A., D. M. Blumenthal, R. Early, E. D. Grosholz, et al. 2012. Global change, global trade, and the next wave of plant invasions. *Frontiers in Ecology and the Environment* 10: 20–28. (4)

Bradshaw, A. D. 1990. The reclamation of derelict land and the ecology of ecosystems. *In* W. R. Jordan III, M. E. Gilpin, and J. D. Aber (eds.), *Restoration Ecol-ogy: A Synthetic Approach to Ecological Research*, pp. 53–74. Cambridge University Press, Cambridge. (10)

Bradshaw, C. J. A. and B. W. Brook. 2014. Human population reduction is not a quick fix for environmental problems. *Proceedings of the National Academy of Sciences* 111: 16610–16615. (1, 4)

Bradshaw, C. J. A., B. M Fitzpatrick, C. C Steinberg, B. W. Brook, et al. 2008. Decline in whale shark size and abundance at Ningaloo Reef over the past decade: the world's largest fish is getting smaller. *Biological Conservation* 141: 1894–1190. (6)

Bradshaw, C. J. A., N. S. Sodhi, and B. W. Brook. 2009. Tropical turmoil: A biodiversity tragedy in progress. *Frontiers in Ecology and the Environment* 7: 79–87. (4)

Bragina, E. V., V. C. Radeloff, M. Baumann, K. Wendland, et al. 2015. Effectiveness of protected areas in the Western Caucasus before and after the transition to post-socialism. *Biological Conservation* 184: 456–464. (8)

Braithwaite, R. W. 2001. Role of Tourism. *In* S. A. Levin (ed.), *Encyclopedia of Biodiversity, Vol. 5*, pp. 667–679. Academic Press, San Diego, CA. (3)

Branton, M. and J. S. Richardson. 2011. Assessing the value of the umbrella-species concept for conservation planning with meta-analysis. *Conservation Biology* 25: 9–20. (6)

Braschler, B. 2009. Successfully implementing a citizen-scientist approach to insect monitoring in a resource-poor country. *BioScience* 59: 103–104. (9)

Braunisch, V., Coppes, J., Bachle, S., and Suchant, R. 2015. Underpinning the precautionary principle with evidence: A spatial concept for guiding wind power development in endangered species' habitats. *Journal for Nature Conservation* 24: 31–40. (3)

Braunisch, V., S.-L. Huang, Y. Hao, S. T. Turvey, et al. 2012. Conservation science relevant to action: a research agenda identified and prioritized by practitioners. *Biological Conservation* 153: 201–210. (8)

Briggs, S. V. 2009. Priorities and paradigms: Directions in threatened species recovery. *Conservation Letters* 2: 101–108. (6)

Brinson, M. M. and S. D. Eckles. 2011. U. S. Department of Agriculture conservation program and practice effects on wetland ecosystem services: A synthesis. *Ecological Applications* 21: S116–S127. (10)

Brodie, J. F., O. E. Helmy, W. Y. Brockelman, and J. L. Maron. 2009. Bushmeat poaching reduces the seed dispersal and population growth rate of a mammal-dispersed tree. *Ecological Applications* 19: 854–863. (8)

Brodin, T. J. Fick, M. Jonsson, and J. Klaminder. 2013. Dilute concentrations of a psychiatric drug alter behavior of fish from natural populations. *Science* 339 (6121): 814–815. (4)

Brook, A., M. Zint, and R. DeYoung. 2003. Landowner's response to an Endangered Species Act listing and implications for encouraging conservation. *Conservation Biology* 17: 1638–1649. (5, 6)

Brooks, J. S. 2016. Design features and project age contribute to joint success in social, ecological, and economic outcomes of community-based conservation projects. *Conservation Letters*. doi: 10.1111/conl.1223. (9)

Brooks, T. M., S. L. Pimm, V. Kapos, and C. Ravilious. 1999. Threat from deforestation to montane and lowland birds and mammals in insular Southeast Asia. *Journal of Animal Ecology* 68(6): 1061–1078. (5)

Brown, M. L., T. M. Donovan, W. Scott-Schwenk, and D. M. Theobald. 2014. Predicting impacts of future population growth and development on occupancy rates of forest dependent birds. *Biological Conservation* 170: 311–320. (1)

Brühl, C. A., T. Schmidt, S. Pieper, and A. Alscher. 2013. Terrestrial pesticide exposure of amphibians: An underestimated cause of global decline? *Scientific Reports 3*, Article 1135. doi: 10.1038/srep01135. (4)

Brundtland, G., M. Khalid, S. Agnelli, S. Al-Athel, et al. 1987. Our Common Future ("Brundtland report"). United Nations Report of the World Commission on Environment and Development. (11)

Brush, S. B. 2004. Growing biodiversity. *Nature* 430: 967–968. (9)

Brush, S. B. 2007. Farmers' rights and protection of traditional agricultural knowledge. *World Development* 35: 1499–1514. (7)

Bryant, D., L. Burke, J. McManus, and M. Spalding. 1998. *Reefs at Risk: A Map-Based Indicator of Threats to the World's Coral Reefs*. World Resources Institute, Washington, DC. (4)

Buchalski, M. R., A. Y. Navarro, W. M. Boyce, T. W. Vickers, et al. 2015. Genetic population structure of Peninsular bighorn sheep (*Ovis canadensis nelsoni*) indicates substantial gene flow across US–Mexico border. *Biological Conservation* 184: 218–228. (6)

Buchholz, R. 2007. Behavioral biology: An effective and relevant conservation tool. *Trends in Ecology and Evolution* 22: 401–407. (7)

Buckley, R. 2009. Parks and tourism. *PLoS Biology* 7: e1000143. (3)

Bull, A. T. 2004. *Microbial Diversity and Bioprospecting*. ASM Press, Washington, DC. (3)

Bulman, C. R., R. J. Wilson, A. R. Holt, A. L. Galvez-Bravo, et al. 2007. Minimum viable metapopulation size, extinction debt, and the conservation of declining species. *Ecological Applications* 17: 1460–1473. (5, 6)

Burgess, M. G., S. Polasky, and D. Tilman. 2013. Predicting overfishing and extinction threats in multispecies fisheries. *Proceedings of the National Academy of Sciences USA* 110: 15943–15948. (4)

Burgess, S. C., K. J. Nickols, C. D. Griesmer, L. A. K. Barnett, et al. 2014. Beyond connectivity: How empirical methods can quantify population persistence to improve marine protected-area design. *Ecological Applications* 24: 257–270. (8)

Burghardt, K. T., D. W. Tallamy, and W. G. Shriver. 2009. Impact of native plants on bird and butterfly biodiversity in suburban landscapes. *Conservation Biology* 23: 219–224. (10)

Burke, L., K. Reytar, M. Spalding, and A. Perry. 2011. *Reefs at Risk Revisited*. World Resources Institute, Washington DC, USA. (4)

Burkhead, N. M. 2012. Extinction rates in North American freshwater fishes, 1900–2010. *BioScience* 62(9): 798–808. (5)

Burrell, J. 2013. Path of the Pronghorn—Leading to New Passages: Part 3. Newswatch, National Geographic. *newswatch.nationalgeographic.com/2013/12/06/path-of-the-pronghorn-leading-to-new-passages-part-3*. (8)

Bussière, E., L. G. Underhill, and R. Altwegg. 2015. Patterns of bird migration phenology in South Africa suggest northern hemisphere climate as the most consistent driver of change. *Global Change Biology* 21(6): 2179–2190. (4)

BusinessWire. 2015. American Burying Beetle–Mitigation Solutions USA Muddy Boggy Endangered Species Bank Expansion in Oklahoma. July 30, 2015. *www.businesswire.com/news/home/20150730006047/en/American-Burying-Beetle-%E2%80%93-Mitigation-Solutions-USA*. (11)

Butchart, S. H. M., M. Clarke, R. J. Smith, R. E. Sykes, et al. 2015. Shortfalls and solutions for meeting national and global conservation targets. *Conservation Letters* 8(5): 329–337. (8, 11)

Butchart, S. H. M., A. J. Stattersfield, and T. M. Brooks. 2006. Going or gone: definingPossibly Extinct'species to give a truer picture of recent extinctions. *Bulletin-British Ornithologists Club* 126: 7. (5)

Butchart, S. H. M., M. Walpole, B. Collen, A. Van Strien, et al. 2010. Global biodiversity: indicators of recent declines. *Science* 328(5982): 1164–1168. (4)

Butler, R. A., L. P. Koh, and J. Ghazoul. 2009. REDD in the red: palm oil could undermine carbon payment schemes. *Conservation Letters* 2: 67–73. (3)

Cafaro, P. 2015. Three ways to think about the sixth mass extinction. *Biological Conservation* 192: 387–393. (12)

Caillouet, C. W., B. J. Gallaway, and A. M. Landry. 2015. Cause and call for modification of the bi-national recovery plan for the Kemp's ridley sea turtle (*Lepidochelys kempii*)—Second Revision. *Marine Turtle Newsletter* 145: 1–4. (1)

Caillouet, C. W., Jr., B. A. Robertson, C. T. Fontaine, T. D. Williams, et al. 1997. Distinguishing captive-reared from wild Kemp's ridleys. *Marine Turtle Newsletter* 77: 1–6. (1)

Cain, M. L., W. D. Bowman, and S. D. Hacker. 2014. *Ecology* 3rd Ed. Sinauer Associates, Sunderland, MA. (2)

Callaway, R. M., G. C. Thelen, A. Rodriguez, and W. E. Holben. 2004. Soil biota and exotic plant invasion. *Nature* 427(6976): 731–733. (4)

Calmy, A., E. Goemaere, and G. Van Cutsem. 2015. HIV and Ebola virus: two jumped species but not two of a kind. *AIDS* 29(13): 1593–1596. (4)

Camacho, A. E. 2007. Can regulation evolve? Lessons from a study in maladaptive management. *UCLA Law Review* 55: 293–358. (6)

Canessa, S., S. J. Converse, M. West, N. Clemann, et al. 2015. Planning for ex situ conservation in the face of uncertainty. *Conservation Biology*. doi: 10.1111/cobi.12613 (7)

Cannon, J. R. 1996. Whooping crane recovery: a case study in public and private cooperation in the conservation of endangered species. *Conservation Biology*10: 813–821. (3)

Carey, M. P., B. L. Sanderson, K. A. Barnas, and J. D. Olden. 2012. Native invaders–challenges for science, management, policy, and society. *Frontiers in Ecology and the Environment* 10(7): 373–381. (4)

Carlson, T. 2013. The Politics of a Tree: How a species became national policy. *In* A. Sher and M. F. Quigley (eds.) *Tamarix: A Case Study of Ecological Change in the American West*. Oxford University Press. (12)

Carnicer, J., M. Coll, M. Ninyerola, X. Pons, et al. 2011. Widespread crown condition decline, food web disruption, and amplified tree mortality with increased climate change-type drought. *Proceedings of the National Academy of Sciences USA* 108: 1474–1478. (4)

Carraro, C., J. Eyckmans, and M. Finus. 2006. Optimal transfers and participation decisions in international environmental agreements. *The Review of International Organizations* 1: 379–396. (6)

Carrière, S. M., E. Rodary, P. Méral, G. Serpantié, et al. 2013. Rio +20, biodiversity marginalized. *Conservation* 6: 6–11. (9, 11)

Carroll, C., R. J. Frederickson, and R. C. Lacy. 2014. Developing metapopulation connectivity criteria from genetic and habitat data to recover the endangered Mexican wolf. *Conservation Biology* 28: 76–86. (6)

Carson, R. 1965. *The Sense of Wonder*. Harper & Row, New York. (3)

Carson, R. 1962. *Silent Spring*. Houghton Mifflin Company. (1)

Carter, I., J. Foster, and L. Lock. 2016. The role of animal translocations in conserving british wildlife: an overview of recent work and prospects for the future. *EcoHealth*. doi: 10.1007/s10393-015-1097-1. (7)

Cassimon, D., D. Essers, and A. Fauzi. 2014. Indonesia's debt-for-development swaps: past, present, and future. *Bulletin of Indonesian Economic Studies* 50(1): 75–100. (11)

Castelletta, M., N. S. Sodhi, and R. Subaraj. 2000. Heavy extinctions of forest avifauna in Singapore: Lessons for biodiversity conservation in Southeast Asia. *Conservation Biology* 14: 1870–1880. (5)

Catlin J., M. Hughes, T. Jones, R. Jones, and R. Campbell. 2013. Valuing individual animals through tourism: Science or speculation? *Biological Conservation* 157:93–98. (3)

Ceballos, G., P. R. Ehrlich, A. D. Barnosky, A. García, R. M. Pringle, and T. M. Palmer. 2015. Accelerated modern human–induced species losses: Entering the sixth mass extinction. *Science Advances* 1(5): e1400253. (5)

Chan, Y. F., K.-P. Chiang, J. Chang, Ø. Moestrup, and C.-C. Chung. 2015. Strains of the morphospecies *Ploeotia costata* (Euglenozoa) isolated from the Western North Pacific (Taiwan) reveal substantial genetic differences. *Journal of Eukaryotic Microbiology* 62: 318–326. (2)

Charky, N. 2014. Congress attempts to change captivity rules for orcas, marine life. *LA Times*, Monday, March 14, 2014. (12)

Chen, X. Y. and F. He. 2009. Speciation and endemism under the model of island biogeography. *Ecology* 90(1): 39–45. (5)

Chen, I., J. K. Hill, R. Ohlemüller, D. B. Roy, and C. D. Thomas. 2011. Rapid range shifts of species associated with high levels of climate warming. *Science* 333: 1024–1026. (4)

Chicago Wilderness. *www.chicagowilderness.org* (8)

Chittaro, P. M., I. C. Kaplan, A. Keller, and P. S. Levin. 2010. Trade-offs between species conservation and the size of marine protected areas. *Conservation Biology* 24: 197–206. (5)

Chivian, E. and A. Bernstein (eds.). 2008. *Sustaining Life: How Human Health Depends on Biodiversity*. Oxford University Press, New York. (3)

Christian, C., D. Ainley, M. Bailey, P. Dayton, et al. 2013. A review of formal objections to Marine Stewardship Council fisheries certifications. *Biological Conservation* 161: 10–17. (4, 12)

Cinner, J. E. and S. Aswani. 2007. Integrating customary management into marine conservation. *Biological Conservation* 140: 201–216. (4)

Clare, S. and I. F. Creed. 2014. Tracking wetland loss to improve evidence-based wetland policy learning and decision making. *Wetlands Ecology and Management* 22: 235–245. (10)

Clark, P. and D. E. Johnson. 2009. Wolf-cattle interactions in the Northern Rocky Mountains. In *Range Field Data 2009 Progress Report. Special Report 1092*, pp. 1–7. Corvallis, OR: Oregon State University, Agricultural Experiment Station. (7)

Clavero, M. and E. García-Berthou. 2005. Invasive species are a leading cause of animal extinctions. *Trends in Ecology and Evolution* 20: 110–110. (4)

Clayton, S. and G. Myers. 2015. *Conservation Psychology: Understanding and Promoting Human Care for Nature*. John Wiley & Sons. (11)

Clements, T., H. J. Rainey, D. An, V. Rours, et al. 2013. An evaluation of the effectiveness of a direct payment for biodiversity conservation: the Bird Nest Protection Program in the northern plains of Cambodia. *Biological Conservation* 157: 50–59. (11)

Clewell, A. F. and J. Aronson. 2006. Motivations for the restoration of ecosystems. *Conservation Biology* 20: 420–428. (10)

Climate Central. 2012. *Global Weirdness: Severe Storms, Deadly Heat Waves, Relentless Drought, Rising Seas, and the Weather of the Future*. Pantheon, New York. (4)

CNN. 2016. SeaWorld's orcas will be last generation at parks. March 16, 2016. (12)

CNN. 2015. Zimbabwean officials: American man wanted in killing of Cecil the lion. July 28, 2015. (3)

Cockle, K. L. and K. Martin. 2015. Temporal dynamics of a commensal network of cavity-nesting vertebrates: increased diversity during an insect outbreak. *Ecology* 96(4): 1093–1104. (2)

Cohen-Shacham, E., T. Dayan, R. de Groot, C., Beltrame, et al. 2015. Using the ecosystem services concept to analyse stakeholder involvement in wetland management. *Wetlands Ecology and Management* 23(2): 241–256. (11)

Colding, J., J. Lundberg, S. Lundberg, and E. Andersson. 2009. Golf courses and wetland fauna. *Ecological Applications* 19: 1481–1491. (9)

Cole, T. 1965. Essay on American Scenery. *In* J. W. McCoubrey (ed.), *American Art, 1700–1960*, pp. 98–109. Prentice-Hall, Englewood Cliffs, NJ. (1)

Cole, I. A., S. M. Prober, I. D. Lunt, and T. B. Koen. 2016. A plant traits approach to managing legacy species during restoration transitions in temperate eucalypt woodlands. *Restoration Ecology.* doi: 10.1111/rec.12334. (10)

Colglazier, W. 2015. Sustainable development agenda 2030. *Science* 349:1048–1050. (11)

Colón-Rivera, R. J., K. Marshall, F. J. Soto-Santiago, D. Ortiz-Torres, and C. E. Flower. 2013. Moving forward: Fostering the next generation of Earth stewards in the STEM disciplines. *Frontiers in Ecology and the Environment* 11: 383–391. (12)

Colwell, R., S. Avery, J. Berger, G. E. Davis, et al. 2012. *Revisiting Leopold: Resource Stewardship in the National Parks.* National Park System Advisory Board, Washington, DC. (8)

Comeau, S., R. C. Carpenter, C. A. Lantz, and P. J. Edmunds. 2015. Ocean acidification accelerates dissolution of experimental coral reef communities. *Biogeosciences* 12(2): 365–372. (4)

Commission for the Conservation of Antarctic Marine Living Resources (CCAMLR). *www.ccamlr.org* (6)

Common, M. and S. Stagl. 2005. *Ecological Economics: An Introduction.* Cambridge University Press, New York. (3)

Connor, E. F. and E. D. McCoy. 2001. Species-area relationships. *In* S. A. Levin (ed.), *Encyclopedia of Biodiversity* 5: 397–412. Academic Press, San Diego, CA. (5)

Conservation International and K. J. Caley. 2008. Biological diversity in the Mediterranean Basin. *In* C. J. Cleveland (ed.), *Encyclopedia of Earth.* Environmental Information Coalition, National Council for Science and the Environment, Washington, DC. *www.eoearth. org/article/Biological_diversity_in_the_Mediterranean_ Basin* (2)

Convention on Biological Diversity. *www.cbd.int.* (11)

Convention on International Trade in Endangered Species of Wild Flora and Fauna (CITES). *www.cites.org* (6)

Cordeiro, N. J., L. Borghesio, M. P. Joho, T. J. Monoski, et al. 2015. Forest fragmentation in an African biodiversity hotspot impacts mixed-species bird flocks. *Biological Conservation* 188: 61–71. (4)

Cork, S. J., T. W. Clark, and N. Mazur. 2000. Introduction: An interdisciplinary effort for koala conservation. *Conservation Biology* 14: 606–609. (8)

Corlett, R. T. 2011. Impacts of warming on tropical lowland rainforests. *Trends in Ecology and Evolution* 26: 606–613. (4)

Corlett, R. T. 2013. Singapore: half full or half empty? *In* N. S. Sodhi, L. Gibson, and P. H. Raven (eds.) *Conservation Biology: Voices from the Tropics*, pp. 142–147. Wiley-Blackwell, Oxford, UK. (8)

Corlett, R. T. 2016. Restoration, reintroduction, and rewilding in a changing world. *Trends in Ecology and Evolution.* doi: 10.1016/j.tree.2016.017. (10)

Corlett, R. T. and R. B. Primack. 2010. *Tropical Rain Forests: An Ecological and Biogeographical Comparison*, 2nd Ed. Blackwell Publishing, Malden, MA. (2, 3, 4)

Corral-Verdugo, V., M. Bonnes, C. Tapia-Fonllem, B. Fraijo-Sing, et al. 2009. Correlates of pro-sustainability orientation: The affinity towards diversity. *Journal of Environmental Psychology* 29: 34–43. (1)

Costa Rica National Parks. *www.costarica-nationalparks. com* (11)

Costanza, R., R. de Groot, P. Sutton, S. van der Ploeg, et al. 2014. Changes in the global value of ecosystem services. *Global Environmental Change* 26: 152–158. (3)

Costello, C., J. M. Drake, and D. M. Lodge. 2007. Evaluating an invasive species policy: Ballast water exchange in the Great Lakes. *Ecological Applications* 17: 655–662. (4)

Costello, M. J. 2015. Biodiversity: The known, the unknown and rates of extinction. Current Biology 25(6): R368–371. (2)

Costello, M. J., R. M. May, and N. E. Stork. 2013. Can we name the Earth's species before they go extinct? *Science* 339: 413–416. (12)

Courchamp, F., J. A. Dunne, Y. Le Maho, R. M. May, et al. 2015. Fundamental ecology is fundamental. *Trends in Ecology and Evolution* 30(1): 9–16. (12)

Cox, P. A. 2001. Pharmacology, biodiversity and. *In* S. A. Levin (ed.), *Encyclopedia of Biodiversity, Vol. 4*, pp. 523–536. Academic Press, San Diego, CA. (3)

Cox, P. A. and T. Elmqvist. 1997. Ecocolonialism and indigenous-controlled rainforest preserves in Samoa. *Ambio* 26: 84–89. (9)

Cox, R. L. and E. C. Underwood. 2011. The Importance of Conserving Biodiversity Outside of Protected Areas in Mediterranean Ecosystems. *PLoS ONE* 6(1): e14508. (9)

Cox, W. A., F. R. Thompson III, and J. Faaborg. 2012. Landscape forest cover and edge effects on songbird nest predation vary by nest predator. *Landscape Ecology* 27(5): 659–669. (4)

Creech, T. G., C. W. Epps, R. H. Monello, and J. D. Wehausen. 2014. Using network theory to prioritize management in a desert bighorn sheep metapopulation. *Landscape Ecology* 29(4): 605–619. (6)

Crees, J. J., A. C. Collins, P. J. Stephenson, H. M. Meredith, et al. 2016. A comparative approach to assess drivers of success in mammalian conservation recovery programs. *Conservation Biology*. In press. doi: 10.1111/cobi.12652. (11)

Christian, C., D. Ainley, M. Bailey, P. Dayton, et al. 2013. A review of formal objections to Marine Stewardship Council fisheries certifications. *Biological Conservation* 161: 10–17. (12)

Costello, M. J., B. Vanhoorne, and W. Appeltans. 2015. Conservation of biodiversity through taxonomy, data publication, and collaborative infrastructures. *Conservation Biology* 29(4): 1094–1099. (12)

Crnokrak, P. and D. A. Roff. 1999. Inbreeding depression in the wild. *Heredity* 83 (1999) 260–270. (5)

Crone, E. E., M. M. Ellis, W. F. Morris, A. Stanley, et al. 2013. Ability of matrix models to explain the past and predict the future of plant populations. *Conservation Biology* 27: 968–978. (6)

Crosmary, W. G., S. D. Côté, and H. Fritz. 2015. The assessment of the role of trophy hunting in wildlife conservation. *Animal Conservation* 18(2): 136–137. (3)

Crowl, T. A., T. O. Crist, R. R. Parmenter, G. Belovsky, and A. E. Lugo. 2008. The spread of invasive species and infectious disease as drivers of ecosystem change. *Frontiers in Ecology and the Environment* 6: 238–246. (4)

Cruickshank, S. S., A. Ozgul, S. Zumbach, and B. R. Schmidt. 2016. Quantifying population declines based on presence-only records for Red List assessments. *Conservation Biology*. doi: 10.1111/cobi.12688. (6)

Cuperus, R., K. J. Canters, H. A. V. de Hars, and D. S. Friedman. 1999. Guidelines for ecological compensation associated with highways. *Biological Conservation* 90: 41–51. (8)

Czech, B. 2008. Prospects for reconciling the conflict between economic growth and biodiversity conservation with technological progress. *Conservation Biology* 22: 1389–1398. (1)

Dahles, H. 2005. A trip too far: Ecotourism, politics, and exploitation. *Development Change* 36: 969–971. (3)

Danielson, F., P. M. Jensen, N. D. Burgess, R. Altamirano, et al. 2014. A multicountry assessment of tropical resource monitoring by local communities. *BioScience* 64: 236–251. (8)

Danielson, F., K. Pirhofer-Walzi, T. P. Adrian, D. R. Kapijimpanga, et al. 2014. Linking public participation in scientific research to the indicators and needs of international environmental agreements. *Conservation Letters* 7: 12–24. (12)

Daru B. H. and P. C. le Roux 2016. Marine protected areas are insufficient to conserve global marine plant diversity. *Global Ecology and Biogeography* 25(3): 324–334. (9)

Daru, B. H., K. Yessoufou, L. T. Mankga, and T. J. Davies. 2013. A global trend towards the loss of evolutionarily unique species in mangrove ecosystems. *PLoS ONE* 8(6): e66686. (4)

Daszak, P., A. A. Cunningham, and A. D. Hyatt. 2000. Emerging infectious diseases of wildlife—threats to biodiversity and human health. *Science* 287: 443–449. (4)

Davenport, C. 2015. National approve landmark climate accord in Paris. *New York Times*. www.nytimes.com/2015/12/13/world/europe/climate-change-accord-paris.html?_r=0. (11)

Davidson, D. J. and J. Andrews. 2013. Not all about consumption. *Science* 339: 1286–1287. (4)

Davis, A. M. and A. R. Moore. 2015. Conservation potential of artificial water bodies for fish communities on a heavily modified agricultural floodplain. *Aquatic Conservation: Marine and Freshwater Ecosystems*. (9)

Davis, M. A. 2009. *Invasion Biology*. Oxford University Press, Oxford, UK. (4)

Dawson, J., F. Patel, R. A. Griffiths, and R. P Young. 2016. Assessing the global zoo response to the amphibian crisis through 20-year trends in captive collections. *Conservation Biology* 30(1): 82–91. (7)

De Grammont, P. C. and A. D. Cuarón. 2006. An evaluation of threatened species categorization systems used on the American continent. *Conservation Biology* 20: 14–27. (6)

Deguise, I. E. and J. T. Kerr. 2006. Protected areas and prospects for endangered species conservation in Canada. *Conservation Biology* 20: 48–55. (9)

De la Riva, I. and S. Reichle. 2014. Diversity and Conservation of the Amphibians of Bolivia. *Herpetological Monographs* 28(1): 46–65. (4)

Dearborn, D. C. and S. Kark. 2010. Motivations for conserving urban biodiversity. *Conservation Biology* 4(2): 432–440. (4)

Denver Post. 2016. Hickenlooper asks EPA for Superfund fix at leaking mines. Feb. 29, 2016. (4)

Dernbach, J. C. and F. Cheever. 2015. Sustainable development and its discontents. *Transnational Environmental Law* 4(02): 247–287. (11)

Devall, B. and G. Sessions. 1985. *Deep Ecology*. Perigrine Smith, Salt Lake, UT. (3)

Diamond, J. 2005. *Collapse: How Societies Choose to Fail or Succeed*. Penguin Books, New York. (3)

Dickman, A. J., E. A. Macdonald, and D. W. Macdonald. 2011. A review of financial instruments to pay for predator conservation and encourage human-carnivore coexistence. *Proceedings Of The National Academy Of Sciences USA* 108: 13937–13944. (9)

Dieter, C. D. and D. J. Schaible. 2014. Distribution and Population Density of Jackrabbits in South Dakota. *Great Plains Research* 24(2): 127–134. (6)

Di Marco, M., B. Collen, C. Rondinini, and G. M. Mace. 2015. Historical drivers of extinction risk: using past evidence to direct future monitoring. *Proceedings of the Royal Society B* 282(1813): 20150928. (5)

Di Minin, E., D. C. Macmillan, P. S. Goodman, B. Escott, et al. 2013. Conservation business and conservation planning in a biological diversity hotspot. *Conservation Biology* 27: 808–820. (11)

Di Minin E., N. Leader-Williams, and C. J. A. Bradshaw. 2016. Banning trophy hunting will exacerbate biodiversity loss. *Trends in Ecology and Evolution* 31(2): 99–102. (9)

Dinerstein, E., K. Varma, E. Wikramanayake, G. Powell, et al. (2013). Enhancing conservation, ecosystem services, and local livelihoods through a wildlife premium mechanism. *Conservation Biology* 27(1): 14–23. (9)

Dinerstein, E., K. Varma, E. Wikramanayake, G. Powell, et al. 2013. Enhancing conservation, ecosystem services, and local livelihoods through a wildlife premium mechanism. *Conservation Biology* 27: 14–23. (9)

Dobson, A. 2005. Monitoring global rates of biodiversity change: Challenges that arise in meeting the Convention on Biological Diversity (CBD) 2010 goals. *Philosophical Transactions of the Royal Society B-Biological Sciences* 360: 229–241. (3)

Dodds, W. K., K. C. Wilson, R. L. Rehmeier, G. L. Knight, et al. 2008. Comparing ecosystem goods and services provided by restored and native lands. *BioScience* 58: 837–845. (10)

Dolman, P. M., N. J. Collar, K. M. Scotland, and R. J. Burnside. 2015. Ark or park: stochastic population modeling to evaluate potential effectiveness of in situ and ex situ conservation for a critically endangered bustard. *Journal of Applied Ecology* 52: 841–850. (7)

Domisch, S., S. C. Jahnig, J. P. Simaika, M. Kuemmerlen, and S. Stoll. 2015. Application of species distribution models in stream ecosystems: the challenges of spatial and temporal scale, environmental predictors and species occurrence data. *Fundamental Applied Limnology* 186(1–2): 45–61. (2)

Donovan, G. H., D. T. Butry, Y. L. Michael, J. P. Prestemon, et al. 2013. The relationship between trees and human health: evidence from the spread of the emerald ash borer. *American Journal of Preventive Medicine* 44: 139–145. (3)

Donlan, J., H. W. Greene, J. Berger, C. E. Bock, et al. 2006. Re-wilding North America. *Nature* 436: 913–914. (7)

Doney, S. C., V. J. Fabry, R. A. Feely, and J. A. Kleypas. 2009. Ocean acidification: the other CO_2 problem. *Annual Review of Marine Science* 1: 169–92. (4)

Doughty, C. E. 2013. Preindustrial human impacts on global and regional environment. *Annual Review of Environment and Resources* 38: 503–527. (4)

Douglas, L. R. and K. Alie. 2014. High-value natural resources: Linking wildlife conservation to international conflict, insecurity, and development concerns. *Biological Conservation* 171: 270–277. (6)

Drayton, B. and R. B. Primack. 1999. Experimental extinction of garlic mustard (*Alliaria petiolata*) populations: Implications for weed science and conservation biology. *Biological Invasions* 1: 159–167. (5)

Drus, G. M., T. L. Dudley, M. L. Brooks, and J. R. Matchett. 2013. The effect of leaf beetle herbivory on the fire behaviour of tamarisk (*Tamarix ramosissima* Lebed.). *International Journal of Wildland Fire* 22(4): 446–458. (4)

Duarte, A., J. S. Hatfield, T. M. Swannack, M. R. Forstner, et al. 2016. Simulating range-wide population and breeding habitat dynamics for an endangered woodland warbler in the face of uncertainty. *Ecological Modelling* 320: 52–61. (6)

Duarte, C. M., K. A. Pitt, C. H. Lucas, J. E. Purcell, et. al. 2013. Is global ocean sprawl a cause of jellyfish blooms? *Frontiers in Ecology and the Environment.* 11: 91–97. (4)

Duchelle, A. E., M. R. Guariguata, G. Less, M. A. Albornoz, et al. 2012. Evaluating the opportunities and limitations to multiple use of Brazil nuts and timber in Western Amazonia. *Forest Ecology and Management* 268: 39–48. (9)

Dudley, N. (ed.) 2008. *Guidelines for Applying Protected Area Management Categories*. IUCN, Gland, Switzerland. (8)

Dudley, N., L. Higgins-Zogib, and S. Mansourian. 2009. The links between protected areas, faiths, and sacred natural sites. *Conservation Biology* 23: 568–577. (1)

Dukes, J. S., N. R. Chiariello, S. R. Loarie, and C. B. Field. 2011. Strong response of an invasive plant species (*Centaurea solstitialis* L.) to global environmental changes. *Ecological Applications* 21: 1887–1894. (4)

Dullinger, S., F. Essl, W. Rabitsch, K. Erb, et al. 2013. Europe's other debt crisis caused by the long legacy of future extinctions. *Proceedings of the National Academy of Sciences USA* 110: 7342–7347. (5)

Dunn, R. R. 2005. Modern insect extinctions, the neglected majority. *Conservation Biology* 19: 1030–1036. (6)

Dvořák, P., A. Poulíčková, P. Hašler, M. Belli M., et al. 2015. Species concepts and speciation factors in cyanobacteria, with connection to the problems of diversity and classification. *Biodiversity and Conservation.* doi: 10.1007/s10531-015-0888-6. (2)

Earnst, S. L., D. S. Dobkin, and J. A. Ballard. 2012. Changes in avian and plant communities of aspen woodlands over 12 years after livestock removal in the northwestern great basin. *Conservation Biology* 26: 862–871. (9)

Ebbin, S. 2009. Institutional and ethical dimensions of resilience in fishing systems: Perspectives from comanaged fisheries in the Pacific Northwest. *Marine Policy* 33: 264–270. (1)

Economist. 2013. Special Report: Biodiversity. *The Economist* September 14, 2013. (10)

Ecosystem Marketplace. 2010. The Katoomba Group. *www.ecosystemmarketplace.com* (9)

Edwards, D. P. and S. G. Laurance. 2012. Green labeling, sustainability and the expansion of tropical agriculture: critical issues for certification schemes. *Biological Conservation* 151: 60–64. (12)

Ehrenfeld, D. W. 1970. *Biological Conservation*. Holt, Rinehart and Winston, New York. (5)

Ehrlich, P. R. and A. H. Ehrlich. 1982. *Extinction: The Causes and Consequences of the Disappearance of Species*. Gollancz, London. (3)

Ehrlich, P. R., P. M. Kareiva, and G. C. Daily. 2012. Securing natural capital and expanding equity to rescale civilization. *Nature* 486(7401): 68–73. (4)

Ehrlich, P. R. and R. M. Pringle. 2008. Where does biodiversity go from here? A grim business-as-usual forecast and a hopeful portfolio of partial solutions. *Proceedings of the National Academy of Sciences USA* 105: 11579–11586. (10)

Elkinton, J. S., D. Parry, and G. H. Boettner. 2006. Implicating an introduced generalist parasitoid in the invasive browntail moth's enigmatic demise. *Ecology* 87: 2664–2672. (4)

Elliott, J. E. and K. H. Elliott. 2013. Tracking marine pollution. *Science* 340: 556–558. (4)

Elrich, P. R. and J. P. Holdren. 1971. Impact of Population Growth. *Science* 171:1212–1217. (4)

Emerson, R. W. 1836. *Nature*. James Monroe and Co., Boston. (1)

Emerton, L. 1999. Balancing the opportunity costs of wildlife conservation for communities around Lake Mburo National Park, Uganda. Evaluating Eden Series Discussion Paper No. 5, International Institute for Environment and Development, London. (3)

Encyclopedia of Life. *eol.org* (2, 12)

Encyclopedia of the Nations. 2009. India. *www.nationsencyclopedia.com/economies/Asia-and-the-Pacific/India.html*. (4)

Engler, M. 2008. Value of the international trade in wildlife. *TRAFFIC Bulletin* 22(1): 4–5. (3)

Equator Initiative. *www.equatorinitiative.org* (9)

Estes, J. A., J. Terbough, J. S. Brashares, M. E. Power, et al. 2011. Trophic downgrading of planet earth. *Science* 333: 301–306. (2)

European Commission. 2008. The EU Biodiversity Strategy to 2020. *ec.europa.eu/environment/nature/info/pubs/docs/brochures/2020%20Biod%20brochure%20final%20lowres.pdf* (10)

Evans, S. R. and B. C. Sheldon. 2008. Interspecific patterns of genetic diversity in birds: Correlations with extinction risk. *Conservation Biology* 22: 1016–1025. (5)

Ewing, S. R., R. G. Nager, M. A. C. Nicoll, A. Aumjaud, et al. 2008. Inbreeding and loss of genetic variation in a reintroduced population of Mauritius kestrel. *Conservation Biology* 22: 395–404. (5)

Fadeeva, N., V. Mordukhovich, and J. Zograf. 2015. New deep-sea large free-living nematodes from marobenthos in the Kuril-Kamchatka trench (the North-West Pacific). *Deep-Sea Research II* 111:95–103. (2)

Faeth, S. H., P. S. Warren, E. Shochat, and W. A. Marussich. 2005. Trophic dynamics in urban communities. *BioScience* 55: 399–407. (2)

Fairtrade International. *www.fairtrade.net* (12)

Faith, D. P. 2008. Threatened species and the potential loss of phylogenetic diversity: conservation scenarios based on estimated extinction probabilities and phylogenetic risk analysis. *Conservation Biology* 22(6): 1461–1470. (6)

Falk, D. A., M. A. Palmer, and J. B. Zedler (eds.). 2006. *Foundations of Restoration Ecology: The Science and Practice of Ecological Restoration*. Island Press, Washington, DC. (10)

Fanshawe, S., G. R. VanBlaricom, and A. A. Shelly. 2003. Restored top carnivores as detriments to the performance of marine protected areas intended for fishery sustainability: a case study with red abalones and sea otters. *Conservation Biology* 17: 273–283. (7)

Farmer, J. R., D. Knapp, V. J. Meretsky, C. Chancellor, and B. C. Fischer. 2011. Motivations influencing the adoption of conservation easements. *Conservation Biology* 25: 827–834. (11)

Farmer, J. R., Z. Ma, M. Drescher, E. G. Knackmuhs, and S. L. Dickinson. 2016. Private landowners, voluntary conservation programs, and implementation of conservation friendly land management practices. *Conservation Letters*. (9)

Fehr-Duda, H. and E. Fehr. 2016. Sustainability: Game human nature. *Nature* 530. (11)

Feist, B. E., E. R. Buhle, P. Arnold, J. W. Davis, and N. L. Scholz. 2011. Landscape Ecotoxicology of Coho Salmon Spawner Mortality in Urban Streams. *PLoS ONE* 6(8): e23424. (4)

Felson, A. J., M. A. Bradford, and T. M. Terway. 2013. Promoting Earth Stewardship through urban design experiments. *Frontiers in Ecology and the Environment* 11: 362–367. (10)

Feng, K., K. Hubacek, and D. Guan. 2009. Lifestyles, technology and CO_2 emissions in China: A regional comparative analysis. *Ecological Economics* 69(1): 145–154. (4)

Ferrer. M., I. Newton, and M. Pandolfi. 2009. Small populations and offspring sex-ratio deviations in eagles. *Conservation Biology* 23: 1017–1025. (5)

Ferro, P. J., M. M. Hanauer, and K. R. E. Sims. 2011. Conditions associated with protected area success in conservation and poverty reduction. *Proceedings of the National Academy of Sciences USA* 108: 13913–13918. (8)

Fischer, J. and D. B. Lindenmayer. 2000. An assessment of published results of animal relocations. *Biological Conservation* 96: 1–11. (7)

Fischer, J., R. Dyball, I. Fazey, C. Gross, et al. 2012. Human behavior and sustainability. *Frontiers in Ecology and the Environment* 10: 153–160. (12)

Fischer, L. K., V. Rodorff, M. von der Lippe, and I. Kowarik. 2016. Drivers of biodiversity patterns in parks of a growing South American megacity. *Urban Ecosystems* 1–19. (9)

Fisher, D. O. and S. P. Blomberg. 2012. Inferring extinction of mammals from sighting records, threats, and biological traits. *Conservation Biology* 26: 57–67. (5)

Fisher, M. 2016. Fall, resurrection and uncertainty: an Arabian tale. *Oryx* 50(01): 1–2. (7)

Fisher, R., R. A. O'Leary, S. Low-Choy, K. Mengersen, K., et al. 2015. Species Richness on Coral Reefs and the pursuit of convergent global estimates. *Current Biology* 25: 500–505. (2)

Fisk, M. R., S. J. Giovannoni, and I. H. Thorseth. 1998. Alteration of oceanic volcanic glass: textural evidence of microbial activity. *Science* 281: 978–980. (2)

Flather, C. H., G. D. Hayward, S. R. Beissinger, and P. A. Stephens. 2011. Minimum viable populations: Is there a "magic number" for conservation practitioners? *Trends in Ecology and Evolution* 26: 307–316. (6)

Flinn, K. M. and P. L. Marks. 2007 Agricultural legacies in forest environments: tree communities, soil properties, and light availability. *Ecological Applications* 17: 452–463. (10)

Flohre, A., C. Fischer, T. Aavik, J. Bengtsson, et al. 2011. Agricultural intensification and biodiversity partitioning in European landscapes comparing plants, carabids, and birds. *Ecological Applications* 21: 1772–1780. (2)

Flory, S. L. and K. Clay. 2009. Effects of roads and forest successional age on experimental plant invasions. *Biological Conservation*. 142: 2531–2537. (4)

Fontaine, B., P. Bouchet, K. Van Achterberg, M. A. Alonso-Zarazaga, et al. 2007. The European Union's 2010 target: Putting rare species in focus. *Biological Conservation* 139: 167–185. (6)

Food and Agriculture Organization of the United Nations (FAO). 2007. FAOSTAT. *faostat.fao.org* (7)

Ford, A. T., A. P. Clevenger, and A. Bennett. 2009. Comparison of methods of monitoring wildlife crossing structures on highways. *Journal of Wildlife Management* 73: 1213–1222. (8)

Forister, M. L., A. C. McCall, N. J. Sanders, J. A. Fordyce, et al. 2010. Compounded effects of climate change and habitat alteration shift patterns of butterfly diversity. *Proceedings of the National Academy of Sciences USA* 107: 2088–2092. (4)

Forsman, A. 2014. Effects of genotypic and phenotypic variation on establishment are important for conservation, invasion, and infection biology. *Proceedings of the National Academy of Sciences* 111(1): 302–307. (5)

Forzza, R. C., J. F. A. Baumgratz, C. E. M. Bicudo, D. A. L. Canhos, et al. 2012. New Brazilian floristic list highlights conservation challenges. *Bioscience* 62: 39–45. (2)

Foster, B. L., K. Kindscher, G. R. Houseman, and C. A. Murphy. 2009. Effects of hay management and native species sowing on grassland community structure, biomass, and restoration. *Ecological Applications* 19: 1884–1896. (10)

Foster, K. R., P. Vecchia, and M. H. Repacholi. 2000. Science and the precautionary principle. *Science* 288: 979–981. (3)

Fox, H. E., M. B. Mascia, X. Basurto, A. Costa, et al. 2012. Reexamining the science of marine protected areas: Linking knowledge to action. *Conservation Letters* 5: 1–10. (8)

Fragkias, M., J. Lobo, D. Strumsky, and K. C. Seto. 2013. Does size matter? Scaling of CO_2 emissions and US urban areas. *PLoS ONE* 8(6): e64727. doi: 10.1371/journal.pone.0064727. (4)

Frankham R. 1995. Effective population size adult population size ratios in wildlife – a review. *Genetics Research* 166: 95–107. (5)

Frankham, R. 2005. Resolving the genetic paradox in invasive species. *Heredity* 94(4): 385–385. (5)

Frankham, R., C. J. A. Bradshaw, and B. W. Brook. 2014. Genetics in conservation management: revised recommendations for the 50/500 rules, Red List criteria and population viability analyses. *Biological Conservation* 170: 56–63. (5)

Frankham, R. 2005. Genetics and extinction. *Biological Conservation* 126: 131–140. (5)

Frankham, R., J. D. Ballou, and D. A. Briscoe. 2009. *Introduction to Conservation Genetics*, 2nd Ed. Cambridge University Press, Cambridge, UK. (2, 5)

Frankham, R., J. D. Ballou, M. R. Dudash, M. D. B. Eldridge, et al. 2012. Implications of different species concepts for conserving biodiversity. *Biological Conservation* 153: 25–31. (2)

Fraser, L. H., W. L. Harrower, H. W. Garris, S. Davidson, et al. 2015. A call for applying trophic structure in ecological restoration. *Restoration Ecology* 23: 503–507. (10)

Gagic, V., I. Bartomeus, T. Jonsson, A. Taylor, et al. 2015 Functional identity and diversity of animals predict ecosystem functioning better than species-based indices. *Proceedings of the Royal Society B* 282: 20142620. (2)

Gagné, S. A., F. Eigenbrod, D. G. Bert, G. M. Cunnington, et al. 2015. A simple landscape design framework for biodiversity conservation. *Landscape and Urban Planning* 136: 13–27. (8)

Gaines, S. D., C. White, M. H. Carr, and S. R. Palumbi. 2010. Designing marine reserve networks for both conservation and fisheries management. *Proceedings of the National Academy of Sciences USA* 107: 18286–18293. (8)

Galbraith, C. A., P. V. Grice, G. P. Mudge, S. Parr, and M. W. Pienkowski. 1998. The role of statutory bodies in ornithological conservation. *Ibis* 137: S224–S231. (1)

Galatowitsch, S. M. *Ecological Restoration*. 2012. Sinauer Associates, Sunderland, MA. (10)

Galetti, M., and R. Dirzo. 2013. Ecological and evolutionary consequences of living in a defaunated world. *Biological Conservation* 163: 1–6. (4)

Gallardo, B., M. Clavero, M. I. Sánchez, and M. Vilá. 2016. Global ecological impacts of invasive species in aquatic ecosystems. *Global Change Biology*. doi: 10.1111/gcb.13004. (4)

Gardiner, S., S. Caney, D. Jamieson, and H. Shue. 2010. *Climate Ethics: Essential Readings*. Oxford University Press, New York. (3)

Garibaldi, L. A., I. Steffan-Dewenter, R. Winfree, M. A. Aizen, et al. 2013. Wild pollinators enhance fruit set of crops regardless of honey bee abundance. *Science* 339(6127): 1608–1611. (3)

Garrard, G. E., F. Fidler, B. C. Wintle, Y. E. Chee, and S. A. Bekessy. 2015. Beyond advocacy: making space for conservation scientists in public debate. *Conservation Letters*. doi: 10.1111/conl.12193. (12)

Garzón-Machado, V., J. M. González-Mancebo, A. Palomares-Martínez, A. Acevedo-Rodríguez, et al. 2010. Strong negative effect of alien herbivores on endemic legumes of the Canary pine forest. *Biological Conservation* 143: 2685–2694. (4)

Gaston, K. J. 2010. *Urban Ecology*. Cambridge University Press, New York. (4)

Gaston, K. J. and J. I. Spicer. 2004. *Biodiversity: An Introduction*, 2nd Ed. Blackwell Publishing, Oxford, UK. (2)

Gates, C. C., P. Jones, M. Suitor, A. Jakes, et al. 2012. The influence of land use and fences on habitat effectiveness, movements and distribution of pronghorn in the grasslands of North America. *In* M. J. Somers and M. W. Hayward (eds.), *Fencing for Conservation*, pp. 277–294. Springer, New York. (4)

Gavin, M. C., J. McCarter, A. Mead, F. Berkes, et al. 2015. Defining biocultural approaches to conservation. *Trends in Ecology and Evolution* 30(3): 140–145. (9)

Geldmann, J., M. Barnes, L. Coad, I. D. Craigie, et al. 2013. Effectiveness of terrestrial protected area in reducing habitat loss and population declines. *Biological Conservation* 161: 230–238. (8)

Gerrodette, T. and W. G. Gilmartin. 1990. Demographic consequences of changing pupping and hauling sites of the Hawaiian monk seal. *Conservation Biology* 4: 423–430. (6)

Gessner, M. O., C. M. Swan, C. I. Dang, B. G. McKie, et al. 2010. Diversity meets decomposition. *Trends in Ecology and Evolution* 25: 372–380. (2)

Gibbs, K. E. and D. J. Currie, D. J. 2012. Protecting endangered species: do the main legislative tools work? *PLoS One* 7(5): e35730. (6)

Gibson, L., A. J. Lynam, C. J. A. Bradshaw, F. He, et al. 2013. Near-complete extinction of native small mammal fauna 25 years after forest fragmentation. *Science* 341: 1508–1510. (4)

Gibson, L., T. M. Lee, L. P. Koh, B. W. Brook, et al. 2011. Primary forests are irreplaceable for sustaining tropical biodiversity. *Nature* 478: 378–381. (4)

Gilbert-Norton, L., R. Wilson, J. R. Stevens, and K. H. Beard. 2010. A meta-analytic review of corridor effectiveness. *Conservation Biology* 24: 660–668. (8)

Gilmour, J. P., L. D. Smith, A. J. Heyward, A. H. Baird, and M. S. Pratchett. 2013. Recovery of an isolated coral reef system following severe disturbance. *Science* 340: 69–71. (4)

Gilpin, M. E. and M. E. Soulé. 1986. Minimum viable populations: Processes of species extinction. *In* M. E. Soulé (ed.), *Conservation Biology: The Science of Scarcity and Diversity*, pp. 19–34. Sinauer Associates, Sunderland, MA. (5)

Gilstad-Hayden, K., L. R. Wallace, A. Carroll-Scott, S. R. Meyer, et al. 2015. Research note: Greater tree canopy cover is associated with lower rates of both violent and property crime in New Haven, CT. *Landscape and Urban Planning* 143: 248–253. (3)

Gliessman, S. R. 2015. *Agroecology: The Ecology of Sustainable Food Systems*. CRC Press, Boca Raton, FL. (7)

Global Environment Facility (GEF): Investing in Our Planet. *www.thegef.org/gef.* (11)

Global Footprint Network: Advancing the Science of Sustainability. 2006. *www.footprint.org* (4)

Godefroid, S., C. Piazza, G. Rossi, S. Buord, et al. 2011a. How successful are plant species reintroductions? *Biological Conservation* 144: 672–682. (7)

Godefroid, S., S. Rivière, S. Waldren, N. Boretos, R. Eastwood, et al. 2011b. To what extent are threatened European plant species conserved in seed banks? *Biological Conservation* 144: 1494–1498. (7)

Godfray, H. C. J., J. R. Beddington, I. R. Crute, L. Haddad, et al. 2010. Food security: The challenge of feeding 9 billion people. *Science* 327: 812–818. (4)

Goetz, S. J., M. Sun, S. Zolkos, A. Hansen, and R. Dubayah. 2014. The relative importance of climate and vegetation properties on patterns of North American breeding bird species richness. *Environmental Research Letters* 9(3): 034013. (6)

González E., A. A. Sher, E. Tabacchi, A. Masip, and M. Poulin. 2015. Restoration of riparian vegetation: A global review of implementation and evaluation approaches in the international, peer-reviewed literature. *Journal of Environmental Management* 158: 85–94. (10)

González-Maya, J. F., L. R. Víquez-R, A. Arias-Alzate, J. L. Belant, and G. Ceballos. 2016. Spatial patterns of species richness and functional diversity in Costa Rican terrestrial mammals: implications for conservation. *Diversity and Distributions* 22: 43–56. (2)

Gore, A. 2006. *An Inconvenient Truth: The Planetary Emergency of Global Warming and What We Can Do About It*. Rodale Books, New York. (4)

Goss, J. R. and G. S. Cumming. 2013. Networks of wildlife translocations in developing countries: an emerging conservation issue? *Frontiers in Ecology and the Environment* 11(5): 243–250. (4)

Gotelli, N. J., A. Chao, R. K. Colwell, W. H. Hwang, and G. R. Graves. 2012. Specimen-based modeling, stopping rules, and the extinction of the ivory-billed woodpecker. *Conservation Biology* 26(1): 47–56. (5)

Granek, E. F., E. M. P. Madin, M. A. Brown, W. Figueira, et al. 2008. Engaging recreational fishers in management and conservation: Global case studies. *Conservation Biology* 22: 1125–1134. (12)

Grant, B. R. and P. R. Grant. 2008. Fission and fusion of Darwin's finches populations. *Philosophical Transactions of the Royal Society. Series B: Biological Sciences* 363: 2821–2829. (5)

Grassle, J. F. 2001. Marine ecosystems. *In* S. A. Levin (ed.), *Encyclopedia of Biodiversity, Vol. 4*, pp. 13–26. Academic Press, San Diego, CA. (2)

Gray, L. K., T. Gylander, M. S. Mbogga, P. Chen, A. Hamann. 2011. Assisted migration to address climate change: Recommendations for aspen reforestation in western Canada. *Ecological Applications* 21: 1591–1603. (7)

Greene, R. M., J. C. Lehrter, and J. D. Hagy. 2009. Multiple regression models for hindcasting and forecasting midsummer hypoxia in the Gulf of Mexico. *Ecological Applications* 19: 1161–1175. (4)

Grenier, M. B., D. B. McDonald, and S. W. Buskirk. 2007. Rapid population growth of a critically endangered carnivore. *Science* 317: 779. (7)

Griffiths, R. A. and L. Pavajeau. 2008. Captive breeding, reintroduction, and the conservation of amphibians. *Conservation Biology* 22: 852–861. (7)

Grman, E., T. Bassett, and L. A. Brudvig. 2014. A prairie plant community data set for addressing questions in community assembly and restoration. *Ecology* 95: 2363. (10)

Groom, M. J., G. K. Meffe, and C. R. Carroll (eds.). 2006. *Principles of Conservation Biology*, 3rd Ed. Sinauer Associates, Sunderland, MA. (2, 3, 4, 5)

Groombridge, B. and M. D. Jenkins. 2010. *World Atlas of Biodiversity; Earth's living resources in the 21st century.* University of California Press, Berkeley. (2)

Gross, L. 2008. Can farmed and wild salmon coexist? *PLoS Biology* 6: e46. (3)

Groves, C., E. Game, M. Anderson, M. Cross, et al. 2012. Incorporating climate change into systematic conservation planning. *Biodiversity and Conservation* 21: 1651–1671. (8)

Grumbine, R. E. 2007. China's emergences and the prospects for global sustainability. *BioScience* 57: 249–255. (4)

Grumbine, R. E. and J. Xu. 2011. Creating a "Conservation with Chinese Characteristics." *Biological Conservation* 144: 1347–1355. (8)

Guerrant, E. O. 1992. Genetic and demographic considerations in the sampling and reintroduction of rare plants. *In* P. L. Fiedler and S. K. Jain (eds.), *Conservation Biology: The Theory and Practice of Nature Conservation, Preservation and Management*, pp. 321–344. Chapman and Hall, New York. (5)

Guerrant, E. O. Jr., K. Havens, and M. Maunder (eds.). 2013. *Ex Situ Conservation: Supporting Species Survival in the Wild.* Island Press, Washington, DC. (7)

Guiry, M. D. 2012 How many species of Algae are there? *Journal of Phycology* 48: 1057–1063. (2)

Gupta, N., R. Raghavan, K. Sivakumar, V. Mathur, and A. C. Pinder. 2015. Assessing recreational fisheries in an emerging economy: Knowledge, perceptions and attitudes of catch-and-release anglers in India. *Fisheries Research* 165: 79–84. (12)

Gusset, M., S. J. Ryan, M. Hofmeyr, G. V. Dyk, et al. 2008. Efforts going to the dogs? Evaluating attempts to re-introduce endangered wild dogs in South Africa. *Journal of Applied Ecology* 45: 100–108. (7)

Gustafsson, L., A. Felton, A. M. Felton, J. Brunet, et al. 2015. Natural versus national boundaries: the importance of considering biogeographical patterns in forest conservation policy. *Conservation Letters* 8(1): 50–57. (12)

Gutiérrez, J. L. and C. Bernstein. 2014. Ecosystem impacts of invasive species. BIOLIEF 2011 2nd World Conference on Biological Invasion and Ecosystem Functioning, Mar del Plata, Argentina, 21–24 November 2011. Acta Oecologica 54: 1–138. (4)

Hall, J. A. and E. Fleishman. 2010. Demonstration as a means to translate conservation science into practice. *Conservation Biology* 24: 120–127. (1)

Halpern, B. S. 2014. Making marine protected areas work. *Nature* 506: 167–168. (1)

Halpern, B. S., C. J. Klein, C. J. Brown, M. Beger, et al. 2013. Achieving the triple bottom line in the face of inherent trade-offs among social equity, economic return, and conservation. *Proceedings of the National Academy of Sciences USA* 110: 6229–6234. (12)

Halpern, B. S., K. A. Selkoe, F. Micheli, and C. V. Kappel. 2007. Evaluating and ranking the vulnerability of global marine ecosystems to anthropogenic threats. *Conservation Biology* 21: 1301–1315. (10)

Halsey, S. J., T. J. Bell, K. McEachern, and N. B. Pavlovic. 2015. Comparison of reintroduction and enhancement effects on metapopulation viability. *Restoration Ecology* 23(4): 375–384. (7)

Hallwass, G., P. F. Lopes, A. A. Juras, and R. A. M. Silvano. 2013. Fishers' knowledge identifies environmental changes in fish abundance trends in impounded tropical rivers. *Ecological Applications* 23: 392–407. (8)

Hambler, C., P. A. Henderson, and M. R. Speight. 2011. Extinction rates, extinction-prone habitats, and indicator groups in Britain and at larger scales. *Biological Conservation* 144: 713–721. (4, 5)

Hammond, P. S., K. Macleod, P. Berggren, D. L. Borchers, et al. 2013. Cetacean abundance and distribution in European Atlantic shelf waters to inform conservation and management. *Biological Conservation* 164: 107–122. (6)

Handel, S. N. 2016. Greens and Greening: Agriculture and Restoration Ecology in the City. *Ecological Restoration* 34(1): 1–2. (10)

Hanley, N., L. P. Dupuy, and E. McLaughlin. 2015. Genuine savings and sustainability. *Journal of Economic Surveys* 29(4): 779–806. (3)

Hanna, E. and M. Cardillo. 2013. Island mammal extinctions are determined by interactive effects of life history, island biogeography and mesopredator suppression. *Global Ecology and Biogeography* 23(4) 395–404. doi: 10.1111/geb.12103. (5)

Hannah, L., P. R. Roehrdanz, M. Ikegami, A. V. Shepard, et al. 2013. Climate change, wine, and conservation. *Proceedings of the National Academy of Sciences USA* 110: 6907–6912. (4)

Hansen, A. J., C. R. Davis, N. Piekielek, J. Gross, et al. 2011. Delineating the ecosystems containing protected areas for monitoring and management. *BioScience* 61: 363–373. (8)

Hansen, M. C., P. V. Potapov, R. Moore, M. Hancher, et al. 2013. High-resolution global maps of 21st-century forest cover change. *Science* 342(6160): 850–853. (4)

Hansen, M. C., S. V. Stehman, P. V. Potapov, B. Arunarwati, et al. 2009. Quantifying changes in the rates of forest clearing in Indonesia from 1990 to 2005 using remotely sensed data sets. *Environmental Research Letters* 4: 034001. (4, 6)

Hanson, T., T. M. Brooks, G. A. B. da Fonseca, M. Hoffmann, et al. 2009. Warfare in biodiversity hotspots. *Conservation Biology* 23: 578–587. (6)

Hanski, I., A. Moilanen, and M. Gyllenberg. 1996. Minimum viable metapopulation size. *American Naturalist* 384: 527–541. (5)

Hardin, G. 1968. The tragedy of the commons. *Science* 162: 1243–1248. (3)

Harding, G., R. A. Griffiths, and L. Pavajeau. 2015. Developments in amphibian captive breeding and reintroduction programs. *Conservation Biology*. doi: 10.1111/cobi.12612. (7)

Hardwick, K. A., P. Fiedler, L. C. Lee, B. Pavlik, et al. 2011. The role of botanic gardens in the science and practice of ecological restoration. *Conservation Biology* 25: 265–275. (7)

Hardy, M. J., J. A. Fitzsimons, S. A. Bekessy, and A. Gordon. 2016. Exploring the permanence of conservation covenants. *Conservation Letters*. doi: 10.1111/conl.12193. (11)

Hare, M. P., L. Nunney, M. K., Schwartz, D. E. Ruzzante, et al. 2011. Understanding and estimating effective population size for practical application in marine species management. *Conservation Biology* 25: 438–449. (5)

Harrington, L. A., A. Moehrenschlager, M. Gelling, R. P. D. Atkinson, et al. 2013. Conflicting and complementary ethics to animal welfare considerations in reintroductions. *Conservation Biology* 27: 486–500. (7)

Harris, J. A., R. J. Hobbs, E. Higgs, and J. Aronson. 2006. Ecological restoration and global climate change. Restoration Ecology 14(2): 170–176. (10)

Harvell, D., R. Aronson, N. Baron, J. Connell, et al. 2004. The rising tide of ocean diseases: Unsolved problems and research priorities. *Frontiers in Ecology and the Environment* 2: 375–382. (4)

Hassett, B, M. Palmer, E. Bernhardt, S. Smith, et al. 2005. Restoring watersheds project by project: trends in Chesapeake Bay tributary restoration. *Frontiers in Ecology and Environment* 3: 259–267. (10)

Hautier, Y., D. Tilman, F. Isbell, E. W. Seabloom, et al. 2015. Anthropogenic environmental changes affect ecosystem stability via biodiversity. *Science* 348: 336–340. (1, 3)

Havens, K., C. L. Jolls, J. E. Marik, P. Vitt, et al. 2012. Effects of a non-native biocontrol weevil, *Larinus planus*, and other emerging threats on populations of the federally threatened Pitcher's thistle, *Cirsium pitcheri*. *Biological Conservation* 158: 202–211. (4)

Haw, J. 2013. Southern California and endangered abalone populations. *Scientific American*, July 3, 2013. (5)

Hayes, T. B., P. Falso, S. Gallipeau, and M. Stice. 2010. The cause of global amphibian declines: a developmental endocrinologist's perspective. *The Journal of Experimental Biology* 213(6): 921–933. (4)

Hayward, M. W. 2009. Conservation management for the past, present, and future. *Biodiversity and Conservation* 18: 765–775. (10)

He, X., M. L. Johansson, and D. D. Heath, D. D. 2016. Role of genomics and transcriptomics in selection of reintroduction source populations. *Conservation Biology*. doi: 10.1111/cobi.12674. (7)

Heber, S. and J. V. Briskie. 2010. Population bottlenecks and increased hatching failure in endangered birds. *Conservation Biology* 24: 1674–1678. (5)

Hedges, S., A. Johnson, M. Ahlering, M. Tyson, and L. S. Eggert. 2013. Accuracy, precision, and cost-effectiveness of conventional dung density and fecal DNA based survey methods to estimate Asian elephant (Elephas maximus) population size and structure. *Biological Conservation* 159: 101–108. (6)

Hedges, S., M. J. Tyson, A. F. Sitompul, M. F. Kinnaird, et al. 2005. Distribution, status, and conservation needs of Asian elephants (*Elephas maximus*) in Lampung Province, Sumatra, Indonesia. *Biological Conservation* 124: 35–48. (5)

Hedrick, P. 2005. Large variance in reproductive success and the N_e/N ratio. *Evolution* 59: 1596–1599. (5)

Heleno, R. H., R. S. Ceia, J. A. Ramos, and J. Memmott. 2009. The effect of alien plants on insect abundance and biomass: a food web approach. *Conservation Biology* 23(2): 410–419. (4)

Helfield, J. M., S. J. Capon, C. Nilsson, R. Jansson, and D. Palm. 2007. Restoration of rivers used for timber floating: Effects on riparian plant diversity. *Ecological Applications* 17: 840–851. (10)

Helm, D. 2015. *Natural Capital: Valuing the Planet*. Yale University Press, New Haven, CT. (3)

Hendricks, S. A., P. R. S. Clee, R. J. Harrigan, J. P. Pollinger, et al. 2016. Re-defining historical geographic range in species with sparse records: Implications for the Mexican wolf reintroduction program. *Biological Conservation* 194, 48–57. (7)

Heschel, M. S. and K. N. Paige. 1995. Inbreeding depression, environmental stress and population size variation in scarlet gilia (*Ipomopsis aggregata*). *Conservation Biology* 9: 126–133. (5)

Hickey, V. and S. L. Pimm. 2011. How the World Bank funds protected areas. *Conservation Letters* 4: 269–277. (11)

Higgs, E. 2003. *Nature by Design: People, Natural Process, and Ecological Restoration*. MIT Press, Cambridge, MA. (10)

Higgs E., D. A. Falk, A. Guerrini, M. Hall, et al. 2014. The role of history in restoration ecology. *Frontiers in Ecology and the Environment* 12(9): 499–506. (10)

Himes Boor, G. K. 2014. A framework for developing objective and measurable recovery criteria for threatened and endangered species. *Conservation Biology* 28: 33–43. (6)

Hinsley, A., D. Verissimo, and D. L. Roberts. 2015. Heterogeneity in consumer preferences for orchids in international trade and the potential for the use of market research methods to study demand for wildlife. *Biological Conservation* 190: 80–86. (4)

Hinz, H., V. Prieto, and M. J. Kaiser. 2009. Trawl disturbance on benthic communities: Chronic effects and experimental predictions. *Ecological Applications* 19: 761–773. (4)

Hinz, H. L., M. Schwarzländer, A. Gassmann, and R. S. Bourchier. 2014. Successes we may not have had: a retrospective analysis of selected weed biological control agents in the United States. *Invasive Plant Science and Management* 7(4): 565–579. (10)

Hitzhusen, G. E. and M. E. Tucker. 2013. The potential of religion for Earth Stewardship. *Frontiers in Ecology and the Environment* 11: 368–376. (1)

Hobbs, R. J. and J. A. Harris. 2001. Restoration ecology: repairing the earth's ecosystems in the new millennium. *Restoration Ecology* 9(2): 239–246. (10)

Hobbs, R. J., D. N. Cole, L. Yung, E. S. Zavaleta, et al. 2010. Guiding concepts for park and wilderness stewardship in an era of global environmental change. *Frontiers in Ecology and the Environment* 8: 483–490. (8, 10)

Hobbs, R. J., E. S. Higgs, and C. M. Hall. 2013. *Novel Ecosystems: Intervening in the New Ecological World Order.* Wiley-Blackwell, Oxford, UK. (4, 10)

Hoeinghaus, D. J., A. A. Agostinho, L. C. Gomes, F. M. Pelicice, et al. 2009. Effects of river impoundment on ecosystem services of large tropical rivers: Embodied energy and market value of artisanal fisheries. *Conservation Biology* 23: 1222–1231. (3)

Hoffman, S. W. and J. P. Smith. 2003. Population trends of migratory raptors in western North America, 1977–2001. *Condor* 105: 397–419. (4)

Hoffmann, M. T. Brooks, G. A. B. Fonseca, and J. M. C. Silva. 2008. Conservation planning and the IUCN Red List. *Endangered Species Research* 6(2): 113–125. (6)

Hoffmann, M., C. Hilton-Taylor, A. Angulo, M. Böhm, et al. 2010. The impact of conservation on the status of the world's vertebrates. *Science* 330: 1503–1509. (5)

Holden, E. and K. G. Hoyer. 2005. The ecological footprint of fuels. *Transportation Research, Part D* 10: 395–403. (4)

Hole, D. G., B. Huntley, J. Arinaitwe, S. H. M. Butchart, et al. 2011. Toward a management framework for networks of protected areas in the face of climate change. *Conservation Biology* 25: 305–315. (4)

Hollings, T., M. Jones, N. Mooney, and H. McCallum. 2014. Trophic cascades following the disease-induced decline of an apex predator, the Tasmanian devil. *Conservation Biology* 28: 63–75. (2)

Hooke, R. L., J. F. Martín-Duque, and J. Pedraza. 2012. Land transformation by humans: a review. *GSA Today* 22(12): 4–10. (4)

Horton, C., T. R. Peterson, P. Banerjee, and M. J. Peterson. 2015. Credibility and advocacy in conservation science. *Conservation Biology* 30(1), 23–32. doi: 10.1111/cobi.12558. (1, 12)

Horwitz, P. and C. M. Finlayson. 2011. Wetlands as settings for human health: incorporating ecosystem services and health impact assessment into water resource management. *BioScience* 61(9): 678–688. (3)

Huffington Post. 2015. Germany to Turn 62 Military Bases into Nature Sanctuary for Birds, Beetles and Bats. *www.huffingtonpost.com/2015/06/19/german-military-bases-nature-reserves_n_7623882.html* (9)

Hughes, A. R., S. L. Williams, C. M. Duarte, K. L. Heck, Jr., and M. Waycott. 2009. Associations of concern: Declining seagrasses and threatened dependent species. *Frontiers in Ecology and the Environment* 7: 242–246. (2)

Hughes, J. B. and J. Roughgarden. 2000. Species diversity and biomass stability. *American Naturalist* 155: 618–627. (5)

Hughes, T. P., H. U. I. Huang, and M. A. Young. 2013. The wicked problem of China's disappearing coral reefs. *Conservation Biology* 27(2): 261–269. (4)

Hughes, T. W. and K. Lee. 2015. The role of recreational hunting in the recovery and conservation of the wild turkey (*Meleagris gallopavo* spp.) in North America. *International Journal of Environmental Studies.* 72(5): 797–809. (7)

Huhta, E. and J. Jokimäki. 2015. Landscape matrix fragmentation effect on virgin forest and managed forest birds: a multi-scale study. *In* J. A. Daniels (ed.), *Advances in Environmental Research, Volume 36*, pp. 95–111. Nova Science Publishers, Inc., Hauppauge, NY. (4)

Hulme, P. E. 2015. Invasion pathways at a crossroad: policy and research challenges for managing alien species introductions. *Journal of Applied Ecology* 52(6): 1418–1424. (4)

Hultine, K. R., D. W. Bean, T. L. Dudley, and C. A. Gehring. 2015. Species introductions and their cascading impacts on biotic interactions in desert riparian ecosystems. *Integrative and Comparative Biology* 55(4): 587–601. (10)

Human Microbiome Consortium. 2012. *www.humanmicrobiome.org/* (2)

Humavindu, M. N. and J. Stage. 2015. Continuous financial support will be needed. *Animal Conservation* 18(1): 18–19. (9)

Humphries, P. and K. O. Winemiller. 2009. Historical impacts on river fauna, shifting baselines, and challenges for restoration. *BioScience* 59: 673–684. (10)

Hunter, M. O., M. Keller, D. Morton, B. Cook, et al. 2015. Structural dynamics of tropical moist forest gaps. *PLoS ONE* 10(7): e0132144. (8)

Hylander, K., S. Nemomissa, J. Delrue, and W. Enkosa. 2013. Effects of coffee management on deforestation rates and forest integrity. *Conservation Biology* 27: 1011–1019. (9)

Ikh Nart Nature Reserve. *IkhNart.com* (12)

Iknayan, K. J., M. W. Tingley, B. J. Furnas, S. R. Beissinger. 2014. Detecting diversity: emerging methods to estimate species diversity. *Trends in Ecology and Evolution.* 29(2): 97–106. (2)

Indigenous Peoples Literature. 2009. *indigenouspeople. net* (9)

Inouye, D. W., M. A. Morales, and G. J. Dodge. 2002. Variation in timing and abundance of flowering by *Delphinium barbeyi* Huth (Ranunculaceae): the roles of snowpack, frost, and La Nina, in the context of climate change. *Oecologia* 130 (4): 543–550. (4)

Intergovernmental Panel on Climate Change (IPCC). 2013. *Climate Change 2013: The Physical Science Basis. Contribution of Working Group I to the Fifth Assessment Report of the Intergovernmental Panel on Climate Change.* T. F. Stocker and 9 others (eds.). Cambridge University Press, Cambridge, UK. (4)

International Joint Commission. 2014. *A Balanced Diet for Lake Erie: Reducing Phosphorus Loadings and Harmful Algal Blooms.* Report of the Lake Erie Ecosystem Priority. (10)

International Plant Protection Convention (IPPC). 2014. R. K. Pachauri and L. A. Meyer (eds.), Climate Change 2014: Synthesis Report. Contribution of Working Groups I, II and III to the Fifth Assessment Report of the Intergovernmental Panel on Climate Change. IPCC, Geneva. (4)

International Union for Conservation of Nature (IUCN) 2015. *The IUCN Red List of Threatened Species. Version 2015–4. www.iucnredlist.org* Downloaded on 19 November 2015. (1)

International Union for Conservation of Nature (IUCN). *www.iucn.org* (6, 8, 12)

International Whaling Commission (IWC). *www.iwc.int.* (6)

International Work Group for Indigenous Affairs (IWGIA). *www.iwgia.org* (9)

International Union for Conservation of Nature (IUCN). 2001. IUCN Red List Categories and Criteria: Version 3.1. IUCN Species Survival Commission. IUCN, Gland, Switzerland. *www.iucnredlist.org/ technical-documents/categories-and-criteria/2001–categories-criteria* (6)

International Union for Conservation of Nature (IUCN). 2004. 2004 IUCN Red List of Threatened Species. *www.iucnredlist.org* (4)

International Union for Conservation of Nature (IUCN). 2015. *2014 Annual Report of the Species Survival Commission and the Global Species Programme. portals.iucn. org/library/node/45591* (1)

Jackson, S. F., K. Walker, and K. J. Gaston. 2009. Relationship between distributions of threatened plants and protected areas in Britain. *Biological Conservation* 142: 1515–1522. (8)

Jacob, J., E. Jovic, and M. B. Brinkerhoff. 2009. Personal and planetary well-being: Mindfulness meditation, pro-environmental behavior and personal quality of life in a survey from the social justice and ecological sustainability movement. *Social Indicators Research* 93: 275–294. (3)

Jacobson, S. K. 2009. *Communication Skills for Conservation Professionals*, 2nd Ed. Island Press, Washington, DC. (12)

Jacobson, S. K., M. D. Mcduff, and M. Monroe. 2007. Promoting conservation through the arts: outreach for hearts and minds. *Conservation Biology* 21(1): 7–10. (1)

Jacobson, S. K., D. Wald, N. Haynes, and R. Sakurai. 2014. Wildlife communication and negotiation. *In* R. McCleery, C. Moorman and N. Peterson (eds.), *Urban Wildlife Science: Theory and Practice*, Springer-Verlag, New York. (12)

Jacquemyn, H., C. Van Mechelen, R. Brys, and O. Honnay. 2011. Management effects on the vegetation and soil seed bank of calcareous grasslands: An 11-year experiment. *Biological Conservation* 144: 416–422. (8)

Jaeger, I., H. Hop, and G. W. Gabrielsen. 2009. Biomagnification of mercury in selected species from an Arctic marine food web in Svalbard. *Science of the Total Environment* 407: 4744–4751. (4)

Jähnig, S. C., A. W. Lorenz, D. Hering, C. Antons, et al. 2011. River restoration success: A question of perception. *Ecological Applications* 21: 2007–2015. (8)

Jäkäläniemi, A., A. H. Postila, and J. Tuomi. 2013. Accuracy of short-term demographic data in projecting long-term fate of populations. *Conservation Biology* 27: 552–559. Many years of data are needed to build a good PVA model. (6)

Jamieson, I. G. 2011. Founder effects, inbreeding, and loss of genetic diversity in four avian reintroduction programs. *Conservation Biology* 25: 115–123. (5)

Janzen, D. H. 2001. Latent extinctions—the living dead. *In* S. A. Levin (ed.), *Encyclopedia of Biodiversity*, pp. 689–700. Academic Press, San Diego, CA. (5)

Jarošík, V., M. Konvička, P. Pyšek, T. Kadlec, and J. Beneš. 2011. Conservation in a city: Do the same principles apply to different taxa? *Biological Conservation* 144: 490–499. (8)

Jarvis, D. I., T. Hodgkin, A. H. Brown, J. Tuxill, et al. 2016. *Crop Genetic Diversity in the Field and on the Farm: Principles and Applications in Research Practices.* Yale University Press. (9)

Jelinski, D. E. 2015. On a landscape ecology of a harlequin environment: the marine landscape. *Landscape Ecology* 30(1): 1–6. (8)

Jenkins, C. N. and L. Joppa. 2009. Expansion of the global terrestrial protected area system. *Biological Conservation* 142: 2166–2174. (8)

Jewell, A. 2013. Effect of monitoring technique on quality of conservation science. *Conservation Biology* 27: 501–508. (6)

Jones, C. G., J. H. Lawton, and M. Shachak. 1996. Organisms as ecosystem engineers. In *Ecosystem Management*, F. B. Samson and F. L. Knopf (eds.), pp. 130–147. Springer, New York. (2)

Jones, H. L. and J. M. Diamond. 1976. Short-time-base studies of turnover in breeding birds of the California Channel Islands. *Condor* 76: 526–549. (6)

Jones, K. E., N. G. Patel, M. A. Levy, A. Storeygard, et al. 2008. Global trends in emerging infectious diseases. *Nature* 451: 990–994. (4)

Jones, L., A. Garbutt, J. Hansom, S. Angus. 2013. Impacts of climate change on coastal habitats. *Marine Climate Change Impacts Partnership: Science Review* 2013: 167–179. (4)

Jones, T. A. 2003. The restoration gene pool concept: beyond the native versus non-native debate. *Restoration Ecology* 11(3): 281–290. (10)

Joppa, L. N., S. R. Loarie, and S. L. Pimm. 2008. On the protection of "protected areas." *Proceedings of the National Academy of Sciences USA* 105: 6673–6678. (8)

Joppa, L. N. and A. Pfaff. 2009. High and far: biases in the location of protected areas. *PLOS ONE* 4: e8273. (8)

Joppa, L. N., D. L. Roberts, and S. L. Pimm. 2011. The population ecology and social behavior of taxonomists. *Trends in Ecology and Evolution* 26: 551–553. (2, 6)

Joppa, L. N., P. Visconti, C. N. Jenkins, and S. L. Pimm. 2013. Achieving the Convention on Biological Diversity's goals for plant conservation. *Science* 341: 1100–1103. (8)

Jorge, M. L. S., M. Galetti, M. C. Ribeiro, and K. M. P. Ferraz. 2013. Mammal defaunation as surrogate of trophic cascades in a biodiversity hotspot. *Biological Conservation* 163: 49–57. (2)

Journey North. Annenberg Learner. *www.learner.org/jnorth* (12)

Junk, W. J., M. T. F. Piedade, L. Lourival, F. Wittmann, et al. 2014. Brazilian wetlands: their definition, delineation, and classification for research, sustainable management, and protection. *Aquatic Conservation* 24(1): 5–22. (3)

Kadoya, T., S. Suda, and I. Washitani. 2009. Dragonfly crisis in Japan: A likely consequence of recent agricultural habitat degradation. *Biological Conservation* 142: 1899–1905. (8)

Kahler, J. S., G. J. Roloff, and M. L. Gore. 2013. Poaching risks in community-based natural resource management. *Conservation Biology* 27: 177–186. (22)

Karanth, K. and R. DeFries. 2011. Nature-based tourism in Indian protected areas: New challenges for park management. *Conservation Letters* 4: 137–149. (3)

Karesh, W. B., R. A. Cook, E. L. Bennett, and J. Newcomb. 2005. Wildlife trade and global disease emergence. *CDC Emerging Infectious Diseases* 11: 1000–1002. (4)

Karl, T. R. 2006. Written statement for an oversight hearing: Introduction to Climate Change before the Committee on Government Reform, U. S. House of Representatives, Washington, DC. (4)

Kautz, R., R. Kawula, T. Hoctor, J. Comiskey, et al. 2006. How much is enough? Landscape-scale conservation for the Florida panther. *Biological Conservation* 130: 118–133. (9)

Keeton, W. S., C. E. Kraft, and D. R. Warren. 2007. Mature and old-growth riparian forests: Structure, dynamics, and effects on Adirondack stream habitats. *Ecological Applications* 17: 852–868. (8)

Keller R. P. and D. M. Lodge. 2007. Species invasions from commerce in live aquatic organisms – problems and possible solutions. *BioScience* 57: 428–436. (4)

Kelm, D. H., K. R. Wiesner, and O. von Helversen. 2008. Effects of artificial roosts for frugivorous bats on seed dispersal in a Neotropical forest pasture mosaic. *Conservation Biology* 22: 733–741. (2)

Keenelyside, K., N. Dudley, S. Cairns, C. Hall, and S. Stolton. 2012. *Ecological restoration for protected areas: principles, guidelines and best practices (Vol. 18)*. IUCN. (10)

Kemp, D. R., H. Guodong, H. Xiangyang, D. L. Michalk, et al. 2013. Innovative grassland management systems for environmental and livelihood benefits. *Proceedings of the National Academy of Sciences USA* 110: 8369–8374. (9)

Keppel, G., A. Naikatini, I. A. Rounds, R. L. Pressey, and N. T. Thomas. 2015. Local and expert knowledge improve conservation assessment of rare and iconic Fijian tree species. *Pacific Conservation Biology* 21(3): 214–219. (6)

Kiesecker, J. M., H. Copeland, A. Pocewicz, N. Nibbelink, et al. 2009. A framework for implementing biodiversity offsets: selecting sites and determining scale. *BioScience* 59: 77–84. (11)

King, A. W., D. J. Hayes, D. N. Huntzinger, T. O. West, et al. 2012. North American carbon dioxide sources and sinks: magnitude, attribution, and uncertainty. *Frontiers in Ecology and the Environment* 10: 512–519. (3)

King, D. I., R. B. Chandler, J. M. Collins, W. R. Petersen, et al. 2009. Effects of width, edge and habitat on the abundance and nesting success of scrub-shrub birds in powerline corridors. *Biological Conservation* 142: 2672–2680. (9)

King, T., C. Chamberlan, and A. Courage. 2014. Assessing reintroduction success in long-lived primates through population viability analysis: western lowland gorillas *Gorilla gorilla gorilla* in Central Africa. *Oryx* 48(02): 294–303. (6)

Kinnaird, M. F. and T. G. O'Brien. 2013. Effects of private land use, livestock management, and human tolerance on diversity, distribution, and abundance of large African mammals. *Conservation Biology* 27: 1026–1039. (9)

Kintisch, K. 2013. Climate study highlights wedge issue. *Science* 339: 128–129. (4)

Kloor, K. 2015. The Battle for the Soul of Conservation Science. *Issues in Science and Technology* 31: 74. (1)

Knapp, R. A., C. P. Hawkins, J. Ladau, and J. G. McClory. 2005. Fauna of Yosemite National Park lakes has low resistance but high resilience to fish introductions. *Ecological Applications* 15: 835–847. (2)

Knoot, T. G., M. Rickenbach, and K. Silbernagel. 2015. Payments for ecosystem services: will a new hook net more active family forest owners. *Journal of Forestry* 113(2): 210–218 (11)

Kobori, H. 2009. Current trends in conservation education in Japan. *Biological Conservation* 142: 1950–1957. (10)

Kobori, H. and R. Primack. 2003. Participatory conservation approaches for Satoyama: The traditional forest and agricultural landscape of Japan. *Ambio* 32: 307–311. (8)

Köhler, H. R. and Triebskorn, R. 2013. Wildlife ecotoxicology of pesticides: can we track effects to the population level and beyond? *Science* 341(6147): 759–765. (4)

Kolby, J. E., K. M. Smith, S. D. Ramirez, F. Rabemananjara, F. et al. 2015. Rapid response to evaluate the presence of amphibian chytrid fungus (*Batrachochytrium dendrobatidis*) and ranavirus in wild amphibian populations in Madagascar. *PLoS ONE* 10(6): e0125330. (7)

Korngold, J. 2008. *God in the Wilderness: Rediscovering the Spirituality of the Great Outdoors with the Adventure Rabbi*. Doubleday, New York. (1)

Krausmann, F., K-H Erb, S. Gingrich, H. Haberi, et al. 2013. Global human appropriation of net primary production doubled in the 20th century. *Proceedings of the National Academy of Sciences* 110(25): 10324–10329. (4)

Krauze-Gryz, D., J. Gryz, and J. Goszczyński, J. 2012. Predation by domestic cats in rural areas of central Poland: an assessment based on two methods. *Journal of Zoology* 288(4), 260–266. (4)

Kroll, G. 2015. An environmental history of roadkill: road ecology and the making of the permeable highway. *Environmental History* 20(1), 4–28. (4)

Kross, S. M., J. M. Tylianakis, and X. J. Nelson. 2012. Effects of introducing threatened falcons into vineyards on abundance of passeriformes and bird damage to grapes. *Conservation Biology* 26: 142–149. (3)

Kubiszewski, I., R. Costanzaa, C. Francob, P. Lawn, et al. 2013. Beyond GDP: Measuring and achieving global genuine progress. *Ecological Economics* 93: 57–68. (3)

Kuebbing, S. E., L. Souza, and N. J. Sanders. 2014. Effects of co-occurring non-native invasive plant species on old-field succession. *Forest Ecology and Management* 324: 196–204. (4)

Kueffer, C. and C. N. K. Kaiser-Bunbury. 2014. Reconciling conflicting perspectives for biodiversity conservation in the Anthropocene. *Frontiers in Ecology and the Environment* 12: 131–137. (10)

Kulkarni, M. V., P. M. Groffman, and J. B. Yavitt. 2008. Solving the global nitrogen problem: It's a gas! *Frontiers in Ecology and the Environment* 6: 199–206. (4)

Kyrpides, N. C., P. Hugenholtz, J. A. Eisen, T. Woyke, et al. 2014. Genomic Encyclopedia of Bacteria and Archaea: Sequencing a Myriad of Type Strains. *PLoS Biology* 12(8): e1001920. doi: 10.1371/journal.pbio.1001920. (2)

Lacher, T. E., L. Boitani, and G. A. B. da Fonseca. 2012. The IUCN global assessments: partnerships, collaboration and data sharing for biodiversity science and policy. *Conservation Letters* 5: 327–333. (6)

Laikre, L., F. W. Allendorf, L. C. Aroner, C. S. Baker, et al. 2010. Neglect of genetic diversity in implementation of the Convention on Biological Diversity. *Conservation Biology* 24: 86–88. (2)

Laikre, L., M. Jansson, F. W. Allendorf, S. Jakobsson, N. Ryman. 2013. Hunting effects on favourable conservation status of highly inbred Swedish wolves. *Conservation Biology* 27: 248–253. (5)

Laikre, L., M. K. Schwartz, R. S. Waples, and N. Ryman. 2010. Compromising genetic diversity in the wild: Unmonitored large-scale release of plants and animals. *Trends in Ecology and Evolution* 25: 520–529. (4)

Land Trust Alliance. *www.landtrustalliance.org* (11)

Langton, M., L. Palmer, and Z. Ma Rhea. 2014. Community-oriented protected areas for indigenous peoples and local communities. *In* S. Stevens (ed.), *Indigenous Peoples, National Parks, and Protected Areas: A New Paradigm Linking Conservation, Culture, and Rights*, pp. 84–107. (8)

Lapeyre, R. 2015. Wildlife conservation without financial viability? The potential for payments for dispersal areas' services in Namibia. *Animal Conservation* 18(1): 14–15. (9)

Larson, C. 2016. Shell trade pushes giant clams to the brink. *Science* 351: 323–324. (4)

Latin American and Caribbean Network of Environmental Funds (RedLAC). *www.redlac.org* (11)

Laurance, W. F. 2008a. Adopt a forest. *Biotropica* 40: 3–6. (10)

Laurance, W. F. 2008b. Theory meets reality: how habitat fragmentation research has transcended island biogeographic theory. *Biological Conservation* 141: 1731–1744. (5)

Laurance, W. F. 2013. Does research help to safeguard protected areas? *Trends in Ecology and Evolution* 28: 261–266. (12)

Laurance, W. F., M. Goosem, and S. G. W. Laurance. 2009. Impacts of roads and linear clearings on tropical forests. *Trends in Ecology and Evolution* 24: 659–679. (4)

Laurance, W. F., H. Koster, M. Grooten, A. B. Anderson, et al. 2012. Making conservation research more relevant for conservation practitioners. Biological Conservation 153: 164–168. (1)

Laurance, W. F., S. G. Laurance, and D. W. Hilbert. 2008. Long-germ dynamics of a fragmented rainforest mammal assemblage. *Conservation Biology* 22(5): 1154–1164. (5)

Laurance, W. F., T. E. Lovejoy, H. L. Vasconcelos, E. M. Bruna, et al. 2002. Ecosystem decay of Amazonian forest fragments: A 22–year investigation. *Conservation Biology* 16: 605–618. (4)

Lavauden L. 1927. Les forêts du Sahara. *Revue des Eaux et Forêts* 7(65): 329–41. (4)

Lawler, J. J., S. P. Campbell, A. D. Guerry, M. B. Kolozsvary, et al. 2002. The scope and treatment of threats in endangered species recovery plans. *Ecological Applications* 12: 663–667. (4)

Lawrence, A. J., R. Afif, M. Ahmed, S. Khalifa, and T. Paget. 2010. Bioactivity as an options value of sea cucumbers in the Egyptian Red Sea. *Conservation Biology* 24: 217–225. (3)

Le Roux, D. S., K. Ikin, D. B. Lindenmayer, G. Bistricer, et al. 2015. Enriching small trees with artificial nest boxes cannot mimic the value of large trees for hollow-nesting birds. *Restoration Ecology*. doi: 10.1111/rec.12303. (8)

Lee, J. S. H., S. Abood, J. Ghahoul, B. Barus, et al. 2014. Environmental impacts of large-scale oil palm enterprises exceed that of smallholdings in Indonesia. *Conservation Letters* 7: 25–33. (4)

Leidner, A. K. and M. C. Neel. 2011. Taxonomic and geographic patterns of decline for threatened and endangered species in the United States. *Conservation Biology* 25: 716–725. (5, 6)

Leopold, A. 1949. *A Sand County Almanac and Sketches Here and There*. Oxford University Press, New York. (1, 3)

Leopold, A. 1953. *Round River: From the Journals of Aldo Leopold*. Oxford University Press, New York. (3)

Leopold, A. C. 2004. Living with the land ethic. *BioScience* 54: 149–154. (1)

Lerner, J., J. Mackey and F. Casey. 2007. What's in Noah's wallet? Land conservation spending in the United States. *BioScience* 57: 419–423. (8)

Levin, S. A. (ed.). 2001. *Encyclopedia of Biodiversity*. Academic Press, San Diego, CA. (2)

Liebhold, A. M., E. G. Brockerhoff, L. J. Garrett, J. L. Parke J. L., and K. O. Britton. 2012. Live plant imports: the major pathway for forest insect and pathogen invasions of the US. *Frontiers in Ecology and the Environment* 10: 135–143. (4)

Lin, J., D. Pan, S. J. Davis, Q. Zhang, et al. 2014. China's international trade and air pollution in the United States. *Proceedings of the National Academy of Science USA*. 111(5) 1736–1741. doi: 10.1073/pnas.1312860111. (11)

Lindenmayer, D. and M. Hunter. 2010. Some guiding concepts for conservation biology. *Conservation Biology* 24: 1459–1468. (1)

Lindenmayer, D. B., G. E. Likens, A. Haywood, and L. Miezis. 2011. Adaptive monitoring in the real world: Proof of concept. *Trends in Ecology and Evolution* 26: 641–646. (8)

Lindenmayer, D. B., A. Welsh, C. Donnelly, M. Crane, et al. 2009. Are nest boxes a viable alternative source of cavities for hollow-dependant animals? Long-term monitoring of nest box occupancy, pest use and attrition. *Biological Conservation* 142: 33–42. (8)

Linder, J. M. and J. F. Oates. 2011. Differential impact of bushmeat hunting on monkey species and implications for primate conservation in Korup National Park, Cameroon. *Biological Conservation* 144: 738–745. (4)

Lindsey, P. A., G. A. Balme, P. J. Funston, P. Henschel, and L. T. Hunter. 2016. Life after Cecil: channelling global outrage into funding for conservation in Africa. *Conservation Letters*. DOI: 10.1111/conl.12224 (9)

Lindsey, P. A., G. Blame, M. Becker, C. Begg, et al. 2013. The bushmeat trade in African savannas: impacts, drivers, and possible solutions. *Biological Conservation* 160: 80–96. (4)

Lindsey, P. A., C. P. Havemann, R. M. Lines, A. E. Price, et al. 2013. Benefits of wildlife-based land uses on private lands in Namibia and limitations affecting their development. *Oryx* 47(01): 41–53. (9)

Linnell, J. D., P. Kaczensky, U. Wotschikowsky, N. Lescureux, and L. Boitani. 2015. Framing the relationship between people and nature in the context of European conservation. *Conservation Biology* 29(4): 978–985. (1)

Liu, H., H. Ren, Q. Liu, X. Wen, M. Maunder, and J. Gao. 2015. Translocation of threatened plants as a conservation measure in China. *Conservation Biology*, 29(6): 1537–1551. (7)

Liu, P., L. Sun, J. Li, L. Wang, et al. 2015. Population viability analysis of *Gloydius shedaoensis* from northeastern China: A contribution to the assessment of the conservation and management status of an endangered species. *Asian Herpetological Research* 1(1): 48–56. (6)

Liu, S. H., K. Li, and D. F. Hu. 2016. The incidence and species composition of Gasterophilus (Diptera, Gasterophilidae) causing equine myiasis in northern Xinjiang, China. *Veterinary Parasitology* 217: 36–38. (7)

Lloyd, P., T. E. Martin, R. L. Redmond, U. Langer, and M. M. Hart. 2005. Linking demographic effects of habitat fragmentation across landscapes to continental source-sink dynamics. *Ecological Applications* 15: 1504–1514. (4)

Lloyd, R. A., K. A. Lohse, and T. P. A. Ferre. 2013. Influence of road reclamation techniques on forest ecosystem recovery. *Frontiers in Ecology and the Environment*. 11: 75–81. (10)

Loke, L. H., R. J. Ladle, T. J. Bouma, and P. A. Todd. 2015. Creating complex habitats for restoration and reconciliation. *Ecological Engineering* 77: 307–313. (9)

Loss, S. R. and R. B. Blair. 2011. Reduced density and nest survival of ground-nesting songbires relative to earthworm invasions in northern hardwood forests. *Conservation Biology* 25: 983–993. (4)

Lotze, H. K., M. Coll, A. M. Magera, C. Ward-Paige, and L. Airoldi. 2011. Recovery of marine animal populations and ecosystems. *Trends in Ecology and Evolution* 26: 595–605. (4)

Loucks, C., M. B. Mascia, A. Maxwell, K. Huy, et al. 2009. Wildlife decline in Cambodia, 1953–2005: Exploring the legacy of armed conflict. *Conservation Letters* 2: 82–92. (4)

Louda, S. M., D. Kendall, J. Connor, and D. Simberloff. 1997. Ecological effects of an insect introduced for the biological control of weeds. *Science* 277: 1088–1090. (4)

Louisiana Coastal Wetlands Conservation and Restoration Task Force and the Wetlands Conservation and Restoration Authority. 1998. *Coast 2050: Toward a Sustainable Coastal Louisiana.* Louisiana Department of Natural Resources, Baton Rouge, LA. (10)

Louv, R. 2005. *Last Child in the Woods: Saving Our Children from Nature-Deficit Disorder.* Algonquin Books, Chapel Hill, NC. (3)

Lovich, J. E., C. B. Yackulic, J. Freilich, M. Agha, et al. 2014. Climatic variation and tortoise survival: Has a desert species met its match? *Biological Conservation* 169: 214–224. (4)

Loyd, K. A. T., S. M. Hernandez, J. P. Carroll, K. J. Abernathy, and G. J. Marshall. 2013. Quantifying free-roaming domestic cat predation using animal-borne video cameras. *Biological Conservation,*160: 183–189. (4)

Lu, Y., K. Wu, Y. Jiang, Y. Guo, et al. 2012. Widespread adoption of Bt cotton and insecticide decrease promotes biocontrol services. *Nature* 487: 362–365. (4)

Luck, G. W., P. Davidson, D. Boxall, and L. Smallbone. 2011. Relations between urban bird and plant communities and human well-being and connection to nature. *Conservation Biology* 25: 816–826. (3)

Lunt, I. D., M. Byrne, J. J. Hellmann, N. J. Mitchell, et al. 2013. Using assisted colonisation to conserve biodiversity and restore ecosystem function under climate change. *Biological Conservation* 157: 172–177. (7)

MacArthur, R. H. and E. O. Wilson. 1967. *The Theory of Island Biogeography.* Princeton University Press, Princeton, NJ. (5, 8)

Mace, G., H. Masundire, J. Baillie, T. Ricketts, et al. 2005. Biodiversity. *In* R. Hassan, R. Scholes, and N. Ash (eds.), *Ecosystems and human well-being: Current state and trends: Findings of the Condition and Trends Working Group,* pp. 77–122. Island Press, Washington, DC. (5)

Mack, R. N., D. Simberloff, W. M. Lonsdale, H. Evans, et al. 2000. Biotic invasions: causes, epidemiology, global consequences, and control. *Ecological Applications* 10(3): 689–710. (4)

MacKay, A., M. Allard, and M. A. Villard. 2014. Capacity of older plantations to host bird assemblages of naturally-regenerated conifer forests: a test at stand and landscape levels. *Biological Conservation* 170: 110–119. (9)

Magera, A. M., J. E. M. Flemming, K. Kaschner, L. B. Christensen, H. K. Lotze. 2013. Recovery trends in marine mammal populations. *PLoS ONE* doi: 10.1371/journal.pone.0077908. (4)

Magnuson, J. J. 1990. Long-term ecological research and the invisible present. *BioScience* 40: 495–501. (6)

Magrach, A., A. R. Larrinaga, and L. Santamaría. 2012. Effects of matrix characteristics and interpatch distance on functional connectivity in fragmented

temperate rainforests. *Conservation Biology* 26: 238–247. (8)

Magurran, A. E. 2013. Measuring Biological Diversity: Frontiers in Measurement and Assessment. Oxford University Press, Oxford, UK. (2)

Malpai Borderlands Group. 2010. *www.malpaiborderlandsgroup.org* (9)

Máñez, K. S., G. Krause, I. Ring, and M. Glaser. 2014. The Gordian knot of mangrove conservation: Disentangling the role of scale, services and benefits. *Global Environmental Change* 28: 120–128. (4)

Manfredo, M. J., T. L. Teel, and A. M. Dietsch. 2016. Implications of human value shift and persistence for biodiversity conservation. *Conservation Biology* 30(2): 287–296. (12)

Mangel, M. and C. Tier. 1994. Four facts every conservation biologist should know about persistence. *Ecology* 75: 607–614. (5)

Marbuah, G., I. M. Gren, and B. McKie. 2014. Economics of harmful invasive species: a review. *Diversity* 6: 500–523. (4)

Maron, M., R. J. Hobbs, A. Moilanen, J. W. Matthews, et al. 2012. Faustian bargains? Restoration realities in the context of biodiversity offset policies. *Biological Conservation* 158: 141–148. (10)

Maron, M., J. R. Rhodes, and P. Gibbons. 2013. Calculating the benefit of conservation actions. *Conservation Letters* 6: 359–367. (1, 3, 8, 12)

Martin, T. G., S. Nally, A. A. Burbidge, S. Arnall, et al. 2012. Acting fast helps avoid extinction. *Conservation Letters* 5: 274–280. (1, 5)

Mascaro, J., R. F. Hughes, R. F., and S. A. Schnitzer. 2012. Novel forests maintain ecosystem processes after the decline of native tree species. *Ecological Monographs* 82(2): 221–228. (10)

Mascia, M. B. and C. A. Claus. 2009. A property rights approach to understanding human displacement from protected areas: The case of Marine Protected Areas. *Conservation Biology* 23: 16–23. (8)

Mascia, M. B. and S. Pallier. 2011. Protected areas downgrading, downsizing, and degazettement (PADD) and its conservation implications. *Conservation Letters* 4: 9–20. (8)

Mascia, M. B., S. Pallier, M. L. Thieme, A., M. C. Bottrill, et al. 2014. Commonalities and complementarities among approaches to conservation monitoring and evaluation. Biological Conservation 169: 258–267. (11)

Mascia, M. B., S. Pallier, R. Krithivasan, V. Roshchanka, et al. 2014. Protected area downgrading, downsizing, and degazettement (PADDD) in Africa, Asia, and Latin America and the Caribbean, 1900–2010. *Biological Conservation* 169: 355–361. (8)

Matteson, K. C., D. J. Taron, and E. S. Minor. 2012. Assessing citizen contributions to butterfly monitoring in two large cities. *Conservation Biology* 26: 557–564. (6)

Matthews, J., G. Van der Velde, A. B. De Vaate, F. P. Collas, et al. 2014. Rapid range expansion of the invasive quagga mussel in relation to zebra mussel presence in The Netherlands and Western Europe. *Biological invasions* 16(1): 23–42. (4)

Matthews, T. J., F. Guilhaumon, K. A. Triantis, M. K. Borregaard, and R. J. Whittaker. 2015. On the form of species–area relationships in habitat islands and true islands. *Global Ecology and Biogeography*. doi: 10.1111/geb.12269. (8)

Maxted, N. 2001. Ex situ, in situ conservation. *In* S. A. Levin (ed.), *Encyclopedia of Biodiversity* 2: 683–696. Academic Press, San Diego, CA. (7)

Maynou, F. and J. E. Cartes. 2012. Effects of trawling on fish and invertebrates from deep-sea coral facies of *Isidella elongata* in the western Mediterranean. *Journal of the Marine Biological Association of the United Kingdom* 92(7): 1501–1507. (4)

McArthur, R. H. and E. O. Wilson. 1967. *The Theory of Island Biogeography.* Princeton University Press, NJ. (5)

McBride, M. F., S. T. Garnett, J. K. Szabo, A. H. Burbidge, et al. 2012. Structured elicitation of expert judgments for threatened species assessment: a case study on a continental scale using email. *Methods in Ecology and Evolution* 3(5): 906–920. (6)

McCaffery, R., C. L. Richards-Zawacki, K. R. Lips. 2015. The demography of *Atelopus* decline: Harlequin frog survival and abundance in central Panama prior to and during a disease outbreak. *Global Ecology and Conservation* 4: 232–242. (6)

McCallum, M. L. 2015. Vertebrate biodiversity losses point to a sixth mass extinction. *Biodiversity and Conservation* 1–23. doi: 10.1007/s10531-015-0940-6. (5)

McCarthy, D. P., P. F. Donald, J. P. W. Scharlemann, G. M. Buchanan, et al. 2012. Financial costs of meeting global biodiversity conservation targets: current spending and unmet needs. *Science* 338(6109): 946–949. (11)

McCarthy, M. A., C. J. Thompson, and N. S. G. Williams. 2006. Logic for designing nature reserves for multiple species. *American Naturalist* 167: 717–727. (8)

McCauley, D. J., M. L. Pinsky, S. R. Palumbi, J. A. Estes, et al. 2015. Marine defaunation: Animal loss in the global ocean. *Science* 347(6219): 1255641. (5)

McCleery, R., J. A. Hostetler, and M. K. Oli. 2014. Better off in the wild? Evaluating a captive breeding and release program for the recovery of an endangered rodent. *Biological Conservation* 169: 198–205. (7)

McClelland, E. K. and K. A. Naish. 2007. What is the fitness outcome of crossing unrelated fish populations? A meta-analysis and an evaluation of future research directions. *Conservation Genetics* 8: 397–416. (5)

McClenachan, L., A. B. Cooper, K. E. Carpenter, and N. K. Dulvy. 2012. Extinction risk and bottlenecks in the conservation of charismatic marine species. *Conservation Letters* 5: 73–80. (5)

McConnachie, M. M., R. M. Cowling, B. W. van Wilgen, and D. A. McConnachie. 2012. Evaluating the cost-effectiveness of invasive plant removal: a case study from South Africa. *Biological Conservation* 155: 128–135. doi: 10.1016/j.biocon.2012.06.006. (4)

McKinley, D. C., M. G. Ryan, R. A. Birdsey, C. P. Giardina, et al. 2011. A synthesis of current knowledge on forests and carbon storage in the United States. *Ecological Applications* 21: 1902–1924. (3)

McNeely, J. A. 1989. Protected areas and human ecology: how national parks can contribute to sustaining societies of the twenty-first century. *In* D. Western and M. Pearl (eds.), *Conservation for the Twenty-first Century*, pp. 150–165. Oxford University Press, New York. (9)

McNeely, J. A., K. R. Miller, W. Reid, R. Mittermeier, et al. 1990. *Conserving the World's Biological Diversity.* IUCN, World Resources Institute, CI, WWF-US, and the World Bank. Gland, Switzerland and Washington, DC. (3)

McWilliams, J. 2013. Fertility treatments. *Conservation* 14: 34–39. (4)

Meffert, P. J. and F. Dziock. 2012. What determines occurrence of threatened bird species on urban wastelands? *Biological Conservation* 153: 87–100. (9)

Meijaard, E., G., G. Albar, Nardiyono, Y. Rayadin, et al. 2010. Unexpected ecological resilience in Bornean orangutans and implications for pulp and paper plantation management. *PLoS ONE* 5(9): e12813. (9)

Meinhardt, K. A., and C. A. Gehring. 2013. *Tamarix* and soil ecology. In A. Sher and M. Quigley (eds.), *Tamarix: A Case Study of Ecological Change in the American West*. Oxford University Press New York. (4)

Meissen, J. C., S. M. Galatowitsch, and M. W. Cornett. 2015. Risks of overharvesting seed from native tallgrass prairies. *Restoration Ecology* 23(6): 882–891. (10)

Melbourne, B. A. and A. Hastings. 2008. Extinction risk depends strongly on factors contributing to stochasticity. *Nature* 454: 100–103. (5)

Menges, E. S. 1992. Stochastic modeling of extinction in plant populations. *In* P. L. Fiedler and S. K. Jain (eds.), *Conservation Biology: The Theory and Practice of Nature Conservation, Preservation and Management*, pp. 253–275. Chapman and Hall, New York. (5)

Merckx, T. and H. M. Pereira. 2015. Reshaping agri-environmental subsidies: From marginal farming to large-scale rewilding. *Basic and Applied Ecology* 16: 95–103. (3)

Meyer, C. K., M. R. Whiles, and S. G. Baer. 2010. Plant community recovery following restoration in temporarily variable riparian wetlands. *Restoration Ecology* 18: 52–64. (10)

Middleton, B. A. 2013. Rediscovering traditional vegetation management in preserves: trading experiences between cultures and continents. *Biological Conservation* 158: 271–279. (8, 9)

Milder, J. C., J. P. Lassoie, and B. L. Bedford. 2008. Conserving biodiversity and ecosystem function through limited development: An empirical evaluation. *Conservation Biology* 22: 70–79. (11)

Millennium Ecosystem Assessment (MEA). 2005. *Ecosystems and Human Well-being*. 4 volumes. Island Press, Covelo, CA. (1, 2, 3, 4, 5, 9)

Miller, B., W. Conway, R. P. Reading, C. Wemmer, et al. 2004. Evaluating the conservation mission of zoos, aquariums, botanical gardens, and natural history museums. *Conservation Biology* 18: 86–93. (7)

Miller, D. C., A. Agrawal, and J. T. Roberts. 2013. Biodiversity, governance, and the allocation of international aid for conservation. *Conservation Letters* 6: 12–20. (11)

Miller, J. K., J. M. Scott, C. R. Miller, and L. P Waits. 2002. The Endangered Species Act: Dollars and sense? *BioScience* 52: 163–168. (6)

Miller, K. A., T. Bell, and J. M. Germano. 2014. Understanding publication bias in reintroduction biology by assessing translocationsof New Zealand's Herpetofauna. *Conservation Biology*. 28(4). doi: 10.1111/cobi.12254. (7)

Miller, K. R. 1996. *Balancing the Scales: Guidelines for Increasing Biodiversity's Chances through Bioregional Management*. World Resources Institute, Washington, DC. (9)

Miller-Rushing, A. J. and R. B. Primack. 2008. Global warming and flowering times in Thoreau's Concord: A community perspective. *Ecology* 89: 332–341. (5)

Mills, S. 2003. *Epicurean Simplicity*. Island Press, Washington, DC. (3)

Mills, M., J. G. Alvarez-Romero, K. Vance-Borland, P. Cohen, et al. 2014. Linking regional planning and local action: towards using social network analysis in systematic conservation planning. *Biological Conservation* 169: 14–53. (8)

Mills Busa, J. H., 2013. Deforestation beyond borders: Addressing the disparity between production and consumption of global resources. *Conservation Letters* 6: 192–199. (4)

Minteer, B. A. and J. P. Collins. 2008. From environmental to ecological ethics: toward a practical ethics for ecologists and conservationists. *Science and Engineering Ethics* 14: 483–501. (3)

Minteer, B. A. and T. R. Miller. 2011. The new conservation debate: Ethical foundations, strategic trade-offs, and policy opportunities. *Biological Conservation* 144: 945–947. (11)

Minuzzi-Souza, T. T. C., N. Nitz, M. B. Knox, F. Reis, et al. 2016. Vector-borne transmission of *Trypanosoma cruzi* among captive Neotropical primates in a Brazilian zoo. *Parasites and Vectors* 9(1): 1. (7)

Miththapala, S. 2008. *Mangroves, Coastal Ecosystems Series*. Ecosystems and Livelihoods Group Asia, IUCN, Colombo, Sri Lanka. (4)

Mitsch, W. J. and J. G. Gosselink. 2015. *Wetlands*. J. Wiley and Sons. New York. (4)

Mittermeier, R. A., P. R. Gil, M. Hoffman, J. Pilgrim, et al. 2005. *Hotspots Revisited: Earth's Biologically Richest and Most Endangered Terrestrial Ecoregions*. Conservation International, Washington, DC. (6)

Mlot, C. 2013. Are Isle Royale's wolves chasing extinction? *Science* 340: 919–921. (6)

Molano-Flores, B. and T. J. Bell. 2012. Projected population dynamics for a federally endangered plant under different climate change emission scenarios. *Biological Conservation* 145: 130–138. (6)

Molnar, J. L., R. L Gamboa, C. Revenga, and M. D. Spalding. 2008. Assessing the global threat of invasive species to marine biodiversity. *Frontiers in Ecology and the Environment* 9: 485–492. (4)

Mora, C. 2009. Degradation of Caribbean coral reefs: focusing on proximal rather than ultimate drivers. Reply to Rogers. *Proceedings of the Royal Society of London B: Biological Sciences* 276(1655): 199–200. (4)

Morandin, L. A. and C. Kremen. 2013. Hedgerow restoration promotes pollinator populations and exports native bees to adjacent fields. *Ecological Applications* 23: 829–839. (10)

Moreno-Mateos, D., V. Maris, A. Béchet, and M. Curran. 2015a. The true loss caused by biodiversity offsets. *Biological Conservation*. 192: 552–559. (10)

Moreno-Mateos, D., P. Meli, M. I. Vara-Rodríguez, and J. Aronson. 2015b. Ecosystem response to interventions: lessons from restored and created wetland ecosystems. *Journal of Applied Ecology* 52(6): 1528–1537. (10)

Moreno-Mateos, D., M. E. Power, F. A. Comín, and R. Yockteng. 2012. Structural and functional loss in restored wetland ecosystems. *PLoS-Biology* 10(1): 45. (4)

Morgan, J. L., S. E. Gergel, and N. C. Coops. 2010. Aerial photography: A rapidly evolving tool for ecological management. *BioScience* 60: 47–59. (8)

Moro, M., A. Fischer, M. Czajkowski, D. Brennan, et al. 2013. An investigation using the choice experiment method into options for reducing illegal bushmeat hunting in western Serengeti. *Conservation Letters* 6: 37–45. (4, 12)

Moro, D., M. W. Hayward, P. J. Seddon, and D. P. Armstrong. 2015. Reintroduction biology of Australian and New Zealand fauna: progress, emerging themes and future directions. *Advances in Reintroduction Biology of Australian and New Zealand Fauna* 178. (7)

Morrell, V. 2014. Science behind plan to ease wolf protection is flawed, panel says. *Science* 343: 719. (7)

Morrison, S. A. 2016. Designing virtuous socio-ecological cycles for biodiversity conservation. *Biological Conservation* 195: 9–16. (12)

Morton, T. A. L., A. Thorn, J. M. Reed, R. Van Driesche, et al. 2015. Modeling the decline and potential recovery of a native butterfly following serial invasions by exotic species. *Biological Invasion* 17:1683–1695. (5)

Moseley, L. (ed.). 2009. *Holy Ground: A Gathering of Voices on Caring for Creation*. Sierra Club Books, San Francisco, CA. (3)

Moss, A., E. Jensen, and M. Gusset. 2016. Probing the Link between biodiversity-related knowledge and self-reported pro-conservation behaviour in a global survey of zoo visitors. *Conservation Letters*. doi: 10.1111/conl.12233. (7)

Mueller, J. G., I. H. B. Assanou, I. D. Guimbo, and A. M. Almedom. 2010. Evalutating rapid participatory rural appraisal as an assessment of ethnoecological knowledge and local biodiversity patterns. *Conservation Biology* 24: 140–150. (6)

Muir, J. 1901. *Our National Parks*. Houghton Mifflin, Boston. (1)

Mukul, S. A. and J. Herbohn. 2016. The impacts of shifting cultivation on secondary forests dynamics in tropics: a synthesis of the key findings and spatio temporal distribution of research. *Environmental Science and Policy* 55: 167–177. (4)

Munawar, S., M. F. Khokhar, and S. Atif. 2015. Reducing emissions from deforestation and forest degradation implementation in northern Pakistan. *International Biodeterioration and Biodegradation* 102: 316–323. (11)

Munilla, I., C. Diez, and A. Velando. 2007. Are edge bird populations doomed to extinction? A retrospective analysis of the common guillemot collapse in Iberia. *Biological Conservation* 137: 359–371. (5)

Munson, L., K. A. Terio, R. Kock, T. Mlengeya, et al. 2008. Climate extremes promote fatal co-infections during canine distemper epidemics in African lions. *PLoS ONE* 3: e2545. (5)

Munson, S. M. and A. A. Sher. 2015. Long-term shifts in the phenology of rare and endemic Rocky Mountain plants. *American Journal of Botany* 102(8): 1268–1276. (4)

Muradian, R., M. Arsel, L. Pellegrini, F. Adaman, et al. 2014. Payments for ecosystem services and the fatal attraction of win-win solutions. *Conservation Letters* 6: 274–279. (11)

Murcia, C. 1995. Edge effects in fragmented forests: implications for conservation. *Trends in Ecology and Evolution* 10(2): 58–62. (4)

Murcia, C., J. Aronson, G. H. Kattan, D., Moreno-Mateos, et al. 2014. A critique of the "novel ecosystem" concept. *Trends in Ecology and Evolution* 29(10): 548–553. (10)

Murray-Smith, C., N. A. Brummitt, A. T. Oliveira-Filho, S. Bachman, et al. 2009. Plant diversity hotspots in the Atlantic coastal forests of Brazil. *Conservation Biology* 23: 151–163. (8)

Musiani, M., C. Mamo, L. Boitani, C. Callaghan, et al. 2003. Wolf depredation trends and the use of fladry barriers to protect livestock in western North America. *Conservation Biology* 17: 1538–1547. (7)

Myers, N., N. Golubiewski, and C. J. Cleveland. 2007. Perverse subsidies. *In* C. J. Cleveland (ed.), *Encyclopedia of Earth*. Environmental Information Coalition, National Council for Science and the Environment, Washington, DC. *www.eoearth.org/article/Perverse_subsidies* (3)

Nabhan, G. P. 2008. *Where Our Food Comes From: Retracing Nicolay Vavilov's Quest to End Famine*. Island Press, Washington, DC. (7)

Naeem, S, A. V. Ingram, T. Agardy, P. Barten, et al. 2015. Get the science right when paying for nature's services. *Science* 347(6227): 1206–1207. (11)

Naess, A. 1989. *Ecology, Community and Lifestyle*. Cambridge University Press, Cambridge, MA. (3)

Naess, A. 2008. *The Ecology of Wisdom: Writings by Arne Naess*. A. Drengson and B. Devall (eds.). Counterpoint, Berkeley, CA. (3)

Naidoo, R., L. C. Weaver, R. W. Diggle, G. Matongo, et al. 2016. Complementary benefits of tourism and hunting to communal conservancies in Namibia. *Conservation Biology*. doi: 10.1111/cobi.12643. (9)

Nakamura, K. and K. Takai. 2015. Indian Ocean Hydrothermal Systems: Seafloor Hydrothermal Activities, Physical and Chemical Characteristics of Hydrothermal Fluids, and Vent-Associated Biological Communities. *In* Ishibashi, J.-i. et al. (eds.), *Subseafloor Biosphere Linked to Hydrothermal Systems*. Springer, Japan. doi: 10.1007/978-4-431-54865-2_12. (2)

Namibia Association of CBNRM Support Organisations (NACSO). 2008. *Namibia's Communal Conservancies: a review of progress in 2008*. NACSO, Windhoek, Namibia. (9)

Namibia Association of CBNRM Support Organisations (NACSO). 2014. *The state of community conservation in Namibia—a review of communal conservancies, community forests and other CBNRM initiatives (2013 Annual Report)*. Windhoek: Namibia Association of CBNRM Support Organizations. doi: 10.1111/j.1523-1739.2012.01960.x. (9)

Naniwadekar, R., Shukla, U., Isvaran, K., Datta, A. 2015. Reduced hornbill abundance associated with low seed arrival and altered recruitment in a hunted and logged tropical forest. *PLoS ONE* 10(3) e0120062. doi: 10.1371/journal.pone.0120062. (2)

Nash, S. 2009. Ecotourism and other invasions. *BioScience* 59: 106–110. (3)

National Geographic. 2015. Graphic Shows Who's Buying and Selling Animals Globally. *news.nationalgeographic.com/2015/06/150615-data-points-infographic-animal-trade* (4)

National Marine Sanctuary of American Samoa. *americansamoa.noaa.gov* (9)

National Oceanic and Atmospheric Administration (NOAA). 2015. National Centers for Environmental Information, State of the Climate: Global Analysis for August 2015. *www.ncdc.noaa.gov/sotc/global/201508* (4)

National Oceanic and Atmospheric Administration (NOAA). 2016. National Centers for Environmental Information, State of the Climate: Global Analysis for January 2016. *www.ncdc.noaa.gov/sotc/global/201601*. (4)

Natural Resource Stewardship and Science. 2015. Office of the Associate Director. Strategic framework for National Park Service Research Learning Centers. National Park Service, Natural Resource Stewardship and Science, Washington, DC. *www.nature.nps.gov/rlc/framework.pdf* (12)

Native Seeds SEARCH: Southwestern Endangered Aridland Resource Clearing House. 2009. *www.nativeseeds.org* (9)

NatureServe Explorer. 2009. *www.natureserve.org/explorer* (6)

NatureServe: A Network Connecting Science with Conservation. 2009. *www.natureserve.org* (6)

Nee, S. 2003. Unveiling prokaryotic diversity. *Trends in Ecology and Evolution* 18: 62–63. (2)

Negrón-Ortiz, N. 2014. Pattern of expenditures for plant conservation under the Endangered Species Act. *Biological Conservation* 171: 36–43. (11)

Nelson, E. J., P. Kareiva, M. Ruckelshaus, K. Arkema, et. al. 2013. Climate change's impact on key ecosystem services and the human well-being they support in the US. *Frontiers in Ecology and the Environment.* 11: 483–493. (4)

Nelson, M. P., J. T. Bruskotter, J. A. Vucetich, and G. Chapron. 2016. Emotions and the Ethics of Consequence in Conservation Decisions: Lessons from Cecil the Lion. *Conservation Letters.* doi: 10.1111/conl.12232 (9)

Nepstad, D., S. Schwartzman, B. Bamberger, M. Santilli, et al. 2006. Inhibition of Amazon deforestation and fire by parks and indigenous lands. *Conservation Biology* 20: 65–73. (4)

Newbold, S. C. and J. V. Siikamäki. 2009. Prioritizing conservation activities using reserve site selection methods and population viability analysis. *Ecological Applications* 19: 1774–1790. (3)

Newmark, W. D. 1995. Extinction of mammal populations in western North American national parks. *Conservation Biology* 9: 512–527. (8)

Newmark, W. D. 2008. Isolation of African protected areas. *Frontiers in Ecology and the Environment* 6: 321–328. (8, 9)

Nicoll, M. A. C., C. G. Jones, and K. Norris. 2004. Comparison of survival rates of captive-reared and wild-bred Mauritius kestrels (*Falco punctatus*) in a re-introduced population. *Biological Conservation* 118: 539–548. (7)

Nieto-Romero, M., A. Milcu, J. Leventon, F. Mikulcak, and J. Fischer. 2016. The role of scenarios in fostering collective action for sustainable development: Lessons from central Romania. *Land Use Policy* 50: 156–168. (11)

Nijman, V., K. A. I. Nekaris, G. Donati, M. Bruford, and J. Fa. 2011. Primate conservation: Measuring and mitigating trade in primates. *Endangered Species Research* 13: 159–161. (4)

Niles, L. J. 2009. Effects of horseshoe crab harvest in Delaware Bay on red knots: Are harvest restrictions working? *BioScience* 59: 153–164. (3)

Nimmo, D. G., L. T. Kelly, L. M. Spence-Bailey, S. J. Watson, et al. 2013. Fire mosaics and reptile conservation in a fire-prone region. *Conservation Biology* 27: 345–353. (8)

Noël, F., N. Machon, and A. Robert. 2013. Integrating demographic and genetic effects of connections on the viability of an endangered plant in a highly fragmented habitat. *Biological Conservation* 158: 167–174. (6)

Nolte, C., A. Agrawal, K. M. Silvus, and B. S. Soares-Filho. 2013. Governance regime and location influence avoided deforestation success of protected areas in the Brazilian Amazon. *Proceedings of the National Academy of Sciences USA* 110: 4956–5961. (8, 11)

Noon, B. R., L. L. Bailey, T. D. Sisk, and K. S. McKelvey. 2012. Efficient species-level monitoring at the landscape scale. *Conservation Biology* 26: 432–441. (6)

Norris, K., A. Asase, B. Collen, J. Gockowski, et al. 2010. Biodiversity in a forest-agriculture mosaic—The changing face of West African rainforests. *Biological Conservation* 143: 2341–2350. (9)

Nuñez, T. A., J. J. Lawler, B. H. Mcrae, D. J. Pierce, et al. 2013. Connectivity planning to address climate change. *Conservation Biology* 27: 407–416. (4, 8)

North American Breeding Bird Survey (BBS). *www.pwrc.usgs.gov/bbs.* (6)

Oakleaf, J. R., C. M. Kennedy, S. Baruch-Mordo, P. C. West, et al. 2015. A World at Risk: Aggregating Development Trends to Forecast Global Habitat Conversion. *PLoS ONE* 10(10): e0138334. (4)

Odell, J., M. E. Mather, and R. M. Muth. 2005. A biosocial approach for analyzing environmental conflicts: A case study of horseshoe crab allocation. *BioScience* 55: 735–748. (3)

O'Donnell, C. F. and J. M. Hoare. 2012. Quantifying the benefits of long-term integrated pest control for forest bird populations in a New Zealand temperate rainforest. *New Zealand Journal of Ecology* 131–140. (4)

Ogden, L. E. 2015. Do Wildlife Corridors Have a Downside? *BioScience* 65(4): 452–452. (8)

Okin, G. S., A. Parsons, J. Wainwright, J. E. Herrick, et al. 2009. Do changes in connectivity explain desertification? *BioScience* 59: 237–244. (4)

Oldekop, J. A., G. Holmes, W. E. Harris, and K. L. Evans. 2015. A global assessment of the social and conservation outcomes of protected areas. *Conservation Biology* 30(1): 133–141. (8)

Olden, J. D., M. J. Kennard, J. J. Lawler, and N. L. Poff. 2011. Challenges and opportunities in implementing managed relocation for conservation of freshwater species. *Conservation Biology* 25: 40–47. (7)

Oldfield, T. E. E., R. J. Smith, S. R. Harrop, and N. Leader-Williams. 2003. Field sports and conservation in the United Kingdom. *Nature* 423: 531–533. (3)

Olivieri, I., J. Tonnabel, O. Ronce, and A. Mignot. 2016. Why evolution matters for species conservation: perspectives from three case studies of plant metapopulations. *Evolutionary Applications* 9(1): 196–211. (7)

O'Meilla, C. 2004. Current and reported historical range of the American burying beetle. U. S. Fish and Wildlife Services, Oklahoma Ecological Services Field Office, OK. (5)

O'Neill, B. C., B. Liddle, L. Jiang, K. R. Smith, K. R., et al. 2012. Demographic change and carbon dioxide emissions. *The Lancet* 380(9837): 157–164. (4)

Oppel, S., B. M. Beaven, M. Bolton, J. Vickery, and T. W. Body. 2011. Eradication of invasive mammals on islands inhabited by humans and domestic animals. *Conservation Biology* 25: 232–240. (4)

Oppenheimer, J. D., S. K. Beaugh, J. A. Knudson, P. Mueller, et al. 2015. A collaborative model for large-scale riparian restoration in the western United States. *Restoration Ecology* 23(2): 143–148. (10)

Ore, J., S. Elbaum, A. Burgin, and C. Detweiler. 2015. Autonomous aerial water sampling. *In* L. Mejias, P. Corke, and J. Roberts (eds.) *Field and Service Robotics*, pp 137–151. Springer International Publishing, Switzerland. (6)

Orgnization for Economic Co-operation and Development. *www.oecd.org* (11)

Orrock, J. L. and E. I. Damschen. 2005. Corridors cause differential seed predation. *Ecological Applications* 15: 793–798. (8)

Österblom, H. and Ö Bodin. 2012. Global cooperation among diverse organizations to reduce illegal fishing in the southern ocean. *Conservation Biology* 26: 638–648. (4)

Osterlind, K. 2005. Concept formation in environmental education: 14–year olds' work on the intensified greenhouse. *International Journal of Science Education* 27: 891–908. (3)

Ostfeld, R. S. 2009. Climate change and the distribution and intensity of infectious diseases. *Ecology* 4: 903–905. (4)

Ottewell, K., J. Dunlop, N. Thomas, K. Morris, et al. 2014. Evaluating the success of translocations in maintaining genetic diversity in a threatened mammal. *Biological Conservation* 171: 209–219. (7)

Packer, C. 2015. *Lions in the Balance: Man-Eaters, Manes, and Men with Guns.* University of Chicago Press. (5)

Packer, C., A. Loveridge, S. Canney, T. Caro, et al. 2013. Conserving large carnivores: Dollars and fence. *Conservation Letters* 16: 635–641. (8)

Paknia, O., H. Rajaei, and A. Koch. 2015. Lack of well-maintained natural history collections and taxonomists in megadiverse developing countries hampers global biodiversity exploration. *Organisms Diversity and Evolution* 15(3): 619–629. (12)

Palomares, F., J. A. Godoy, J. V. López-Bao, A. Rodriguez, et al. 2012. Possible extinction vortex for a population of Iberian lynx on the verge of extirpation. *Conservation Biology* 26: 689–697. (5)

Pan, Y., R. A. Birdsey, J. Fang, R. Houghton, et al. 2011. A large and persistent carbon sink in the world's forests. *Science* 333: 988–993. (3)

Pandolfi, J. M., S. R. Connoly, D. J. Marshall, and A. L. Cohen. 2011. Projecting coral reef futures under global warming and ocean acidification. *Science* 333: 418–442. (4)

Papadopoulou, A., D. Chesters, I. Coronado, G. De La Cadena, et al. 2015. Automated DNA-based plant identification for large-scale biodiversity assessment. *Molecular Ecology Resources* 15: 136–152. (2)

Papworth, S. K., J. Rist, L. Coad, and E. J. Milner-Gulland. 2009. Evidence for shifting baseline syndrome in conservation. *Conservation Letters* 2: 93–100. (6)

Pardini, R., S. M. de Souza, R. Braga-Neto, and J. P. Metzger. 2005. The role of forest structure, fragment size and corridors in maintaining small mammal abundance and diversity in an Atlantic forest landscape. *Biological Conservation* 12: 253–266. (8)

Parlato, E. H. and D. P. Armstrong. 2013. Predicting post-release establishment using data from multiple reintroductions. *Biological Conservation* 160: 97–104. (7)

Pawlowski, J., Audic, S., Adl, S., Bass, D., et al. 2012. CBOL protist working group: barcoding eukaryotic richness beyond the animal, plant, and fungal kingdoms. *PLoS Biology* 10(11): e1001419. (2)

PBL Netherlands Environmental Assessment Agency. 2012. Trends in Global CO_2 Emissions 2012 Report. PBL Netherlands Environmental Assessment Agency The Hague The Netherlands, European Commission Joint Research Centre Institute for Environment and Sustainability. doi: 10.2788/33777. (4)

Peace Parks Foundation. *www.peaceparks.org* (11)

Peakall, R., D. Ebert, L. J. Scott, P. F. Meagher, and C. A. Offord. 2003. Comparative genetic study confirms exceptionally low genetic variation in the ancient and endangered relictual conifer, *Wollemia nobilis* (Araucariaceae). *Molecular Ecology* 12: 2331–2343. (5)

Pe'er, G., M. A. Tsianou, K. W. Franz, Y. G. Matsinos, et al. 2014. Toward better application of minimum area requirements in conservation planning. *Biological Conservation* 170: 92–102. (5, 6)

Peery, M. Z., S. R. Beissinger, S. H. Newman, E. B. Burkett, and T. D. Williams. 2004. Applying the declining population paradigm: Diagnosing causes of poor reproduction in the marbled murrelet. *Conservation Biology* 18: 1088–1098. (5)

Peh, K. S. H., J. de Jong, N. S. Sodhi, S. L. H. Lim, et al. 2005. Lowland rainforest avifauna and human disturbance: persistence of primary forest birds in selectively logged forests and mixed-rural habitats of southern Peninsular Malaysia. *Biological Conservation* 123: 489–505. (9)

Peres, C. A. and M. Schneider. 2012. Subsidized agricultural resettlements as drivers of tropical deforestation. *Biological Conservation* 151: 65–68. (4)

Perring, M. P., P. Audet, and D. Lamb. 2014. Novel ecosystems in ecological restoration and rehabilitation: Innovative planning or lowering the bar? *Ecological Processes* 3(1): 1–4. (10)

Perry, G. and D. Vice. 2009. Forecasting the risk of brown tree snake dispersal from Guam: A mixed transport-establishment model. *Conservation Biology* 23: 992–1000. (4)

Peterson, M. J., D. M. Hall, A. M. Feldpausch-Parker, and T. R. Peterson. 2010. Obscuring ecosystem function with the application of the ecosystem services concept. *Conservation Biology* 24: 113–119. (3)

Phelps, J., and E. L. Webb. 2015. "Invisible" wildlife trades: Southeast Asia's undocumented illegal trade in wild ornamental plants. *Biological Conservation* 186: 296–305. (3)

Phelps, J., L. R. Carrasco, E. L. Webb, L. P. Koh, et al. 2013. Agricultural intensification escalates future conservation costs. *Proceedings of the National Academy of Sciences USA* 110: 7601–7606. (9, 11)

Philpott, S. M., P. Bichier, R. Rice, and R. Greenberg. 2007. Field-testing ecological and economic benefits of coffee certification programs. *Conservation Biology* 21: 975–985. (9)

Phua, M. H., S. Tsuyuki, N. Furuya, and J. S. Lee. 2008. Detecting deforestation with a spectral change detection approach using multitemporal Landsat data: A case study of Kinabalu Park, Sabah, Malaysia. *Journal of Environmental Management* 88: 784–795. (4)

Pikesley, S. K., B. J. Godley, H. Latham, P. B. Richardson, et al. 2016. Pink sea fans (*Eunicella verrucosa*) as indicators of the spatial efficacy of Marine Protected Areas in southwest UK coastal waters. Marine Policy 64: 38–45. (9)

Pilgrim, J. D., S. Brownlie, J. M. M. Ekstrom, T. A. Gardner, et al. 2013. A process for assessing the offsetability of biodiversity impacts. *Conservation Letters* 6: 376–384. (11, 12)

Pimentel, D., R. Zuniga, and D. Morrison. 2005. Update on the environmental and economic costs associated with alien-invasive species in the United States. *Ecological Economics* 52: 273–288. (4)

Pimm, S. L, C. N. Jenkins, R. Abell, T. M. Brooks, et al. 2014. The biodiversity of species and their rates of extinction, distribution, and protection. *Science* 344: 1246752. (1, 5)

Pimm, S. L. and C. Jenkins. 2005. Sustaining the variety of life. *Scientific American* 293(33): 66–73. (4, 5)

Pimm, S. L. and L. N. Joppa. 2015. How Many Plant Species are There, Where are They, and at What Rate are They Going Extinct? *Annals of the Missouri Botanical Garden* 100(3): 170–176. (5)

Pinchot, G. 1947. *Breaking New Ground*. Harcourt Brace, New York. (1)

Pittman, S. E., M. S. Osbourn, and R. D. Semlitsch. 2014. Movement ecology of amphibians: A missing component for understanding population declines. *Biological Conservation* 169: 44–53. (6)

Pocock, M. J., S. E. Newson, I. G. Henderson, J. Peyton, et al. 2015. Developing and enhancing biodiversity monitoring programmes: a collaborative assessment of priorities. *Journal of Applied Ecology* 52(3): 686–695. (8)

Polidoro, B. A., K. E. Carpenter, L. Collins, N. C. Duke, et al. 2010. The loss of species: Mangrove extinction risk and geographic areas of global concern. *PLoS ONE* 5(4): e10095. (4)

Polishchuk, L. V., K. Y. Popadin, M. A. Baranova, and A. S. Kondrashov. 2015. A genetic component of extinction risk in mammals. *Oikos* 124(8): 983–993. (5)

Pooley, S. P., J. A. Mendelsohn, and E. J. Milner-Gulland. 2014. Hunting down the chimera of multiple disciplinarity in conservation science. *Conservation Biology* 28: 22–32. (1)

Porensky, L. M. and T. P. Young. 2013. Edge effect interactions in fragmented and patchy landscapes. *Conservation Biology* 27: 509–519. (4)

Porszt, E. J., R. M. Peterman, N. K. Dulvy, A. B. Cooper, et al. 2012. Reliability of indicators of decline in abundance. *Conservation Biology* 26: 894–904. (6)

Posey, D. A. and M. J. Balick (eds.). 2006. Human Impacts on Amazonia: The Role of Traditional Ecological Knowledge in Conservation and Development. Columbia University Press, New York. (8, 9)

Possingham, H. P., M. Bode, M. and C. J. Klein, C. J. 2015. Optimal conservation outcomes require both restoration and protection. *PLoS Biology* 13(1): e1002052–e1002052. (8)

Poudel, R. C., M. Möller, J. Liu, L.-M. Gao, et al. 2014. Low genetic diversity and high inbreeding of the endangered yews in Central Himalaya: implications for conservation of their highly fragmented populations. *Diversity and Distributions* 20(11): 1270–1284. (5)

Powell, K. I., J. M. Chase, and T. M. Knight. 2013. Invasive plants have scale-dependent effects on diversity by altering species-area relationships. *Science* 339: 316–318. (3)

Power, M. E., D. Tilman, J. A. Estes, B. A. Menge, et al. 1996. Challenges in the quest for keystones. *BioScience* 46: 609–620. (2)

Power, T. M. and R. N. Barret. 2001. *Post-Cowboy Economics: Pay and Prosperity in the New American West*. Island Press, Washington, DC. (3)

Prescott-Allen, C. and R. Prescott-Allen. 1986. *The First Resource: Wild Species in the North American Economy*. Yale University Press, New Haven, CT. (3)

Primack, R. B. and A. J. Miller-Rushing. 2012. Uncovering, collecting, and analyzing records to investigate the ecological impacts of climate change: A template from Thoreau's Concord. *BioScience* 62: 170–181. (4)

Primack, R. B., A. J. Miller-Rushing, and K. Dharaneeswaran. 2009. Changes in the flora of Thoreau's Concord. *Biological Conservation* 142: 500–508. (5)

Protected Planet. *www.wdpa.org* (8)

Purvis, A., J. L. Gittleman, G. Cowlishaw, and G. M. Mace. 2000. Predicting extinction risk in declining species. *Proceedings of the Royal Society of London B: Biological Sciences* 267(1456): 1947–1952. (5)

Quammen, D. 1996. *The Song of the Dodo: Island Biogeography in an Age of Extinctions*. Scribner, New York. (5)

Quayle, J. F., L. R. Ramsay, and D. F Fraser 2007. Trend in the status of breeding bird fauna in British Columbia, Canada, based on the IUCN Red List Index method. *Conservation Biology* 21(5): 1241–1247. (6)

QUINTESSENCE Consortium. 2016. Networking our way to better ecosystem service provision. *Trends in Ecology and Evolution* 31(2): 105–115. (12)

Radeloff, V. C., S. I. Stewart, T. J. Hawbaker, U. Gimmi, et al. 2010. Housing growth in and near United States protected areas limits their conservation value. *Proceedings of the National Academy of Sciences USA* 107: 940–945. (9)

Raghavan, R., N. Dahanukar, M. F. Tlusty, A. L. Rhyne, et al. 2013. Uncovering an obscure trade: threatened freshwater fishes and the aquarium pet markets. *Biological Conservation,*164: 158–169. (4)

Rai, N. D. and K. S. Bawa. 2013. Insering politics and history in conservation. *Conservation Biology* 27: 425–428. (9)

Ramakrishnan, U., J. A. Santosh, U. Ramakrishnan, and R. Sukumar. 1998. The population and conservation status of Asian elephants in the Periyar Tiger Reserve, southern India. *Current Science India* 74: 110–113. (5)

Ramsar Convention Secretariat. *www.ramsar.org* (11)

Rands, M. R., W. M. Adams, L. Bennun, S. H. M. Butchart, et al. 2010. Biodiversity conservation: Challenges beyond 2010. *Science* 329: 1298–1303. (11)

Randolph, J. and G. M. Masters. 2008. *Energy for Sustainability: Technology, Planning, Policy.* Island Press, Washington, DC. (4)

Raup, D. M. 1988. Diversity Crises in the Geological Past. *In* E. O. Wilson and F. M Peter (eds.), *Biodiversity,* Chapter 5. National Academies Press, Washington, DC. (5)

Raup, D. M. and Sepkoski, J. J. 1982. Mass extinctions in the marine fossil record. *Science* 215(4539): 1501–1503. (5)

Reading, R. 2015. *Mongolia Field Program Update.* Association of Zoos and Aquariums (AZA) 2015 Mid-year Meeting, Columbia, SC. (12)

Reading, R. R., T. J. Weaver, J. R. Garcia, R. Elias Piperis, et al. (eds.). 2011. Lake Titicaca's Frog (*Telmatobius culeus*) Conservation Strategy Workshop. December 13–15, 2011. Bioscience School, Universidad Nacional del Altiplano, Puno, Peru. Conservation Breeding Specialist Group CBSG/(SSC/IUCN) Mesoamerica. (4)

Redford, K. H. 1992. The empty forest. *BioScience* 42: 412–422. (2, 4)

Redford, K. H. and W. M. Adams. 2009. Payment for ecosystem services and the challenge of saving nature. *Conservation Biology* 23: 785–787. (3)

Redford, K. H., G. Amato, J. Baillie, P. Beldomenico, et al. 2011. What does it mean to successfully conserve a (vertebrate) species? *BioScience* 61: 39–48. (6)

Redford, K. H., B. J. Huntley, D. Roe, T. Hammond, et al. 2015. Mainstreaming Biodiversity: Conservation for the Twenty-First Century. *Frontiers in Ecology and Evolution* 3: 137. (12)

Redpath, S. M., J. Young, A. Evely, W. M. Adams, et al. 2013. Understanding and managing conservation conflicts. *Trends in Ecology and Evolution* 28: 100–109. (9, 12)

Reed, D. H., J. J. O'Grady, J. D. Ballou, and R. Frankham. 2003. The frequency and severity of catastrophic die-offs in vertebrates. *Animal Conservation* 6(02): 109–114. (5)

Reed, J. M. 1999. The role of behavior in recent avian extinctions and endangerments. *Conservation Biology* 13: 232–241. (5)

Reed, J. M., C. S. Elphick, E. N. Ieno, and A. F. Zuur. 2011. Long-term population trends of endangered Hawaiian waterbirds. *Population Ecology* 53: 473–481. (6)

Reed, J. M., C. S. Elphick, A. F. Zuur, E. N. Ieno, and G. M. Smith. 2007. Time series analysis of Hawaiian waterbirds. *In* A. F. Zuur, E. N. Ieno, and G. M. Smith (eds.), *Analysis of Ecological Data*, pp. 613–632. Springer-Verlag, the Netherlands. (6)

Reed, S. E., J. A. Hilty, and D. M. Theobald. 2014. Guidelines and incentives for conservation development in local land-use regulations. *Conservation Biology* 28: 258–268. (11)

Reeve, R. 2014. *Policing International Trade in Endangered Species: The CITES Treaty and Compliance.* Routledge, London. (6)

Régnier, C., B. Fontaine, and P. Bouchet. 2009. Not knowing, not recording, not listing: Numerous unnoticed mollusk extinctions. *Conservation Biology* 23: 1214–1221. (5, 6)

Regos, A., M. D'Amen, N. Titeux, S. Herrando, A. Guisan, and L. Brotons. 2016. Predicting the future effectiveness of protected areas for bird conservation in Mediterranean ecosystems under climate change and novel fire regime scenarios. *Diversity and Distributions*, 22(1): 83–96. (8)

Regueira, R. F. S. and E. Bernard. 2012. Wildlife sinks. Quantifying the impact of illegal bird trade in street markets in Brazil. *Biological Conservation* 149: 16–22. (4)

Riehl, B., H. Zerriffi, and R. Naidoo. 2015. Effects of community-based natural resource management on household welfare in Namibia. *PLoS ONE* 10(5). (9)

Reiker, J., B. Schulz, V. Wissemann, and B. Gemeinholzer. 2015. Does origin always matter? Evaluating the influence of nonlocal seed provenances for ecological restoration purposes in a widespread and outcrossing plant species. *Ecology and Evolution* 5(23): 5642–5651. (10)

Reiter, N., J. Whitfield, G. Pollard, W. Bedggood, et al. 2016. Orchid re-introductions: an evaluation of success and ecological considerations using key comparative studies from Australia. *Plant Ecology* 1–15. (7)

Reyers, B., R. Biggs, G. S. Cumming, T. Elmqvist, et al. 2013. Getting the measure of ecosystem services: a social-ecological approach. *Frontiers in Ecology and the Environment* 11: 268–273. (3)

Reyers, B., D. J. Roux, R. M. Cowling, A. E. Ginsburg, et al. 2010. Conservation planning as a transdisciplinary process. *Conservation Biology* 24: 957–965. (1)

Ribeiro, J., G. R. Colli, J. P. Caldwell, and A. M. Soares. 2016. An integrated trait-based framework to predict extinction risk and guide conservation planning in biodiversity hotspots. *Biological Conservation* 195: 214–223. (9)

Ricciardi, A. 2003. Predicting the impacts of an introduced species from its invasion history: An empirical approach applied to zebra mussel invasions. *Freshwater Biology* 48: 972–981. (4)

Richardson, C. J. and N. J. Hussain. 2006. Restoring the Garden of Eden: an ecological assessment of the marshes of Iraq. *BioScience* 56: 477–489. (10)

Richardson, S. J., R. Clayton, B. D. Rance, H. Broadbent, et al. 2015. Small wetlands are critical for safeguarding rare and threatened plant species. *Applied Vegetation Science* 18(2): 230–241. (8)

Ricketts, T. H., E. Dinerstein, D. M. Olson, C. J. Loucks, et al. 1999. *Terrestrial Ecoregions of North America: A Conservation Assessment*. Island Press, Washington, DC. (2)

Ricklefs, R. E., and F. He. 2016. Region effects influence local tree species diversity. *Proceedings of the National Academy of Sciences USA*, 113: 674–679. (2)

Riehl, B., H. Zerriffi, and R. Naidoo. 2015. Effects of community-based natural resource management on household welfare in Namibia. *PLoS ONE* 10(5): e0125531. (9)

Rife, A. N., B. Erisman, A. Sanchez, and O. Aburto-Oropeza. 2013. When good intentions are not enough … Insights on networks of "paper park" marine protected areas. *Conservation Letters* 6: 200–212. (8)

Rinella, M. F., B. D. Maxwell, P. K. Fay, T. Weaver, and R. L. Sheley. 2009. Control effort exacerbates invasive-species problem. *Ecological Applications* 19: 155–162. (4)

Ripple, W. J. and R. L. Beschta. 2012. Trophic cascades in Yellowstone: The first 15 years after wolf reintroduction. *Biological Conservation* 145: 205–213. (2)

Ripple, W. J., J. A. Estes, R. L. Beschta, C. C. Wilmers, et al. 2014. Status and ecological effects of the world's largest carnivores. *Science* 343: 1241484. doi: 10.1126/science1241484. (2, 5, 11)

Riskas, K. A., M. M. Fuentes, and M. Hamann. 2016) Justifying the need for collaborative management of fisheries bycatch: a lesson from marine turtles in Australia. *Biological Conservation* 196: 40–47.

Rissman, A. R. and V. Butsic. 2011. Land trust defense and enforcement of conserved areas. *Conservation Letters* 4: 31–37. (11)

Rissman, A. R., J. Owley, M. R. Shaw, and B. B. Thompson. 2015. Adapting conservation easements to climate change. *Conservation Letters* 8(1): 68–76. (11)

Riverá Ortíz, F. A., R. Aguilar, M. D. C. Arizmendi, M. Quesada, and K. Oyama. 2014. Habitat fragmentation and genetic variability of tetrapod populations. *Animal Conservation* 18(3): 249–258. (5)

Robertson, G. C. Moreno, J. A. Arata, S. G. Candy, et al. 2014. Black-browed albatross numbers in Chile increase in response to reduced mortality in fisheries. *Biological Conservation* 169: 319–333. (4)

Robertson, M. M. 2006. Emerging ecosystem service markets: Trends in a decade of entrepreneurial wetland banking. *Frontiers in Ecology and the Environment* 4: 297–302. (10)

Robinson, J. G. 2011. Ethical pluralism, pragmatism, and sustainability in conservation practice. *Biological Conservation* 144: 958–965. (3)

Robinson, J. G. 2012. Common and conflicting interests in the engagements between conservation organizations and corporations. *Conservation Biology* 26: 967–977. (11, 12)

Robinson, R. A., H. P. Q. Crick, J. A. Learmonth, I. M. D. Maclean, et al. 2008. Travelling through a warming world: Climate change and migratory species. *Endangered Species Research* 7: 87–99. (4)

Robles, M. D., C. H. Flather, S. M Stein, M. D. Nelson, and A. Cutko. 2008. The geography of private forests that support at-risk species in the conterminous United States. *Frontiers in Ecology and the Environment* 6: 301–307. (9)

Rochefort, L. and E. Lode. 2006. Restoration of degraded boreal peatlands. *In* R. K. Wieder and D. H. Vitt (eds.), *Boreal Peatland Ecosystems,* p. 381–423. Springer Berlin Heidelberg. (10)

Roe, D., E. Y. Mohammed, I. Porras, and A. Giuliani. 2013. Linking biodiversity conservation and poverty reduction: De-polarizing the conservation–poverty debate. *Conservation Letters* 6: 162–171. (9, 11)

Rolston, H., III. 1994. 2012. *Environmental Ethics*. Temple University Press. (3)

Roman, J., Dunphy-Daly, D. W. Johnston, and A. J. Read. 2015. Lifting baselines to address the consequences of conservation success. *Trends in Ecology and Evolution* 30.6: 299–302. (1)

Roman, J. and S. R. Palumbi. 2003. Whales before whaling in the North Atlantic. *Science* 301(5632): 508–510. (4)

Romero, G. Q., T. Gonçalves-Souza, C. Vieira, and J. Koricheva. 2015. Ecosystem engineering effects on species diversity across ecosystems: a meta-analysis. *Biological Reviews* 90(3): 877–890. (2)

Rompré, G., W. D. Robinson, A. Desrochers, and G. Angehr. 2009. Predicting declines in avian species richness under nonrandom patterns of habitat loss in a Neotropical landscape. *Ecological Applications* 19: 1614–1627. (5)

Rood, S. B., S. Kaluthota, K. M. Gill, E. J. Hillman, et al. 2015. A twofold strategy for riparian restoration: combining a functional flow regime and direct seeding to re-establish cottonwoods. *River Research and Applications*. doi: 10.1002/rra.2919. (7)

Rosa, I. B. D., C. Souza Jr., and R. M. Ewers. 2012. Changes in size of deforested patches in the Brazilian Amazon. *Conservation Biology* 26: 932–937. (4)

Roscher, C., J. Schumacher, B. Schmid, E-D. Schulze. 2015. Contrasting effects of intraspecific trait variation on trait-based niches and performance of legumes in plant miztures. PloS ONE 10(3) e0119786. doi: 10.1371/journal. pone.0119786. (2)

Rosenzweig, M. L. 2003. Win-win ecology: how the Earths species can survive in the midst of human enterprise. Oxford University Press, New York. (9, 10)

Ruane, J. 2000. A framework for prioritizing domestic animal breeds for conservation purposes at the national level: A Norwegian case study. *Conservation Biology* 14: 1385–1393. (7)

Ruscoe, W. A., P. J. Sweetapple, M. Perry, and R. P. Duncan. 2013. Effects of spatially extensive control of invasive rats on abundance of native invertabrates in mainland New Zealand forests. *Conservation Biology* 27: 74–82. (4)

Russell, J. C., J. G. Innes, P. H. Brown, A. E. Byrom. 2015a. Predator-free New Zealand: conservation country. *BioScience*, biv012. (4)

Russell, J. C., H. P. Jones, D. P. Armstrong, F. Courchamp, et al. 2015b. Importance of lethal control of invasive predators for island conservation. *Conservation Biology*. (4)

Rust, N. A. 2015. Media Framing of Financial Mechanisms for Resolving Human–Predator Conflict in Namibia. *Human Dimensions of Wildlife* 20(5): 440–453. (3)

Ryan, M. E. J. R. Johnson, B. M. Fitzpatrick, L. J. Lowenstine, et al. 2013. Lethal effects of water quality on threatened California salamanders but not on co-occurring hybrid salamanders. *Conservation Biology* 27: 95–102. (2)

Saarinen, K., A. Valtonen, J. Jantunen, and S. Saarino. 2005. Butterflies and diurnal moths along road verges: does road type affect diversity and abundance? *Biological Conservation* 123: 403–412. (9)

Saccheri, I., M. Kuussaari, M. Kankare, P. Vikman, et al. 1998. Inbreeding and extinction in a butterfly metapopulation. *Nature* 392: (6675): 491–494. (5)

Sachs, J. D. 2008. *Common Wealth: Economics for a Crowded Planet*. Penguin Press, New York. (3)

Sæther, B. E. and S. Engen. 2015. The concept of fitness in fluctuating environments. *Trends in Ecology and Evolution* 30(5): 273–281. (6)

Sagoff, M. 2008. On the compatibility of a conservation ethic with biological science. *Conservation Biology* 21: 337–345. (3)

Sairam, R., S. Chennareddy, and M. Parani. 2005. OBPC Symposium: Maize 2004 and Beyond—plant regeneration, gene discovery, and genetic engineering of plants for crop improvement. *In Vitro Cellular and Developmental Biology—Plant* 41: 411. (3)

Sajeva, M., C. Augugliaro, M. J. Smith, and E. Oddo. 2013. Regulating internet trade in CITES species. *Conservation Biology* 27: 429–430. (6)

Sale, P. F. 2015. Coral reef conservation and political will. *Environmental Conservation* 42(02): 97–101. (12)

Sanderson, E., M. Jaiteh, M. A. Levy, K. H. Redford, et al. 2002. The human footprint and the last of the wild. *BioScience* 52: 891–904. (4)

Sandom, C., S. Faurby, B. Sandel, and J. C. Svenning. 2014. Global late Quaternary megafauna extinctions linked to humans, not climate change. *Proceedings of the Royal Society of London B: Biological Sciences* 281(1787): 20133254. (5)

Sanford, E., B. Gaylord, A. Hettinger, E. A. Lenz, et al. 2014. Ocean acidification increases the vulnerability of native oysters to predation by invasive snails. *Proceedings of the Royal Society of London B: Biological Sciences* 281(1778): 20132681. (4)

Sanjayan, M., L. H. Samberg, T. Boucher, and J. Newby. 2012. Intact faunal assemblages in the modern era. *Conservation Biology* 26: 724–730. (5)

Sauer, J. R., J. E. Hines, J. E. Fallon, K. L. Pardieck, et al. 2014. The North American Breeding Bird Survey, Results and Analysis 1966–2013. Version 01.30.2015. USGS Patuxent Wildlife Research Center, Laurel, MD. (6)

Sauer, J. R., W. A. Link, J. E. Fallon, K. L. Pardieck, and D. J. Ziolkowski, Jr. 2013. The North American breeding bird survey 1966-2011: summary analysis and species accounts. *North American Fauna* 79(79): 1–32. (6)

Sayre, N. F., R. R. McAllister, B. T. Bestelmeyer, M. Moritz, et al. 2013. Earth Stewardship of rangelands: coping with ecological, economic, and political marginality. *Frontiers in Ecology and the Environment* 11: 348–354. (4, 12)

Scariot, A. 2013. Land sparing or land sharing: the missing link. *Frontiers in Ecology and the Environment* 11: 177–178. (9)

Schaider, L. A., R. A. Rudel, J. M. Ackerman, S. C. Dunagan, and J. G. Brody. 2014. Pharmaceuticals, perfluorosurfactants, and other organic wastewater compounds in public drinking water wells in a shallow sand and gravel aquifer. *Science of the Total Environment* 468: 384–393. (4)

Scheckenbach, F., K. Hausmann, C. Wylezich, M. Weitere, and H. Arndt. 2010. Large-scale patterns in biodiversity of microbial eukaryotes from the abyssal sea floor. *Proceedings of the National Academy of Sciences USA* 107: 115–120. (2)

Scheffer, M., S. Barrett, S. R. Carpenter, C. Folke, C., et al. 2015. Creating a safe operating space for iconic ecosystems. *Science* 347(6228): 1317–1319. (8)

Schlacher, T. A., and L. Thompson. 2012. Beach recreation impacts benthic invertebrates on ocean-exposed sandy shores. *Biological Conservation* 147: 123–132. (3)

Schleuning, M. and D. Matthies. 2009. Habitat change and plant demography: Assessing the extinction risk of a formerly common grassland perennial. *Conservation Biology* 23: 174–183. (5)

Schmidt, C., Kucera, M., and Uthicke, S. 2014. Combined effects of warming and ocean acidification on coral reef Foraminifera *Marginopora vertebralis* and *Heterostegina depressa*. *Coral Reefs* 33(3): 805–818. (4)

Schmitz, P., S. Caspers, P. Warren, P., and K. Witte. 2015. First steps into the wild–exploration behavior of

European bison after the first reintroduction in Western Europe. *PLoS ONE* 10(11). (7)

Schofield, G., R. Scott, A. Dimadi, S. Fossette, et al. 2013. Evidence-based marine protected area planning for a highly mobile endangered marine vertebrate. *Biological Convservation.* 161: 101–109. (8)

Schonewald-Cox, C. M. 1983. Conclusions: guidelines to management: a beginning attempt. *In* C. M. Schonewald-Cox, S. M. Chambers, B. MacBryde and L. Thomas (eds.), *Genetics and Conservation: A Reference for Managing Wild Animal and Plant Populations,* pp. 414–445. Benjamin/Cummings, Menlo Park, CA. (6)

Schuur, E. A. G., A. D. McGuire, C. Schädel, G. Grosse, et al. 2015. Climate change and the permafrost carbon feedback. *Nature* 520(7546): 71–179. (4)

Schwartz, M. W. 2008. The performance of the Endangered Species Act. *Annual Review of Ecology, Evolution, and Systematics* 39: 279–299. (6)

Schwenk, W. S. and T. M. Donovan. 2011. A multispecies framework for landscape conservation planning. *Conservation Biology* 25: 1010–1021. (8)

Schwitzer, C., R. A. Mittermeier, S. E. Johnson, G. Donati, et al. 2014. Averting lemur extinctions amid Madagascar's political crisis. *Science* 343: 842–843. (5)

Scott, J. M., B. Csuti, and F. Davis. 1991. Gap analysis: An application of Geographic Information Systems for wildlife species. *In* D. J. Decker, M. E. Krasny, G. R. Goff, C. R. Smith, and D. W. Gross (eds.), *Challenges in the Conservation of Biological Resources: A Practitioner's Guide,* pp. 167–179. Westview Press, Boulder, CO. (8)

Scott, J. M., F. W. Davis, R. G. McGhie, R. G. Wright, et al. 2001. Nature reserves: do they capture the full range of America's biological diversity? *Ambio* 11: 999–1007. (8)

Scott, J. M., D. D. Goble, J. A. Wiens, D. S. Wilcove, et al. 2005. Recovery of imperiled species under the Endangered Species Act: The need for a new approach. *Frontiers in Ecology and the Environment* 3: 383–389. (6)

Scott, J. M., F. L. Ramse, M. Lammertink, K. V. Rosenberg, et al. 2008. When is an "extinct" species really extinct? Gauging the search efforts for Hawaiian forest birds and the ivory-billed woodpecker. *Avian Conservation and Ecology* 3(2): 3. (5)

Seastedt, T. R. 2015. Biological control of invasive plant species: a reassessment for the Anthropocene. *New Phytologist* 205(2): 490–502. (10)

Sebastián-González, E., J. A. Sánchez-Zapata, F. Botella, J. Figuerola, et al. 2011. Linking cost efficiency evaluation with population viability analysis to prioritize wetland bird conservation actions. *Biological Conservation* 144: 2354–2361. (7)

Seddon, P. J., C. J. Griffiths, P. S. Soorae, and D. P. Armstrong. 2014. Reversing defaunation: Restoring species in a changing world. *Science* 345(6195): 406–412. (7)

Şekercioğlu, Ç. H., R. B. Primack, and J. Wormworth. 2012. The effects of climate change on tropical birds. *Biological Conservation* 148(1): 1–18. (4, 5)

Selier, S. A. J., B. R. Page, A. T. Vanak, and R. Slotow. 2014. Sustainability of elephant hunting across international borders in southern Africa: A case study of the greater Mapungubwe Transfrontier Conservation Area. *The Journal of Wildlife Management* 78(1): 122–132. (3)

Selomane, O, B. Reyers, R. Biggs, H. Tallis, and S. Polasky. 2015. Towards integrated social-ecological sustainability indicators: Exploring the contribution and gaps in existing global data. *Ecological Economics* 118: 140–146. (11)

Sethi, P. and H. F. Howe. 2009. Recruitment of hornbill-dispersed trees in hunted and logged forests of the eastern Indian Himalaya. *Conservation Biology* 23: 710–718. (5)

Shackelford, N., R. J. Hobbs, N. E. Heller, L. M. Hallett, et al. 2013. Finding a middle-ground: The native/non-native debate. *Biological Conservation* 158: 55–62. (4)

Shafer, C. L. 1997. Terrestrial nature reserve design at the urban/rural interface. *In* M. W. Schwartz (ed.), *Conservation in Highly Fragmented Landscapes,* pp. 345–378. Chapman and Hall, New York. (8)

Shafer, C. L. 1999. History of selection and system planning for U. S. natural area national parks and monuments: beauty and biology. *Biodiversity and Conservation* 8: 189–204. (8)

Shafer, C. L. 2001. Conservation biology trailblazers: George Wright, Ben Thompson, and Joseph Dixon. *Conservation Biology* 15: 332–344. (1)

Shaffer, M. L. 1981. Minimum population sizes for species conservation. *BioScience* 31: 131–134. (6)

Shafroth, P. B., V. B. Beauchamp, M. K. Briggs, K. Lair, et al. 2008. Planning riparian restoration in the context of Tamarix control in western North America. *Restoration Ecology,* 16(1): 97–112. (10)

Shanley, P. and L. Luz. 2003. The impacts of forest degradation on medicinal plant use and implications for health care in eastern Amazonia. *BioScience* 53: 573–584. (3)

Sharma, N., M. D. Madhusudan, and A. Sinha. 2014. Local and landscape correlates of primate distribution and persistence in the remnant lowland forests of the Upper Brahmaputra Valley, northeastern India. *Conservation Biology* 28: 95–106. (5)

Sher, A. Introduction to the Paradox Plant. 2013. *In* A. Sher and M. F. Quigley (eds.), *Tamarix: A Case Study of Ecological Change in the American West,* pp. 1–20. Oxford University Press, New York. (10)

Sher, A. A. and L. A. Hyatt. 1999. The disturbed resource-flux invasion matrix: a new framework for patterns of plant invasion. *Biological Invasions* 1(2–3): 107–114. (4)

Shutt, K., M. Heistermann, A. Kasim, A. Todd, et al. 2014. Effects of habituation, research and ecotourism on faecal glucocorticoid metabolites in wild western lowland gorillas: implications for ecotourism. *Biological Conservation* 172: 72–79. (3)

Shwartz, A., A. Turbé, L. Simon, R. Juliard. 2014. Enhancing urban biodiversity and its influence on city dwellers. *Biological Conservation* 171: 82–90. (10)

Siikamäki, J. 2011. Contributions of the US state park system to nature recreation. *Proceedings of the National Academy of Sciences USA* 108: 14031–14036. (3)

Simaika, J. P., M. J. Samways, J. Kipping, F. Suhling, et al. 2013. Continental-scale conservation prioritization of African dragonflies. *Biological Conservation* 157: 245–254. (2, 8)

Simberloff, D. S., J. A. Farr, J. Cox, and D. W. Mehlman. 1992. Movement corridors: conservation bargains or poor investments? *Conservation Biology* 6: 493–505. (8)

Simberloff, D., C. Murcia, and J. Aronson, J. 2015. Novel ecosystems are a Trojan horse for conservation. *Ensia. ensia.com/voices/novel-ecosystems-are-a-trojan-horse-for-conservation.* (10)

Simmons, R. E. 1996. Population declines, variable breeding areas and management options for flamingos in Southern Africa. *Conservation Biology* 10: 504–515. (6)

Siraj, A. S., M. Santos-Vega, M. Bouma, D. Yadeta, et al. 2014. Altitudinal changes in malaria incidence in highlands of Ethiopia and Colombia. *Science* 343: 1154–1158. (4)

Skelly, D. K., S. R. Bolden, and L. K. Freidenburg. 2014. Experimental canopy removal enhances diversity of vernal pool amphibians. *Ecological Applications* 24: 340–345. (8)

Smith, A. 1909. An Inquiry into the Nature and Causes of the Wealth of Nations. *In* J. L. Bullock (ed.), *The Harvard Classics*, P. F. Collier and Sons, New York. (3)

Smith, S., A. A. Sher, and T. A. Grant. 2007. Genetic Diversity in Restoration Materials and the Impacts of Seed Collection in Colorado's Restoration Plant Production Industry. *Restoration Ecology* 15: 369–374. (10)

Smyser, T. J., S. A. Johnson, K. Page, C. M. Hudson, et al. 2013. Use of experimental translocations of allegheny woodrat to decipher casual agents of decline. *Conservation Biology,* 27: 752–762. (7)

Soanes, K., M. C. Lobo, P. A. Vesk, M. A. McCarthy, et al. 2013. Movement re-established but not restored: inferring the effectiveness of road-crossing mitigation for a gliding mammal by monitoring use. *Biological Conservation* 159: 434–441. (8)

Soares-Filho, B., P. Moutinho, D. Nepstad, A. Anderson, et al. 2010. Role of Brazilian Amazon protected areas in climate change mitigation. *Proceedings of the National Academy of Sciences USA* 107: 10821–10826. (9)

Society for Conservation Biology. 2016. What is SCB? Organizational Values. *conbio.org/about-scb/who-we-are* (1)

Society for Ecological Restoration. *www.ser.org* (10)

Sodhi, S. N., R. Butler, W. F. Laurance, and L. Gibson. 2011. Conservation successes at micro-, meso- and macroscales. *Trends in Ecology and Evolution* 26: 585–594. (1)

Sogge, M. K, E. H. Paxton, C. Van Riper. 2013. *Tamarix* in riparian woodlands: a bird's eye view. *In* A. Sher and M. F. Quigley (eds.), *Tamarix: A Case Study of Ecological Change in the American West*, pp. 189–206. Oxford University Press, New York. (10)

Solow, A., W. Smith, M. Burgman, T. Rout, T. et al. 2012. Uncertain sightings and the extinction of the ivory-billed woodpecker. *Conservation Biology* 26(1): 180–184. (5)

Sorice, M. G., C-O. Oh, T. Gartner, M. Snieckus, et al. 2013. Increasing participation in incentive programs for biodiversity conservation. *Ecological Applications* 23: 1146–1155. (11)

Soulé, M. E. 1985. What is conservation biology? *BioScience* 35: 727–734. (1)

Soulé, M. E. and D. Simberloff. 1986. What do genetics and ecology tell us about the design of nature reserves? *Biological Conservation* 35: 19–40. (8)

Spalding, M. D., L. Fish, and L. J. Wood. 2008. Towards representative protection of the world's coasts and oceans—Progress, gaps, and opportunities. *Conservation Letters* 1: 217–226. (8)

Spielman, D., B. W. Brook, and R. Frankham. 2004. Most species are not driven to extinction before genetic factors impact them. *Proceedings of the National Academy of Sciences USA* 101: 15261–15264. (5)

Sponberg, A. F. 2009. Great Lakes: Sailing to the forefront of national water policy? *BioScience* 59: 372. (10)

Srinivasan, U. T., S. P. Carey, E. Hallstein, P. A. T. Higgins, et al. 2008. The debt of nations and the distribution of ecological impacts from human activities. *Proceedings of the National Academy of Sciences USA* 105: 1768–1773. (6)

Stankey, G. H. and B. Shindler. 2006. Formation of social acceptability judgments and their implications for management of rare and little-known species. *Conservation Biology* 20(1): 28–37. (6)

Staricco, J. I. and S. Ponte. 2015. Quality regimes in agrofood industries: A regulation theory reading of Fair Trade wine in Argentina. *Journal of Rural Studies* 66–76. (12)

Steadman, D. W. 1995. Prehistoric extinctions of Pacific island birds: biodiversity meets zooarchaeology. *Science* 267(5201: 1123–1131. (5)

Stein, B. A., L. S. Kutner, and J. S. Adams (eds.). 2000. *Precious Heritage: The Status of Biodiversity in the United States.* Oxford University Press, New York. (4)

Stein, B. A., C. Scott, and N. Benton. 2008. Federal lands and endangered species: the role of the military and other federal lands in sustaining biodiversity. *BioScience* 58: 339–347. (9)

Stocks, A. 2005. Too much for too few: problems of indigenous land rights in Latin America. *Annual Review of Anthropology* 34: 85–104. (12)

Stokes, E. J., S. Strindberg, P. C. Bakabana, P. W. Elkan, et al. 2010. Monitoring great ape and elephant abundance at large spatial scales: Measuring effectiveness of a conservation landscape. *PLoS ONE* 5(4): e10294. (9)

Stokstad, E. 2009. Obama moves to revitalize Chesapeake Bay restoration. *Science* 324: 1138–1139. (10)

Stouffer, P. C., E. I. Johnson, R. O. Bierregaard Jr, and T. E. Lovejoy. 2011. Understory bird communities in Amazonian rainforest fragments: Species turnover through 25 Years post-isolation in recovering landscapes. *PLoS ONE* 6(6): e20543. (4)

Strain, D. 2011. 8.7 million: A new estimate for all the complex species on earth. *Science* 333: 1083. (2)

Strayer, D. L. 2009. Twenty years of zebra mussels: lessons from the mollusk that made headlines. *Frontiers in Ecology and the Environment* 7: 135–141. (4)

Sullivan, B. L., J. L. Aycrigg, J. H. Barry, R. E. Bonney, et al. 2014. The eBird enterprise: An integrated approach to development and application of citizen science. *Biological Science* 169: 31–40. (12)

Sun, Z., Q. Huang, C. Opp, T. Hennig, T., and U. Marold. 2012. Impacts and implications of major changes caused by the Three Gorges Dam in the middle reaches of the Yangtze River, China. *Water Resources Management* 26(12): 3367–3378. (4)

Sutherland, W. J., W. M. Adams, R. B. Aronson, R. Aveling, et al. 2009. One hundred questions of importance to the conservation of global biological diversity. *Conservation Biology* 23: 557–567. (12)

Sutherland, W. J., S. Bardsley, L. Bennun, M. Clout, et al. 2011. Horizon scan of global conservation issues for 2011. *Trends in Ecology & Evolution* 26: 10–16. (12)

Svenning, J. C., P. B. Pedersen, C. J. Donlan, R. Ejrnæs, et al. 2015. Science for a wilder Anthropocene: Synthesis and future directions for trophic rewilding research. *Proceedings of the National Academy of Sciences* 113(4): 898–906. (10)

Swanson, F. J., C. Goodrich, and K. D. Moore. 2008. Bridging boundaries: scientists, creative writers, and the long view of the forest. *Frontiers in Ecology and the Environment* 6: 499–504. (3, 12)

Sweka, J. A. and T. C. Wainwright. 2014. Use of population viability analysis models for Atlantic and Pacific salmon recovery planning. *Reviews in Fish Biology and Fisheries* 24(3): 901–917. (6)

Szabo, J. K., S. H. M. Butchart, H. P. Possingham, and S. T. Garnett. 2012. Adapting global biodiversity indicators to the national scale: a Red List index for Australian birds. *Biological Conservation* 148: 61–68. (6)

Szlávik, J. and M. Füle. 2009. Economic consequences of climate change. American Institute of Physics Conference Proceedings, Sustainability 2009: *The Next Horizon* 1157: 73–82. (12)

Tallis, H., R. Goldman, M. Uhl, and B. Brosi. 2009. Integrating conservation and development in the field: implementing ecosystem service projects. *Frontiers in Ecology and the Environment* 7: 12–20. (9)

Tallis, H., J. Lubchenco, V. M. Adams, C. Adams-Hosking, et al. 2014. Working together: A call for inclusive conservation. *Nature* 515(7525): 27–28. (12)

Tallis, H. and S. Polasky. 2009. Mapping and valuing ecosystem services as an approach for conservation and natural-resource management. *Annals of the New York Academy of Sciences* 1162: 265–283. (6)

Tamarisk Coalition. 2016. Tamarisk beelte (*Diorhabda* spp.) in the Colorado River basin: synthesis of an expert panel forum. Tamarisk Coalition, Grand Junction, CO. (10)

Taylor, M. F. J., K. F. Suckling, and J. J. Rachlinski. 2005. The effectiveness of the Endangered Species Act: A quantitative analysis. *BioScience* 55: 360–366. (6)

Temperton, V. M. and R. J. Hobbs. 2004. The search for ecological assembly rules and its relevance to restoration ecology. In V. M. Temperton, R. J. Hobbs, T. J. Nuttle, and S Halle (eds.), *Assembly Rules And Restoration Ecology—Bridging The Gap Between Theory And Practice*, pp. 34–54. Island Press, Washington, DC. (10)

Temple, S. A. 1991. Conservation biology: New goals and new partners for managers of biological resources. *In* D. J. Decker, M. Krasny, G. R. Goff, C. R. Smith, and D. W. Gross (eds.), *Challenges in the Conservation of Biological Resources: A Practitioner's Guide*, pp. 45–54. Westview Press, Boulder, CO. (1)

Templeton, A. R. 1986. Coadaptation and outbreeding depression. *In* M. E. Soulé (ed.), *Conservation Biology: The Science of Scarcity and Diversity*, pp. 105–116. Sinauer Associates, Sunderland, MA. (5)

Tende, T., B. Hansson, U. Ottosson, M. Åkesson, and S. Bensch. 2014. Individual identification and genetic variation of lions (Panthera leo) from two protected areas in Nigeria. *PLoS ONE* 9(1): e84288. (5)

Thatcher, C. A., F. T. van Manen, and J. D. Clark. 2009. A habitat assessment for Florida panther population expansion into central Florida. *Journal of Mammalogy* 90: 918–925. (9)

Theobald, D. M. 2004. Placing exurban land-use change in a human modification framework. *Frontiers in Ecology and the Environment* 2: 139–144. (9)

Thiollay, J. M. 1989. Area requirements for the conservation of rainforest raptors and game birds in French Guiana. *Conservation Biology* 3: 128–137. (6)

Thogmartin, W. E., C. A. Sander-Reed, J. A. Szymanski, P. C. McKann, et al. 2013. White-nose syndrome is likely to extirpate the endangered Indiana bat over large parts of its range. *Biological Conservation* 160: 162–171. (4)

Thomsen P. F. and W. Willerslev. 2015. Environmental DNA—an emerging tool in conservation for monitoring past and present biodiversity. *Biological Conservation* 183: 4–18. (6)

Thondhlana, G. S. Shackleton, and J. Blignaut. 2015. Local institutions, actors, and natural resource governance in Kgalagadi Transfrontier Park and surrounds, South Africa. *Land Use Policy* 47: 121–129 doi: 10.1016/j.landusepol.2015.03.013. (11)

Thoreau, H. D. 1854. *Walden; or, Life in the Woods*. Ticknor and Fields, Boston. (3)

Thorp, J. H., J. E. Flotemersch, M. D. Delong, A. F. Casper, et al. 2010. Linking ecosystem services, rehabilitation, and river hydrogeomorphology. *BioScience* 60: 67–74. (3)

Tilman, D., R. M. May, C. L. Lehman, and M. A. Nowak, M. A. 1994. Habitat destruction and the extinction debt. *Nature* 371: 65–66. (5)

Timmer, V. and C. Juma. 2005. Biodiversity conservation and poverty reduction come together in the tropics: Lessons learned from the Equator Initiative. *Environment* 47: 25–44. (9)

Tingley, R., R. A. Hitchmough, and D. G. Chapple. 2013. Life-history traits and extrinsic threats determine extinction risk in New Zealand lizards. *Biological Conservation* 165: 62–67. (5)

Tittensor, D. P., C. Mora, W. Jetz, H. K. Lotze, et al. 2010. Global patterns and predictors of marine biodiversity across taxa. *Nature* 466: 1098–1101. (2)

Tlusty, M. F., A. L. Rhyne, L. Kaufman, M. Hutchins, et al. 2013. Opportunities for public aquariums to increase the sustainability of the aquatic animal trade. *Zoo Biology* 32: 1–12. (7)

Tognelli, M. F., M. Fernández, and P. A. Marquet. 2009. Assessing the performance of the existing and proposed network of marine protected areas to conserve marine biodiversity in Chile. *Biological Conservation* 142: 3147–3153. (8)

Toledo, V. M. 2001. Indigenous peoples and biodiversity. *In* S. A. Levin (ed.), *Encyclopedia of Biodiversity* 3: 451–464. Academic Press, San Diego, CA. (9)

Towne, E. G., D. C. Hartnett, and R. C. Cochran. 2005. Vegetation trends in tallgrass prairie from bison and cattle grazing. *Ecological Applications* 15: 1550–1559. (8)

Traill, L. W., B. W. Brook, R. R. Frankham, and C. J. A. Bradshaw. 2010. Pragmatic population viability targets in a rapidly changing world. *Biological Conservation* 143: 28–34. (6)

Tran, P. M. and L. Waller. 2013. Effects of landscape fragmentation and climate on Lyme disease incidence in the northeastern United States. *Ecohealth* 10(4): 394–404. (4)

Tranquilli, S., M. Abedi-Lartey, F. Amsini, L. Arranz, et al. 2012. Lack of conservation effort rapidly increases African great ape extinction risk. *Conservation Letters* 5: 48–55. (8)

Tree of Life Web Project. *tolweb.org* (12)

Triantis, K., D. Nogues-Bravo, J. Hortal, A. V. Borges, et al. 2008. Measurements of area and the (island) species-area relationship: new directions for an old pattern. *Oikos* 117: 1555–1559. (5)

Troupin, D. and Y. Carmel. 2014. Can agro-ecosystems efficiently complement protected area networks? *Biological Conservation* 169: 158–166. (9)

Tscharntke, T., Y. Clough, T. C. Wanger, L. Jackson, et al. 2012. Global food security, biodiversity conservation and the future of agricultural intensification. *Biological Conservation* 151: 53–59. (9)

Turner, I. M. 2001. *The Ecology of Trees in the Tropical Rain Forest.* Cambridge University Press. (4)

Turvey, S. 2008. *Witness to Extinction: How We Failed to Save the Yangtze River Dolphin.* Oxford University Press. (5)

Twilley, R. R. and J. W. Day. 2012. Mangrove wetlands. *In* J. W. Day et al. (ed.), *Estuarine Ecology*, 2nd Ed., pp. 165–202. Wiley-Blackwell, Hoboken, NJ. (4)

U. S. Census Bureau. *www.census.gov* (1)

U. S. Energy Information Administration. 2015. International Energy Statistics 2008–2011. *www.eia.gov/tools/faqs/faq.cfm?id=85&t=1* (1)

U. S. Fish and Wildlife Service (USFWS). 2003. Guidance for the Establishment, Use, and Operation of Conservation Banks. *www.fws.gov/endangered/esa-library/pdf/Conservation_Banking_Guidance.pdf* (11)

U. S. Fish and Wildlife Service (USFWS). 2013. Federal and State Endangered and Threatened Species Expenditures Fiscal Year 2013. *www.fws.gov/endangered/esa-library/pdf/2012.EXP.FINAL.pdf*, p. 5. (6)

U. S. National Park Service. 2009. National Park of American Samoa. *www.nps.gov/npsa/index.htm* (9)

U. S. National Park Service. *www.nps.gov* (9)

UNESCO World Heritage Centre. 2010. *whc.unesco.org* (11)

United Nations Development Programme. 2006. *www.undp.org* (4)

United Nations Environment Programme (UNEP). 2014. Protected Planet Report. *www.unep-wcmc.org/resources-and-data/protected-planet-report-2014* (8)

United Nations Environment Programme World Conservation Monitoring Centre (UNEP-WCMC). 2014. Report on achievements for the year 2014. Cambridge, UK. (8)

United Nations Environment Programme Mediterranean Action Plan (UNEPMAP). 2016. Countries Agree Ambitious Conservation Measures for Mediterranean. *www.unep.org/Documents.Multilingual/Default.asp?DocumentID=27058&ArticleID=35947&l=en* (9)

United Nations. 1993a. *Agenda 21: Rio Declaration and Forest Principles.* Post-Rio Edition. United Nations Publications, New York. (11)

United Nations. 1993b. *The Global Partnership for Environment and Development.* United Nations Publications, New York. (11)

Uriarte, M., M. Pinedo-Vasquez, R. S. DeFries, K. Fernandes, et al. 2012. Depopulation of rural landscapes exacerbates fire activity in the western Amazon. *Proceedings of the National Academy of Sciences* 109(52): 21546–21550. (4)

Valiela, I. and P. Martinetto. 2007. Changes in bird abundance in eastern North America: Urban sprawl and global footprint? *BioScience* 57: 360–370. (4)

van Swaay, C., D. Maes, S. Collins, M. L. Munguira, et al. 2011. Applying IUCN criteria to inverte-brates: How red is the Red List of European butterflies? *Biological Conservation* 144: 470–478. (6)

Van Teeffelen, A. J. A., C. C. Vos, and P. Opdam. 2012. Species in a dynamic world: consequences of habitat network dynamics on conservation planning. *Biological Conservation* 153: 239–253. (8)

Van Vugt, M. 2009. Averting the tragedy of the commons: using social psychological science to protect the environment. *Current Directions in Psychological Science* 18: 169–173. (11)

Van Wilgen, B. W., G. G. Forsyth, D. C. Le Maitre, A. Wannenburgh, et al. 2012. An assessment of the effectiveness of a large, national-scale invasive alien plant control strategy in South Africa. *Biological Conservation* 148: 28–38. (4)

Vanbergen and the Insect Pollinator Initiative. 2013. Threats to an ecosystem service: pressures on pollinators. *Frontiers in Ecology and the Environment.* 11: 251–259. doi: 10.1890/120126. (3)

Vandermeer, J., I. Perfecto, and S. Philpott. 2010. Ecological complexity and pest control in organic coffee production: Uncovering an autonomous ecosystem service. *BioScience* 60: 527–537. (9)

Varma, V., J. Ratnam, V. Viswanathan, A. M. Osuri, et al. 2015. Perceptions of priority issues in the conservation of biodiversity and ecosystems in India. *Biological Conservation* 187: 201–211. (12)

Venevsky, S. and I. Venevskaia. 2005. Hierarchical systematic conservation planning at the national level: Identifying national biodiversity hotspots using abiotic factors in Russia. *Conservation Biology* 124: 235–251. (7)

Venter, O., J. E. M. Watson, E. Meijaard, W. F. Laurance, and H. P. Possingham. 2010. Avoiding unintended outcomes from REDD. *Conservation Biology* 24: 5–6. (3)

Verstraete, M. M., R. J. Scholes, and M. S. Smith. 2009. Climate and desertification: Looking at an old problem through new lenses. *Frontiers in Ecology and the Environment* 7: 421–428. (4)

Vianna, G. M. S., M. G. Meekan, D. J. Pannell, S. P. Marsh, and J. J. Meeuwig. 2012. Socio-economic value and community benefits from shark-diving tourism in Palau: A sustainable use of reef shark populations. *Biological Conservation* 145: 267–277. (3)

Viblanc, V. A., C. Saraux, N. Malosse, and R. Groscolas. 2014. Energetic adjustments in freely breeding–fasting king penguins: does colony density matter? *Functional Ecology* 28(3): 621–631. (6)

Vidal, O., J. López-García, and E. Rendón-Salinas. 2014. Trends in deforestation and forest degradation after a decade of monitoring in the Monarch Butterfly Biosphere Reserve in Mexico. *Conservation Biology* 28: 177–186. (8)

Vincent, A. C., Y. J. Sadovy de Mitcheson, S. L. Fowler, and S. Lieberman. 2014. The role of CITES in the conservation of marine fishes subject to international trade. *Fish and Fisheries* 15(4): 563–592. (4)

Vranckx, G., H. Kacquemyn, B. Muys, and O. Honnay. 2012. Meta-analysis of susceptibility of woody plants to loss of genetic diversity through habitat fragmentation. *Conservation Biology* 26: 228–237. (5)

Vredenburg, V. T. 2004. Reversing introduced species effects: Experimental removal of introduced fish leads to rapid recovery of a declining frog. *Proceedings of the National Academy of Sciences USA* 101: 7646–7650. (4)

Wackernagel, M. and W. Rees. 1996. *Our Ecological Footprint: Reducing Human Impact on the Earth (New Catalyst. Bioregional Series).* New Society Publishers, Gabriola Island, BC. (4)

Wade, L. 2013. Gold's dark side. *Science* 341: 1448–1449. (4)

Wagner, K. I., S. K. Gallagher, M. Hayes, B. A. Lawrence, et al. 2008. Wetland restoration in the new millennium: do research efforts match opportunities? *Restoration Ecology* 16: 367–372. (10)

Waterman, C., L. Calcul, J. Beau, W. S. Maet al. 2016. Miniaturized Cultivation of Microbiota for Antimalarial Drug Discovery. *Medicinal Research Reviews* 36(1): 144–168. (3)

Wake, D. B. and V. T. Vredenburg. 2008. Are we in the midst of the sixth mass extinction? A view from the world of amphibians. *Proceedings of the National Academy of Sciences USA* 105: 11466–11473. (5)

Waldron, A., R. Justicia, L. Smith, and M. Sanchez. 2012. Conservation through chocolate: a win-win for biodiversity and farmers in Ecuador's lowland tropics. *Conservation Letters* 5: 213–221. (9)

Walker, R., E. Arima, J. Messina, B. Soares-Filho, et al. 2013. Modeling spatial decisions with graph theory: logging roads and forest fragmentation in the Brazilian Amazon. *Ecological Applications* 23(1): 239–254. (5)

Wallace K. J. 2007. Classification of ecosystem services: problems and solutions. *Biological Conservation* 139(3–4): 235–246. (3)

Waller, D. M. 2015. Genetic rescue: a safe or risky bet? *Molecular Ecology* 24(11): 2595–2597. (5)

Wallmo, K. and D. K. Lew. 2012. Public willingness to pay for recovering and downlisting threatened and endangered marine species. *Conservation Biology* 26: 830–839. (3)

Wanger, T. C., K. Darras, S. Bumrungsri, T. Tscharntke, et al. 2014. Bat pest control contributes to food security in Thailand. *Biological Conservation* 171: 220–223. (3)

Ward, P. 2004. The father of all mass extinctions. *Conservation* 5: 12–17. (5)

Warkentin, I. G., D. Bickford, N. S. Sodhi, and C. J. A. Bradshaw. 2009. Eating frogs to extinction. *Conservation Biology* 23: 1056–1059. (4)

Watanabe, M. E. 2015. The Nagoya Protocol on Access and Benefit Sharing: international treaty poses challenges for biological collections. *BioScience* doi: 10.1093/biosci/biv056. (11)

Watson, J. 2013. Endangered species thrive on US military ranges. Associated Press. August 10. *bigstory. ap.org/article/endangered-species-thrive-us-military-ranges.* (9)

Watson, J. E., E. S. Darling, O. Venter, M. Maron, et al. 2016. Bolder science needed now for protected areas. *Conservation Biology.* (9)

Watson, J. E., N. Dudley, D. B. Segan, and M. Hockings. 2014. The performance and potential of protected areas. *Nature* 515: 67–73 (8)

Watson, J. E. M., H. Grantham, K. A. Wilson, and H. P. Possingham. 2011. Systematic Conservation Planning: Past, Present and Future. *In* R. Whittaker and R. Ladle (eds.), *Conservation Biogeography,* pp. 136–160. Wiley-Blackwell, Oxford. (6)

Wearn, O. R., D. C. Reuman, and R. M. Ewers. 2012. Extinction debt and windows of conservation opportunity in the Brazilian Amazon. *Science* 337(6091): 228–232. (5)

Wedekind, C. G. Evanno, T. Székely, M. Pompini, et al. 2013. Persistent unequal sex ratio in a population of grayling (Salmonidae) and possible role of temperature increase. *Conservation Biology* 27: 229–234. (5)

Weis, J. S. and C. J. Cleveland. 2008. DDT. *In* C. J. Cleveland (ed.), *Encyclopedia of Earth*. Environmental Information Coalition, National Council for Science and the Environment, Washington, DC. *www.eoearth.org/article/DDT* (4)

Weiser, E. L., C. E. Grueber, and I. G. Jamieson. 2013. Simulating retention of rare alleles in small populations to assess management options for species with different life histories. *Conservation Biology* 27(2): 335–344. (5)

West, P. C., G. R. Narisma, C. C. Barford, C. J. Kucharik, and J. A. Foley. 2011. An alternative approach for quantifying climate regulation by ecosystems. *Frontiers in Ecology and the Environment* 9: 126–133. (3)

Western and Central Pacific Fisheries Commission (WCPFC). *www.wcpfc.int* (6)

Western, D., R. Groom, and J. Worden. 2009. The impact of subdivision and sedentarization of pastoral lands on wildlife in an African savanna ecosystem. *Biological Conservation* 142: 2538–2546. (9)

Whisenant, S. 1999. *Repairing damaged wildlands: a process-orientated, landscape-scale approach (Vol. 1)*. Cambridge University Press, Cambridge, UK. (10)

White, P. S. 1996. Spatial and biological scales in reintroduction. *In* D. A. Falk, C. I. Millar, and M. Olwell (eds.), *Restoring Diversity: Strategies for Reintroduction of Endangered Plants*, pp. 49–86. Island Press, Washington, DC. (6)

White, T. H., Jr., N. J. Collar, R. J. Moorhouse, V. Sanz, et al. 2012. Psittacine reintroductions: common denominators of success. *Biological Conservation* 148: 106–115. (7)

Whitman, W. B., T. Woyke, H. P. Klenk, Y. Zhou, et al. 2015. Genomic Encyclopedia of Bacterial and Archaeal Type Strains, Phase III: the genomes of soil and plant-associated and newly described type strains. *Standards in Genomic Sciences* 10(1): 1. (2)

Whittington, R. J. and R. Chong. 2007. Global trade in ornamental fish from an Australian perspective: the case for revised import risk analysis and management strategies. *Preventive Veterinary Medicine* 81(1): 92–116. (4)

Wibbels, T. and E. Bevan. 2015. New Riddle in the Kemp's Ridley Saga. In *State of the World's Sea Turtles Report*. Oceanic Society. (1)

Wiens, J. J. 2007. Species delimitations: new approaches for discovering diversity. *Systematic Biology* 56(6): 875–878. (2)

Wiersma, Y. F., T. D. Nudds, and D. H. Rivard. 2004. Models to distinguish effects of landscape patterns and human population pressures associated with species loss in Canadian national parks. *Landscape Ecology* 19: 773–786. (8)

Wikramanayake, E., E. Dinerstein, J. Seidensticker, S. Lumpkin, et al. 2011. A landscape-based conservation strategy to double the wild tiger population. *Conservation Letters* 4: 219–227. (8)

Wikström, L., P. Milberg, and K. Bergman. 2008. Monitoring of butterflies in semi-natural grasslands: Diurnal variation and weather effects. *Journal of Insect Conservation* 13: 203–211. (6)

Wilcove, D. S. and L. L. Master. 2005. How many endangered species are there in the United States? *Frontiers in Ecology and the Environment* 3: 414–420. (4, 6)

Wilcove, D. S. and M. Wikelski 2008. Going, going, gone: Is animal migration disappearing. *PLoS Biology* 6(7): e188. doi: 10.1371/journal.pbio.0060188. (5, 8)

Wildt, D. E., P. Comizzoli, B. Pukazhenthi, and N. Songsasen. 2009. Lessons from biodiversity—The value of nontraditional species to advance reproductive science, conservation, and human health. *Molecular Reproduction and Development* 77: 397–409. (7)

Wilhere, G. F. 2012. Inadvertent advocacy. *Conservation Biology* 26: 39–46. (1)

Willi, Y., M. van Kleunen, S. Dietrich, and M. Fischer. 2007. Genetic rescue persists beyond first-generation outbreeding in small populations of a rare plant. *Proceedings of the Royal Society B* 274: 2357–2364. (5)

Willis, C. G., B. R. Ruhfel, R. B. Primack, A. J. Miller-Rushing, and C. C. Davis. 2008. Phylogenetic patterns of species loss in Thoreau's woods are driven by climate change. *Proceedings of the National Academy of Sciences USA* 105: 17029–17033. (4)

Wilson, E. O. 1989. Threats to biodiversity. *Scientific American* 261(3): 108–116. (5)

Wilson, E. O. 1992. *The Diversity of Life*. Harvard University Press, Cambridge. (10)

Wilson, E. O. 2010. Within one cubic foot: Miniature surveys of biodiversity. *National Geographic* 217(2): 62–83. (2)

Wilson, H. B., E. Meijaard, O. Venter, M. Ancrenaz, and H. P. Possingham. 2014. Conservation strategies for orangutans: reintroduction versus habitat preservation and the benefits of sustainably logged forest. *PLoS ONE* 9(7): e102174. (7)

Wilson, M. C., X. Y. Chen, R. T. Corlett, R. K. Didham, et al. 2016. Habitat fragmentation and biodiversity conservation: key findings and future challenges. *Landscape Ecology* 31(2): 219–227. (4)

Winker, K. 2009. Reuniting phenotype and genotype in biodiversity research. *BioScience* 59(8): 657–665. (2)

With, K. A. 2015. How fast do migratory songbirds have to adapt to keep pace with rapidly changing landscapes? *Landscape Ecology* 30(7) 1351–1361. (4)

Wofford, J. E., R. E. Gresswell, and M. A. Banks. 2005. Influence of barriers to movement on within-watershed genetic variation of coastal cutthroat trout. *Ecological Applications* 15(2): 628–637. (5)

World Bank. *www.worldbank.org* (1, 11)

World Resources Institute (WRI). 2003. *World Resources 2002–2004: Decisions for the Earth: Balance, voice, and power*. World Resources Institute, Washington, DC. (11, 12)

World Resources Institute (WRI). 2005. *World Resources 2005: The Wealth of the Poor—Managing Ecosystems to Fight Poverty*. World Resources Institute, Washington, DC. (4).

World Wildlife Fund (WWF) Global. *wwf.panda.org* (6, 9, 11)

World Wildlife Fund (WWF) International. *www.worldwildlife.org* and *www.panda.org* (6, 9, 11)

World Wildlife Fund (WWF) and M. McGinley. 2009. Central American dry forests. *In* C. J. Cleveland (ed.), *Encyclopedia of Earth*. Environmental Information Coalition, National Council for Science and the Environment, Washington, DC. *www.eoearth.org/article/Central_American_dry_forests* (4)

Wright, A. J., D. Veríssimo, K. Pilfold, E. C. M. Parsons, et al. 2015. Competitive outreach in the 21st century: Why we need conservation marketing. *Ocean & Coastal Management* 115: 41–48. (12)

Wright, G., K. Andersson, C. Gibson, and T. Evans. 2015. What incentivizes local forest conservation efforts? Evidence from Bolivia. *International Journal of the Commons* 9(1). (12)

Wright, S. 1931. Evolution in Mendelian populations. *Genetics* 16: 97–159. (5)

Wright, C. K. and M. C. and Wimberly. 2013. Recent land use change inthe western corn belt threatens grasslands and wetlands. *PNAS*. doi: 10.1073/pnas.1215404110. (4)

Wright, H. L., I. R. Lake, and P. M. Dolman. 2012. Agriculture—A key element for conservation in the developing world. *Conservation Letters* 5: 11–19. (9)

Wright, S. J., G. A. Sanchez-Azofeifa, C. Portillo-Quintero, and D. Davies. 2007. Poverty and corruption compromise tropical forest reserves. *Ecological Applications* 17: 1259–1266. (8)

Wright, S. J., H. Zeballos, I. Domínguez, M. M. Gallardo, et al. 2000. Poachers alter mammal abundance, seed dispersal, and seed predation in a Neotropical forest. *Conservation Biology* 14: 227–239. (8)

Wu, J. and R. J. Hobbs (eds.). 2009. *Key Topics in Landscape Ecology*. Cambridge University Press, Cambridge, UK. (8)

Wu, R., S. Zhang, D. W. Yu, P. Zhao, et al. 2011. Effectiveness of China's nature reserves in representing ecological diversity. *Frontiers in Ecology and the Environment* 9: 383–389. (8)

Wuerthner, G., E. Crist, and T. Butler (eds.) 2015. *Protecting the Wild: Parks and Wilderness*. The Foundation For Conservation. Island Press. (8)

Wunder, S. 2013. When payments for environmental services will work for conservation. *Conservation Letters* 6: 230–237. (9, 11)

Wünscher, T. and S. Engel. 2012. International payments for biodiversity services: review and evaluation of conservation targeting approaches. *Biological Conservation* 152: 222–230. (9)

Xu, H., X. Pan, Y. Song, Y. Huang, M. Sun, and S. Zhu. 2014. Intentionally introduced species: more easily invited than removed. *Biodiversity and Conservation* 23(10): 2637–2643. (4)

Yamaoko, K., H. Moriyama, and T. Shigematsu. 1977. Ecological role of secondary forests in the traditional farming area in Japan. *Bulletin of Tokyo University* 20: 373–384. (8)

Yamaura, Y., T. Kawahara, S. Iida, and K. Ozaki. 2008. Relative importance of the area and shape of patches to the diversity of multiple taxa. *Conservation Biology* 22: 1513–1522. (8)

Yang, S. L., J. D. Milliman, K. H. Xu, B. Deng, et al. 2014. Downstream sedimentary and geomorphic impacts of the Three Gorges Dam on the Yangtze River. *Earth-Science Reviews* 138: 469–486. (4)

Yen, S. C., K H. Chen, Y. Wang, and C. P. Wang. 2015. Residents' attitudes toward reintroduced sika deer in Kenting National Park, Taiwan. *Wildlife Biology* 21(4): 220–226. (7)

Yi, B. L. Islamic clerics issue a fatwa against poachers in Indonesia and Malaysia. Public Radio International. *www.pri.org/stories/2016-01-07/islamic-clerics-issue-fatwa-against-poachers-indonesia-and-malaysia* (3)

Yochim, M. J. and W. R. Lowry. 2016. Creating conditions for policy change in national parks: contrasting cases in Yellowstone and Yosemite. *Environmental Management* 1–13. (7)

Yodzis, P. 2001. Trophic levels. *In* S. A. Levin (ed.), *Encyclopedia of Biodiversity, Vol. 5*, pp. 695–700. Academic Press, San Diego, CA. (2)

Young, T. P. 1994. Natural die-offs of large mammals: implications for conservation. *Conservation Biology* 8: 410–418. (5)

Young, T. P., T. M. Palmer, and M. E. Gadd. 2005. Competition and compensation among cattle, zebras, and elephants in a semi-arid savanna in Laikipia, Kenya. *Biological Conservation* 112: 251–259. (9)

Zander, K. K. and S. T. Garnett. 2011. The economic value of environmental services on indigenous-held lands in Australia. *PLoS ONE* 6(8): e23154. (3)

Zaradic, P. A., O. R. W. Pergams, and P. Kareiva. 2009. The impact of nature experience on willingness to support conservation. *PLoS ONE* 4: e7367. (11)

Zellweger, F., V. Braunisch, F. Morsdorf, A. Baltensweiler, et al. 2015. Disentangling the effects of climate, topography, soil and vegetation on stand-scale species richness in temperate forests. *Forest Ecology and Management* 349: 36–44. (2)

Zhu, L. and Y. C. Zhao. 2015. A feasibility assessment of the application of the Polluter-Pays Principle to ship-source pollution in Hong Kong. *Marine Policy* 57: 36–44. (12)

Zhu, Y. Y., Y. Y. Wang, H. R. Che, and B. R. Lu. 2003. Conserving traditional rice varieties through management for crop diversity. *BioScience* 53: 158–162. (9)

Zimmer, C. 2013. Bringing them back to life: The revival of an extinct species is no longer a fantasy. But is it a good idea? *National Geographic* 223(14). (7)

Zimmermann, A., M. Hatchwell, L. Dickie, and C. D. West (eds.) 2008. *Zoos in the 21st Century: Catalysts for Conservation*. Cambridge University Press, Cambridge. (7)

Zimmermann, N. E., T. C. Edwards, Jr., G. G. Moisen, T. S. Frescino, and J. A. Blackard. 2007. Remote sensing-based predictors improve distribution models of rare, early successional and broadleaf tree species in Utah. *Journal of Applied Ecology* 44: 1057–1067. (7)

Zomer, R. J., J. Xu, M. Wang, A. Trabucco, and Z. Li. 2015. Projected impact of climate change on the effectiveness of the existing protected area network for biodiversity conservation within Yunnan Province, China. *Biological Conservation* 184: 335–345. (8)

Zydelis, R., B. P. Wallace, E. L. Gilman, and T. B. Werner. 2009. Conservation of marine megafauna through minimization of fisheries bycatch. *Conservation Biology* 23: 608–616. (4)

Zylberberg, M., K. A. Lee, K. C. Klasing, and M. Wikelski. 2013. Variation with land use of immune function and prevalence of avian pox in Galapagos finches. *Conservation Biology* 27: 103–112. (4)

Index

The letter *f* after a page number indicates that the entry is included in a figure; *t* indicates that the entry is included in a table.